Modern Philosophy

An Introduction and Survey

Roger Scruton was educated at Jesus College, Cambridge. He has taught for twenty years in the departmen[illegible] College, Unive[illegible] Professor of Ph[illegible] sity, Massachus[illegible] *Salisbury Review.* [illegible] *Imagination*, *The Aesthetics of Architecture*, *The Meaning of Conservatism*, *Sexual Desire* and *The Philosopher on Dover Beach*. He has written four works of fiction, most recently the highly acclaimed *Xanthippic Dialogues*. When in the UK, he lives in Wiltshire.

Other books by
Roger Scruton
include

NON-FICTION

Art and Imagination
The Aesthetics of Architecture
The Meaning of Conservatism
Sexual Desire
The Philosopher on Dover Beach

FICTION

Fortnight's Anger
Francesca
A Dove Descending
Xanthippic Dialogues

Modern Philosophy

A Survey

Roger Scruton

Published in the United Kingdom in 1997 by
Arrow Books

7 9 10 8 6

First published in the United Kindom in 1994 by Sinclair-Stevenson Ltd
This edition first published in 1996 by Mandarin Paperbacks

Arrow Books
Random House UK Ltd
20 Vauxhall Bridge Road, London SW1V 2SA

Random House Australia (Pty) Limited
20 Alfred Street, Milsons Point, Sydney,
New South Wales 2061, Australia

Random House New Zealand Limited
18 Poland Road, Glenfield, Auckland 10, New Zealand

Random House South Africa (Pty) Limited
Endulini, 5a Jubilee Road, Parktown 2193, South Africa

Random House UK Limited Reg. No. 954009

A CIP catalogue record for this book
is available from the British Library

Papers used by Random House UK Limited
are natural, recyclable products made from wood grown in sustainable forests. The manufacturing processes conform to the environmental regulations of the country of origin

Printed and bound in Great Britain
by Cox & Wyman Ltd, Reading, Berkshire

ISBN 0 7493 1902 X

Contents

Note to the Reader

The main text is without footnotes or other scholarly apparatus. I have therefore provided a Study Guide, which covers the material chapter by chapter, sometimes at length, sometimes cursorily, depending on the subject. In the Study Guide the reader will find the following:

(a) precise references to the works discussed or mentioned in the text;
(b) suggested preliminary reading;
(c) elucidation of specific difficulties;
(d) expansion of the topic, where appropriate, in order to show how it is now conceived;
(e) bibliography;
(f) occasional topics for discussion.

The Study Guide is far from comprehensive; but I hope that it will enable the reader, whether student or layman, to use the main text to the best advantage.

Introduction

This book originated in a series of lectures, delivered first at Birkbeck College, London, and subsequently at Boston University, Massachusetts. I have expanded the material where necessary, and added a Study Guide to compensate for the absence of references and footnotes in the main text. The purpose is to guide the reader who may know no philosophy as far as possible towards the frontier of the subject, without becoming bogged down in the minute controversies of the academy.

I reject the notion that there are 'central' questions of philosophy; the agenda of this book is therefore drawn far more widely than is normal in introductory texts. I also have serious reservations concerning the utility of much that passes for 'research' in modern philosophy, while recognising, nevertheless, that the subject has been irreversibly changed as a result of Frege and Wittgenstein, and must be understood from the most modern perspective if it is to be understood at all. It is by no means easy to convey this modern perspective in language accessible to the common reader; but the attempt is, I believe, worthwhile, and not only for the layman's sake. The technocratic style of modern philosophy – and in particular that emerging from the Anglo-American universities – is in danger of killing all interest in the subject, and of severing its connection to humane education. Only if philosophers can rediscover the simplicity and directness of a Frege, a Russell or a Wittgenstein, so as to express the problems of the head in the language of the heart, will they really know what they are doing in the realm of abstract ideas. Part of my motive in preparing these lectures for publication has therefore been to rediscover the subject, by presenting it in the language that seems most clear and natural to me.

The book begins slowly, from specific questions, and referring to well-known texts. As the argument develops, however, I explore original lines of thinking, in order to show the subject not merely as it is, but as, in my opinion, it ought to be. Every now and then, therefore, the material will become controversial; I have tried to indicate where this is so, either explicitly, or by an appropriate change of style. The reader is not asked to agree with my more controversial claims, but only to find arguments against them.

Aristotle observed that you should not attempt to impose more exactitude on a study than the matter permits. Likewise, you should not strive to give easy versions of ideas that are inherently difficult. The best that the reader can hope for is that the difficulties are intrinsic to the subject matter, and not generated by the author's style. When the subject becomes truly technical, however, I have tried to circumvent the difficulty, while giving a sufficient idea of its nature. My hope is that the curious reader will be able, at the end of this book, to find his way through most of the recent literature, and all the classics, of philosophy.

I have discussed the various drafts of this work with several friends and colleagues. I am particularly grateful to Robert Cohen, Andrea Christofidou, Dorothy Edgington, Fiona Ellis, Sebastian Gardner and Anthony O'Hear, whose advice and criticism have helped me to avoid many errors of thought and presentation.

1

The Nature of Philosophy

The purpose of this work is to acquaint the reader with the principal arguments, concepts and questions of modern philosophy, as this subject is taught in English-speaking universities. Sometimes words like 'analytic' are used to describe this kind of philosophy, though that implies a greater unity of method than really exists. Let us say merely that contemporary English philosophy is modern in the true sense of the word – the sense in which science, mathematics and the common law are modern. It attempts to build on past results and, where they are inadequate, to supersede them. Hence English philosophy pays scrupulous attention to arguments, the validity of which it is constantly assessing; it is, like science, a collective endeavour, recognising and absorbing the contributions of many different workers in the field; its problems and solutions too are collective, emerging often 'by an invisible hand' from the process of debate and scholarship.

The word 'modern' is used in other ways, of which two are important:

(1) To denote the modern, as opposed to the ancient or medieval, era of our civilisation. The modern era is held to be contemporaneous with the rise of natural science, and the decline of the centralising tendency in Christendom. Hence Descartes is described as a modern philosopher, while Aquinas is not. Within the modern period certain cultural and intellectual episodes are marked out as particularly important – notably the Enlightenment, by which is meant the irresistible current of secularisation, scepticism and political aspiration which began in the seventeenth century (maybe in Descartes's time), and which culminated in the profoundly unenlightened follies of the French Revolution.

(2) To mean 'modernist', as in 'modern art'. A modernist is *committed* to the modern age, believing that traditions must be overthrown or redefined in order to do justice to the new forms of experience. For a modernist it is intellectually, morally or culturally necessary to manifest one's modernity, to 'challenge' what resists it, and to pour scorn on those who take refuge in the values and habits of a superseded age. (Since these people are in short supply, a vast modernist industry is devoted to inventing them. They are the 'bourgeois', targeted in the writings of Sartre, Foucault, Habermas and Adorno.)

English-speaking philosophy is modern, but not modernist. French philosophy in our time (the work of Foucault and Derrida especially) is modernist, without being particularly modern – i.e., without basing itself in the assessment of arguments, or in the desire to build on established truths. Since the main fear of a modernist is that he may be unwittingly behind the times, he tends to affirm himself as resolutely ahead of them. Hence modernists have invented the label 'post-modern' to define their latest position. This label has been adopted by several French thinkers, notably the sociologist Jean Baudrillard and the philosopher J.F. Lyotard, though it is perhaps more familiar to English readers through the writings of the architectural critic Charles Jencks. Exactly what is meant by 'post-modern' and 'post-modernism' is a question that I take up in the Study Guide. Modernism is often a tenable position – although just how long it can be maintained is a matter of dispute. (A modernist needs to define himself *against* something, so that the very success of his enterprise threatens to undermine it.) No one can doubt the contribution that modernism has made to music and poetry in our century. In architecture, politics and philosophy, however, its contribution has not always been so welcome. The least that can be said is that modernists will not enjoy this book, while post-modernists will probably hate it.

Many historical philosophers are known for their speculative systems, in which a complete account of reality is promised or attempted. Hegel is one of the most accomplished of the system-builders, though his close rival Schopenhauer is equally ambitious and rather more agreeable to read. Modern philosophers are not system-builders in general: or, at least, their systems are peculiarly bare and unconsoling, in a manner foretold in the title of a work by the influential logical positivist Rudolf Carnap: *The Logical Construction of the World*. Since the turn of the century, philosophical problems and arguments have usually been introduced through articles, often devoted to some minute work of logical analysis, and sparking off debates which to an outsider may seem extremely arid and in any case pointless when set beside the aching questions of the human spirit. Learning to take an insider's view of these debates, and to discover that they are not arid at all but, on the contrary, addressed to the most important of human questions, is an exciting intellectual adventure. But it is hard work, and nothing can be learned without the patient study of difficult texts. The one mercy is that – with few exceptions – the greatest works of modern philosophy are short.

So much by way of context-setting. I now turn to the first topic, which is philosophy itself.

1. What is Philosophy?

There is no simple answer to this question: indeed, it is in one respect the main question of philosophy, whose history is a prolonged search for its own definition. Nevertheless, a kind of answer can be given, in terms that explain what follows. Philosophy involves the attempt to formulate, and also to answer, certain questions. These questions are distinguished by their *abstract* and *ultimate* character:

(a) *Abstraction:* We recognise many questions about concrete things: What is that noise? Why did she fall? What does this sentence mean? And we have methods for answering those questions – experiment, theory-building, analysis. But, as every child knows, questions can be repeated, in order to discover grounds for their answers. As they are repeated they tend to emancipate themselves from the circumstances that gave rise to them, and to take on an abstract form – by which I mean that they lose all reference to concrete circumstances, and apply generally, to the world as such. What is a noise? Why do things fall? What is meaning? Abstraction is a matter of degree, and a wholly abstract question is phrased by means of concepts which do not seem to be special cases of any wider category, but, on the contrary, to be themselves the widest categories that we have. Thus the concept of a fall is a special case of the concept of an event. But it is hard to see what general category events belong to, other than the category of events. (I have used the term 'category' to refer to these most abstract of concepts, since that was the term introduced into philosophy by Aristotle, and adopted by Kant, precisely for this purpose.) If you ask why things fall, you are still asking a scientific question. But if you ask why there are *events*, you stand at the threshold of philosophy.

(b) *Ultimacy:* We are rational creatures, and seek for explanations, reasons and causes. All our search for knowledge is based on the tacit assumption that the world can be given a rational explanation. And we have no difficulty in finding explanations in the first instance. 'She fell because she was drunk'. The second instance too causes no anxiety: 'She was drunk because she had consumed three bottles of wine'. And the third instance: 'She had drunk the wine because she was unhappy'; and so on. But each step in the chain demands a further explanation, and if it is not forthcoming, everything that depends on that step is 'ungrounded'. Is there, then, some ultimate point in the chain of causes, some final resting place, so to speak, where the cause of everything is found? If there is, then what is its explanation? Thus we arrive at the *ultimate* question: Is there a first cause? And this question will, by its very form, have set itself beyond science, *beyond* the methods that we normally use to answer questions about causes. Is it a real question? There are philosophers who say not: but that, too, is a philosophical position, a response to the ultimate question.

As an example, consider one of Kant's arguments about cosmology, in the *Critique of Pure Reason*, which is without question the greatest work of modern philosophy. Each event in time is preceded by some other event. The series of events stretches into the past and also into the future, with ourselves buoyed up on the apex of the present. But what of the series as a whole? Did it have a beginning in time? To put it another way: Did the world have a beginning in time? This is a philosophical question. It is wholly abstract (using only the categories of 'world' and 'time'). It is also ultimate: it arises only when all questions about actual events have been answered or discarded. Hence, Kant argued, it cannot be answered by any scientific method. Suppose then that the world has a beginning – at time t. What was happening at the preceding moment $t - \delta$? It must have been a very *special* moment, if it had power to generate from itself a whole world. But if it had that power, something was *true* at that moment. Something already existed: namely the power to produce the astonishing thing that succeeded it. In which case the world did not begin at t, but was already in existence at $t - \delta$. So there is a contradiction involved in the idea that the world began at time t. (And t here is *any* time.) So let us take the alternative view, that the world had no beginning in time. In that case an infinite sequence of times must have elapsed up to the present: an infinite series must have *come to an end*. And that, Kant suggested, is *also* a contradiction. Hence whichever view we take – that the world had a beginning in time, or that it had not – we arrive at a contradiction. So there must be *something wrong with the question*. Perhaps ultimate questions are not questions at all.

I don't say that Kant's argument is valid. But the example is interesting in showing one characteristic response of philosophy to its own problems: namely, to argue that the very process which gives rise to them, also shows them to be unreal. The task of philosophy is to diagnose what went wrong.

(c) *The interest in truth:* There is another feature that characterises philosophy, and distinguishes it from neighbouring modes of thought. Not all human thinking is directed towards truth. In art and myth we allow ourselves the free use of fictions. Truth lurks within those fictions, and in the case of myth a kind of revelation may advance behind the veil of falsehood. But neither art nor myth is assessed on the grounds of its literal truth, and neither is discarded merely because it presents no valid arguments. In philosophy, however, truth is all-important, and determines the structure of the discipline. Validity, indeed, is normally defined in terms of truth, a valid argument being one in which the premises could not be true while the conclusion is false. Even those who believe that philosophical questions have no answers, assert this to be true; and the 'discovery' that they have no answers is made only by the attempt to find a true one.

We need to interpret our experiences, and frequently make use of fictions in doing so. Sometimes these fictions are obscurely intertwined with truths, as in the mysteries of religion. When interpreting the world our purpose is not merely to know it – perhaps not even to know it – but to establish and justify our place within its boundary. Many things assist us in this task besides religion: art, story-telling, symbolism, ritual, along with common morality and rules of conduct. When we refer to the philosophy of the Talmud, or the philosophy of John Keats, we have this kind of thing in mind. Fragments of this 'philosophy' are aimed at abstract truth and so are philosophical in the true sense; but most are a matter of religious, aesthetic and moral interpretation, whose primary goal is not truth but consolation.

There are philosophers who have repudiated the goal of truth – Nietzsche, for example, who argued that there are no truths, only interpretations. But you need only ask yourself whether what Nietzsche says is true, to realise how paradoxical it is. (If it is true, then it is false! – an instance of the so-called 'liar' paradox.) Likewise, the French philosopher Michel Foucault repeatedly argues as though the 'truth' of an epoch has no authority outside the power-structure that endorses it. There is no trans-historical truth about the human condition. But again, we should ask ourselves whether that last statement is true: for if it is, it is false. There has arisen among modernist philosophers a certain paradoxism which has served to put them out of communication with those of their contemporaries who are merely modern. A writer who says that there are no truths, or that all truth is 'merely relative', is asking you not to believe him. So don't.

2. What is the Subject-matter of Philosophy?

Is there any special domain which is the domain of philosophy? Or can philosophical questions arise in any context? Roughly speaking there are three standard answers to this question:

(a) Philosophy studies another realm of being, to which it gains access through its own procedures. The purpose of the ultimate question is therefore to open the gates into that other realm. This was the view adopted by Plato, and argued for in some of the most inspiring and beautiful of all philosophical writings. The 'Platonic' philosophy postulates another and higher realm of being, which is the true reality and to which the human soul is attached by its reasoning powers. This idea was taken up by many of the thinkers of antiquity – the most notable being the pagan Plotinus, and the early Christian father Clement of Alexandria.

The problem for the Platonist is this: How do you know that this other realm exists? The philosopher must establish that our reasoning powers really can give access to it. The main task of philosophy then becomes the critical assessment of the human intellect. The description of the 'higher' realm takes second place to the analysis of reason. Moreover, the fact that Plato is able to give glimpses of this 'higher' realm only through elaborate story-telling and extended metaphor (i.e. only by using language in a way that deviates from the goal of literal truth), might lead us to suspect that philosophy, conceived in Plato's way, is really impossible: that we cannot rise above the world of our day-to-day thinking to the point of view of reason.

(b) Philosophy studies *anything*. Philosophical questions arise at any juncture and concern any kind of thing. There are philosophical questions about tables – for example, what makes this table the *same* as the table I encountered yesterday? (the problem of 'identity through time') – about people, works of art, political systems: in short, about anything that exists at all. Indeed, there are philosophical questions about fictions, and about impossible objects too – even existence is not a requirement that the subject-matter of philosophy must meet; if it is ultimate questions that interest you, then non-existence is as good a starting-point as any.

The main problem for such a view is this: How do we define a philosophical (as opposed to a scientific, artistic, moral or religious) question? I have given an answer in section 1; but it is not an answer that every thinker would accept.

(c) Philosophy studies *everything*: it tries to provide a theory of the *whole* of things. In contrast to the 'bittiness' of science, philosophy attempts an integrated account of the world, in which all truth will be harmonised. Usually the philosopher promises to console the rational being who is fortunate enough to grasp the totality that he offers. (The modern suspicion that the truth about the world may be unutterably dreadful gives us a motive to stick to the details: but the 'leap into the totality' remains a permanent temptation, too, for the very same reason.)

Again there is a problem: *Can* we understand the whole of things? There are those who argue that we can have a theory of anything, but not a theory of everything. Such a theory would have to be *too* general, taking up a standpoint *outside* the world, and so rendering itself unintelligible to those who are *in* the world. Once again, as in (a), the question of the whole of things tends to get postponed, in favour of this more urgent question, as to whether reason is capable of knowing it in any case. Advocates of philosophical holism include most of the great idealists: for instance, Hegel, Schelling, and F.H. Bradley.

3. Does Philosophy Have a Distinctive Method?

Even if we rule out the idea of a special subject-matter for philosophy we are still confronted by the question of philosophical method. Are there techniques that we should apply, some set of assumptions or procedures, when confronted with the ultimate questions? The search for method has been a constant preoccupation of philosophers down the centuries; here are four important living options:

(a) *Thomism*. Named after its originator, St Thomas Aquinas (1226–74), Thomism was an attempt to synthesise the results of philosophical reflection in so far as these recommended themselves to the 'natural light of reason', and in so far as they could be reconciled with the teachings of the Church. Drawing heavily on the newly discovered works of Aristotle, and the commentaries of Muslim, Christian and Jewish theologians, Aquinas described philosophy as the highest of the sciences, the discipline that explores the ultimate 'ground' or explanation of everything. Each special science inquires about the sphere that defines it: biology about life, physics about matter, psychology about mind, and so on. But each makes assumptions that it cannot justify. The task of philosophy is to explore how the world must be, if those assumptions are to be valid. Hence philosophy cannot base its results on experience. Its conclusions are established by reason alone, and it describes the 'highest principles' of things. The sphere of such a study is universal: *everything* falls under it, since every subsidiary science depends upon philosophy for its ultimate credentials. However, although philosophy applies to everything, it does not ask the same questions as those asked by the specific sciences. For these sciences deal with the 'secondary causes' of things – i.e. they explain one contingent thing in terms of another – while philosophy deals with 'first causes', which explain all contingent things in terms of the ultimate nature of reality.

Jacques Maritain, whose *Introduction to Philosophy* is a succinct and accessible defence of the Thomist position, defines his subject thus:

> Philosophy is the science which by the natural light of reason studies the first causes or highest principles of all things – is, in other words, the science of things in their first causes, in so far as these belong to the natural order.

Can there be such a 'science'? If modern writers are sceptical, it is because they are uncertain as to the nature and possibility of *a priori* argument (see below). In contrast to the vast ambition of Thomism, therefore, we find the studied minimalism of:

(b) *Linguistic or 'conceptual' analysis*. These labels, briefly popular, are no longer in general use. But what was meant by them has a certain plausibility. The idea is this: philosophical questions arise at the *end* of science, when all particular inquiries about matters of fact have been exhausted. So what is there left to ask about? Not the world, since we have said what we could about it. About human thought, then? But *what* about it? In so far as thought is part of the world, there is a science of thought too, which will be no more concerned with ultimate questions than the science of worms or galaxies. But there is still the question of the *interpretation* of thought: What do we *mean* by this or that thought? This will not be a scientific question; for it is settled by the *analysis* of a thought, rather than an explanation of why the thought occurred. Such an analysis must concern either the words used to express the thought, or the concepts that compose it. The fundamental philosophical question is therefore the question of *meaning*. We must analyse the meanings of our terms (i.e. the concepts expressed by them), in order to answer the questions of philosophy. This explains why the results of philosophy are not merely scientific results, and also why they seem to have a kind of eternal or necessary truth. When a philosopher asks 'What is a person?' he does not seek the particular facts about particular people, nor the scientific truth about people in general. He wants to know what it *is* to be a person: what makes something a *person* rather than a mere animal, say. Hence he is asking what the word 'person' means. If his answer is that 'person' means 'rational agent', then the assertion that a person is a rational agent will be, for him, not just true, but necessarily true. It will not denote a contingent truth (a mere 'matter of fact') but a truth about the very nature of the thing described.

On this view, philosophical results attain the dignity of necessity, only by losing their air of substantiality. If it is merely a matter of the *meaning* of the word 'person' that persons are rational agents, why is this result important? Moreover, what *are* meanings, and how do we *decide* that this or that is the meaning of a word? These too are philosophical questions, and the method of linguistic analysis has left people seriously puzzled by them.

(c) *Critical philosophy*. This expression was introduced by Kant, to denote the task of philosophy as he conceived it. For Kant too philosophy is concerned with the analysis of thought. But this means something more than the analysis of words or their meanings. Philosophy must rise above thought, so as to set limits to its legitimate activity. It must tell us which procedures tend towards the truth, which patterns of argument are valid, and which employments of our reasoning powers are not illusory. Philosophy must 'define the limits of the thinkable', and its method involves a 'second order' reflection on reflection itself. Sometimes Kant calls this method 'transcendental', meaning roughly that it involves getting behind (transcending) thought, so as to describe the conditions that make thinking possible. The results of critical philosophy tell us what the world must be like if it is to be thinkable; since the world is thinkable, we can know by purely philosophical argument what the world is like.

Not surprisingly, this 'critical' stance raises as many questions as it answers. How, asks Hegel, can philosophy set limits to thought without at the same time transcending them? (There is no such thing as a one-sided boundary.) Already, for Kant himself, the possibility of critical philosophy (and of 'transcendental arguments' in particular) had become the major philosophical question.

(d) *Phenomenology*. Literally 'phenomenology' means the study of appearances, i.e. the study of the world as it appears to consciousness. Appearances may be deceptive; they may also be revealing, without being identical with the non-mental reality that is known through them. (Consider the face in the picture: this is an appearance, which is genuinely and objectively there to the conscious observer. But is it part of physical reality?) To understand the world as it appears is certainly part of the task of philosophy: the most important things in life (goodness, beauty, love and meaning) are grounded in appearance. For the phenomenologists, however, appearances are the *primary* subject-matter of philosophy. And since appearances are dependent on the subject who observes them, phenomenology involves a study of consciousness itself. So argued Edmund Husserl, the Moravian founder of the discipline, who wrote during the early decades of this century. (The term 'phenomenology' was introduced in the eighteenth century by the mathematician J. H. Lambert, and was also used by Hegel to describe the general theory of consciousness.)

According to Husserl, the aim of philosophy is to study the contents of consciousness, seen from the point of view of the subject. Although philosophy must begin from a study of consciousness, however, it does not, according to Husserl, end there. On the contrary, it has another and more ambitious goal, which is to understand the 'essences' of things. We understand the world because we bring it under concepts, and each concept presents an essence: the essence of man, of matter, of unicorns, and so on. These essences are not discovered by scientific inquiry and experiment, which merely studies their *instances*. But they are 'revealed to', 'posited in', consciousness, where they can be grasped by an intuition. The problem is to clear the mind of all the junk that prevents the intuition from forming. Our minds are cluttered with beliefs about the contingent and the inessential; we can approach essences, therefore, only if we 'bracket' those beliefs, and study what is left as the object of a pure inner awareness. The method of 'bracketing' – also described as 'phenomenological reduction' – will be discussed in later chapters.

Phenomenology, like linguistic analysis, proposes *meaning* as its primary subject-matter. But it is not the narrow species of meaning that resides in language; it is the meaning of life itself – the process whereby we relate to the world and make it our own. This explains the appeal of phenomenology, especially to those who are looking for the answer to moral, aesthetic or religious questions. On the other hand, phenomenology has never succeeded in justifying itself to its critics' satisfaction. In particular, it has never shown how a study of what is 'given' to consciousness can lead us to the essence of anything at all.

All of (a), (b), (c) and (d) suppose philosophy to be deeper than, and prior to, science. Moreover, on all four views, philosophy *leaves science as it is*. Hence no scientific theory can prove or disprove a philosophical theory. To understand why this is so, we need to explore an important distinction, which will occupy us in later chapters:

4. The *a priori* and the Empirical

Philosophy is said to be an *a priori* inquiry, although precisely what this means is a matter of controversy. While science proceeds by experiment, and tests all its theories against the evidence, philosophy reaches its results by thought alone, and makes no reference to experience in doing so. Something like this must be so, as we can deduce from our discussion. For philosophical questions arise at the *end* of science. They ask whether the methods employed in science are valid; whether experience is a guide to reality; whether the world as a whole is intelligible. Such questions cannot be answered by science, which *presupposes* a positive answer to them. No experience can bear on the question whether experience is a guide to reality: this is a question for thought alone. In calling it an *a priori* question, philosophers mean that it is prior to experience, and must be settled by thinking if it is to be settled at all.

Philosophers therefore try to justify the idea of an *a priori* discipline, and to show that *a priori* knowledge is possible. Indeed, Kant thought, this just is the principal question of philosophy.

5. Branches of Philosophy

Philosophy has acquired certain recognised branches: they are not as clearly divided from one another as they may appear, and it is arguable that you cannot really understand any part of philosophy without having some inkling of the whole. (This, indeed, is the principal weakness of anglophone philosophy: not that it is too narrow or analytical, but that it is too specialised. When someone can call himself a philosopher, while entertaining *no views whatsoever* on aesthetics, political philosophy, morality or religion, something has gone wrong with his conception of the subject.)

It is useful to divide pure philosophy from applied philosophy. In the first of these, philosophy generates its *own* questions and answers. In the second, it reaches out to explore the foundations of disciplines whose subject-matter it does not control.

(a) *Pure philosophy*. Four branches are generally recognised:
(i) Logic: the study of reasoning. Which forms of argument are valid, and why? What follows from what, and what does 'follows from' mean? What are the 'laws of thought', or are there none? What is the distinction between necessary and contingent truth? And so on.
(ii) Epistemology: the theory of knowledge. What can I know, and how? Does perception provide knowledge? What guarantee do I have of judgements based on memory? Is there knowledge of the past, of universal laws, of the future? Can knowledge reach beyond experience? And so on.
(iii) Metaphysics: the theory of being. (Named after a book of Aristotle's, which coming 'after the *Physics*' (*meta ta phusika*), thereby gave a name to its subject-matter.) What exists? What is existence? Does God exist? What are the basic items in the world? Do properties exist, as well as the individuals that possess them? And so on. Certain branches of metaphysics are so important as to be treated separately – notably the philosophy of mind.
(iv) Ethics and aesthetics: the theory of value. Is there a real distinction between those things, actions, affections which are good and those

which are bad or evil? Can we justify the belief that we ought to do this rather than that? What is virtue, and why should we cultivate it? What is beauty, and why should we pursue it? And so on.

(b) *Applied philosophy*. There are as many branches of this as there are occasions for human folly. Of particular importance are the following:

(i) Philosophy of religion. This is sometimes taken to include theology, although it would be more accurate to say that its subject is really the *possibility* of theology.

(ii) The philosophy of science: a branch of epistemology (concerned with the validity of scientific method), together with a branch of metaphysics (concerned with the existence of the entities postulated by science, many of which – quanta and quarks, for example – are metaphysically highly problematic).

(iii) The philosophy of language – concerned with the understanding of meaning and communication. This increasingly important branch of the subject now threatens to engulf the remainder, since so many philosophical questions can be rephrased as questions about meaning.

(iv) Political philosophy. The oldest branch of applied philosophy, and the theme of the first indisputable masterpiece of Western philosophy, Plato's *Republic*.

(v) Applied ethics. A growing branch of the subject, involving the application of philosophical argument to specific moral problems; for example, to sexual conduct, business ethics, abortion and euthanasia.

In addition to studying most of those, it is normal to study the history of philosophy. But this raises an interesting question: What is meant by the history of philosophy, and why is it important? If philosophy is really the 'modern' subject that I have supposed it to be, what need does it have of its own history? Why has that history not been superseded, as the history of science has been superseded through its own success? A physicist may with impunity ignore all but the recent history of his subject and be none the less expert for that. Conversely, someone with only a very inadequate grasp of physics (of the system of physics which is currently accepted as true), may nevertheless be a competent historian of the subject, able to explore and expound the assumptions and historical significance of many a dead hypothesis. (That is why science and the history of science are separate academic disciplines.)

One answer is this. Philosophical questions are ultimate: hence they lie at the limits of the human understanding. It is difficult to know whether we have truly grasped them. In order to grasp them, therefore, we study the works of the great thinkers who have wrestled with them, and whose superior intellect, even when cluttered by outmoded beliefs and discredited conceptions, guides us more truly to the heart of the subject than we should ever be guided by our own capacities. Such a study does not only present us with the highest reaches of human thought; it helps us to uncover our prejudices and to see their roots.

6. History of Philosophy and History of Ideas

What then is the distinction between the history of philosophy and the history of ideas? Or is there no distinction?

The history of philosophy is treated as a contribution to philosophy. That is to say, philosophers of the past are studied on account of their contribution to *present* problems, or to problems that can be *made* present. The same concern for truth that animates the study of philosophy, disciplines our conception of its history. It is a precondition of entering the thought of historical philosophers that one does not regard the issues that they discussed as 'closed'. The map of philosophical history is therefore less a map of influences and continuities than a patchwork of figures who, regardless of their place in the temporal order, can be understood as our contemporaries. The history of philosophy is taught according to a philosophical agenda, and figures whose influence in the realm of ideas far outstrips the influence of the great philosophers (Rousseau, for example, or Schelling) may have a comparatively minor part to play in philosophical history. An historian of ideas may be a bad philosopher (bad, that is, at assessing the validity of arguments or the truth of their conclusions); but he must be a good historian. For he is concerned to describe, and if possible to explain, the influence of ideas in history, regardless of their philosophical merit.

This is why the history of philosophy is so full of gaps. There are important thinkers whose problems have been either solved or stated better by another writer – Malebranche, for example, or Lotze. These are seldom encountered by students of the subject. There are others, however, who have given model expression to problems that remain with us – Plato, Aristotle, Descartes, Kant and Hume being among them. Every now and then a thinker of the past is rediscovered as a great philosopher, and then makes the transition from the history of ideas to the history of philosophy. This happened recently with

Adam Smith. Sometimes the journey goes the other way, as it did with Lotze. And there are philosophers – Hegel being the principal among them – who every ten years are dismissed as merely influential only to be 'rediscovered' as great philosophers ten years further on.

Does this mean that there is no progress in philosophy? It is a tacit belief of 'modern' philosophy that progress is possible. There is a truth to be uncovered; but it lies at the limit of our understanding. Hence we must hold on to those philosophical achievements which serve as our paradigms of argument. Bit by bit, the terrain of philosophy has been illuminated; and while ignorance threatens always to re-engulf us, we can certainly claim a better understanding today than Descartes had, of the problems that he discovered.

2

Scepticism

Modern philosophy began, not with Descartes exactly, but with the thing that Descartes made famous: systematic doubt. Descartes's position in the history of philosophy is certainly not identical with his position in the history of ideas. Indeed, he was part of a long tradition of attempts to formulate and answer the sceptical question. (See the impressive study by R.H. Popkin, *The History of Scepticism from Erasmus to Spinoza*.) It is this that placed epistemology for the first time at the centre of philosophy (a position that it is now beginning to lose). Scepticism is often presented in terms of the concept of knowledge – a concept to which I return in Chapter 22. But this concept, of such vital concern to the Greeks, is not one that we need to use, in order to state the sceptical problem. It is sufficient to ask the question whether we have adequate grounds for our everyday beliefs, in order to see the force of sceptical arguments. At once, however, we come across a problem of language, and it is well to clear this up at the outset.

1. Sentence, Proposition, Statement, Thought, Belief

Suppose I utter the sentence 'The cat is on the mat' ('*p*' for short). Not only have I uttered the sentence '*p*'; I have expressed the proposition that *p*, which is in turn identical with the *thought* that *p*. The proposition is the meaning of the sentence. In expressing this proposition I may also be making a *statement*: the statement that *p*. I don't *always* make this statement by uttering the sentence, however: I am not making the statement *now*, since I am not telling you or anyone that the cat *is* on the mat, but merely holding up the proposition for your attention. (Contrast my answer to your anxious question 'Where is the cat?') In making the statement I may in turn be expressing my *belief* that *p*. So here we have four (possibly five) things that can be identified by '*p*': the sentence '*p*'; the proposition (thought) that *p*; the statement that *p* and the belief that *p*. These belong to different categories: the sentence is a piece of language, the proposition is what is *meant* by the sentence, the statement what is *done* with it, and the belief a mental state *expressed* by it. But they all have an important feature in common: they can be true or false (they have a 'truth-value', as modern philosophers put it). Moreover they are made true by the very same state of affairs: the state of affairs that *p* (so there is another thing that '*p*' identifies). Sceptical arguments can be expressed, therefore, in a variety of ways: as arguments about the justification of sentences, of propositions, of statements and of beliefs. For our purposes we do not need to distinguish these; though the question which comes first (which is 'fundamental', in the sense of defining and identifying the others) is important in logic and the philosophy of mind.

2. The Structure of a Sceptical Problem

Scepticism begins by identifying some set of beliefs which are basic to our view of the world, and whose truth we do not question. It then identifies all the *grounds* for those beliefs: not the actual grounds that this or that person may have, but all the *possible* grounds. It proceeds to show that those grounds do not justify the beliefs. Mild scepticism argues that they do not *prove* the beliefs conclusively; radical scepticism argues that they offer no reason for believing *at all*. It is radical scepticism that provides the greatest stimulus to philosophy, since, if we have no answer to it, we have no grounds for thinking that our ordinary beliefs are true.

Here we need to make a distinction among beliefs. Some beliefs are, so to speak, epistemological luxuries. I could give them up without losing my conception of the world and my place within it. Consider the belief in God: it may be morally and emotionally difficult for me to abandon this belief. But if the sceptic showed that I could not conceivably have any grounds for it, he would not undermine my conception of the world. I could retain my scientific beliefs, for example, while admitting the non-existence of God. In attacking these epistemological luxuries the sceptic may actually confirm my more ordinary convictions.

But there are epistemological *necessities* too. First among them is the belief that I inhabit an objective world that is distinct from me, and whose existence does not depend upon my thinking. If I gave up *this* belief, all my scientific knowledge, and indeed all my common-sense judgements, would have to be abandoned. I would find it hard, perhaps impossible, even to formulate a conception of what I am; maybe even my ability to think and speak rationally would be put in doubt. It is against epistemological necessities of this kind that Descartes's argument in the *Meditations of First Philosophy* is directed. And interestingly his own *answer* to scepticism involves resuscitating the belief in God, which ceases, therefore, to be the mere luxury that modern people assume it to be.

3. Descartes's Argument

Descartes begins by putting his own beliefs (those in the epistemologically 'necessary' category) in doubt. His initial procedure is to show that my ordinary grounds for those beliefs are compatible with their falsehood. I have had *these* beliefs, on *these* grounds, and turned out to be mistaken. So how do I know that I am not mistaken now? For example, sense experience, which is my normal ground for beliefs about the physical world, is notoriously prone to error. I may suffer illusions, hallucinations, sensory aberrations; and these are qualitatively indistinguishable from what, on other occasions, I may hold to be 'veridical' (i.e. true) perceptions. But if the ground for my belief that I am sitting at my desk is the way things seem, and this ground is compatible with my not sitting at my desk, it is not a *sufficient* ground.

From this point Descartes moves to two more radical arguments, designed to show that my experience offers no ground *at all*.

(i) The dreaming argument. Often, in dreams, I have had the experiences that I have in waking life. And there is nothing logically absurd in the supposition that my experience while dreaming should be exactly the experience that I have now, sitting by the fire and meditating. So how do I know that I am not dreaming? Dreams provide no foundation for the beliefs that occur in them; and yet they could in principle manifest just the order and connectedness of waking experience. So what grounds have we for trusting our waking experience?

This may seem to be merely a form of 'mild' scepticism. Someone may reply that it is in fact very unlikely that I *am* dreaming; hence even if I cannot *prove* that I am awake, I have very good grounds for the *belief* that I am awake. But such a reply misses the point. We judge what is likely in terms of long-term connections. It is likely that the dog will not be hungry when I get home, since generally my wife arrives first, and feeds him. The likelihood of *q* given *p* is something that we know, because we have grounds for associating *q* with *p* in general. But if it is possible that I am dreaming now, it is equally possible that I am dreaming on every occasion – that even my experiences of waking have been dreams. I cannot establish the long-term connection between the nature of my experience, and the fact of being awake, which will enable me to deduce that it is likely that, when my experience is like this, I am not dreaming.

Other philosophers reply by saying that the concept of dreaming must nevertheless get its *sense* from the contrast with waking, and that the very possibility of making the contrast supposes that I have some criterion upon which I can rely. (Cf. Norman Malcolm's argument in *Dreaming*.) Without getting into technicalities, we can at least wonder whether Descartes's invocation of the concept of dreaming does not automatically compel him to retain the idea that, if he is dreaming now, this is because he exists in a world where he is sometimes awake. On the other hand, even if there is a criterion for distinguishing waking from dreaming, could I not merely *dream* that I have applied it? And how did I learn about the existence of this criterion? Maybe I merely dreamed of its existence.

Descartes himself did not regard the dreaming argument as conclusive. He recognised that, since the ideas in dreams must come from somewhere, and since he is not their creator (they being involuntary), he is entitled to suppose that he lives in a world which has the power to produce those ideas. He can believe *something* about an objective reality. He therefore turns his attention to the second and more powerful argument.

(ii) The demon. Descartes now imagines that his experience is exactly as it is, except that it has been produced in him by a demon – 'an evil genius' – who is powerful enough, and perverse enough, to generate in his victims the steady illusion of an objective reality. If this hypothesis is coherent, then it could be that there is no such reality, indeed that there is nothing in the world, besides me and the demon who deceives me. So, how do I know that the hypothesis is not true?

Various moves might be attempted in reply to this. But we can no longer rely on the distinction between dreaming and waking; nor can we rely on the distinction between veridical and illusory perception.

Suppose there *were* some mark of veridical perception, whereby we habitually distinguish the true from the false among our sensory experiences. Could not the demon manufacture experience which displayed that mark? Could he not allow me to make all the distinctions I presently make, and to separate quite reasonably the true from the false among my perceptions, to construct just the very picture of the world that presently persuades me, to distinguish waking from dreaming and being from seeming – and yet, in all this, to be no more anchored in an objective reality than if I were the only thing that existed?

Someone might argue that, even so, the 'demon' is at best an *hypothesis*, and therefore no better than a *rival explanation* of my experience. The ordinary explanation – that things seem as they do because I inhabit an objective world which corresponds to my opinion – is equally good, if not better. Indeed, some philosophers, relying on what has been called 'inference to the best explanation', have suggested that we do have grounds, and good grounds, for inferring the truth of our common-sense view of things, since this provides the *best explanation* of our experience. (Gilbert Harman, 'The Inference to the Best Explanation': see Study Guide.) But even if such philosophers are able to say what is meant by the 'best' explanation, it is arguable that the demon is a better explanation than the one afforded by common sense. Instead of supposing the existence of a complex world, with a multiplicity of objects, whose laws we barely understand, the demon hypothesis proposes just one object (the demon) operating according to a principle (the desire and pursuit of deception) that we are intimately acquainted with. The hypothesis is both simpler, and more intelligible, than the doctrine of common sense. Maybe it *is* the best explanation!

4. Scepticism Generally

The purpose of such arguments is to show that the grounds for our common-sense beliefs may be satisfied, even though those beliefs are false; and to show in addition that we have no better reason for supposing the truth of our common-sense beliefs than for supposing the truth of some rival view. The demon recurs in many guises, and many are the arguments that have seemed, to this or that philosopher, to bottle him up in some harmless container. But always he escapes: the whole purpose of inventing him was to give him the power to do so. His latest incarnation is through an argument due to Hilary

Putnam (*Reason, Truth and History*, pp. 4–7), which suggests that my experience could be just as it is, even though I am in fact nothing but a 'brain in a vat', at the mercy of some malign scientist who stimulates me with his electrodes. (Modern people are happier with mad scientists than evil demons; but this too is the work of the devil: see Chapter 30.)

Descartes used the expression 'hyperbolical doubt' to describe the position in which he was put by his radical arguments. Rather than establish doubt by contrasting doubt with certainty, his arguments infect every belief and every certainty. They leave us without the contrast between the known and the unknown, upon which our world-view seems to depend.

This radical scepticism is not the only kind. In each area of epistemology there is a *local* scepticism, which serves to challenge the objectivity of our beliefs, while leaving the rest of knowledge unaffected. Here are some examples:

(a) God. Beliefs about God are based on beliefs about the world (for example, the belief that the world is harmonious with our desires). But these beliefs about the world could be true, and yet God not exist.

(b) Other minds. Beliefs about other minds are based on beliefs about behaviour and bodily circumstances. But these beliefs about behaviour could be true, even though there were no other mind.

(c) Values. Beliefs about values (though are they exactly beliefs?) are based in beliefs about the world. But etc. . . .

(d) 'Theoretical' entities in science (e.g. electrons, photons, quanta). Beliefs about theoretical entities are based on beliefs about observable entities. But etc. . . .

And so on, through all the realm of epistemology. The possibility of local scepticism is precisely what defines an epistemological problem; and one of the questions, to be touched on in the next chapter, is whether we need to answer scepticism case by case, or whether there might be, on the contrary, some global solution to a problem that always seems to take the same repeatable form.

5. Appearance and Reality

One way of interpreting scepticism is through the distinction between appearance and reality. Descartes's arguments seem to show that the reality of the world is distinguishable from the appearance of the world. The problem is to derive the reality from the appearance. Can it be done?

Many philosophers have answered in the following way: Yes, it can be done, provided the reality is not too *distant* from the appearance. Furthermore, if we are to know what we mean by 'reality', the reality *cannot* be distant. A good example of such a philosopher is Berkeley, whose *Three Dialogues Between Hylas and Philonous* should be read at the first opportunity. Berkeley was attacking a view that he associated (perhaps wrongly) with Locke. According to this view, the world is composed of 'material substance' or (more familiarly) matter, which we know through experience – i.e. through the way it appears. But this idea, Berkeley suggested, leads to contradictions. If material substance is really independent of us, it must possess properties regardless of its appearance. But which properties? We have no conception of what properties are, except through the experiences that define them – experiences of colour, shape, hot and cold, etc. So is matter really hot, say, square or green? What grounds do we have for saying such a thing? Only the fact that 'material substance' *appears* hot, square or green. We say this object is hot because it feels hot. Plunge your hand in hot water and withdraw it: the object will no longer appear hot at all. So it appears both hot and not hot. Which appearance shows the reality? There is, Berkeley suggested, no answer to this question. Either both appearances are true – in which case the very idea of material substance involves a contradiction; or neither are true – in which case properties cannot be attributed to material substance, but only to the appearances which supposedly represent it. Whichever solution we take, material substance drops out of consideration, and the appearance comes in its place.

Should we then say that reality is appearance? Or that it is constructed from appearance? Or that it is 'reducible' to appearance? To understand these questions, as they have developed since Berkeley, we need to define a few common philosophical positions.

3

Some More –isms

It is easier to understand the force of scepticism if one has some grasp of the extent and variety of the attempts to combat it. In this chapter I shall survey the most important among traditional responses to the demon, and thereby develop a brief history of modern philosophy.

1. Idealism

The term 'idealism' is used of a variety of positions. I shall begin from that of Berkeley.

Berkeley's concern was to show that we have no grounds for believing in anything, save the existence of 'ideas' and whatever 'perceives' or 'conceives' them. 'To be is to be perceived'. By 'idea' Berkeley meant any mental state, whether perception, thought or sensation – in short anything of the kind that I can discover in myself by 'looking inwards'. His arguments involve the standard sceptical moves, together with certain additions of his own, directed specifically at Locke. The outcome was, he believed, quite simple: we fall into error and confusion, just as soon as we postulate a world of 'material substances'. So long as we talk only of that which is known to us – namely, of the ideas that enter our consciousness, and whatever may be rightly inferred from them – we are safe. Moreover, this is well understood by ordinary people, who do not really mean anything by their words, other than the ideas that are denoted by them.

Berkeley may seem to be invulnerable to the demon. About my ideas I cannot be deceived; and if this is all that I refer to when I speak of a material world, then I cannot be deceived about that either. However, things are not quite so simple. When I speak of tables and chairs, Berkeley concedes, I am not referring to single ideas, but to what he calls 'collections' of ideas – meaning, roughly, the totality of the experiences that lead me to deploy these concepts. I certainly cannot be deceived into thinking that *this* idea, that I have now, is other than it seems. But I can mistakenly believe it to belong to a certain 'collection'. And I may be mistaken in my memories of previous ideas. Hence the picture that I presently entertain of the world may be entirely misleading; and maybe I have no means available, to arrive at the *true* picture. Perhaps talk of a true picture is not really meaningful. Perhaps I ought to refer *only* to my present ideas. But in such a case can I really *refer* to them? (See Chapter 5.)

Berkeley is unique among idealists in being entirely honest about what he is trying to say. In effect he is arguing that there is no physical world, and that nothing exists except minds – yours, mine and God's. (It is because God is always there to 'conceive' things, that they do not disappear when I turn my back on them.) This honesty earned him the label 'subjective idealist' from Schelling and Hegel. He was prepared to say, quite naïvely, that everything is 'composed' of mental states, and to define those states (the 'ideas') through their 'subjective' aspect – i.e. through their 'inner' nature, their way of being 'given' to consciousness. But there is another kind of idealism, which is called (again following Schelling and Hegel) 'objective idealism'. The objective idealist believes that reality is in some sense *independent* of mind: it is objective in relation to the subject who perceives it. But he also believes that reality is 'organised mentally': it gains its character through the process whereby it is known (the process of 'bringing under concepts'). On this view it is impossible to say that physical objects are 'composed of' ideas, or of any other mental stuff. What we must say instead is that physical objects are the *objects* of mental states, and are endowed with their nature by the process whereby they are 'posited' in the observing intellect. Furthermore, mental states are not merely 'subjective': they are not 'given' to the subject in the way envisaged by Berkeley. They too acquire their nature by the process which gives them objective (and not merely subjective) reality. Mental states are 'realised' in the objective world, which is in turn realised through the process of knowledge.

The details of that Hegelian position are complex and far from clear (I return to them in Chapter 12). To make matters worse there is a third kind of idealism, namely the 'transcendental' idealism of Kant (which was the main inspiration for Hegel's philosophy). According to Kant, the world is independent of us, but also 'conforms to' our faculties. The world is the way it *is* because that is the way it *seems* – even though being is more than seeming. The way it seems is the way we order it, and the way we *must* order it if we are to have objective knowledge. We also have the idea of a 'transcendental' world – a world unconstrained by the requirement that we should know it. But it is *only* an idea, which can be translated into no knowledge of a transcendental *reality*.

Kant's position is extremely subtle – so subtle, indeed, that no commentator seems to agree with any other as to what it is.

2. Verificationism

Berkeley is often compared to the 'verificationists' or 'logical positivists'. Verificationism arose in Vienna between the wars, as part of the 'culture of repudiation' whereby central Europe threw away its inheritance and committed moral suicide. The ostensible inspiration was Wittgenstein's *Tractatus*; but it drew on other sources, notably on British empiricism, and on a profound respect for science and 'scientific method' as the one proven route to knowledge. The root idea is the 'verification principle', which says that the meaning of a sentence is given by the procedure for verifying it (for establishing its truth). You know what you mean by '*p*' just so long as you know how to find out whether *p* is true. This implies that many things we are disposed to say are meaningless, since there exists no actual or possible procedure for establishing their truth. Most of metaphysics, for example, is meaningless – a result which the verificationists enthusiastically welcomed.

Verificationism was brought to Britain by A.J. Ayer, who made his reputation with the book – *Language, Truth and Logic* – in which he expounded it. This book is a kind of classic, and should be read if possible, provided it is read *quickly* and *inattentively*. The details of the argument are preposterous; but the outline shows clearly what motivated verificationism and why it was so influential. It was one of the first systematic attempts to recast philosophical problems as problems about *meaning*, and to propose the abolition of philosophy as the aim of philosophy. It also gave an answer to scepticism. If the evidence for *p* is *q*, and that is the only evidence there is or can be, then '*p*' *means q*. Hence there is no gap between evidence and conclusion, and the sceptical problem does not arise. (Here is the connection with Berkeley.) Thus if the only evidence I have for my statements about physical objects consists in the truth of other statements about experience, that is what I *mean* by my statements about physical objects. Any residual worries about the 'physical object in itself', the item 'beyond' experience, are meaningless: they are philosophical inventions, which would disappear were we to use language according to its proper logic. Philosophy creates its own problems: they are solved by showing them not to be problems at all.

Verificationists were also behaviourists. Since the only evidence I have for my statements about your mind are observations of your behaviour, that is all I could mean by referring to your mental processes. But what about my own mind? Surely I don't mean to refer to my *behaviour* when I say that I am in pain or thinking. If that is what I meant, then I could be mistaken! Here lies one of the many difficulties that verificationism has encountered. Indeed, verificationism is studied now largely for these difficulties. The verificationists, like Berkeley, made the mistake of being honest. They fell victim to their own naïveté, by making it possible to refute every single thing they said, including the verification principle itself. How, after all, would you verify the principle? There seems to be no answer. In which case, the principle, by its own reckoning, is meaningless.

One legacy of logical positivism remains, however. This is the view that language has a systematic underlying structure, which may not be immediately revealed in its apparent structure. We may have to *discover* what we mean, by analysing our language, and laying bare the logical relations between our sentences. This idea was derived by the positivists from Russell. But the use they made of it served to project it into every sphere of philosophical inquiry.

3. Reductionism

Verificationism is also a form of reductionism. It asks questions about 'problematic entities' – for example, physical objects, minds – and answers them by reducing those objects to the evidence which leads us to believe in them. The principle here is that of 'Ockham's razor', named after the medieval English empiricist, William of Ockham (sometimes written Occam) who probably never referred to it. Ockham's razor says that 'entities are not to be multiplied beyond necessity'. We should suppose the existence only of those things that we *need* to assume, in order to explain our experience. Entities which are not needed should be excised from our world view, or else 'reduced' to other necessary things. An example is society: does a society (say British society) exist, over and above the individuals who compose it? Mrs Thatcher notoriously said there is *no such thing* as society; meaning that a society is composed of individuals and nothing else. Does it follow that there are no *facts* about societies which are not facts about individuals? And if that follows, are we entitled to say that societies do not exist? The reductionist will reply that we *can* say that societies exist, provided we have 'reduced' societies to their members. Reductionists are ontologically parsimonious. Their opponents – including 'objective idealists' like Hegel – will argue for a richer 'ontology'. Hegel, for example, would say that when people come together in society a new *entity* – civil society – is born out of their mutual agreements. And this entity changes the nature of the individuals who compose it. People outside society are a *different kind of thing* from people who have associated. So how can we say that a society is composed of non-social individuals? We have a new, organic whole, which cannot be reduced to the cells that compose it, since their nature depends upon their participation in this whole.

Verificationists espoused a particularly modern form of reductionism. When people say that a society is 'composed' of individuals, they are, according to the verificationists, speaking obliquely. What they should really say is that 'sentences about society are equivalent to other sentences about individuals', or 'sentences about society can be translated without loss of meaning into other sentences about individuals'. 'Equivalent' here means *logically* equivalent (necessarily having the same truth-value). To put the point in other words: societies are not composed of individuals: they are 'logical constructions' out of individuals. The idea of a logical construction – due to Russell and the early Wittgenstein – was thenceforth to play an important role in philosophy. It is therefore worth making a few brief remarks about it.

We frequently have cause to refer to the average man. But only someone in the grip of philosophical madness would say that the average man *exists*. The reason why we do not infer his existence is that we know that our references to him are shorthand: they are convenient summaries of facts about men. If we say that the average man has 2.3 children we certainly do not mean that there is something that has 2.3 children. We mean that the number of children of men, divided by the number of men, is 2.3. That is an example of what the verificationists have in mind, when they say that sentences about one thing are *equivalent* to, or *translatable* into, sentences about another. There are two ways of expressing this idea. We can speak about *sentences* involving the phrase 'the average man', and say that they are translatable into sentences about men. Or we can speak about the average man himself, and say that he is a logical construction out of men. Although these say the same thing, one is about words, and the other is about a putative *entity* to which those words 'refer'. Carnap, whose book *The Logical Construction of the World* carried verificationism to its limit, wrote that the first way of talking is in the 'formal' mode, while the second is in the 'material' mode. Philosophical problems arise when we become too attached to the material mode, and forget the formal mode of discourse which supplies its meaning. A philosophical problem might start from the premise that we refer to the average man; *therefore* the average man exists. Translation into the formal mode shows that problem to be nonsense.

The verificationists rephrased Berkeley's theory that tables and chairs are collections of ideas, arguing instead that they are *logical constructions* out of 'sense data'. Sense data are what are 'given' in sensory experience.

The term 'reductionism' is sometimes used in another way and another context, to denote the attempt to rid the human world of the values, myths and superstitions that supposedly encumber it. For instance, a sexologist of the Kinsey variety may describe human sexual behaviour without reference to the thoughts of the participants, and 'reduce' it to a biological function, or a pleasurable sensation in the sexual parts. A Marxist may describe the legal system of a country without reference to the rights and duties that it defines, 'reducing' it instead to the power relations which are enforced by it. Here we encounter a peculiar use of the phrase 'nothing but'. Sex is 'nothing but' the means whereby our genes perpetuate themselves; or is 'nothing but' the pleasure felt in the sexual parts; justice is 'nothing but' the power requirements of the ruling class; gallantry is 'nothing but' the means we have devised for reminding women of their servitude. (See the argument of Thrasymachus in the first book of Plato's *Republic*.) Reductionism of this kind does not merely involve a host of philosophical confusions. It is essentially anti-philosophical, based in the desire to simplify the world in favour of some foregone conclusion, whose appeal lies in its ability to disenchant and so demean us. The reductionist 'opens our eyes' on to the truth of our condition. But of course, it is not the truth at all, and is believed to be true only because it is shocking. There is, here, a contempt for truth and for human experience that a philosopher should do his best to overcome. Even if a genetic strategy *explains* human sexual behaviour, this does not entitle us to conclude that the thing explained is identical with, reducible to, or 'nothing but' the thing that explains it. (After all, a genetic strategy explains our belief in mathematics.) One task for philosophy in our time is to teach people to resist this kind of vulgar reductionism. Unfortunately, modern universities devote a great amount of their energy to endorsing it.

4. Empiricism

Empiricism, named after Sextus Empiricus (*c* 200 AD), who advocated its main principles, is the view that all knowledge, and all understanding, have their roots in experience – particularly in the experience we obtain through the senses. The theory has two parts: it affirms first, that experience is the basis of our *knowledge*; secondly, that experience is the basis of our *understanding* – i.e. of the concepts used to formulate that knowledge. It is fairly obvious why these should go together, although there are philosophies which try to distinguish them (for example, that of Kant). Modern empiricists tend to emphasise the second part of the theory, which is typified by verificationism (see above). They argue that the meaning of what we think and say is given by the experiences which 'verify' or 'ground' it. Locke had a similar thought in mind, when he argued that all our ideas are *derived* from the senses, though his way of putting the point is unattractive to modern philosophers, since it seems to imply that empiricism is merely a theory about the *cause* of our 'ideas' (or concepts) and not about their content. (The same defect can be observed in Berkeley and Hume, who are usually classified, along with Locke, as the great modern empiricists.)

Empiricism of a kind is implied by a medieval scholastic principle, 'nothing in the intellect that was not first in the senses', accepted even by St Thomas Aquinas. But it is not clear how far Thomas, or his contemporaries, were prepared to take the principle. Those medieval philosophers who would definitely be counted as empiricists – most notably William of Ockham – were thought to be distinctly heretical, even dangerous, at the time.

Empiricism has a common-sensical air; but it is also, once accepted, fairly iconoclastic, since it implies that every claim to knowledge must be put to the test of experience and, if found wanting, rejected. Authority, tradition, revelation are all put in doubt. But what is meant by 'the test of experience'? About this there is much disagreement. The most sophisticated position makes reference to 'scientific method', arguing that the test of experience involves experimental procedures, induction (see Chapter 15) and a host of other intellectual devices, whose authority must in turn be validated by their ability to pass the test of experience. (This sounds like a vicious circle, but, as I said, the idea is sophisticated, and will be accompanied by arguments purporting to prove that the circle is not vicious at all.)

The greatest obstacles to empiricism are mathematics and metaphysics. It seems impossible to dismiss mathematical knowledge; yet it is surely not founded on experience; nor do mathematical concepts (number, addition, etc.) derive their sense from experiences in which we 'ground' them. The empiricist's attempt to show that mathematics is therefore merely a matter of definition has led to some of the most important results of modern philosophy. As for metaphysics, one empiricist response is that given by the verificationists – to dismiss it as nonsense. But this is too simple. For empiricists are forced to make metaphysical assumptions, and must therefore give a procedure for distinguishing the true from the false among them. (After all, isn't empiricism itself a metaphysical theory?)

Reflection on this problem-area lies at the root of much argument about 'necessary truth'. Mathematical and metaphysical propositions are, if true, necessarily true – true whatever the course of experience, however the world may 'turn out to be'. What does that mean, how can it be so, and – most importantly for an empiricist – how can we know that it is so? These are matters that I shall discuss in Chapter 13.

5. Rationalism

Rationalism was first distinguished from empiricism by Kant, who argued, in the *Critique of Pure Reason*, that empiricism and rationalism represent two comprehensive options, and that the philosophers of his day were drawn respectively to one or other of them. (Kant went on to argue for a third option, his own, which incorporated, as he saw it, what was true in both, while avoiding their errors.) Rationalism is founded in a thorough-going scepticism about experience. Experience delivers, the rationalist argues, only obscure and ambiguous results. In particular, it can only show the appearance of the world, and is silent about the reality. Hence it gives no grounds for *knowledge*. If we know anything it is because we can obtain certainty; and certainty comes only by rational reflection, based in self-evident principles. True knowledge is therefore *a priori* knowledge – knowledge whose justification can be delivered by reasoning alone.

Rationalism leads almost inexorably to the conclusion that reason is superior to the senses wherever it is in competition with them (and also to the equally important conclusion that it is reason which *decides* whether it is in competition with the senses). This in turn leads to a re-casting of the distinction between appearance and reality. *All* beliefs derived from experience, whether 'illusions', fantasies or well-founded scientific results, are beliefs about appearances. What is *really* real is known only to reason.

Rationalism has a long tradition. It is at least as old as Plato, who is often taken as a paradigm case of it. In modern times it includes (according to standard typology) Descartes, Leibniz, Spinoza and, in all probability, Hegel. However, one should beware of attaching too much importance to a label which is itself an invention of a philosopher – moreover, a philosopher (Kant) with an over-systematising mind. Modern commentators like to emphasise the 'empiricist' aspects of Descartes and Leibniz (even the empiricist side of Spinoza), all of whom were powerful *scientific* thinkers, whose philosophy arose in part from the problem of explaining the world as they observed it. Nevertheless, for the sake of convenience the division between empiricists and rationalists has become generally accepted.

One question that the rationalist has to face, which parallels the questions posed to the empiricist, is how to account for *contingent* truths – truths which 'might have been otherwise'. The results of rational reflection always have an air of necessity. We can prove them, but only by proving that this is how things *must* be, as in mathematics. If the really real is known by reasoning, then are there any contingent truths about it? Only Spinoza was prepared categorically to answer 'no'.

6. Realism

This 'ism' has become particularly important in recent decades, for complicated historical reasons that need not concern us. Roughly speaking, you are a 'realist' about *x* if you think that *x* exists independently of our thoughts about it, our experience of it, and so on. In this sense most people are realists about tables and chairs, but not about characters in myth and fiction. Verificationism might be seen as a thorough-going *anti*-realism, and the term 'anti-realism' has recently been generalised to cover a variety of theories, among which verificationism is a special case. Previously philosophers might have contrasted (as Kant did) realism with *idealism*. Reluctance to call 'verificationism' a form of idealism has led to the use of the broader label.

This topic has become extremely complex, since 'realism' is now used to describe a general position in the theory of meaning. I return to it in Chapter 19, by which time it will be possible to explore some of the subtle positions that lie between realism and idealism, for which 'anti-realism' is only one possible name.

7. Relativism

This is frequently adopted by those impatient with the burden of sceptical argument, for it seems to cut through the whole dispute, leaving the individual sovereign over his little opinions. It tells us that there is no such thing as objective truth, since all truth is 'relative'. In argument about moral problems, relativism is the first refuge of the scoundrel. 'That is your opinion,' says the relativist, 'and you are welcome to it. But it is not my opinion, and *I* am welcome to *mine*. No opinion has authority apart from the point of view which adopts it. I have as much right to believe that adultery is right as you have to believe that it is wrong. Neither right nor wrong exist, apart from the opinions which we entertain about them.'

Relativism involves accepting a certain kind of scepticism. It has secured a foothold not only in morality, but also in the philosophy of science. Thomas Kuhn's *Structure of Scientific Revolutions* purports to show that scientific thinking always involves the adjustment of a theoretical 'paradigm', which is without justification since it defines the terms of the debate. Under the influence of this idea relativism enjoyed an enormous following in the seventies.

But the debate is far older than Kuhn. Indeed, it goes back to Plato, whose *Theaetetus* contains a sustained response to the relativism of Protagoras. Plato begins by asking whether a relativist can provide an account of expertise, teaching, discussion and debate, without introducing the notion of objective truth. Plato does not define expertise; instead he asks 'what do people *think* about expertise?' (170ab). The answer is that they think some people to be knowledgeable, others to be ignorant, and that the ignorant make false judgements. But Protagoras holds that there can be no false judgements. So either people are right to think there is false judgement, in which case Protagoras is refuted; or they are wrong, in which case Protagoras is also refuted, since their judgement is an example of a false one.

The relativist might reply that the argument *assumes* what is in issue, namely the concept of absolute truth. Maybe it is true *for me* that there are false judgements; but it does not follow that it is true *for Protagoras*. As Plato shows, however, the objectivity of our beliefs is not jeopardised, if relativism is true only for the *relativist*. Moreover in asserting that relativism is true for *him*, the relativist asserts that it is true for him absolutely. He is committed to absolute truth by the very practice of assertion, which has absolute truth as its goal.

The dispute here has not ended. But two things are certain: *vulgar* relativism has no hope of surviving outside the minds of ignorant rascals; *sophisticated* relativism has to be so sophisticated as barely to deserve the name.

4

Self, Mind and Body

Let us return now to Descartes's *Meditations*, and in particular to the second Meditation, in which Descartes ventures his first answer to the hyperbolical doubt. You will remember that Descartes had a variety of arguments, two of which in particular seemed to convince him that he could not trust the evidence of the senses. Descartes's own way of responding to these arguments was with a question: Of what can I be certain? Or, to put it in another way: Is there something which I cannot doubt? If there is such a thing then it, and it alone, can provide foundations for knowledge.

Descartes's answer to this question is often known as the '*cogito*', from an earlier statement of the argument, summarised in the words '*cogito ergo sum*': I think therefore I am. (See *Discourse on Method*. The formulation is in fact due to Descartes's friend Jean de Silhon.) This, however, is not what Descartes says in the *Meditations*. Here he says rather that the proposition 'I am, I exist' is necessarily true each time I pronounce it, 'or inwardly affirm it'.

The phrase 'necessarily true' must be read carefully. After all, the proposition that I exist is *not* necessarily true. What is necessarily true is another proposition – namely, that if I think then I exist. (The necessity of this is, however, no different from the necessity of 'If I eat then I exist'; hence the inadequacy of Descartes's original formulation of his argument.) It is vital to recognise that the propositions which Descartes proposes as immune from doubt include propositions like this – 'I am' – which are *contingently* true. Unlike a truth of mathematics, this proposition might have been false. Being contingent, it promises to offer foundations for a theory of the world as it *happens to be*. I know the world contains me; it might not have done, but it does. Here, then, is the starting point for my theory of what the world in fact contains.

Is Descartes right in thinking that 'I exist' is immune from doubt? The modern consensus is that the truth of this sentence is guaranteed by the rule that 'I' refers to the speaker. The proposition that I exist can be expressed (whether overtly or in thought) only by someone who thereby refers to himself. The successful utterance of the proposition therefore presupposes the existence of the person uttering it. The proposition cannot be entertained without being true. So Descartes was right.

But what follows? Descartes was aware that he could not really build on the proposition that he exists: to say that something exists is to say nothing, if we cannot also say *what* exists. I need to know what kind of thing I am, if I am to advance a theory of the world that contains me. Descartes entertains, one by one, various propositions *about himself* – that he is a reasonable animal, that he has hands and face and arms, that he has a body which he can move at will; and so on. But all of these he believes he can doubt. The demon could be deceiving him into the belief that he has arms, face and a body. The same is true of any physical attribute – of anything that 'pertains to the nature of body'. Only if we explore the soul do we begin to find some refuge from the demon. However much I doubt the truth of propositions about my body, I cannot doubt that I am thinking:

> What of thinking? I find there that thought is an attribute that belongs to me; it alone cannot be separated from me. I am, I exist, that is certain. But how often? Just when I think; for it might possibly be the case if I ceased entirely to think, that I should cease altogether to exist.

Descartes goes on to conclude that he is a thinking thing, in other words 'a thing which doubts, understands, conceives, affirms, denies, wills, refuses, which also imagines and feels'.

Descartes is using the term 'thought' to cover a great many things, as that sentence shows. In fact, thought includes every *present mental state of the subject*. Concerning all such states, I am immune to error, and the demon cannot deceive me. This is true not only of the 'intellectual' mental states, but of sensory perception too. As Descartes puts it:

> I am the same who feels, that is to say, who perceives certain things, as by the organs of sense, since in truth I see light, I hear noise, I feel heat. But it

will be said that these phenomena are false and that I am dreaming. Let it be so; still it is at least certain that it seems to me that I see light, that I hear noise and that I feel heat. That cannot be false; properly speaking it is what is in me called feeling (*sentire*): and used in this precise sense that is no other thing than thinking.

Descartes's argument involves a move beyond the original immunity from doubt of 'I exist'. The rule that 'I' refers to the speaker does not guarantee the truth of every proposition involving 'I'. Consider 'I walk'. I could say this, and think it, even though it is not true (because I am dreaming). Even 'I am speaking' fails to be self-validating. For though this proposition is (necessarily) true whenever uttered aloud, I may mistakenly believe that I am uttering it (as opposed to thinking it or dreaming it). Only *inner* speech (i.e. thought) gives me the full immunity from doubt that Descartes is seeking. But it seems that *everything* that could be described as thought (every present mental state) is immune from doubt.

I can doubt that I am speaking, but not that I am thinking. Nor can I doubt that I am in pain, or having a certain visual experience, wanting to go home; etc. There seems to be something about my present mental states which puts them beyond doubt. But propositions about my bodily states – walking, speaking, weighing 12 stone, etc. – can be doubted. From this idea arises another: namely, that the 'immunity to error' of my beliefs about my own mind, contrasted with the error-prone nature of my beliefs about the physical world, derives from a fundamental difference between the mental and the physical – an 'ontological divide'. My mental states are more truly *part* of me than my physical states, and that is why I stand to them in this privileged relation, which guarantees my claims to knowledge.

In the sixth Meditation, Descartes goes further, attempting to establish a 'real distinction' between soul and body. His argument is roughly this: I can doubt all propositions about my body; but I cannot doubt propositions about my soul (i.e., about my present mental states). In particular, I cannot doubt that I am a thinking thing. If anything is true of me, *this* is true. But since I can conceive of the falsehood of all propositions about my body (including the proposition that I have a body) while being unable to conceive of the falsehood of the proposition that I am a thinking thing, then the former could be false, even though the latter must be true, just so long as I exist at all. From this it seems to follow that I am *essentially* a thinking thing (a soul), but only *accidentally* (or contingently) a body.

Moreover, there seems to be nothing wrong with the hypothesis that the soul could exist without the body.

The argument for the 'real distinction' is deep and tricky. (For an excellent discussion, see Bernard Williams, in *Descartes: the Project of Pure Enquiry*.) I mention it here, so as to indicate the direction in which Descartes's thought is tending: towards the view now known as the 'Cartesian' theory of the mind. According to this view the mind is a non-physical entity (or 'substance'), which is transparent to its own awareness, and connected only contingently with the world of physical objects. Each subject (each 'I') is identical with such a mind, over which it exerts a kind of epistemological sovereignty. The relation between mind and body is therefore problematic. (For example: what makes my body *my* body?)

Descartes did not present the theory in quite those terms. In particular he did not use words like 'mental' and 'physical'. For him all mental states were modes of 'thought' (he also used the word 'idea' to refer indifferently to perceptions, thinkings, willings, and sensings). Likewise all physical states were modes of 'extension' (which corresponds to our idea of 'matter in space'). He concludes the second Meditation with a famous argument (about a piece of wax) designed to show, first that extension is the *essence* of ordinary (physical) objects (just as thought, he later argues, is the essence of the soul); and secondly, that we can know this fact by rational reflection, but not by empirical observation. (The argument is therefore authority for the view that Descartes is a rationalist.)

Cartesian Approaches to the Mind

Descartes's argument was designed to roll back the tide of scepticism. But its influence has not been confined to that sphere. More important, historically, has been the Cartesian theory of mind, as this emerges from the anti-sceptical argument. This theory has been adopted in one or another form by empiricists, rationalists and phenomenologists. It is worth reviewing what each of these schools has made of it.

A. Empiricism

Classical empiricism maintains the following positions:
(i) All claims to knowledge are based on knowledge of experience: experience is the 'foundation' of knowledge.

(ii) Experience can provide such a foundation only if my beliefs about my present experiences are immune to error. (Otherwise they too need a foundation.)
(iii) But my experience does provide a foundation, since the realm of experience is 'set apart' from the physical world: it is a realm of 'privileged access', where I (and I alone) am sovereign. Your beliefs about my mind will never be as well founded as mine, since mine require no further foundation. They are self-founding.

While this sequence of thoughts does not lead inexorably to the Cartesian theory of mind, it has certainly persuaded Locke, Berkeley and others to adopt that theory, and so to be landed, like Descartes, with a problem concerning the relation between mind and body. The problem is particularly acute for an empiricist. For is it not strange that my knowledge of the physical world should be based on my knowledge of something that is not even a part of it? Hence Berkeley's idealism.

B. Rationalism

Descartes's 'real distinction' was derived from rationalist considerations concerning the essence of soul and body, as these are understood by reason. Subsequent philosophers took the Cartesian division of the world into thought (the attribute of soul) and extension (the attribute of body) very seriously. Even if they did not accept the details of Descartes's theory of mind, the great 'ontological divide' between thought and extension survived in their philosophies. Spinoza tried to heal this divide, by making thought and extension into attributes of a single 'substance'. But even in Spinoza the belief persists that thought is fundamentally 'apart from' the physical world, and to be studied in its own terms. In Leibniz, too, the Cartesian theory survives in vestigial form; indeed souls for Leibniz have more fundamental reality than bodies.

Like Descartes, rationalists had little difficulty in entertaining the view that the soul can survive the body. Most also suggested that the soul has no spatial properties – that it exists 'nowhere'. Both rationalists and empiricists tended to follow Descartes in using one comprehensive word to denote the objects of all mental states – the word 'idea'. For the rationalist the paradigm of an idea is a concept; for the empiricist the paradigm is a sensory experience. (Hume made a certain advance by distinguishing these two things, calling the first idea, the second impression. He then went on to spoil the achievement, by describing ideas as faded impressions.) In general, theories of mind were, during the early years of modern philosophy, hampered by an inability to make systematic distinctions among the various kinds of mental object.

C. Phenomenology

The most recent revival of the Cartesian approach to the mind – phenomenology – has escaped that disability. Phenomenology arose from a desire to analyse the 'given' in its full complexity, and is described by its principal founder, Edmund Husserl, in his significantly named *Cartesian Meditations*, as an *a priori* 'egology' – an *a priori* study of the ego. Just as Descartes had begun by doubting everything that he could not prove, so should the study of the mind begin, according to Husserl, by 'bracketing' all that is not given to consciousness. For example, suppose that I am angry at the political slogans that have been scrawled on the blackboard. It is not part of my *anger* that those slogans are really there. I may be suffering from a fit of paranoia which causes me to see what does not exist. Hence the slogans are to be 'bracketed', as not belonging to the mental state of the subject. This is the method of phenomenological 'reduction', which leads to the pure mental content, and to the 'essence' contained in it. In the course of the reduction, the entire physical world is bracketed, leaving only a Cartesian remainder, a subject who is 'apart from' the world on which he meditates.

But Husserl does not stop there. This meditating subject is, he says, a merely 'empirical self', part of the everyday furniture of the world. It too must be bracketed, since it is not an object of consciousness. The phenomenological reduction abstracts even from the thinker, leaving only a pure consciousness, which Husserl calls the 'transcendental ego'. It is difficult to determine the status of this transcendental ego. But there is no doubt that Husserl's method has led him into a view of the mind that is largely Cartesian: whatever the mind is, it is through 'inner' awareness that the mind is known; and it is doubtful that it even makes sense to describe the self – whether 'empirical' or 'transcendental' – as a part of physical reality.

The distinction between 'empirical' and 'transcendental' self was in fact introduced by Kant, and it is interesting to note that it was used by Kant precisely to cast doubt, not only on the Cartesian theory of the mind, but also on the whole idea that there is a realm of 'inner' awareness, from the study of which a knowledge of essences can be derived. The very process which leads us to retreat to that inner realm, Kant argues, deprives us of every object of knowledge, and leaves us at last empty-handed, invulnerable to doubt precisely because there is nothing left to doubt:

> We can lay at the foundation of psychology nothing but the simple and in itself perfectly contentless presentation *I*, which cannot even be called a conception, but merely a consciousness which accompanies all conceptions. But of this I, or he, or it, who or which thinks, nothing more is presented than a transcendental subject of thought = *x*, which is cognised only by means of the thoughts that are its predicates, and of which, apart from these, we cannot form the least conception.

That passage (from the second edition of the *Critique of Pure Reason*, B404), is part of a sustained argument against the Cartesian theory, anticipating the argument that I shall discuss in the following chapter. One of the most exasperating aspects of Husserl is that, when pillaging Kant for his language, he overlooked so many of the arguments.

Even if the world is 'bracketed' by our reduction, the *reference* to the world remains. I cannot describe my anger, and distinguish it from other mental states, if I neglect to mention that it is anger about those ghastly political slogans. The 'aboutness' of the mental state is intrinsic to it. My anger contains a 'slogan-shaped' hole, so to speak, which must be filled by an appropriate entity in the world if the emotion is to be 'well-founded'. This feature of 'aboutness' is characteristic of many (according to some phenomenologists, of all) of our mental states. The feature is usually called 'intentionality' (from the Latin *intendere*, to aim). (Husserl was probably the inventor of this particular technical term; his master, Brentano, had written instead of 'intentional inexistence', a phrase that he borrowed from the medieval scholastics.) In addition to its intentionality, a mental state will be analysable in terms of its component parts and non-detachable 'moments', such as its intensity. Other phenomenologists distinguish mental states in terms of their relation to time (some states can be precisely dated, others not); whether they are 'subject to the will' (you can command me to imagine something, but not to believe it); and so on. All these interesting ideas led the phenomenologists to attempt comprehensive divisions of the mental realm, into the various species that inhabit it. Yet the results are delivered by a study of the 'first-person' case: of how the mind is presented to consciousness. Hence phenomenology tends inevitably to endorse the Cartesian distinction between mental and physical, inner and outer, that which is really and essentially *me*, and that which is just standing in my environment (if environment is the right word).

Consciousness and Self-consciousness

Doubt is brought to an end, for Descartes, by the study of the 'first person': the subject of doubt himself. In the course of the argument, however, the first person has acquired a character. He is not merely conscious, but self-conscious. He is able to use the word 'I', in order to describe his own condition and to distinguish himself from those things which are other than himself. The distinction between self and other is built into the very basis of his world-view, by the attempt to answer the sceptical question. This may mean that the Cartesian theory is not a theory of the mind as we normally understand it, but a theory of the self-conscious mind.

Not all minds are self-conscious. Animals, for example, are normally credited with minds, however rudimentary. At least they have mental *states*. For example, they see things, desire things, even think things. A dog may think there is a cat in the garden; he may smell a rat, see his master, hear his neighbour barking; he may want his supper, fear the postman and so on. Yet in none of this does he make a distinction between self and other. Certainly, he never attributes these mental states to *himself* as their subject. Why, then, do we say he has these states? Neither he nor anyone else seems to have the 'privileged access' to them – the immunity to error – on which the Cartesian theory of consciousness is based. It is true that the dog makes no mistakes about his mental states; but that is because he makes no judgements about them at all. They *are* his judgements, so to speak.

Descartes inclined to the view that animals do *not* have minds. They are a peculiar kind of living machine, like us in many respects, but without that crucial thing – the soul – which distinguishes us among the works of nature. Most people find this very implausible; but most people also find it hard to prove that animals *do* have minds, since they remain instinctively wedded to the Cartesian view that the mind is essentially the self, revealed to itself in the act of introspection. On the other hand, they are also attached to the view that the mind is not 'apart from' the physical world, but a real and active component of it. Why, after all, do we attribute mental states to animals? Is it not because this is a very good way of explaining their behaviour? And how could mental states explain behaviour if they did not cause it?

Causation, as normally understood, is a relation between physical things. Should we therefore conclude that the mind is physical? The problem becomes acute when we turn back to the human case. For here too we invoke mental states in order to explain behaviour. ('He killed her out of jealousy'; 'She smiled because she was happy to see him'; 'He resigned because he was affronted'; and so on.) One of the greatest difficulties facing the Cartesian is that of reconciling the theory of the mind as a non-physical entity, revealed to introspection alone, with the view that the mind acts on, and is acted upon by, the physical world.

The Unconscious

Someone might wonder how the Cartesian theory can be reconciled with Freudian psychology, and with the many philosophical and psychological theories that have postulated an unconscious component in the human mind. For Descartes, to speak of an unconscious thought is close to nonsense. Leibniz was more flexible, arguing that much mental activity escapes the purview of self-consciousness (which he called 'apperception'), since it occurs either too rapidly or too minutely to be observed. On the other hand, recent writers seem to be committed to the existence of an 'unconscious' which is more deeply hidden than that implies: something not just outside the reach of self-consciousness, but not conscious at all.

Now if you thought that the mind is a physical thing – for example, the brain – you would surely be prepared to admit that it has unconscious states: physical states which simply are never the objects of awareness, like the level of some bacterium in the bloodstream. But that is not what people have in mind when they refer to the unconscious. They mean that I might have some state, which is just like this belief, desire, or perception before my mind, except that it is *not an object of consciousness*. Having introduced mental states solely as ideas (objects of consciousness) and conscious thoughts it is hard for Descartes to allow such a suggestion. But maybe his is only *one* way of characterising the mental. And maybe there are other ways, which do not have the implication that every mental state is also fully known to the subject who has it.

Of course, if we admit the existence of unconscious mental states, we must still recognise the reality of *consciousness*. And indeed, since it is this which defines my view on the world, it is open for Descartes to argue that, in providing an account of consciousness, he has described what I essentially am. The self and the mind, he will argue, are one, and all mental states are therefore conscious by definition. Reference to the unconscious is perfectly legitimate, provided that the unconscious is conceived as non-mental, or pre-mental, in the manner of a physical condition that may, at some time, intrude into consciousness, by becoming the cause of mental states. But that is of no philosophical interest.

However, the matter is not so simple. For what is this 'self' which is the seat of consciousness? Is it ever the object of consciousness? Or is it a pure subject, a transcendental ego, as Husserl claimed? And if we can never observe it (since it exists only as observer, and never as observed), why are we so sure that it can have no unconscious states? Those vertiginous questions may seem unanswerable (and indeed, Hume was inclined to say that the self is a kind of illusion, as are all the conundrums which derive from it). However there are those who have regarded them as the starting point of philosophy. Fichte, for example, argued that there can be no knowledge at all, unless the self knows itself as object. But in doing so, the self becomes an object and so alien to itself: the self as known is therefore a 'not-self'. From such paradoxical beginnings Fichte derived a metaphysical system in which, he supposed, all possible knowledge was contained and justified.

Rather than speak of the 'self' and the 'subject', however, we should look at the core idea of Descartes's *cogito*, which can be expressed without reference to the concept of consciousness at all.

Immunities to Error

Consider the following propositions:

(1) I exist.
(2) I think.
(3) I am dreaming.
(4) I am in pain.
(5) I want a drink.
(6) I am six-foot tall.
(7) I have a body.
(8) I am identical with Roger Scruton.

Some of these I can doubt, some not. Which of them is immune to doubt, and why? And is immunity to doubt the same as immunity to error?

The first proposition is immune to doubt, according to Descartes, since the mere fact that I think it, is sufficient grounds for its truth; moreover, I can know that this is so. Likewise with the second proposition. And here the grounds for the truth of the proposition are contained in the very *fact* of doubting. Doubting is thinking: to doubt this proposition is not just to give grounds for its falsehood: it is to *falsify* it. The third proposition is also interesting. For I *can* doubt that I am dreaming, even when I am dreaming. (I might dream that I am dreaming: but then I really do dream.) And I might think that I am dreaming when I am not. The explanation lies in the fact that to describe a thought or experience as a dream is to say something about its external circumstances, and about those, which are outside the reach of consciousness, I may be in error.

The fourth proposition is interesting for yet another reason. To doubt that I have a pain is not to have a pain: nor is it to give proof that I have one. So there is no way in which we can construe this proposition as self-confirming, either in the manner of (1) or in the manner of (2). Yet it seems to be immune from error. If I think that I am in pain, then I am in pain. Moreover, if I am in pain, I know it. Some philosophers argue, therefore, that sensations, and other similar mental states, are 'incorrigibly known', and also 'self-intimating' (i.e., if they are present, I know that they are). While this position goes naturally with the Cartesian view of thought, it cannot be explained as the *cogito* is explained. Indeed, its explanation is one of the unsolved questions of modern philosophy. (See Chapter 31.)

Maybe something similar is true of (5). For desires seem to be objects of awareness in just the way that sensations are; and the supposition that I can make mistakes about them is fraught with difficulty. On the other hand, the fact that philosophers are tempted to endorse the theory of unconscious desires, even when virtually no philosopher has ever countenanced unconscious sensations, suggests that things are not so simple. Furthermore, I might find myself saying 'I don't know whether I want a drink or not': Could this ever mean that I *do* want a drink, but just don't know it? Or does it mean something else?

About (6) I can be in error. I can also doubt it. This is Descartes's paradigm of a proposition which touches on what is accidental and external to the self. But what about (7)? I can certainly *doubt* that I have a body: after all, Descartes did so. But can I really be in error? Imagine a philosophical argument for the anti-Cartesian view that nothing can think without a body. If that is so, then I can deduce from the fact that I think that I have a body: moreover, I can never be in error when I believe that I have a body. Only embodied beings can believe anything. This exemplifies a strategy made famous by Kant. What is presupposed, Kant asked, in my ability to entertain the thought 'I think'? Whatever it is, we know that it is *true*.

The final proposition is a real teaser. Surely I cannot doubt that I am identical with the person who I am! On the other hand, there may be another sense in which I may make a mistake as to who that person is. I may believe that I am Napoleon: or even, in my delusions of grandeur, that I am Roger Scruton. Such crazy mistakes do occur. At the same time, I always successfully refer to just the very person that I am, whenever I refer to myself in the first person. I cannot be mistaken in identifying myself as 'I', any more than I can be mistaken in identifying the time as 'now', or the place where I am as 'here'. This makes it look as though immunities to error arise from grammar. That is the deep and spooky suggestion that we find in Wittgenstein, who uses it to overthrow the entire Cartesian picture.

5

The Private Language Argument

Descartes took refuge from the demon in the first person. The self is the place where doubt expires; it is the one absolutely certain thing; it is also metaphysically distinct from everything else over which my beliefs may range. It is not only the anchor of epistemology; it is also the starting point of metaphysics.

Descartes's own response to the hyperbolical doubt does not, however, rely on features of self-consciousness. He turns his attention instead to his proof of his own existence, and asks what enabled him to be persuaded by it. The answer is that the proof was composed of 'clear and distinct ideas': ideas which are grasped completely by the intellect, and which are unmixed with any others that might confuse them. (Very roughly, this corresponds to our notion of a 'self-evident' argument.) He deduced from this that clearness and distinctness are marks of truth, and thereby acquired a criterion for distinguishing true ideas from false ones. Any proof composed of clear and distinct ideas must lead to true conclusions. He produces two such proofs for the existence of a supremely benevolent God, and so exorcises the demon.

I shall return to Descartes's proof of God in a later chapter; what is important here is to grasp Descartes's strategy. He begins from the subjective sphere, and the beliefs that he entertains there; he establishes his own existence, and then the existence of God. By virtue of God he rises out of the subjective sphere (the sphere of pure *seeming*) and attains a point of view on the world which shows the world as it *is*. By reflecting on the nature of God he can deduce that the world is as it seems to him, since God would not deceive him. If he makes mistakes, it is because he is not using his faculties as God intended.

This argument exhibits a pattern that occurs elsewhere. It begins from the subject, and the sphere of seeming where he is sovereign. It then argues outwards to an 'objective' viewpoint. From that viewpoint it establishes the existence of an objective world, and the sphere of being is constructed from the result. Such a pattern of argument is typical of the epistemological position known as 'foundationalism'. This tries to justify our beliefs by providing certain (or indubitable) foundations for them. Since indubitability attaches to first-person knowledge (to the sphere of seeming), such knowledge tends to become the foundation of all other beliefs.

Recent philosophers have been impressed by the fact that, once this approach is adopted, with the sphere of seeming as the premise, it becomes actually impossible to emerge into the objective world. The hope of constructing a bridge to the objective viewpoint, and of establishing justified claims about a sphere of being outside the self, rapidly dwindles, as we explore the vast assumptions that have to be made in the course of such a construction. Descartes could only make the journey via a proof of God's existence which few would regard as valid. Hume never made the journey at all.

But maybe there is something wrong with the premise? Maybe Descartes is wrong in assuming that he is entitled to be certain about the subjective sphere (the sphere of seeming)? That is the striking suggestion that emerges from a series of modern 'anti-Cartesian' arguments, among which Wittgenstein's argument against the possibility of a private language is the most famous.

1. Background

In the last chapter I briefly surveyed the manner in which the Cartesian theory of the mind has been either overtly or tacitly endorsed by modern philosophers. Its presence is felt in verificationism and empiricism; it underlies much traditional epistemology, through its endorsement of the contrast between inner and outer; and it is assumed by phenomenology, with its emphasis on the 'transcendental subject' and the 'given'. But what exactly does the theory *say*?

This question is not the same as 'What did *Descartes* say about the theory?' The 'Cartesian theory' of mind has evolved since the day of its supposed invention, and what now goes by that name is a synthesis of various views, many of which were indeed held by Descartes, brought together in a way of which he may not have approved. Here then is the theory in its modern form – the form under which it is usually attacked.

The mind is a non-physical entity whose states are essentially 'inner' – that is, connected only contingently (by no necessary connection) with 'outer' circumstances. The subject himself is identical with his mind, and has a peculiar 'privileged' view of it. In particular, his own present mental states are known to him indubitably. If he thinks he has them, then he does; and if he has them he is aware of it. Moreover, he cannot make mistakes about them. (This applies only to his *present* mental states: for clearly he can make mistakes about his past and future states.) He has no such privileged view of his own physical states – and indeed this is one of the grounds for thinking that mind and body are distinct things. Nor does he have such a privileged view of other minds. 'First-person privilege' is a distinguishing mark of the mental and attaches to every mind, or at any rate, to every self-conscious mind.

The picture has two parts: first that the mental is non-physical, separate from any 'outer' process. Secondly that each subject has 'first-person privilege' in relation to his present mental states. The second thesis is supposed to imply the first, but does it? When people attack the Cartesian theory, they usually seek to reject its first part – the theory that the mind is non-physical. To attack the second part is more difficult; since, with elaborate qualifications, this part is true.

In addition, Descartes held that the mind is also a *substance*. This claim – whose precise sense is far from clear, and which I shall consider at a later point – is not commonly regarded as essential to the 'Cartesian' theory, as this is now discussed.

2. Epistemology versus Anthropology

The Cartesian theory is associated with a particular approach to philosophy, which, because it emphasises the first person (the self) and his predicament, inevitably centres on questions of epistemology. I can be certain of my own states of mind. But this certainty seems to set me apart from the rest of reality. So how can I know that there is anything *besides* myself? Only if I can use my indubitable knowledge of my own experience as a *basis* for my beliefs about a larger world. To show that I can do so is the primary task of epistemology.

Suppose, however, that we were to forget the self for a moment, and think about the other. Surely, I could ask the question 'How does *he* know?' But here nothing that I say by way of an answer would be indubitable to me. I would be describing his epistemological capacities, without either raising or answering the sceptical question. All the same, there may be some philosophical sense in pursuing the matter. It may be useful to give a philosophical description of the 'epistemological predicament' from a point of view outside it. For instance, we could imagine a philosophy which argued for the following conclusions:

(a) Only a being of a certain kind can have knowledge of an objective world (where objective means 'independent of the knower', or, 'not dependent upon the knower for its nature or existence').

(b) Only a being of a certain kind can have knowledge of the grounds of its own knowledge. (Maybe *we* have that knowledge, but dogs do not.)

(c) Only a being of a certain kind can have knowledge of its own mental states.

(d) Only a being of a certain kind can have 'first-person privilege'.

(e) Only a being of a certain kind can have epistemological (and specifically sceptical) problems. My horse has never had problems of that kind; when he refuses to jump, it is never because he doubts the *existence* of the world that lies before him.

Such conclusions could form part of what Kant called a 'philosophical anthropology', by which he meant an *a priori* description of our condition, which will show at the same time what has to be true if we are to have beliefs and experiences. This philosophical anthropology gives precedence to the third-person viewpoint. It also by-passes the epistemological problems that bother the Cartesian. Instead of asking 'how do I know?' it asks, for example, 'what kind of a creature can ask the question "how do I know?"?'

Suppose, however, that our philosophical anthropologist came up with the following conclusion: 'Only a creature who inhabited a world of physical objects could doubt that he did so'. Surely that would offer us a way out of the epistemological impasse. For I could at once deduce, from the very fact that I have the epistemological question about the world, that the question can be answered in the affirmative. I do not *base* my conclusion about the physical world on my knowledge of my experience. It follows simply from the fact that I have the question. I can answer the question in the affirmative, without proposing 'foundations' for my beliefs, and without giving any special priority to my own experience.

Our anthropologist could come up with even more surprising conclusions (if he could find the arguments!). For example, he might be able to drive a wedge between the two parts of the Cartesian theory of consciousness. He might find reasons for saying that only a creature whose mind was *part* of the physical world could have 'first-person privilege'. Only if the Cartesian theory of the mind is false do we have the privileged access to our own mental states which disposes us to believe it. That, roughly, is the conclusion of the 'Private Language Argument'.

Kant initiated the shift away from epistemology towards philosophical anthropology. It was taken further by Hegel, whose theory of 'objective spirit' is really an extended and subtle description of the human condition from the third-person perspective. Unfortunately, Hegel's paradoxical language and ebullient self-confidence led to his being ignored by philosophers once the initial shock had died down. Only with the later Wittgenstein has this line of reasoning come into its own, and epistemology begun to take second place on the philosophical agenda. When philosophers, following W.V. Quine, refer to a 'naturalised' epistemology (or epistemological 'naturalism') it is by way of endorsing a particular development of the anthropological viewpoint. (See W.V. Quine, 'Epistemology naturalized', in *Ontological Relativity & Other Essays*, 1962.) A naturalised epistemology is precisely an epistemology that has been demoted to second place. The debate is now about what should occupy the *first* place: Physics? Metaphysics? Pataphysics?

3. Wittgenstein's Argument

The Private Language Argument is scattered among sections 243 to 351 of the *Philosophical Investigations*. It has many parts, and is not fully intelligible outside the context of Wittgenstein's later philosophy of language. Some of its broader conceptions may, however, be understood at this stage, and they will suffice to present a powerful challenge to the Cartesian theory of the mind.

According to the Cartesian, mental states are private, in the quite specific sense of being accessible, and therefore knowable, only to the person who has them. Another way of putting this is to say that they are separate from, or only 'contingently connected with', the public world. By 'public world' I mean a world whose contents are equally accessible, equally knowable, to *more* than one person. So let us say that the Cartesian holds that mental states are 'private entities' or 'private objects' in this sense. The question is whether that view is true: i.e. whether mental states are private objects.

Well, what do we mean by 'mental state'? Wittgenstein's paradigm is sensation: something that we feel, and to which we have privileged access. Why do we call this mental? What precisely do we *mean* by 'mind'? This is a question that Wittgenstein does not directly answer. But we can already suggest *one* part of an answer to it. The mind is the thing upon which I have a 'first person perspective'. My mental states are those which are transparent – which are what I believe them to be, and which are always revealed to me. Wittgenstein does not like the normal ways of describing this 'first-person privilege'. In particular, he thinks it is nonsense to say that I *know* that I am in pain – since that implies the possibility of error. (See section 246.) He prefers to put the point this way: 'it makes sense to say about other people that they doubt whether I am in pain; but not to say it about myself'. (246). But he also has another and more suggestive way of putting the point, which is this: that in my own case I do not need a criterion in order to identify this, that I have now, as a pain. I have no 'criterion of identity' for my own sensations. (See section 288, which contains the Private Language Argument in a nutshell.)

Why is this? Wittgenstein does not exactly say. But he suggests that we are dealing with a fact of 'grammar': a fact about self-reference in our language. Our public language is so constructed that, if someone were to apply procedures and to make 'mistakes' in applying the word 'pain' to himself, that would merely show that he did not understand the word 'pain'. We learn the words for sensations in a way that ensures their privileged first-person use. Imagine the opposite case. Imagine a person who, whenever asked 'are you in pain?' 'are you thinking?' and so on, replied 'I am not sure, let me see', and set about to study his own behaviour. Could you make sense of this? He might continue in this way: 'I am behaving like someone in acute pain, certainly. Or at least, so it seems to me. Or rather does it seem that way to me? I am not sure. Actually I am not sure whether I am not sure, since the behavioural evidence for saying that I am not sure is far from conclusive. Or at least, it seems to me far from conclusive. Although maybe I shouldn't say that – it is so difficult in such a case to say what things seem like to me.' And so on. Clearly something has gone wrong here. The first-person use of mental terms has not even got off the ground. You would be tempted to say that this person has simply not understood the crucial difference between 'he' and 'I': he does not have a grasp of the idea of self. Maybe he *has* no self. In which case, who is speaking?

The Cartesian position is this: that the privileged access that we have to our own mental states is explained by their privacy. It is because only I can know, that I really *know*. (Putting it that way already suggests that there might be something wrong with the Cartesian position.) But suppose sensations really were 'private objects'. Would we have the kind of 'privileged access' that guarantees our first-person reference to them?

Wittgenstein imagines the following case:

> Suppose everyone had a box with something in it: we call it a 'beetle'. No one can look into anyone else's box, and everyone says he knows what a beetle is only by looking at *his* beetle. Here it would be quite possible for everyone to have something different in his box. One might even imagine such a thing constantly changing. But suppose the word 'beetle' had a use in these people's language. If so it would not be used as a name of a thing. The thing in the box has no place in the language-game at all; not even as a *something*: for the box might even be empty. No, one can 'divide through' by the thing in the box; it cancels out, whatever it is. (Section 293.)

The phrase 'language-game' is Wittgenstein's shorthand for a rule-guided practice in which many people can join. (One of his contentions is that language-games are of different kinds, and that we go wrong in philosophy by trying to understand one game in terms of the rules of another.) His argument is meant to show that you can never introduce into a public language a word which refers (in that language) to a private object. For just look at the case. Someone uses the word 'beetle' just as I do, even though in his case there is *nothing* in the box. Yet we both agree that he has a beetle, and that 'beetle' is the correct description of the thing that he has. So it cannot be the 'private object' that we refer to. It 'drops out of consideration as irrelevant'.

To this one is tempted to reply as follows: Maybe words in a public language cannot refer to private objects. But the words of our public language are used also in another way. Each of us has his own *private* language, in which he refers to what can be known only to himself – his 'private objects'. And this is a language that no one else can understand, since no one else can know whether I am using the words of that language correctly. Is this a coherent idea? Wittgenstein says not. We think that it is coherent, because we believe that we have 'privileged access' to our private objects: we just know when they occur, and cannot be mistaken about it. So there is no problem in inventing a language whose words latch on to these things. I can always be sure that I am using the words of such a language properly, since I can be sure, without having to check up on the matter, that the thing referred to is really there.

Here is the subtle part. Wittgenstein asks us to look back at our public language, in which, as he repeatedly says, we *do* refer to sensations. This already casts doubt on the view that *sensations* (as we normally refer to them) are private objects. It is part of the 'grammar' of sensation words that we do not make mistakes when using them of ourselves, provided that we understand them. That is what our first-person certainty amounts to. The inventor of a private language is supposing that just the same guarantee is available to *him*. But it was a guarantee provided by the grammar of our public language; so he is not entitled to assume that he has it. For him the question can arise: how do I know that this, that I have now, is a private object (or is the same private object that I had when last I used the word 'S')? Once the question has arisen, we find that there is no answer to it. There is no criterion to be found, which would guarantee to the user of the private language, that he really has identified a *private* object, and really has succeeded in endowing his words with a private reference, or indeed with any reference at all. He could be wrong about this, and yet never know that he is wrong. And if he is right, he could not know that either. Since nobody else knows anything about the matter, he cannot appeal to other speakers. So there simply *is* no criterion of correctness, no rule that attaches his terms to intrinsically private entities. The private language is impossible. We imagine that the private-language user has a guarantee of reference, because we wrongly describe his situation in terms of the public language: in terms of the 'first-person privilege' which is part of the *public* grammar of sensation words. This public grammar is tied to the process whereby sensation words are taught and learned – through the expression of sensations in behaviour. But 'if I assume the abrogation of the normal language-game with the expression of sensation, I need a criterion of identity for the sensation [i.e. the private object]; and then the possibility of error exists.' (Section 228.)

There are many ramifications to Wittgenstein's argument, and I have given only a very brief summary of one central strand in it. Naturally, the whole matter is immensely controversial. Many disagree with the argument, and many think that it is valid; but nobody thinks it can be ignored. If it is valid, it seems to have the following conclusion: first that we cannot refer to private objects in a public language; secondly that we cannot refer to them in a private language; and hence that we cannot refer to them. The Cartesian theory of the mind is therefore certainly false, since if true it would have the conclusion that we could not even *speak* of the mind.

Descartes's flight into the privileged region of the 'soul' is not a flight from the demon at all. On the contrary, his private refuge has already been eaten by an invisible worm.

The point is this: *if* I allow the demon to destroy my confidence, and if I search for foundations in what can be known only to me, then the belief that I have discovered those foundations will also fall victim to the demon. Having retreated into a realm which is insulated from physical reality (being wholly 'apart from' it), I still believe that I can think. I believe, in other words, that I know what I mean by 'I' and 'think'. But this is precisely what I can no longer assume. If the demon has pushed me into this private corner, it is only to slay me there.

On the other hand, the private-language argument offers an answer to the demon. It says: Stop looking for foundations for your beliefs, and step out of the first-person viewpoint. Look at your situation from outside, and ask how things must be, if you are to suffer from these doubts and uncertainties. You will see that one thing at least is true: that you speak a language. And if that is true, it must be possible for others, too, to learn your language. If you can think about your thinking, then you must be speaking a public language. In which case, you must be part of some 'public realm', in which other people could wander. This public realm is no fiction of the demon, but the fundamental reality.

4. Kant and the Noumenon

Wittgenstein's argument has many interesting features, besides the one to which I have already drawn attention, of placing anthropology on the throne. Equally remarkable is its peculiar conclusion, that a certain kind of thing – a 'private object' – *cannot be referred to*. Is that not a paradox? For have I not already referred to it, in saying it cannot be referred to?

To understand Wittgenstein's position it is helpful to study a parallel position of Kant's. In the *Critique of Pure Reason*, Kant argues as follows: we live in a world of observable things – of phenomena, as he calls them. Our concepts only make sense when applied to phenomena – i.e. to objects that are knowable through experience. On the other hand, the very structure of our thought tempts us towards the idea of a world 'beyond' experience, a world of 'things-in-themselves' whose nature is quite independent of appearance. This idea makes sense only to thinking: we have no experience that corresponds to it. The thing-in-itself is, as he puts it, not a phenomenon at all. It is a 'noumenon': something that is given only in thought.

Concepts acquire their significance, Kant argued, in acts of judgement, when we use them to describe and explain the world. But the only world that we can describe and explain is the world that appears to us: the world of phenomena. Hence the concept of a noumenon can never be employed in an act of judgement – it can never be used to say *how the world is*. It has no 'positive' employment, as Kant put it. We can use the concept only *negatively*: only in order *to forbid its own application*. I can say that if the soul existed, it would have to be a thing-in-itself, in which case I could not make judgements about it. Therefore, since I do refer to the mind, the mind is not a soul.

Kant produced a variety of arguments to show that many 'philosophical' entities are merely noumena. As soon as you try to make judgements about them, you find that you are speaking of something else – of the observable phenomena. The noumenon 'drops out of consideration as irrelevant', like the soul in my example. This exactly parallels Wittgenstein's argument about the private object. In effect, Wittgenstein has mounted a challenge to the Cartesian, to show how the private realm which he postulates could really be an object of reference, either to the self or to another. The Cartesian has no means of showing that he has successfully 'picked out' the thing to which he is trying to refer. It always and necessarily eludes his conceptual grasp, just as the noumenon does for Kant. In which case, as Wittgenstein laconically puts it, 'a nothing would do as well as a something about which nothing can be said'. The private language argument is not so much an examination of the private object as a systematic rejection of 'the grammar which tries to force itself on us', when we talk about the mental. (See section 304.)

There are grounds for thinking that Kant and his followers accepted arguments similar to Wittgenstein's. Both Kant and Hegel, for example, believed that the 'subject' of experience is given only through its interaction with the world of 'objects'. There is no such thing as a realm of pure subjectivity, as Descartes had envisaged it. The subject makes sense of himself only through applying concepts; and concepts derive their sense from their *primary* application in the realm of objects – the realm where there is a real distinction between applying them correctly and making a mistake (a distinction between being and seeming). Hegel added that I can have knowledge of myself only through a complex process of 'self-realisation', in which I become a member of an objective and interpersonal order, recognising others and conceding their rights. (The famous 'master' and 'slave' argument in the *Phenomenology of Spirit*: see Chapter 20.) All such arguments tend towards Wittgenstein's conclusion, that the first-person case is not the starting point of philosophy, still less the foundation of knowledge, but the by-product of reference and activity in a public realm.

5. Strawson and the Modern Approach

Another, and closely related, answer to the sceptic is given by Sir Peter Strawson, in the first three chapters of his book *Individuals*. Strawson is concerned, like Wittgenstein, to show how our language works, and what has to be true if we are to have thoughts about our condition. He *says* that he is merely describing our language, and the metaphysical assumptions that happen to be made in it. (He calls *Individuals* an 'essay in descriptive metaphysics'.) In fact, however, he comes closer to Wittgenstein, in suggesting that the way we *do* use language, is the way we *have* to use it, if we are to refer at all. For Strawson reference requires a 'frame' of enduring objects. These are the individuals of his title, and provide the constancy through time that is needed, if we are to mean anything consistently. I attach my language to these individuals, by primary acts of reference which serve to anchor me in an objective realm. Strawson tries to prove that this realm is not only objective, but also spatial: without a spatial dimension, he believes, I cannot 'reidentify' objects at different times, and therefore cannot arrive at a conception of an enduring individual.

Individuals divide into two fundamental kinds: material objects and persons. Persons are distinguished by the fact that certain predicates can be applied to them which cannot be meaningfully applied to material objects. Strawson confusingly calls these P-predicates (though he might have said mental predicates). And he proceeds against the Cartesian in the following way: predicates like 'blue', 'man', 'is thinking' are general terms. They apply to indefinitely many instances (potentially, at any rate). I cannot understand them without being prepared to apply them in this way. A predicate that I just did not know *how* to apply to any instance other than the one used to teach it, would not be a genuine predicate: it would be something like a *name*. It could not be used to say anything about that instance – anything that might be true or false – no more than I could describe an object simply by calling it Henry.

If the Cartesian is right, then I learn mental predicates like 'is in pain', 'is thinking', from my own case. But I can encounter no *other* case. The procedure for applying these predicates to myself (in terms of the inner character of my mental life) cannot be duplicated outside myself. So I could not learn to apply the predicates to another instance. They would not *be*, for me, predicates. Hence I could not use them to describe even my *own* condition. The Cartesian conception of the mind is self-defeating.

What is the rival position? Strawson again approximates to Wittgenstein's picture, arguing that I can understand mental predicates only if I am prepared to recognise other instances where they would be correctly used. This means that I must be prepared to *identify* other 'subjects of consciousness' or 'persons' in my world. And for Strawson to *identify* means also to 'reidentify through time'. So I must have some procedure for identifying and reidentifying persons in my world, if I am to have knowledge of my own mental states. (This part of the argument is not yet demon-proof. For I could identify and reidentify persons in the hallucinatory world that the demon offers me. Still, it may *become* demon-proof, just as soon as we block the retreat to the first-person case, and reject the Cartesian picture of the mind.)

Strawson's argument, like Wittgenstein's, has attracted much commentary. It is fairly slippery in places: the notion of 'identifying' the object of reference is notoriously vague. So too is the notion of a mental predicate (a P-predicate). (Do animals have P-predicates?) Some (e.g. Ayer) object to the root idea, that I must be able to identify 'other instances' of pain, if I am to atribute pain to myself. Surely I can admit that there could be 'other instances', without being able to recognise them or pick them out? And that would be enough to establish that the word 'pain' had, for me, the sense of a predicate.

Still, the argument serves to reinforce the new direction that philosophy has taken during the latter part of our century. Sceptical problems are no longer given the priority that they used to receive. Nor are philosophers so keen to find 'foundations' for our knowledge. Nor does the Cartesian theory of the mind, or the idea of a private 'inner' realm, have very much appeal for a modern philosopher. All three shifts are part of a single comprehensive move away from epistemology towards the study of the human condition. And the primary subject of this study is language.

6

Sense and Reference

The last chapter rushed forward into difficult territory, and some of the terms that I used have not yet been explained. I spoke of reference, predicates, meaning and rules, while assuming that an intuitive understanding of these things would suffice. To become clearer we must now go back to Frege, the German philosopher who created the philosophy of language as we know it, and whose work was carried forward in Wittgenstein's *Philosophical Investigations*. Frege wrote little, and everything that he wrote is absolutely first rate. Thanks to the writings of Michael Dummett the main ideas of Frege's philosophy are now in free circulation, and nobody can doubt their importance. There is one paper of Frege's that everyone interested in philosophy *has* to read, which is the paper that I shall discuss in this chapter. In their new edition of the philosophical writings Geach and Black translate it under the title 'On Sense and Meaning', but it is rightly called 'On Sense and Reference' by everyone else. Apart from this article, two other works of Frege's are particularly relevant to modern philosophy: the essay called 'The Thought' (printed in P.F. Strawson, *Oxford Readings in Philosophical Logic*), and the short masterpiece called *The Foundations of Arithmetic*.

It is a commonplace that logic, which ought to be the most scientific part of philosophy, is in many ways the most controversial and also the slowest to change. Aristotle summarised and classified the valid 'syllogisms', and gave a subtle account of truth and inference. But nobody built on his achievement until modern times. Although Leibniz made some important advances, the knowledge of logic among philosophers actually *declined* during the nineteenth century. The greatest nineteenth-century philosopher – Hegel – wrote a book called *Logic* which consists only of formally invalid arguments, and contains nothing pertinent to logic as we know it. Various empiricists did rather better, and John Stuart Mill's *System of Logic* is one of the few pre-Fregean works that repay serious study. But the real changes came only at the end of the nineteenth century, when Frege and Russell simultaneously began research into the foundations of mathematical thinking. I shall consider one of Russell's contributions in the next chapter; much of what Frege discovered, Russell discovered too.

1. The Structure of a Sentence

Ancient philosophers were familiar with the ordinary subject-predicate sentence, such as 'John is bald', which they divided into three parts: the subject term ('John'), the copula ('is'), and the predicate term ('bald'). They had great problems, however, in fitting other sentences into this mould. For example, sentences like 'All swans are white' (universal sentences), 'John exists' (existential sentences), and 'It is necessary that you come' (modal sentences). The subject-predicate sentence was fundamental for Aristotle, partly because it suggested an all-important metaphysical division, between the entity picked out by the subject term, and the entity picked out by the predicate term. There seems to be no way of thinking without the use of subject-predicate sentences. And this seems to suggest an ultimate division in reality between substances and their properties. Substances are particular, properties are universal; substances exist unproblematically in space and time (some of them, at any rate); properties have only a problematic existence. (Plato said that they exist not in space and time, but eternally.) This division led therefore to two metaphysical ideas, which I shall discuss in Chapter 8: substance, and universals.

The 'Fregean revolution' led to the modern theory of sentence structure. According to this theory the copula 'is' has nothing to do with the 'is' of existence. Indeed, we must distinguish subject-predicate sentences from existential sentences, which have a quite different logic. For the purposes of logic, moreover, we should divide the subject-predicate sentence into *two* parts, not three: 'John', and 'is bald'. The sentence predicates baldness of John. As for the existential sentence, this cannot be expressed in subject-predicate form at all. We need the mathematical idea of a *variable* if we are to make sense of it. The sentence 'John exists' tells us that there is something that is John – in other words, that there is an x, such that x is identical with John.

Likewise, if I say that a red thing exists, I mean that there is something that is red – or, there is an x such that x is red. Logicians would write these two sentences in the following way: $(\exists x)(x = \text{John})$, and $(\exists x)(R(x))$. The importance of this symbolism will become clear as we proceed; but it is by no means necessary to the argument. The subject-predicate sentence would be expressed as 'B(John)' – indicating that baldness is being 'predicated of' John.

The introduction of the variable into logic made it possible, for the first time, to understand the logic of existence. Existence is not a predicate, just as Kant had argued. It is to be represented by a 'quantifier', which tells us *how many* things possess the predicate: namely, at least one. This quantifier 'binds' the variable in the sentence $R(x)$; otherwise the variable is 'free'. (Sentences with free variables are called 'open sentences'.) We can understand universal sentences in a similar way, as involving a variable bound by a quantifier. Here the quantifier says that *everything* has a certain property. 'Everything is green' is written: $(x)(G(x))$, or alternatively $(\forall x)\ (G(x))$ – for every x, green x. And then comes the first interesting result, namely that the two quantifiers can be defined in terms of each other, by means of negation. To say that everything is green is to say that it is not the case that something is not green – in symbols $\sim(\exists x)(\sim G(x))$; and to say that something is green is to say that it is not the case that everything is not green – in symbols: $\sim(x)(\sim G(x))$. Not surprisingly, therefore, Frege and Russell felt that they had found the clue to mathematics. Maybe we could develop the logic of quantification until all the numbers were defined in terms of it.

2. Singular Terms and Identity

The idea of a predicate is not too hard to understand: a predicate is a term that 'predicates' something of an object; alternatively, it 'applies a concept' to that object, or assimilates the object to a certain *kind*. But what about the subject-term of a sentence? What does this do?

Here Frege proposed a new terminology. The idea of a 'subject-term' is extremely misleading. For the very same term can occur as what used to be called the 'object' of a sentence. 'John kicks Mary' contains a term – 'Mary' – which surely functions exactly like the term 'John', in picking out an item in the world. But it is not part of the subject of the sentence. For Frege 'John' and 'Mary' are names, and names are special cases of what we should now call singular terms. The category of 'singular terms' also includes what Russell called 'descriptions', whether 'indefinite' such as 'a man', or 'definite' such as 'the queen of England' and 'the morning star'. All such terms refer, according to Frege, to 'objects'. What are objects? This is a question for metaphysics. But we can at least say two things about them: first, that they are the bearers of properties (which are 'predicated' of them), and secondly that they are the subject-matter of identity-statements. Sentences of the form 'a = b' are meaningful, so long as 'a' and 'b' are singular terms (names) – i.e. so long as they refer to objects.

What then do we mean by 'reference'? The answer is not so much stated by logic as *shown* by it. Reference is the relation that holds between a singular term and the object that corresponds to it – between the name 'John' and John himself. We all have an intuitive idea of this relation. The idea becomes more precise, Frege thinks, as we discover that reference is not confined to singular terms, but describes a dimension of meaning in general. All terms refer; indeed, the *purpose* of language is to refer, to pick out things in the world and to make true statements about them. That's why we invented language in the first place.

3. Sense and Reference

But reference (*Bedeutung*) is not the whole of meaning. Consider an identity statement, 'a = b'. What makes this true is the fact that 'a' and 'b' refer to the same object. If the meaning of 'a' consisted in the object that it referred to, then anyone who understood 'a' and understood 'b' would know immediately the truth of 'a = b', just as we all know the truth of 'a = a'. Frege gives an example. Nobody needs to know anything about the world in order to know the truth of 'The morning star is identical with the morning star': this is self-evident, and its truth is guaranteed simply by the meanings of the words. But you might understand the sentence 'The morning star is identical with the evening star' and *not* know that it is true. This suggests that it does not have the same meaning as 'The morning star is identical with the morning star'. On the other hand, the terms in the two sentences refer to the same things – namely the planet Venus, and identity. It follows that there must be more to meaning than reference.

This is one of the arguments that Frege offers for his distinction between sense (*Sinn*) and reference (*Bedeutung*). When you understand a word, you grasp its 'sense', and you can do this without knowing that it refers to an object that some *other* word also picks out. The sense is what you understand when you understand a word. But if language is to perform its referential function, the sense must in some way *fix* the reference. It must be the case that the person who understands the word is, so to speak, *pointed towards* its reference, just as understanding 'the morning star' you are pointed in the direction of Venus, the evening star. The sense of a term contains the information necessary to determine its reference. Frege sometimes calls the sense 'the mode of presentation' of the reference. (For recent doubts about this argument, see Chapter 19, section 6.)

The sense of a word must be distinguished from the 'ideas' associated with it. For example, I may associate with the word 'Venus' various images (Botticelli's *Birth of Venus*), certain words (the Temple of Venus in *The Knight's Tale*), certain sounds ('Venus' from Holst's *Planets*), and certain ideas (beauty, classical perfection, the power of love): all these are mere 'ideas' for Frege; that is to say, subjective associations, which are peculiar to me, and which are not part of the shared, public meaning of the word – the meaning which you and I understand, and which leads us to use the word with the same *sense*. To put this in another, and more Wittgensteinian way, the sense of a word is given by the *rules* governing its public use; the associated 'ideas' are local consequences of those rules. This is the origin of Wittgenstein's later view that meaning is essentially public.

4. Predicates and Sentences

So much for singular terms. What about predicates? And what about the complete sentences that are formed from them? Presumably these have sense; and they must have reference too, otherwise the reference of singular terms would simply be inert, no part of the language in which those terms occur.

This is where the theory gets difficult. Frege believed that 'it is only in the context of a sentence that a word refers to (*bedeutet*) anything'. In other words, reference occurs only in the complete sentence and, in assigning a reference to its parts (as I have done with singular terms), we are speaking obliquely. Really we should see the reference of a word as a *contribution* made by that word to the reference of a sentence. Likewise with sense: the sense of a word is its *contribution* to the sense of the sentence in which it occurs. We can speak of *the* sense, and *the* reference of a word, because the contribution is always the same: it is systematic. Language works because the rules for composing words into sentences also serve to compose the sense of sentences out of the senses of words, and the reference of sentences out of the references of words.

Consider a sentence, 'John is bald'. What does this refer to? To answer this question we have to consider again what the point of language is. We wish our sentence to *correspond to reality*, in the way that 'John' corresponds to John. A successful sentence is one that represents the world *as it is*, one which *leads* us to the reality. In other words it is a *true* sentence. Truth is the all-important mark of success. Frege argued that there are just two 'truth-values' as he called them: the true and the false. He therefore suggested that a sentence will refer to one or other of two things: truth (the true) or falsehood (the false). At first sight this looks ridiculous – for it seems to imply that all the infinitely many possible sentences in a language say one of only two things. But that is not the right conclusion. What Frege meant is that a sentence is related to its truth-value as a name is related to its object. And you can prove this in another way. For suppose John is identical with Mr Smith. It follows that 'John' and 'Mr Smith' have the same reference. In which case you could substitute 'Mr Smith' for 'John' in any sentence, and expect the reference to remain the same. Well, *what* remains the same, when 'Mr Smith' is substituted for 'John' in sentences containing 'John'? The answer is: the truth-value. If it is true that John is bald, then it is true that Mr Smith is bald. If it is false that John is married, it is false that Mr Smith is married; and so on. This suggests that the truth-value is to the sentence what the object is to the name.

(There are problems for this argument, however, and Frege discusses them. Sometimes you *cannot* substitute terms with the same reference without a change of truth-value. For example, take the sentence 'Mary believes that John is bald'. This might be true, even though it is not true that Mary believes that Mr Smith is bald, since Mary does not know that John is Mr Smith. Frege argues that you have to treat these sentences as *special cases*. Most other philosophers agree with him. The consensus is that we should distinguish the straightforward cases, in which terms may be substituted for one another provided they have the same reference, from deviant, oblique, or 'opaque' contexts, in which substitution is impossible. We try to understand the 'direct' contexts first, and explain the indirect in terms of them. See Chapter 12.)

So what now of predicates? To what do *they* refer? Intuitively, one would say 'a concept', or something similar. A predicate applies a concept to an object, and, if the object 'falls under' the concept, then the resulting sentence is true. Up to a point, Frege agrees with this. But, he suggests, it does not really show how language *works*. It does not explain the way in which the reference of the predicate 'latches on' to that of the singular term, so as to generate the reference of the sentence (the truth-value). The matter becomes clearer, Frege suggests, if we borrow another mathematical idea – that of a 'function'. A function is an operation which systematically transforms one mathematical object into another, for each object of the appropriate kind. For example the function $\sqrt{}$ gives, for each number, the square root of that number. It is a 'function from a number to its root'. Likewise a concept determines a function, from an object to a truth-value. Insert an object into the function, and it delivers one of two results: true or false. This is only a technical way of saying what we all know already; but it has proved very useful in the philosophy of language. We shall encounter another use of the term 'function' shortly.

If predicates and sentences have reference, do they have a sense? Indeed they do, says Frege, who analyses the sense of a sentence in just the same way. The sense of a sentence stands to its reference (the truth-value) as the sense of a singular terms stands to *its* reference (the object). As we have seen, the sense of a singular term is an operation we *understand* in understanding the term, and which *fixes* its reference. By understanding the term, we can know what it refers to. Likewise for a sentence. Understanding the sentence is grasping the *thought* expressed by it, which in turn is identified by the truth-conditions of the sentence. When we can grasp these truth-conditions we acquire what we need in order to understand the sentence. The truth-conditions determine the reference (the truth-value): if they are fulfilled, the sentence is true; if not, false.

As for the sense of the predicate, this once again looks after itself. The sense of a predicate is a function from the sense of singular terms to the sense of subject-predicate sentences. To put it in another and more common way: the sense of any term is to be understood as its contribution to the truth-conditions of any sentence of which it is a part.

5. The Synoptic Picture

Here, then, in rough outline, is Frege's theory of language, together with one or two philosophically important conclusions that may be derived from it.

(a) Sense and reference. There are two dimensions of meaning. Every

term, and every meaningful combination of terms, has both a sense and a reference. The sense *determines* the reference; it is also what we understand in understanding the term. By understanding a term, we acquire a 'route' to its reference: we have what we need, in order to relate the term to the world.
(b) The reference of a singular term is an object, while that of a sentence is a truth-value. The predicate refers to a concept, which is to be understood as determining a function from object to truth-value.
(c) The sense of a sentence is the thought expressed by it, which is given by the conditions for its truth. The sense of every term in a language can be seen as a (systematic) contribution to the truth-conditions of the sentences which include it.
(d) The true unit of meaning is the sentence: the completed expression of a thought. It is only in the context of a sentence that the meaning and reference of a term are fully realised. (This is sometimes called the 'context principle'.)
(e) At the same time, sentences are built up from the terms that compose them: each term makes a systematic contribution to the truth-conditions. This means that, if we know the rules governing the use of individual terms, we can derive the truth-conditions of all the infinitely many sentences that can be built from them.

The last point is extremely important, since it endorses an observation that it is frequently necessary to make, both in the philosophy of language and in linguistics. (It is an observation associated also with the work of the linguist Noam Chomsky.) It seems that our understanding of language is creative. We understand sentences that we have never heard before; and there seems to be no limit to the number of new sentences that we can form. Yet we are finite creatures, with only finite capacities. So how is this possible? It is possible because we can build indefinitely many sentences from a finite vocabulary, by means of repeated applications of structural principles (the rules of syntax); and because the sense of each sentence so formed is determined by the senses of the terms of which it is composed, by another set of rules (semantic rules). Furthermore, because sense determines reference, we acquire, through understanding the senses and the semantic rules that combine them, a procedure for assigning not only objects to singular terms, but also truth-values to sentences. Our language is connected to the world by the rules that govern it.

Frege's argument is an impressive achievement. Without recourse to anything more than common intuition, he is able to outline a comprehensive theory of language and our way of understanding it, which is also pregnant with metaphysical suggestions. Almost all his conclusions derive from the simple observation that language has the primary goal of expressing and communicating what is true. We immediately see that it is in terms of the concept of truth that the relation between words and the world must be represented, and also that there is an unbreakable connection between meaning and truth-conditions. Those two ideas alone account for much of subsequent philosophy. And, in deriving his account of 'deep structure', Frege makes no assumption that his opponent would not also be forced to make, once he had learned to look language in the face.

6. The Revolution in Logic

Logic studies valid inferences, and tries to give rules for distinguishing the valid from the invalid. If the inference from *p* to *q* is valid, someone might conjecture, this is because the meaning of '*p*' 'contains' that of '*q*': *q* is comprised in what we understand by '*p*'. On this view the primary subject-matter of logic would be Frege's sense: that which we understand when we understand a sentence.

The revolution in logic occurred when philosophers realised that what matters is *reference*. Logic does not describe – as Cartesian philosophers supposed – the relations among 'ideas'. It describes the relation between language and the *world*, in its most abstract and systematic outline. An inference is valid, just so long as the premises cannot be true without the conclusion being true. We can therefore understand the validity of arguments in terms of the 'truth-functions' whereby propositions are combined.

The idea is this: each sentence in a language can have one of two references, true or false. The sentences are *evaluated*, by assigning one of those 'truth-values' to them. Until evaluated, a sentence is simply an 'uninterpreted sign'. Hence the first step in logic is to envisage a language of 'primitive sentences', represented by propositional signs *p*, *q*, *r* and so on. These signs are treated as variables, to be evaluated by assigning to each of them one or other of the values T or F. We cannot do much with this language yet, since it has no structure. So we need to *combine* propositions into complexes, by introducing propositional 'connectives', such as *and* and *not*, usually represented by the symbols &, and ~. We understand these connectives by assigning to them a 'value'. Obviously, they are not true or false in themselves. However, they would conform to our general idea of reference if we could think of them as standing for *functions*, which yield new truth-values, depending on the values assigned to the propositions that they join. Consider the connective & (and). We could assign to this the following function: p & q has the value true, if p has value true and q has value true. Otherwise it has the value false. Sometimes this is represented in a truth-table as follows:

p	&	q
T	T	T
F	F	T
T	F	F
F	F	F

The middle column represents the value that must be assigned to p & q, when p and q each has the value given in the columns beneath the symbols 'p' and 'q'.

We can also define ~, negation, in the same way: ~p has the value false when p has the value true, true when p has the value false. There are sixteen possible truth-functions of two propositional variables, not all of them interesting. One, however, has caught the attention of logicians, since it captures some part of our ordinary idea of implication, the function given by the table:

p	$\supset$	q
T	T	T
F	T	T
T	F	F
F	T	F

$\supset$ is called 'material implication'. And you can see by inspection, that from the propositions p and $p \supset q$ you can deduce q: whenever the first two have the value true, the third must also have the value true. That is an elementary proof of the 'rule of inference' known in logic as *modus ponendo ponens* or *modus ponens*, which says, from p and p implies q, infer q.

Although there are sixteen truth-functions of two propositions, we do not need sixteen symbols in order to introduce them, since they can be defined in terms of one another. $p \supset q$, for example, can be defined as ~(p & ~q). You can make do with only one connective, though the result looks rather alarming.

7. Formal Languages

Modern logic, having defined its starting point in that way, advances through the study of formal languages. A formal language is an artificial language, which duplicates some, but not all, of the features of a natural language, while strictly obeying the requirements of the person who creates it. The simplest formal languages are designed to display the relations between elementary sentences and their combinations, while permitting systematic interpretation by the assignment of truth-values. Such a language will contain:

(a) A vocabulary; say, a list of primitive propositional variables p, q, r, s (which stand for truth-values, just as sentences do, according to Frege, in a natural language), and a list of 'constants' – usually connectives such as &, $\sim$, $\supset$.

(b) Syntactic rules – rules defining what is called a 'well-formed formula'; for example: if p is a wff then so is $\sim p$.

(c) Rules of interpretation, which assign values to the constants, so that one can interpret the complex sentences on the basis of the values assigned to their parts. These values are truth-functions.

Such a language duplicates some of the features of natural language. It permits infinitely many sentences to be constructed from finitely many parts; and the reference of every meaningful sentence is completely determined by the reference of its parts. It contains, however, nothing corresponding to Frege's idea of sense. Nor do we need this idea, in order to explore the domain of logic. The ideas of proof and logical truth, for example, can already be studied by means of our simplified language. We need only add 'rules of inference'. For example:

(d) Rules of inference: (i) From p and $p \supset q$, infer q. (*Modus ponens.*) (ii) If p and q have the same truth-value, then p may be substituted for q in every formula where q occurs. (Rule of substitution.)

The language has now 'come to life'. Put down any proposition, and infinitely many more will be generated by the system. If we add axioms to our language, we can proceed to deduce theorems, and so *use* our language to construct a theory. Our choice of axioms will be dictated by the purpose of the theory. In logic, the most interesting axioms are those which are logically true: i.e. which cannot be false, given the interpretation assigned to the 'constants'. A logical truth comes out true whatever the value of the variables contained in it. For example $(p \mathbin{\&} (p \supset q)) \supset q$ is a logical truth. Whatever the values of p and q it will come out true. Such formulae are sometimes called 'truth-functional tautologies'. If we make the right choice of rules of inference (such as the two I have given), we shall have a theory which generates *only* logical truths from logically true axioms. We might even be able to generate *all* logical truths by repeated application of our rules. (In this case the theory will be 'complete' with respect to logical truth.) How can we know that a theory is complete? There we have a neat and interesting mathematical question, which requires us to construct a theory about our theory – a 'meta-theory' as it would be called. Much modern logic consists in the construction of meta-theories: proofs about proofs, which tell us where we can arrive by means of this or that procedure. (See Chapter 26.)

What bearing has all this on natural language? Obviously our formal language is not very rich. Moreover, it models only a small part of the structure of natural language – the part which concerns connectives, like 'and', 'if' and 'not'. We need a far richer language if we are to penetrate *inside* the structure of sentences, so as to explore the syntactic and semantic properties of their parts. This is what the 'predicate calculus' is designed to do, by taking as its primitives signs representing the *parts* of sentences: predicate signs, *F*, *G*, etc.; signs for singular terms *a*, *b*, *c*; and signs for variables *x*, *y*, *z*, and for the quantifiers that bind them. Thanks to the work of Tarski, logicians know how to construct this kind of language on Fregean principles, and to 'assign values' to its sentences on the basis of the values of their parts. The theory of reference has been gradually expanded in this way, through the development of formal languages which come ever closer to the natural paradigm. And always we seem to be able to articulate the logical features of such a language, while considering only the reference and never the sense of its formulae.

What bearing has all this on philosophy? In what way has the new approach to logic altered the philosophical agenda? In the next chapter I answer by way of an example.

7

Descriptions and Logical Form

Russell's paper 'On Denoting', which first appeared in *Mind* 1906, is perhaps the most famous paper in the whole of modern philosophy, a model of understated iconoclasm, and an inspiration for much that happened in Anglo-American philosophy for the next fifty years. Its primary subject-matter is the word 'the'; and readers were not slow to recognise that, if a whole philosophy can be implied in giving the meaning of this tiny word, something momentous must have happened to philosophy. Much of what Russell says is now rejected; all is disputed. Yet there is a core of irresistible good sense which still repays study. Moreover, it is impossible to understand modern philosophy without understanding Russell's argument. In the brief period of Oxford 'ordinary-language' philosophy during the fifties, Strawson wrote a reply to Russell (the paper entitled 'On Referring'). This too is important; in retrospect, however, it has dated far more rapidly than the conceptions that it criticises.

1. Russell's Problem

Russell's thought, like Frege's, is rooted in the study of logic and the foundations of mathematics, and the two men share the credit for the theory of mathematics that was eventually to be spelled out in Russell's and Whitehead's *Principia Mathematica*. However, Russell did not accept the two-tiered theory of meaning that Frege advanced in 'On Sense and Reference', largely because he half saw (what I briefly summarised at the end of the last chapter) that the theory of 'reference' is all that logic requires. We can build our logic on the idea that words 'stand for' things: names for objects, sentences for truth-values, and general terms for 'classes' (things which have other things as their 'members'). Thus the sentence 'John is bald' stands for a truth-value (true, say); and it says that the thing named 'John' (i.e. John) is a member of the class denoted by 'is bald' (the class of bald things). Russell has caused much confusion by describing John as the *meaning* of 'John', while assuming that the meaning of a *sentence* is a proposition. This is to confuse Frege's two levels of meaning. It is the truth-value, not the proposition, that stands to the sentence as John stands to his name. But Russell did not understand Frege's motive, which has been misunderstood throughout the period of Frege's influence. For Russell, therefore, John becomes a component of every proposition about him: he himself stands there, in the middle of a proposition, an intruder from the real world, weirdly trapped in logical space. Ridding Russell's theory of this confusion is no easy matter; what I shall say will therefore not use Russell's words, and will often stray from his argument, in order to capture the real force of it.

There is another aspect of Russell's theory that must also be set aside if we are to understand its present-day significance. He is addressing *two* underlying questions: first, what is the meaning of a phrase; secondly, how do we *know* its meaning? In the case of names, the meaning is the object referred to; we know the meaning, according to Russell, by being acquainted with its object. This idea became associated in Russell's philosophy with a theory of 'knowledge by acquaintance' which had the paradoxical result that we are not actually acquainted with anything that we normally refer to, so that most of the names in our language are not names at all.

Russell's problem is this. The expressions that Frege had called names (the singular terms) are of two radically different kinds: names, and 'denoting phrases', the latter being either indefinite ('a man') or definite ('the man'). Obviously 'John' is a different kind of term from 'the morning star'. For we cannot assume 'John' to be a name at all, unless there is someone to whom it refers. But we use such denoting phrases as 'the morning star' quite happily, without being sure that there is anything denoted by them. For example, we may speculate about the golden mountain, saying 'the golden mountain is made of gold', 'the golden mountain is hidden from view', 'the golden mountain does not exist'. But suppose (as is surely true) that there is *no* golden mountain. How then do we evaluate these sentences? Are they true or false? Frege would say that they contain 'empty' names: names that do not refer to anything, and which are therefore not really names at all. But if 'the golden mountain' acts as a name, while lacking a reference, then all the sentences in which it occurs *also* lack a reference. (The reference of the sentence is determined only when the references of its parts are determined.) Just as 'the golden mountain' lacks a reference, therefore, the sentence 'the golden mountain is hidden' lacks a truth-value. It is neither true nor false. If Frege is right, our language will be full of such sentences, which we use quite happily, without being able to assign a truth-value to them at all. In which case, our logic – which is based on the relation between sentences and truth-values – goes out of the window. Indeed matters are worse. For we can use these denoting phrases without even *knowing* whether there is anything that corresponds to them. Indefinitely many sentences play an active part in our thought and inference, even though we cannot say whether they have a real place in it at all, or what that place might be. This, to Russell, was intolerable.

Russell was not the only philosopher to worry about this problem. I have already referred to Husserl, founder of modern phenomenology, whose murky *Logical Investigations* were being composed at this time. Husserl belonged to a circle of Austrian philosophers, who frequented the universities of Vienna and Prague at the turn of the century, and who specialised in the border regions between logic and the philosophy of mind. One such was Alexius Meinong, known today largely as the object of Russell's scorn, but of great interest in his own right. Austrian culture was, at this period, rich in imaginative achievements. Mahler, Freud, Klimt and Rilke were at the height of their powers; Adolf Loos had not yet begun his war on architecture, nor Schoenberg his revolt against music; even Musil still believed that there are men with qualities. Meinong's contribution to this world of fertile illusion was to suggest that logic should study the objects of *thought*. Even if there is no golden mountain, we have the power to *think* of it, and this is what anchors 'the golden mountain' in the haven of reference. We have no difficulty in understanding 'the golden mountain is hidden', because 'the golden mountain' refers to a mental object. This object does not exist in the real world, but 'subsists' in the realm of thought.

Russell probably did not understand Meinong very well. At any rate, he took him to be saying that, just *because* sentences containing the phrase 'the golden mountain' are meaningful, there must in some sense *be* a golden mountain. Otherwise those sentences would not be about anything. Every denoting phrase therefore brings some entity into the realm of 'subsistence'. But what about 'the round square'? Does that subsist as well? What a peculiar realm this must be, if it contains impossible objects! Surely, Russell conjectured, we should start again, and examine the logic of denoting phrases without our Fregean shackles. Maybe we have misunderstood their logic altogether.

2. Russell's Theory

Russell begins from an idea of definition. Normally, when I define a term, I give an equivalent term: a term that can replace the one defined in all contexts where it occurs, without changing meaning or truth-value. (The sign for this in logic is '=df'.) A definition says that the one term can be substituted for another, so that, by convention, they always denote the same thing.

Such definitions Russell called 'explicit' definitions. They eliminate terms from our language, by substituting other terms with the same function. But no explicit definition could get rid of 'the golden mountain' without introducing a new term which gave rise to exactly the same logical and metaphysical problems. What we need, Russell suggested, is an *implicit* definition, by which he meant a procedure for replacing the term in every context of its use, that would show exactly how to evaluate the resulting sentence. Take the phrase 'The present king of France'. Since there *is* no present king of France (assuming that we subscribe, as Russell did, to the republican heresy), we cannot understand definitions of the form 'The king of France = x': the concept of identity has no clear function here. Nevertheless, the sentence 'The present king of France is bald' is meaningful; it ought therefore to have a truth-value. So how do we evaluate the denoting phrase? The answer is that we evaluate it by reconstructing the whole *sentence*, so as to settle the question of its truth-value. We give an implicit definition of the phrase, by showing how sentences in which it occurs can be replaced by other sentences with the same truth-value, in which the phrase itself does *not* occur. This is the best we can do, and all we need to do. So how is it done?

The answer lies in the word 'the'. In describing the present king of France as bald, Russell reasonably argued, I am implying first that there *is* a king of France, secondly that there is at most *one* king of France, and thirdly that whatever is a king of France is bald. Those three ideas capture the meaning of the sentence, for a reason that Frege would have endorsed: namely, they capture what has to be true if the sentence is to be true. They 'give the truth-conditions' of the original sentence. (This was not Russell's way of putting the point, but that does not matter for our purpose.)

Here then is Russell's analysis of 'the present king of France is bald':

'There is a king of France; there is at most one king of France; and everything that is a king of France is bald.'
To put it another way:
'There is an x, such that x is a king of France, x is bald, and for every y, y is a king of France only if y is identical with x.' In symbols:
$(\exists x)(K(x) \,\&\, B(x) \,\&\, (y)(K(y) \supset (y = x)))$.

We have translated the sentence into forms recognised in the 'predicate calculus', and shown that the original sentence has exactly the same truth conditions as a tripartite conjunction, in which the phrase 'the king of France' does not occur. Moreover, we can see that the original sentence *does* have a truth-value after all, and the value is (on the republican assumption): false.

This process can be repeated for every denoting phrase in our language, whether indefinite ('a man') or definite ('the man'). The denoting phrases that most interested Russell were those composed from the definite article and a description: such as 'the king of France', 'the golden mountain', 'the morning star'. He called these 'definite descriptions'. Hence the name for his theory: Russell's theory of descriptions. And he believed that he had not only given a general rule for eliminating all such phrases from our language, but also said all that needs to be said about the meaning of the word 'the'.

3. Logical Form

Certain features of Russell's theory deserve emphasis.
(i) He gives the meaning of a sentence by stating the conditions for its truth. In doing so, he shows not only that the sentence has a truth value, but also how that truth value is determined.
(ii) The sentence analysed is a subject-predicate sentence (speaking grammatically). It has a subject-term ('the king of France') and a predicate term ('is bald'), and attributes the predicate to the subject. But the analysis is not a subject-predicate sentence at all. It is a composite sentence, which is *existential* in form. It is actually *saying* that something exists, and then predicating various properties of it.
(iii) The grammatical form of the original sentence caused us to mistake its logic: it was this that led us to suppose, either that the sentence has no truth-value (Frege's option), or that the subject-term

refers (Meinong's option). If Russell is right, the grammatical form is systematically misleading.

This led Russell to an interesting conclusion. The grammar of ordinary language, he suggested, may, and often does, conceal the true *logical form* of the thoughts expressed in it. The logical form is given by the role of a sentence in inferences: we understand the sentence when we understand what can be inferred from it, and how its truth-value is determined. Moreover, Russell assumed, we have in the new logic a perfect instrument for representing logical form. The language of logic is designed precisely to capture the role of sentences as bearers of truth-values, and to show how the references of the parts of a sentence determine the reference of the whole. The elementary logical operations (truth-functions, quantification, variables and so on) are perfectly understood, since they are defined in *terms* of their logical role – their role in composing the truth-values of sentences from the references of their parts. If we can assume that the language of logic is complete – that it captures all the ways in which truth-values can be assigned and all the ways in which inferences may be judged for their validity – we shall be tempted like Russell to say that logic takes precedence over natural grammar. Logic tells us what we really mean by our ordinary language: or at least what we ought to mean. The first step in philosophy is to give the logical form of the sentences that trouble us; then we shall have a clue to what we can and cannot say by means of them.

Those ideas were immensely influential on the early Wittgenstein and the logical positivists. Even now we find philosophers arguing in terms of them. Donald Davidson is one prominent recent example. (See his collection *Inquiries into Truth and Interpretation*, 1984.) Ordinary language creates philosophical problems, such philosophers suggest, because we do not understand the truth-conditions of the thoughts expressed in it. When assigning truth-conditions, we should use an ideal language (a logical language) whose operations are entirely understood. What makes a language ideal for Davidson is that there exists what he calls a 'theory of truth' for it. By this he means a theory which shows, in a systematic way, how all the sentences of the language can be evaluated, given an evaluation of their parts. In practice, however, the result is similar to Russell's. For Davidson doubts that a theory of truth can be provided for natural language, and seems at times to agree with Russell in thinking the 'predicate calculus' to be the best language that we have for representing reality.

Three vital ideas, therefore, have come out of Russell's theory: that philosophy proceeds by giving the truth-conditions of problematic sentences; that truth-conditions are given by the 'logical form' of a sentence, represented in some logically transparent language; and perhaps that, for each sentence, there is one and only one correct analysis (or logical form).

4. Mathematical Implications

For Russell, there was an even more interesting application of his theory in the philosophy of mathematics. Consider the sentence: 'Three is greater than two'. It contains two names: 'three' and 'two'. To what do those names refer? The obvious answer – to numbers – opens the door to the inflated metaphysics of Plato, who believed that the numbers really exist, but that they do not exist in space and time. For thinkers like Plato the numbers, existing eternally and immutably, objects of our most certain claims to knowledge, occupy a privileged position in the metaphysical hierarchy, serenely smiling on the ruin of mortal things which their deification engenders. If we are to avoid such metaphysical extravagance, Russell thought, we must try to show that the logical form of sentences like 'Three is greater than two' is not what it seems: we must try to rewrite sentences involving numerical expressions, so that they say the same thing, but without using those expressions as names – preferably, without using numerical expressions at all.

In fact much of what we say about numbers can be said using their *adjectival* form, as in 'There are three apples in the basket'; most ordinary activities of counting and calculating tacitly assume that there is something being counted, some subject-matter other than numbers themselves, which is the real object of our attention. A first step towards understanding mathematics would be taken, therefore, if we could understand numerical expressions in their adjectival use. And the theory of descriptions shows how. For we already have an analysis of the expression 'one' in terms of purely logical ideas, such as existence and identity. To say that there is one king of France, is to say that there exists a king of France, and everything that is a king of France is identical with that thing. In symbols:

$(\exists x)(K(x) \;\&\; (y)(K(y) \supset . y = x))$

In that sentence there occurs no numerical expression. Yet we have said exactly what we wished to say by using the term 'one'. We can do the same now for 'zero' and 'two'. 'There is no king of France' reads:

It is not the case that there is a king of France. 'There are two kings of France' reads: There is an *x*, and there is a *y*, such that *x* is a king of France, and *y* is a king of France, and for all *z*, *z* is a king of France only if *z* is identical with *x* or identical with *y*. In symbols:

$$(\exists x)(\exists y)(K(x) \;\&\; K(y) \;\&\; (z)(K(z) \supset. \;(z = x) \vee (z = y)))$$

Once again, we have eliminated a numerical expression, using only logical ideas. (Note the introduction of two more logical symbols: 'v', to mean *or*, defined as a truth-function, and the punctuation mark '.' after the implication sign, which tells us to take all the signs that succeed it (other than brackets opened earlier in the sentence) as lying within the 'scope' of the implication.)

Of course, we are a long way yet from giving definitions, explicit or implicit, of the numbers themselves. And our translations are not only cumbersome, but as yet divorced from any laws of arithmetical procedure that would enable them to play a part in calculation. Nevertheless, it is surely of immense philosophical interest that we can replace mathematics by logic in this particular application. Even these simple results have a seismic effect on the Platonic super-sphere, and give encouragement to those who believe, as Russell believed, that mathematics is, in the last analysis, merely logic in another guise.

Now you begin to see that there has been no word more important in the history of philosophy – with the exception of the verb 'to be' – than the definite article. From his analysis of 'the', Russell derived a theory of language, a philosophy of being, a method of analysis, and the first steps towards a derivation of arithmetic from logic. And all of these have been as influential as anything else in the modern history of the subject.

5. Meinong's Jungle

What I have just discussed is rather difficult and involves a leap forward into matters that will concern us later in this work. But the underlying point is important, both historically and philosophically, so I shall elaborate a little by considering the Russellian approach to 'ontology'. (The historical ground over which I am racing is explored more meticulously in J.A. Passmore's worthy but somniferous volume entitled *A Hundred Years of Philosophy*.) Traditional philosophy is remarkable for introducing entities into the world of which we hitherto had no suspicion: Plato's Forms, the Cartesian self, the Hegelian Absolute Idea, Schopenhauer's will, 'material substance', Kant's 'thing in itself', and so on. Berkeley, as we have seen, was particularly vexed by the idea of 'material substance', arguing that we have and could have no conceivable evidence for its existence, and that we could in any case say everything that we wanted to say without referring to it. By the time we come to Meinong, however, it seems (or at any rate, it seemed to Russell), that philosophy had become capable of spawning indefinitely many metaphysical objects – just as many as there are meaningful phrases to 'refer' to them. The philosopher enters a universe oppressively over-populated by his own creations, without any weapon to cut them down. What is the way out of this jungle? Or are we lost there for ever?

Russell's answer is simple and is, in fact, a re-writing of Berkeley's answer, in terms of modern logic. Speak, he says, about what you have to speak, using sentences whose meaning and structure you understand. You may not say exactly the same as you say in ordinary language: but at least you know what you mean. Moreover, there is no need to mean anything else. There is nothing that can be clearly said that is not capable of being said in the language of logic; and by confining yourself to this language you will see that no philosopher has, or can have, any grounds for asserting the existence of the metaphysical entities to which his theories seem to commit him.

We have already seen an application of this idea in the theory of logical constructions (which is a generalisation of the theory of descriptions as Russell presented it), and in the philosophy of mathematics. We find similar thoughts too in the American philosopher W.V. Quine, who summarises his metaphysics in the slogan 'to be is to be the value of a variable'. In other words, if your beliefs commit you to affirming sentences of the form 'There is an x such that $F(x)$', then, and only then, are you committed to the existence of something that is F. Everything that you are obliged to 'quantify over' in order to express the truth exists. Nothing else exists. That way, by displaying logical form, we get rid of the metaphysical entities lurking in the 'ontological slums' of language.

It is because it pointed in this direction that F.P. Ramsey, a pupil and disciple of Russell's, called the article 'On Denoting' a 'paradigm of philosophy'. By careful analysis of language it seems to dispose of a philosophical problem, and of all the ghostly entities that had been invoked in solving it.

6. Strawson's Criticism

It is on this point, however, that Strawson's objections to Russell's theory turn. Strawson was rebelling against the tradition that arose out of Russell's article, and which placed logical form before natural grammar. He did not understand Russell's theory in quite the terms I have used. For he was under the influence of other thinkers, who had taken Ramsey's praise to heart. He was exasperated by the assumption that formal logic is somehow *nearer* to reality than natural language, and that if something cannot be said in a formal language then it cannot be said at all. He believed that there is a logic of natural language, different from that of the formal languages presented by Frege and Russell, but nevertheless just as capable of delivering true results from true premises, while being rather more sensitive to the principal purpose of language, which is to communicate with rational beings.

First of all, he argues, Russell assumes that sentences in a language have meaning in themselves, so that we can understand them even when they are neutrally displayed upon a page. But language does not *have* meaning: it *acquires* meaning, through our use of it. Furthermore, it is not sentences that are true or false, but the *statements* that we make by means of them. Making a statement is an activity, which requires a context; and built into this context are presuppositions that are not stated by the speaker.

Russell's analysis proceeds by supposing that sentences containing denoting phrases imply various other sentences: for example 'The king of France is bald' implies that the king of France exists. But sentences do not imply anything at all, until they are put to use in statements. And, if the king of France does not exist, there is no statement to be made about him. The best we can say is that the statement that the king of France is bald *presupposes* the existence of the king of France, since the one who makes such a statement is in a position to do so, only when the king exists. *If* the king exists, then we can refer to him; if not, we can't. Referring is something that *we* do; 'denoting' is a mere fabrication of the logic-besotted intellect, and certainly nothing that a *phrase* can do. Phrases *do* nothing.

The logic of ordinary language, Strawson goes on, cannot be captured in the structures of formal logic. It is, for all that, a logic, laying constraints on our speech, and obliging us to aim in just the direction that Russell wishes us to aim: towards the truth about the world. It is absurd to say that 'the king of France is bald' *means* 'There is one and only one *x* such that *x* is a king of France and every king of France is bald'. That is not what the sentence *means* at all, even if the statement that a person would make by means of it presupposes the existence of the king of France.

What then *does* the sentence mean? Unfortunately, Strawson does not favour us with an answer. Perhaps he should say that the meaning of the sentence is given by the statement that it is standardly used to make. But what is that statement? Here we enter mysterious territory. We know what sentences are. But do we know what statements are? Do we know how to count them, distinguish them, recognise them? Could they, too, have stepped from the ontological slums? Such would be Quine's conclusion.

In any case, the disagreement between Russell and Strawson is far less radical than it seems. Strawson is saying that a sentence does not have a truth-value; only statements have truth-values. The statement that the king of France is bald may be true or false, and it can be true or false only when *made*. However, it cannot be made unless its presuppositions are fulfilled. One of these presuppositions is the existence of the king of France. If the king of France does not exist, it is neither true nor false that the king of France is bald. Strawson has, in effect, retreated to the position occupied by Frege.

Perhaps too, the best way of identifying a statement is in terms of the circumstances under which it would be correct to make it. The statement that the king of France is bald is correctly made – when? *One* answer (though not the only one) is that it is correctly made when true: in other words, when there is a king of France and only one king of France, and he is bald. But then, we have identified the statement in terms of its truth-conditions. And they are the very same truth-conditions that Russell described. We are not so very far from 'logical form' after all.

There is more to the dispute than that, but it serves to show how difficult it is to reject the Russellian analysis entirely. As for the promise of a 'logic of ordinary language', the least that can be said is that Strawson does not fulfil it. Maybe it cannot be fulfilled. (This connects with a persistent question, which is: Why is logic so *difficult*? Should it not be the easiest thing in the world, to decide what follows from what and why? The fact is, however, that the truths that are most evident are also the hardest to explain.)

7. Moving on

Strawson's objections are not entirely wrong, although nobody now would make them in the way that he chooses. There is a growing sense that language functions in a variety of different ways, and that the 'regimentation' (as Quine calls it) required in order to display its supposed 'logical form' does damage to our ordinary ways of thinking. Moreover, philosophers (with the exception of Davidson), are unsatisfied with the view that simple logical languages have the power either to represent the sum of human thought or (for the same reason) to describe the structure of reality.

Definite descriptions provide, in fact, a good illustration of this. When I say 'George is the horse who won', I may mean one of two things, depending on the context. You may be inspecting George with a view to buying him, and I utter this sentence as a recommendation. Alternatively you may have heard the outcome of the race, and want to know just who was the genius, to use language abhorrent to Musil, who won it. In the first case my concern is to describe George's qualities; in the second case my concern is to identify the winning horse. One and the same sentence can function both as an 'attribution' of qualities and as an assertion of identity. Surely it does not have the same logic in these two uses? In a distinguished recent contribution to the theory of descriptions Keith Donnellan explores the distinction between these two uses of definite descriptions, and shows that the Russellian theory does not capture the whole truth about either of them. (See K. Donnellan, 'Reference and definite descriptions'.)

But this brings us back to metaphysics. For whether we take logic or ordinary language as our focus of concern, it has become clear that the structure of what we say has considerable bearing on the structure of the world. It could even be that these two things are different aspects of one thing: so the Hegelians tell us.

8

Things and Properties

Return for a moment to Descartes's 'cogito': I am thinking. Why was he certain of this? Because even if he doubted it, that merely proved that someone doubted. But has he not gone too far? The eighteenth-century German aphorist Lichtenberg suggested that Descartes was entitled only to assert the existence of thought: not 'I am thinking', but 'it is thinking', to be understood on the analogy with 'it is raining'. Certainly, there is thinking going on; but why assume the existence of a thinker? Many people have been impressed by this objection. For, in supposing a thinker, you have introduced a powerful metaphysical hypothesis. By the private language argument, the hypothesis can rescue the entire world from the demon.

But is this objection really tenable? Surely, there cannot be thinking unless *someone* is thinking? Thoughts do not exist without an owner. Unlike rain, thought is a state or property, not an independent item in the world. To say that thinking is going on already implies the existence of a thinker. Otherwise we could be in the condition of the dying Mrs Gradgrind (*Hard Times*), who said that there is a pain in the room somewhere, but she couldn't rightly say that it was *hers*!

The point is general. Properties require an object to instantiate them. There cannot be squareness in the world without something square. This is an impressive fact, which has troubled philosophers since Plato. For it seems that we unerringly divide reality into two kinds of entity – properties, and the things that bear them. And we are already in possession of a vast store of knowledge about the two. We know that properties inhere in things, but not things in properties; that a thing can change in respect of its properties but no property ever changes at all; that a thing can go out of existence but a property never; that things must be known through their properties, but properties can be known in themselves. And so on.

These observations have led philosophers to ask what the division between thing and property amounts to. We have already seen that the division in language between subject and predicate arises spontaneously and is carried over, in one form or another, into the logical analysis of a sentence. Aristotle took this division between subject and predicate as fundamental: there cannot be truth, unless a property is 'predicated of' something. Predication is part of the fundamental structure of thought. Moreover, it is by virtue of predication that we obtain truth: a true sentence is one that predicates of its subject a property that in fact belongs to it. So reality must mirror this structure in our thinking. Indeed that is *why* thought has the subject-predicate structure. It has this structure because reality divides into two: substances on the one hand, and attributes on the other.

Until now I have spoken of things and their properties; others refer to objects and qualities, substances and attributes, individuals and concepts. And into each pair of terms there is built a wealth of metaphysical theory which forbids us from saying that these are several versions of the same distinction. Behind all of the theories, however, lies an attempt to understand a fundamental divide, which is that between particulars and universals. Particulars are what we identify by names and definite descriptions. Universals are *instantiated* by particulars, and shared by them, as the property of blueness is shared by everything blue.

1. Particulars

Particulars come in several guises:

(a) Concrete particulars, or things in space and time. Tables and chairs, animals and persons, atoms and galaxies – all these are 'things in space and time'. They can be individuated, counted and described. There is much discussion as to whether some are more *basic* than others. Could we really identify atoms and galaxies, for example, if we could not identify tables and chairs? (See section 6, below.) Philosophers also wonder which particulars are truly *substantial*. For example, a heap of sand seems more 'arbitrary' than an animal or a person. Its being *one* thing (rather than two or many) does not flow from its nature, as does the oneness of a cat. In looking for particulars, we are looking for the things that must be enumerated, when 'taking stock' of the world. A universe containing a heap of sand is little different from a universe containing the same heap divided. But a universe containing a cat is very different from one containing two halves of a cat. Empiricism tends to the view that only *concrete* particulars are *really* real. To give an inventory of the universe is to identify all that is contained in space and time. Rationalism tends to the view that reality is not so obvious.

(b) Abstract particulars. For do we not also refer to and describe things like numbers, classes, possibilities and fictions? Numbers especially are the source of much philosophy, as we have already seen in discussing Russell. They are 'objects' in Frege's sense: that is, we give them names, and strive to discover the truth about them. Yet it is absurd to say that they exist in space and time: as though there were some place where the number nine could at last be encountered. Should we therefore say that they too are particulars, but of a very special kind?

Empiricists worry about these abstract particulars for a variety of reasons. For one thing, numbers cannot be known through the senses, but only through thought. Moreover, they do not *act* on anything, so as to produce results. They are 'powerless' in the natural world, and leave no trace there. (When I put two apples in the balance and it swings, it is the apples that cause the movement, not the number two, which takes no part in the process. If I thought that the number 2 was involved in this process, then I should have to think the same of the number 4 – there being four half-apples in the pan. And so on for all the numbers. If any number participates in a causal process, therefore, they all do. Which means that the 'presence' of a particular number is not a distinguishing condition of any process – and therefore not part of its cause.) In which case, how do the numbers affect our thought, and why do we say that, by thinking, we gain *knowledge* of them? Many empiricists therefore try to construe numbers and other abstract objects as 'creations of the mind', with no independent reality.

(c) Mixed cases. There are also some rather puzzling cases which seem to hover between abstract and concrete. An interesting instance is that of a 'type'. If I refer to the Ford Cortina, I do not refer to one particular car, but to a type of car. The individual Cortinas are 'tokens' of this type. At the same time, logically speaking, the type behaves like a particular. The Ford Cortina has *properties* (namely, those that are shared by its undamaged tokens); it is to be described and explained in terms of concrete processes in the spatio-temporal world. Nevertheless, there is no place where the Ford Cortina *is*. It remains aloof from the world of its tokens, just as numbers do.

(d) Problem cases. Much of metaphysics is concerned with 'ontology': what exists and what we must assume to exist in order to achieve a cogent description of reality. Problems of ontology are as real for empiricists as for rationalists, and they are as hotly debated today as they were by the pre-Socratics. Problematic entities are frequently thrown up by philosophical discussions, and we may wonder whether they are particulars, rather than properties or states of something else. Examples include: events, facts, propositions, states of affairs, sights and sounds, works of art (where is Beethoven's fifth symphony?), and so on.

2. Universals

Universals inhere in, are instantiated by, particulars and the same universal may inhere in indefinitely many particulars. Universals are of several kinds:

(a) *Properties.* When I say that the glass is green, I attribute a property to the glass. This property corresponds to, is expressed by, the predicate 'is green'. Sometimes the attribution of a property uses another kind of language. For example, when I say that John is a man, I ascribe a property to him: however, it is not an adjective but a noun that performs the function.

Properties are of many kinds, and certain interesting philosophical divisions have been made among them through the ages. One of the most important has proved to be the distinction between 'primary' and 'secondary' qualities, discussed by Locke, in his *Essay of the Human Understanding* (Bk II, Ch. 8), though not invented by him. We intuitively feel that some properties belong to an object, regardless of how it may appear to us, while others owe their nature to our capacity to perceive them. A thing is square, heavy, solid, whether or not it looks so. But is it bitter whether or not it tastes so, red whether or not it looks so, loud whether or not it sounds so? Locke suggested that bitterness, redness or loudness must be defined in terms of the power to produce experiences in the normal observer. Things have colours *because* they appear coloured to the normal observer. Colour, therefore, is a 'secondary' quality. Locke's distinction belongs to a wider enterprise, which is that of separating the world as science discovers it to be, from the world as it appears in our everyday perceptions.

Further distinctions are sometimes made between enduring properties and temporary states, between dispositions (courage, fragility) and occurrences (being in pain, being green), between accidental and essential properties. Some of these distinctions I shall discuss later. All that needs to be understood for the moment is the vastness and diversity of the class of properties. Whatever can be truly 'predicated' of a particular is a property: there are as many varieties of properties as there are varieties of fact.

(b) *Relations*. These caused great problems to the traditional (Aristotelian) logic. A relation has instances; but it requires more than one particular in which to inhere. Some philosophers – notably Leibniz and Spinoza – found this paradoxical: it makes the truth about *one* particular dependent upon the truth about another, even when there is no causal or metaphysical connection between them. Modern logic dismisses such worries as self-engendered, and treats relations simply as properties of ordered sets of objects. Thus the relation expressed by 'taller than' is a property of *pairs* of things, the relation expressed by 'between' a property of *triples*. Saying that these pairs, triples or whatever are 'ordered' is a way of recognising the fact that in saying that John is taller than Mary, we are not also saying that Mary is taller than John. The relation 'taller than' applies to the pair John and Mary *in that order*. To some extent the mystery contained in the idea of relation is retained in this new concept of ordering. Who does the ordering and how?

(c) *Kinds*. We could not understand the world if we did not learn to assign particulars to their kinds: to identify this as a tiger, that as a piece of gold, this as a human being. Kinds are universals (they have instances), but there is a dispute as to whether they are merely that. In identifying something as a tiger, we have said a vast amount more than we have said when describing it as yellow. Nor is this 'vast amount more' merely additional properties of the yellow variety. We have said something about the nature of a thing, in identifying it as a tiger. I shall return to this point below.

(d) *Sortals*. Locke distinguished properties that could be used to count their instances, from those that could not. Take the general term 'man'. This gives me a means to count things, not merely a means to describe them. 'How many men are there in this room?' is a determinate question. Contrast 'How many green things?' This has no determinate answer: it all depends how I divide the areas of green. Is a green pullover one bit of green or three (a trunk and two sleeves)? Is a green picture-frame one bit of green or four? And so on. The concept of a sortal has become significant, for reasons that I touch on below.

(e) *Mass terms*. Finally, just as we should recognise terms that feature in answer to the question 'How many?', so should we recognise other general terms, which feature in the answer to 'How much?' How much water is there in the bottle? How much snow fell last night? And so on. These terms also express universals: but universals of a very peculiar kind, which often seem to behave, logically speaking, as though they were particulars. Although each drop of water instantiates the universal water, we speak of the universal as prior to its instances. We enumerate the properties of water, describing it by means of subject-predicate sentences in which 'water' is the subject-term. This too has baffled philosophers, and 'mass nouns', as they are called, now attract a special chapter in the literature.

3. The Problem of Universals

We could continue to make distinctions among universals, and the result would not be without interest. But it will not answer the question that has puzzled philosophers since Plato: namely, what *are* they? To *what* do general terms refer?

Plato was impressed by the fact that universals are abstract, like numbers. The colour blue has instances in space and time, but in itself is nowhere and nowhen. Moreover, the knowledge we have of it – for example that blue is a colour, that nothing can be blue and red at the same time, that blue things are visible to the eye, and so on – is knowledge of *necessary* truths. Perhaps *all* the truths about universals are necessary, just like the truths of mathematics. Nor do universals change in any respect: the colour blue has no history, and is what it is for all time and unconditionally. In other words, a universal behaves exactly like an abstract particular. So maybe that is what it *is*.

This theory gains credence if you think that terms in a language all function in one way, by standing for things (i.e., as 'names'). 'John' stands for John; 'this chair' stands for this chair; so what does 'brave' stand for? The answer is that it stands for an entity which is *universal*: an Idea, as Plato sometimes describes it. (You then encounter a formidable problem, concerning what the word 'is' stands for in a sentence like 'John is brave'; this – the classical question of Being – is a matter to which I return in Chapter 12.)

There is no doubt that Plato was disposed to think of language in something like the way that I have just described: as a string of names, each separately standing for something, and combined purely by their sequential order. In considering Frege's theory of the function we have already encountered an important objection to that way of seeing things. The parts of a sentence have separate roles, and make sense only in combination, and by performing the tasks that distinguish them.

What, then, are universals, according to Plato? He does not exactly say, and indeed constantly changes his mind. But in certain dialogues, notably the *Parmenides*, the *Phaedo* and the *Republic*, he introduces his famous theory of Forms, which was for a long time assumed to be an answer to the problem of universals. (Modern commentators believe this interpretation to be too simple.) Take a general term, like 'bed'. How do we come to apply it? The answer is simple. 'Bed' denotes the abstract Form of bed. We compare the particular object with the Form, and discover the fitness between them. Then we apply the term 'bed' to the object. (The example of 'bed' is that considered by Plato in the *Republic*; but did he intend it seriously? See the Study Guide.)

Plato was unhappy with the argument and proposed an objection (the so-called 'third man argument') which need not concern us here (but see again the Study Guide). The force of Plato's objection can be understood from considering another, due to Bertrand Russell. If you say that you apply the term 'bed' by virtue of a relation between the bed and the Form, then what about the relation? Is it not a universal too? Do you not need to recognise that the bed and the Form instantiate this relation? But it too must have its Form; so that, relating the object to the Form of bed, you must first relate it to the Form that relates it to the Form of bed. But what about this new relation? Is this too not a Form? We are at the beginning of a vicious infinite regress.

Other things too made Plato unhappy. If we say that every general term denotes a Form, then the sphere of Forms becomes densely populated with far from pleasant things: the Form of evil, the Form of pain and so on. How can these coexist with their opposites so easily? Moreover, evil and pain acquire the sacrosanct and eternal reality that Plato had been hoping to reserve for the Good and the Beautiful. He therefore continued to revise the theory in obedience to other preoccupations.

Some people have wondered whether there can really *be* a theory of universals, as Plato envisaged it. For what is such a theory supposed to do? Is it supposed to give a general account of how we can apply general terms? But then it must *use* general terms, and therefore assume that we already understand them. A theory that says 'General terms can be applied because the world is thus and so', applies the predicate 'thus and so': and what does this refer to?

This argument is very powerful, and you find it repeated in a variety of forms in the later writings of Wittgenstein, and in those of Quine and Nelson Goodman. But it does not remove our worry: if anything, it deepens it. For while we can make a first shot at identifying particulars, and giving theories of language which show how to refer to them, we seem to be unable to do the same for universals. So what *is* happening, when we describe a book as red? Surely, we have said something more about the world in saying that the book is red – more than is implied in saying that the book exists. Redness is part of reality. What part?

4. Realism and Nominalism

Traditionally philosophers have divided into two camps in answer to that question. The 'realists' say that universals really exist, independently of our thought. The nominalists say that universals are brought into being by thought – specifically, that there is no more to the reality of universals than our use of general terms. ('Nominalism', because the theory gives precedence to the *name* of the property over the property itself. For the sake of simplicity I ignore the theory of conceptualism, which is like nominalism, except that it emphasises general *thoughts* in the place of general *terms*. See the Study Guide.)

Realists are of two kinds. First, there are those like Plato, who postulate a separate realm where universals reside – the realm of abstract things – and who try to understand general terms as standing for items in this realm. Secondly, there are those, like Aristotle, who believe that universals really exist, but only *in and through* their instances. There is no special realm were redness resides: it exists in the here and now, and consists in the redness of particular things. At the beginning of the *Nicomachean Ethics*, Aristotle summarises his objections to the theory of Forms: the principal one being that it completely mistakes the function of the very general term (the term 'good') that it was introduced to explain. (Does the Form of the Good play a role in the meaning of 'good flower-pot', 'good frying-pan', or 'good guillotine'? And is it always the *same* role?) Aristotle's point in this important passage is to remind us that general terms acquire their meaning in the description of particular things. If we learned them by applying them in the abstract realm, then our thoughts would stay in that realm, and never descend to earth. But it is only in their concrete application that terms like 'red' can be understood. And their role is quite different from the one implied by Plato.

Aristotle is sometimes said to have believed in universals *in rem* (in the thing). The suggestion is that this kind of commonsensical realist admits that redness exists, but only *in* red things. But that does not seem to be a satisfactory theory, for two reasons. First, what does the word 'in' mean? Is that not part of what we were trying to explain? Redness is not in the book as I am in this room. And how is it that redness can be in this book and that book, when we know that it is impossible for me to be in this room and that one simultaneously? Are we not just refusing to answer the real question, when we take up this position?

Secondly, what about universals that have no instances? Suppose no red things had existed. Would that have entailed the non-existence of redness? Certainly, we do think of certain universals as closely tied to the destiny of their instances: the dinosaur went out of existence with the last of its instances. (Or did it?) But much of our thought is devoted to exploring universals that may have no instances at all. Consider the universal expressed by 'square'. In geometry we define this and develop all kinds of theorems about it. But is there anything in the world that is actually square, as the geometers define the term? Many things are *approximately* square – and there are some very close approximations. But is anything actually *square*? Maybe not. Yet we know more about this universal (thanks to geometry) than we do about any of those that we encounter in the street. That was the kind of argument that impressed Plato.

Maybe an Aristotelian should say, in response to this, that a universal exists so long as it *could* have instances. 'Redness exists' means 'red things are possible'. But what does 'possible' mean? Does this not denote another universal – and of a highly abstract kind? Some philosophers think that we understand possibility through the study of 'possible worlds' – abstract entities that are not so very far from Plato's Forms, at least in the metaphysical qualms to which they give rise.

So what does the nominalist say about all this? There are two parts to nominalism. First, the nominalists argue that only particulars exist, and that nothing else exists. If we say that universals exist this only means that we apply general terms to particulars. Abstract entities (and this goes for abstract particulars too) are simply shadows cast by language. We imagine that, because we apply the predicate 'green' to many different things, there is a single entity – greenness – which is common to them. But all that they have in common is that we call them 'green'.

The second part of nominalism is more subtle. What, the nominalist asks, is really being said by someone who argues that greenness exists? Presumably, that there are green things. But what does that mean? Presumably, that some sentences of the form 'x is green' are true. In other words there are instances where it is correct to apply the predicate 'green'. So what makes it correct to say that x is green? The realist says: the fact that x *is* green, and so goes round in a circle. The nominalist says something quite different. We regard the use of the word 'green' to describe x as a correct use. This is a fact, but it is a fact about *us*. According to the rules of our language, x is correctly classified as green. But we could have classified x in a quite different way. Classifications merely gather individuals under a common label, and at some level, all classifications arise from our *decisions*.

The realist is dissatisfied, and wishes to anchor our language in reality, to say that there are *real properties* of things which justify our descriptions. But in arguing thus, he produces new descriptions, and so begs the question. You do not anchor our classifications in reality, merely by replacing one classification by another. The realist, like the nominalist, is bound to use words. But he is under the illusion that his words show him a reality beyond the one they create. And for this he can have no evidence.

Two very important arguments have been developed along these lines in recent philosophy: Wittgenstein's argument about rule-following, in *Philosophical Investigations*, and Nelson Goodman's argument about predication (sometimes called 'Goodman's paradox') in *Fact, Fiction and Forecast*. They are difficult, and it is not necessary to understand them at this stage. (See Chapters 14 and 19.) Each argument tends to the conclusion, as Wittgenstein would put it, that you cannot use language to get between language and the world. Our attempt to explain how we classify things will always involve classifying things: we cannot, in language, step outside language so as to confront an unconceptualised reality. At some point we are forced to accept, therefore, the fact that we use words as we do.

The realist will not be satisfied with this. He will argue that our use of language is not arbitrary, but is *constrained* by reality. If we use words as we do, it is because the world constrains our communications. The world contains universals, like redness, which we try to capture in our use of general terms. Moreover, the realist will argue, the nominalist has embarked on a dangerous journey. He seems to tell us that the world depends on language: that we *make* the world by speaking. Indeed that is what the more radical nominalists (Nelson Goodman for instance, in his *Ways of Worldmaking*) explicitly *say*. Is this not to lose the 'robust sense of reality' that Russell recommended? In the wrong hands, nominalism can be an intellectual disaster. If I believe that language users form reality through their conceptions then I am likely to be suspicious of your conceptions. Why do you want to build the world in your way, if not for purposes of your own? Do I not have a duty to liberate myself from your conceptions? Look at the writings of Michel Foucault and you will see how corrosive this approach becomes. Indeed a kind a vulgar nominalism underlies many of the modernist outlooks that have colonised the academic world: feminist criticism and deconstruction are examples.

5. Substance

Is it true, as I have assumed, that we at least understand particulars? Vivid though the image may be, of those cats and dogs and saucepans, can we really say what it is that makes these into particular things? Or does this fact lie at the limit of language, as incapable of further explanation as the fact that we call green things 'green'?

Two features of particulars have attracted the attention of modern philosophers: unity and identity. The saucepan is *one* thing: it is counted once in the inventory of the world. It is also identical with itself, which means that it can be the subject of 'identity questions': is this the same saucepan as the one I saw yesterday? And so on. The two features belong together, and have to do with the fact that particulars can be *counted*.

Aristotle, however, was dissatisfied with many of our ways of counting particulars. For *how* we count them depends on how we classify them. And there are ways of classifying which are near to nature, others which are artificially devised. If I ask you to count the heaps that are in this room, you would hardly know how to begin, and in any case your answer would be arbitrary: you can divide a heap into two heaps for the sake of counting it, without doing violence to reality. If, on the other hand, I ask you to count the number of desks, your answer is not so arbitrary. And if I ask you to count the number of cats, your answer is not arbitrary at all.

Furthermore there are things which seem like particulars, but which, on inspection, are more like universals. Consider again the Fort Cortina. Certainly, it has properties: but only because individual cars have those properties. They are instances of the *type*, which is 'predicated of' them. Or consider the average man. The argument that he is a logical construction establishes also that the average man is 'predicated of' men. For the truths about him depend upon the truths about men, and there is no additional truth which is purely about *him*. Likewise, when I speak of the 'Class of '91', I am really referring to its members. Although I seem to be describing a particular, this particular is 'predicated of' the individuals who compose it.

Putting together those two ideas, we approach the highly charged notion for which Aristotle is famous: that of substance or *ousia*. Attributes are 'predicated of' substances; but substances are predicated of nothing. Moreover, particular substances are those particulars which *must* be counted if the contents of the world are to be identified; and the way of counting them is fixed by their nature.

Is a heap a substance? Surely not. For the truths about the heap are really truths about the items that compose it. It can be 'broken down' without remainder. Moreover, we can summarise those truths equally well however the heap is counted: as one thing, two, three or as many as it has members. Matters are very different with Moggins the cat. The truths about Moggins are truths about *her*, and not about the parts of her. Divide her into parts and she ceases to exist. And the relation between those parts is a natural relation: it is a fact of nature that these parts are joined to form *one* self-contained thing. Moggins must be counted once and once only in the inventory of the world.

When it comes to the desk or the saucepan, matters are not so simple. Can we not divide a desk into a desk-top and four legs? Could we not saw it in half, so as to count it, now as two things, now as one? Is it not a desk only in so far as we *use* it as such – so that its nature is, so to speak, lent to it by our interests? In which case should we not say that the facts about a desk are really facts about the people who use it, so that the desk is *predicated* of those people?

Here we are in deep water, and it is best to jump out quickly. The distinction between cats and desks is that between 'natural' and 'functional' kinds. The examples tell us that, in our pursuit of those non-arbitrary particulars which are ultimate bearers of our predications, we are forced to divide the world into *kinds*, and to distinguish the arbitrary from the non-arbitrary among them. Some universals describe the *nature* of the things to which they apply; others describe their passing condition. Moggins could cease to be black without ceasing to be Moggins; but she could not cease to be a cat without ceasing to be. Aristotelians would say that her being a cat belongs to her *essence*, while her colour is an 'accident'. The ultimate inventory of the world describes things in terms of their essences. And if the essence determines a way of counting (if it is a 'sortal'), then each of its instances is an individual substance.

The idea of substance was taken up by later philosophers and bent in several directions. In the philosophy of the rationalists it was to play a vital role. Descartes's conclusion about himself is not that he thinks, but that he is a thinking *substance*. And it is very important to know what he means by this. His immediate successors, Leibniz and Spinoza, inherited his idea of substance, and tried to iron out its inconsistencies. Leibniz came to the conclusion that there are infinitely many substances (the monads) which together make up the world (though Leibniz's world is very different from the world as we normally perceive it); Spinoza came to the conclusion that there is only one substance, which is God or Nature, depending how it is conceived.

Empiricists have been more scathing. If we mean by a substance that which is the bearer of qualities, Locke argued, then this is a bare 'substratum': it is what remains when we peel all the qualities away, the onion without its layers. But what is that? If we say something about it, we simply describe one of its qualities. Nothing in our language enables us to identify the substance *itself*. It is a 'something we know not what'. The same is true of the 'real essences' which supposedly guide our predications. They too are unknown (though perhaps not unknowable). Our language is attached not to real essences but to *nominal* essences – classifications which we invent for convenience' sake, and which may have no other foundation in reality. (Here you see a very good instance of how a philosopher's stand on the question of particulars goes hand in hand with a view about universals.)

6. Individuals

Discussions about substance faded away from modern philosophy, and it is only thanks to the patient work of David Wiggins that they have regained their old importance. In the meantime another set of arguments dominated the field – argument about identity which have their origin in Frege, but which are more frequently associated with the work of Quine and Strawson. It is worth summarising the argument in the first chapter of the latter's *Individuals*, since it has shaped the way in which modern philosophers approach the questions we have been discussing.

Strawson argues that all discourse depends on identifying (referring to) particulars, and predicating properties of them (bringing them under concepts). How do we identify a particular? Well, one way is to identify it in terms of other particulars: as the Class of '91 is identified through its members. But obviously that merely postpones the answer to our question. How do we identify the members?

Strawson argues that there are *basic* particulars, which have to be identified if we are to identify *anything*. (The basic particular is Strawson's substitute for Aristotle's substance.) These will be the true 'individuals' that compose our universe: the things that are counted, in the inventory of the world. What makes them basic is that our linguistic practices are anchored in them: it is by referring to them that we attach our discourse to the world. Hence basic particulars cannot include, for example, the micro-physical components of reality. For those are beyond our ken. They must be ordinary 'medium-sized dry goods', as J.L. Austin put it, the kind of things we see and bump into as we negotiate our passage through the world. The metaphysics of ordinary language therefore gives pride of place to the very things – tables and chairs – which have been first to disappear under the interrogating gaze of the metaphysicians.

If we are to identify something, in order to refer to it, then we need also to make statements of identity. We have to answer such questions as 'Is this the same table as yesterday?' 'Is that one chair or two I see against the wall?' And so on. In common with many philosophers, Strawson argues that identity, in the case of physical objects, includes *identity through time*. Reference to a physical object is possible only if it can be 're-identified' through change – whether change in it or change in me. Otherwise I should have no guarantee that the thing I am referring to lasts long enough for the reference to be completed, or to be understood by you. Strawson puts this point by saying that we need 'criteria of identity' for the basic particulars in our universe.

Strawson further argues that such criteria of identity are available only if we locate our basic particulars in space. Space provides the enduring 'frame of reference' which enables me to trace an object's history, and to say that this table is the one that was here yesterday. It is therefore no accident that the basic particulars in our world-view are medium-sized objects in space and time. The concrete particular is vindicated as the foundation of reference – the thing which makes discourse possible.

Strawson leads us to the 'reidentifiable particular' as the metaphysically privileged entity on which our world-view depends. And he places the following notions at the heart of philosophy: space, time, and identity. I shall survey these notions in future chapters. The time has come, however, to explore two other concepts of metaphysics: truth and reality. For I have assumed that we know what we mean by each of those terms, and it is probable that we do not know what we mean by either.

9

Truth

In one sense, truth has been the subject of the last three chapters, which have treated the abstract relation between language and the world. But the theories I have considered relate to the place of truth in argument – to the *logic* of truth. And they are neutral with respect to the *metaphysical* nature of truth. What is it, for something to be true or false? Intuitively truth is a relation – between the thing that is true, and the thing that makes it so. But both terms of the relation are in dispute, as is the relation itself. Philosophers differ as to whether the 'truth-bearer' is a sentence, a proposition, a thought, a statement, a belief, or some other entity, whether linguistic or mental. They differ too as to what truth consists in. Some speak of correspondence – but with what? (And here again there are a variety of positions, summarised in the terms 'fact', 'situation', 'reality' and 'state of affairs'.) Others replace correspondence with some other relation: coherence, for example. Others still reject the whole idea of truth as a *relation*, regarding it instead as an intrinsic property of whatever possesses it. There are even those who argue that truth is neither a property nor a relation, and that the concept is merely *redundant*. In this chapter I shall explore some of these theories, which take us to the heart of metaphysics.

1. Reality

The 'robust sense of reality' recommended to us by Russell sounds a warning against *hubris*: do not suppose, it tells us, that you are the only thing in the world, or even the centre of the world as you know it. Everything you think or say stands to be assessed in terms of what is *not* you. The measure of thought is *reality*, and reality is neither created by thought nor controlled by it. Reality is *objective*: its being is distinct from its seeming: what it is does not depend on what we *think* it to be. Our thought is aimed at reality, and when it hits the target, then and only then can we speak of truth.

All that is common sense. But it contains a great many metaphysical assumptions that are hard to clarify, harder still to prove. For one thing, we often speak in terms of truth, falsehood and validity, in contexts where we are not at all sure whether there is some reality that is targeted by our judgements. Consider ethics. You may express the view that eating people is wrong; and I may agree with your judgement, by describing it as true. And I may go on to assert that your belief that eating animals is wrong is *false*. This is not just an expression of arbitrary taste. On the contrary, it is an opinion over which we argue. You at once reply that I am mistaken, since it is wrong to take the life of an animal if it can be avoided (say, by eating some ghastly concoction out of soya beans). I in turn argue that it is right to eat animals if that gives them their only chance to live. The argument builds on a point of agreement (roughly, that it is good that animals should live) – but this too could be questioned; would it be a good thing that people should live, if they were merely bred to be eaten? Such debates are familiar to us. They are also immensely important. For although we are in broad agreement about fundamental moral judgements (the ten (give or take a few) commandments), we can find our way through the human jungle only with the aid of sober advice and reasoned casuistry. Where should we be in this, if we could not avail ourselves of the ideas of truth and valid argument?

Yet do we wish to say that there is a moral *reality*, which underpins our moral judgements and guarantees their truth? Some philosophers have *argued* for such a view ('moral realism'). But it is very much a philosophical position, neither endorsed by all who have considered it, nor implied by common sense. Our moral arguments could have the structure that they have, even if there were no moral reality, but only a striving for agreement – the 'common pursuit of true judgement' as T.S. Eliot called it, whose sole and sufficient reward is social harmony. If the example of moral judgement does not convince you, then consider arguments in aesthetics. It is *obvious* that St Paul's Cathedral is beautiful, and the new Lloyd's building repulsive; but is there some 'aesthetic reality' that makes these judgements true?

Those questions have led philosophers to try to define truth without assuming realism. To speak of truth and its role in discourse is one thing; but we should not *assume* that, in doing so, we are referring to an independent reality.

Nevertheless, we might agree on certain broad platitudes about truth, in terms of which to frame our theory. Here are some of them:
(i) If the sentence '*p*' is true, then so is the sentence ' "*p*" is true'; and vice versa.
(ii) In asserting a proposition, making a judgement, etc., we *aim* at truth.
(iii) Our judgements are not true simply because we decide to call them true.
(iv) Whatever is true has *conditions* for its truth: conditions whose fulfilment confers truth upon it.
(v) A true proposition is consistent with every other true proposition: no truth is contradicted by another.

2. The Correspondence Theory

Those platitudes lead naturally, though not inevitably, to the correspondence theory. This is often introduced by some words of Aristotle's: 'To say of what is, that it is, and of what is not, that it is not, is to speak truly'. But this gnomic utterance stands in need of interpretation – especially as it would be accepted both by those who defend the correspondence theory and by those who oppose it. So let's start again.

The basic idea is this: truth consists in correspondence between the thing that is true, and the thing that makes it true. To which you naturally reply: Between *what* and *what*? Things which can be true include: sentences, statements, propositions, beliefs and thoughts. Which do we choose? The simple answer is that it does not matter. If we can say what it is for a sentence to be true, then we can extend our theory to the proposition that it expresses, the statement that it makes, the belief that it identifies, and so on. So let's stick to the proposition – the abstract entity which captures what is *said* by a sentence, what is *believed* by a believer, what is *stated* by a statement, and so on.

What is the other term of the relation? To what do true propositions correspond? One answer is reality. But we have already seen that this leads towards metaphysical commitments that we may wish to avoid. Another answer is 'things', which is scarcely an improvement. Moreover, things are not enough, if we mean by this merely the array of objects that clutter the world. The proposition that my car will not start is not made true by my car, but by the fact that my car won't start. We need something more abstract than a thing if we are to reach the underpinning of the proposition: something like a fact, a state of affairs, or a situation. Which? And does it matter? Maybe it doesn't matter. For facts, states of affairs and situations have one important property in common, namely, that they are 'individuated' (identified as the particular facts, states of affairs or situations that they are) by means of a clause beginning with 'that': the fact that my car won't start; the state of affairs that my car won't start; the situation that my car won't start. Sometimes we can make do with another idiom: the situation, fact, etc. of my car's not starting. But the point remains the same, namely, that facts, states of affairs etc., are fully identified only through relative clauses.

In other words, you can only identify a fact through a *proposition*: the proposition that my car won't start, which can be 'nominalised' into the phrase 'my car's not starting', but which is, for all that, the very same entity that occurs on the other side of the truth relation. To say that a proposition is true if and only if it corresponds to a fact sounds illuminating. But when you come down to brass tacks, and confess that the proposition that *p* is true if and only if it corresponds to the fact that *p*, your theory sounds far less impressive. For the very same entity – the proposition that *p* – seems to occur on both sides of the equation. And it *has* to be that way: which is the reason for Aristotle's gnomic utterance (or rather the reason for its gnomery).

Does this matter? Defenders of the correspondence theory think not; defenders of the coherence theory think it matters very much indeed. We cannot get out of the problem by shifting from facts to states of affairs, situations or truth-conditions. These notions help us in suggesting something about the *way* in which the truth-value of a sentence is determined by the facts. But it still remains true that to identify the thing that makes a proposition true, we must offer a proposition. Moreover, only if it is the same proposition do we have any *a priori* certainty that our theory is a theory of *truth*.

So let us stick to the facts. Is anything really said by the theory that the truth of the proposition that *p* consists in its correspondence with the fact that *p*? The defender of the theory will argue that something *is* said: a proposition, he will claim, is a mental or linguistic entity, whereas a fact is a thing in the world. The proposition that *p* might never have been formulated; but still the *fact* that *p* would exist. In comparing our propositions with the facts, therefore, we are comparing them with something other than themselves. In a similar way a map of my village shares the contours of the village: but it can be compared with the village for accuracy, and shown to be true or false.

However, the analogy with the map shows the weakness of the reply. The features of the village are reproduced on the map. But it is not by the map that we first identify them. On the contrary, I could point to them, walk over them, measure them and describe them without the aid of the map. They are 'separately identifiable'; hence something substantial is said by the statement that they correspond to features of the map. By contrast, we seem to have no other way of identifying facts, save through the propositions that they are supposed to 'anchor'. Why are we so sure that we have *two* things, when both are identified in the very same way?

But can't we point to the facts? Are we *forced* to identify them through propositions? Does there not come a point where our words are attached to the *world*, and is not pointing one of the ways by which this attachment is achieved?

3. The Coherence Theory

Pointing is a gesture, and its meaning must be understood. Suppose I point my finger at a picture. What leads you to suppose that I am pointing at the *picture*, rather than the mirror behind my shoulder? After all, you could have read the gesture in another way, from the finger backwards to the shoulder. The simple answer is that we read the gesture as we do because there is a *convention* that governs it. This is how we understand it. Moreover, the convention says only that the gesture points to the *thing* in front of me; further conventions have to be invoked, in order to know which *fact* about the thing I am singling out for your attention. Pointing belongs to language, and leans on language for its precision. It is only when we are able to read the gesture as an expression of thought that we can use it to anchor our words in reality. But that raises the question *what* thought? Why, the thought that *p*! We are back with that wretched proposition. Indeed, no other thought would do: only this would serve to convey the fact that we have in mind, when referring to whatever makes it true that *p*.

Arguments like that have been given by Wittgenstein in defence of a kind of nominalism, and by Hegel in defence of the coherence theory of truth. So what does the coherence theory say? The basic idea is this: try as you may, you will not be able to step outside thought and lay hold of some independent realm of facts. To say what makes one thought true is to express a thought: usually the *same* thought. We can anchor our thoughts, but only to other thoughts. No thought bears a logical relation to something that is not a thought – unless it be another of those 'truth-bearing' entities (sentences, propositions, etc.) which are, for the sake of this discussion, in just the same metaphysical boat. If we believe that a thought or proposition is made true by its relation to something outside it, that thing must also be a thought or proposition.

Truth, therefore, is a relation among propositions. What relation? The answer usually given is 'coherence'. A false view of the world is one that does not cohere; a true one is one that 'hangs together' in some way, with each of its components supporting and supported by every other.

There is something undeniably attractive in this picture. For it overcomes the disobliging 'atomism' of the correspondence theory. It helps us to see why the search for truth links all our thoughts in a common endeavour, and why every one of our beliefs is exposed to refutation by the trial of every other. It makes science and theory-building intelligible, shows their place at the heart of knowledge, while at the same time endowing them with some of the majesty of art.

The problem is that the relation of coherence proves hard to define. A first shot at a definition might be in terms of consistency. For we know that every true proposition is consistent with every other. (This was the fifth of our initial platitudes.) In fitting our propositions into a total view of the world, we are building a system of mutually consistent thoughts. The problem with this is that it gives a necessary condition for truth, but not a sufficient condition. Take the totality of true contingent propositions and negate them. The result will also be a mutually consistent system. But every one of its components will be false. (The objection holds only in so far as we confine our attention to contingent truths.)

One response for the defender of the coherence theory, who wishes to take consistency as the truth-relation, is to argue that all truths are in any case necessary truths. And the negations of these will form an *inconsistent* system. (That, roughly speaking, is the line taken by Hegel, following Spinoza.) But then he has given himself another criterion of truth, namely necessity. Truth does not consist in a relation between propositions at all, but in an internal property possessed by each proposition in itself: the property of necessary truth. But what is *that* property? The prospects of defining it without relying on the idea of truth seem pretty slender.

Various alternatives have been considered. One very promising alternative relies on the idea of evidence. The true system is the one in which each proposition provides evidence for some or all of the others, or makes the others more probable, or is connected to the others through chains of evidential support. But here too there is a difficulty. For there can be conflicting systems – perhaps infinitely many – which exemplify this relation of mutual support, and they cannot all be true. (Another consequence of our fifth platitude.) So the idea of evidence needs strengthening if this line is to yield the result that the defender of the coherence theory is seeking. Perhaps he should speak instead of the *best* evidence: for it is possible that there is only one set of propositions which is such that each provides the *best* evidence for some other in the system. But how do we define 'best' evidence? Surely, *p* is the best evidence for *q* just in the case where *p* entails *q*: i.e. where *p* cannot be true without *q* being true. Once again we are driven to define coherence in terms of truth. Moreover we are approaching by another route the Spinozist view, that the system of true beliefs is one in which there is no place for contingency.

The defender of the correspondence theory will say that, however we define coherence, either it will fail to capture our basic intuitions about truth (including the five platitudes given earlier); or, if it does capture them, this is because the definition surreptitiously introduces a notion of *comparison* between thought and reality, between language and the world. The defender of the coherence theory will retort that there *is* no such comparison. We can compare thought with reality only by bringing reality under concepts: and then we compare thought with thought. (See Ralph Walker, *The Coherence Theory of Truth*.)

Although the argument seems deadlocked, that may be no more than an appearance. For it is open to the defender of correspondence to identify some feature of reality other than a fact as the entity that confers truth on a proposition. In a series of heroic papers, Fred Sommers has argued that we do not need facts, or any other 'language-shaped' entities, as 'truth-makers'. What makes it true that brown dogs exist is a property of the world – namely, its containing brown dogs. From the point of view of logic, a fact and a property of the world belong to wholly different categories, and the wicked arguments for the coherence theory therefore miss the point. Of course, the defender of that theory will not be satisfied by this: he will argue that we have merely changed the terms of the debate, but that in comparing sentences with properties of the world we are still really comparing sentences with sentences. But perhaps there is a reply to that. For either the defender of the coherence theory is making the trivial observation that we have to use words in order to describe the world – an observation which surely cannot have the consequence that an extra-linguistic reality *does not exist*! – or he is saying that we have to identify reality through true sentences; in which case he is saying something false.

4. Pragmatism

There is one theory that tries to incorporate the best of the two that we have so far discussed. According to this theory, our beliefs constitute a system, bound together by logical relations such as entailment and presupposition. We can amend or reject any proposition in the system, provided we adjust all the others and that they cohere with (are consistent with) it. But there is an *external* requirement that the system must meet. The system must match our experience: it must be possible to derive from it an account of experience that is actually confirmed. Our beliefs 'confront the tribunal of experience as a whole', as Quine puts it. As usually expressed (e.g. by Quine, its principal defender) this theory does not really differ from Hegel's, except in its adoption of empiricist categories. For Hegel would argue that experience only confirms or refutes our beliefs because it *includes* beliefs. Experience is a mode of 'concept application', and in testing the body of our thought against experience we are once again merely testing thought against thought. And the test can only be the old one, of coherence. The most we have achieved by Quine's move is to include a new range of thoughts (experiential thoughts) into the total system.

Quine himself is not a Hegelian. He identifies his philosophy with 'pragmatism', a peculiar American tradition, founded in the last century by C.S. Peirce, and passed on to William James, John Dewey and Quine. (See the essays in Quine's *From a Logical Point of View*.) Pragmatism is the view that 'true' means useful. A useful belief is one that gives me the best handle on the world: the belief which, when acted upon, holds out the greatest prospect of success.

It is very difficult to persuade a pragmatist to say anything as clear as what I have just said. Simple definitions of truth in terms of utility seem transparently absurd. More complex positions tend to become indistinguishable from the coherence theory (Quine) or the correspondence theory (Peirce), by virtue of the system into which they are incorporated. Obviously, if we say that a belief is true when useful, we must know what we mean by 'useful'. Anyone seeking a career in an American university will find feminist beliefs useful, just as Marxist beliefs were useful to the university apparatchik in the Soviet Union (not to speak of Britain or Italy). But this hardly shows those beliefs to be true. So what do we mean by 'useful'? One suggestion is: part of a successful scientific theory. But what makes a theory successful? (Marxism was successful, if you mean that it was spread to a large number of believers.) Some say that a successful theory leads to true predictions. But if we take that line, we end by defining utility in terms of truth. Indeed, it is hard to find a plausible pragmatism that does not come to this conclusion: that a true proposition is one that is useful in the way that *true* propositions are useful. Impeccable, but vacuous.

For this reason, modern pragmatists tend to retreat, like Richard Rorty (who ought to be described, perhaps as 'post-modern') to the coherence theory of truth. Relying on arguments similar to those given above, Rorty decides that we must reject the idea that our discourse 'represents' or 'matches' an independent reality. We cannot aim our thought beyond language, and the attempt to do so will merely return us to our language in another guise. Here is how Rorty defines his position:

> [Pragmatists] view truth as, in William James's phrase, what is good for *us* to believe. So they do not need an account of a relation between beliefs and objects called 'correspondence', nor an account of human cognitive abilities which ensures that our species is capable of entering into that relation. They see the gap between truth and justification not as something to be bridged by isolating a natural and trans-cultural sort of rationality which can be used to criticise certain cultures and praise others, but simply as the gap between the actual good and the possible better. From a pragmatist point of view, to say that what is rational for us now to believe may not be *true*, is simply to say that somebody may come up with a better idea . . . For pragmatists, the desire for objectivity is not the desire to escape the limitations of one's community, but simply the desire for as much intersubjective agreement as possible, the desire to extend the reference of 'us' as far as we can . . . (*Objectivity, Realism and Truth*, pp. 22–3.)

The pragmatist, on this view, thinks that his views are better than those of his opponent (whom Rorty calls the 'realist'), but he does not think that his views correspond to the nature of things. Pragmatism, in other words, wraps together the claims of nominalism, with those of the coherence theory of truth, and tells us that 'we' are the test of truth, and the final court of appeal in all our scientific judgements. Since truth, on this view, is indistinguishable from widespread agreement (or at least, widespread agreement among 'us'), the pursuit of truth – science – is nothing but the attempt to spread agreement as far as we can.

It is hard to argue with such a view, just as it is hard to argue with the nominalist or the idealist: a pragmatist, so defined, will always remain in the seat that he has carved for himself, provided he never drops his guard. As Rorty himself acknowledges (p. 24) the pragmatist does not really have a theory of truth. He is proposing that we drop this concept, with its futile suggestion that we can compare our beliefs with language-shaped features of reality; instead we should concentrate on the business of living. If there is a choice to be made here, it is a choice of life-styles – the open and inclusive style of the free-thinking democrat, versus the closed and exclusive style of the tribesman or the priest.

This is, however, a remarkable conclusion to have drawn from so slender a premise. For the sole ground for Rorty's position is the trivial truth that we cannot describe the world except by means of the language and the procedures that are ours. Do we not, nevertheless, make a distinction between those procedures that reveal reality, and those which merely fortify our place in it? The Islamic Ummah was and remains the most extended consensus of opinion that the world has ever known. It expressly recognises consensus (*ijma'*) as a criterion of truth, and is engaged in a never-ceasing endeavour to include as many as possible in its comprehensive first-person plural. Moreover, whatever Rorty or James mean by 'good' or 'better' beliefs, the pious Muslim must surely count as having some of the very best: beliefs that bring security, stability, happiness, a handle on the world, and a clear conscience as one eliminates one's enemy. Yet still, is there not a nagging feeling somewhere, that those beliefs might not be true, and that the enervated opinions of a post-modern atheist might possibly have the edge on them? Rorty, who blithely asserts that God does not exist, must believe that *this* belief is in some way a better foundation for the consensual community that he envisages. Yet how does he know that? Certainly not by applying the methods that are 'ours': since history points quite in the opposite direction. (Look at the two great attempts at an atheist community: Nazi Germany and Soviet Russia.) It is quite clear that, when it comes to the point of believing something, Rorty is as prepared as the rest of us to look beyond the consensus, and to evaluate beliefs on some other grounds than their 'goodness' or utility. Indeed, if he rejects the belief that God exists, it is because, like any atheist, he is convinced that there is, in reality, nothing to which that belief corresponds. Surely, when the founding father of the pragmatist school espoused 'fallibilism' – meaning that none of our beliefs should be regarded as beyond the reach of questioning (C.S. Peirce, *Collected Papers*, vol. 1) – he did not mean to deprive us of the tests whereby we decide that the *bad* beliefs are also *false*. Even if we are never entitled to declare ourselves certain, we still reject our old beliefs by describing them as untrue, and usually by accepting new beliefs that contradict them. Indeed, the method of refutation is so fundamental to science, that it is hard to imagine any assertions that do not presuppose it. And in availing ourselves of concepts like falsehood and contradiction, we are covertly reaffirming our commitment to truth. Moreover, there seems to be no clear account of the convergence at which science aims, and which it also seems (unlike religion, for example) to achieve, than the supposition that it is convergence on the *truth*. A Muslim, a Copt, a Druze and a Buddhist disagree on many things: but, if they have thought about the matter at all, they agree about the laws of physics.

As I say, that cannot be the last word; for pragmatists are just what their name suggests – wily casuists like Protagoras who, armed with a few impregnable arguments, can always accuse their accusers of begging the question. (See Study Guide.) And we shall encounter these arguments again, when discussing the nature of scientific theories, and the concept of meaning. For present purposes, the question is this: On whom does the onus lie? Does the pragmatist have to show how we can dispense with the idea of a language-independent reality? Or does his opponent have to show that this idea is – well, true? The wisest response to these questions is that given by Kant, whose 'transcendental idealism' delivers the following schematic answer: we can know the world only from the point of view that is ours. We cannot step outside our concepts so as to know the world 'as it is in itself', from no point of view. Nevertheless, our concepts are shaped by the belief that judgements are representations of reality: our concepts are concepts of objectivity, and apply to the realm of 'objects'. Without that underlying belief we could not begin to think.

At the same time, the belief in an objective order generates the idea of a world seen from no perspective: the world 'in itself', as God knows it. We cannot attain God's perspectiveless view of things; but the thought of it inhabits our procedures as a 'regulative idea', exhorting us always along the path of discovery.

Kant's view (which has been re-presented as 'internal realism' by Hilary Putnam) is not the last word in the matter. But it is the best word that has ever been uttered. And it shows that the question of the nature of truth is really not settled by arguments like Rorty's. The fact that we cannot step outside our concepts does not determine whether this particular concept – the concept of truth – has the representation of the world, rather than the expression of life, as its fundamental purpose. (For remember, the concept of the world is also one of our concepts.)

5. The Redundancy Theory

This theory, due to F.P. Ramsey ('Facts and Propositions'), is a radical response to the puzzles that we have been considering. Suppose that there are only two truth-values, true and false. For any proposition *p*, it will then be the case that *p* is logically equivalent to the proposition that *p* is true. The two sentences '*p*' and '*p* is true' necessarily have the same truth-value. (This is not one of our platitudes, since it depends on the controversial theory that every proposition is either true or false.)

If we accept the equivalence of '*p*' and '*p* is true', however, we might be tempted to argue for the conclusion that we do not *need* the concept of truth. We can say everything we want to say without it. Instead of saying that *p* is true, it suffices to say that *p*. All that we add with the word 'true' is a reaffirmation of the original proposition.

This may seem like an attempt to avoid the issue. But it is based in two subtle ideas, first expounded by Kant:

(1) The philosophical problem of truth arises from an illegitimate generalisation. For each proposition we can raise the question, what makes it true: and we answer by giving the truth-conditions. We can also give theories which 'pair' each proposition with its truth-conditions. But these theories give a different result for each proposition. If, instead of asking the question what makes this or that proposition true, I ask the question what makes *any* proposition true, then I can find no answer: the question is over-generalised. (Compare 'How much does this book weigh?' with 'How much does anything weigh?') Philosophers speak of 'correspondence' and 'coherence', but these are mere words. We can meaningfully compare each propo-

sition with its truth-conditions. But there is no general truth about truth in the sense required by the traditional theories.

(2) The difficulties that we have encountered stem from the attempt to survey our thought as a whole, and to judge its credentials. But we could do this only if we could achieve a point of view outside human thought, so as to assess the totality for its coherence, or so as to compare it with reality. However, there is no such point of view to be adopted. We are locked in our thought; but this is not a limitation, since there is no other viewpoint to which we might have escaped. When we use a word like 'true', we use it from within our language: it does not have the magic property which nothing *could* have, of leading us out of language, into some 'direct' or 'transcendental' encounter with the world. Hence the word 'true' does not add anything that could satisfy the metaphysicians. It can be dropped from the language.

The redundancy theory has not satisfied many philosophers. It requires considerable contortions if it is to account for all the things we say about truth. (Consider 'the truth about Mozart's death'; 'the pursuit of truth'; 'a story which is largely true'. How could you remove the words 'truth' and 'true' from these phrases, on Ramsey's theory?) The theory is also deeply unsatisfying: we *do* have an intuitive idea of what the classical theories are saying, and we recognise the choice between them as not only a real one, but the most fundamental choice in the whole of metaphysics. Nevertheless, philosophers have often returned to Ramsey's thoughts in this matter, partly because they exemplify a new and increasingly popular approach to the problem.

6. Minimalist Theories

In a famous article, to which I shall return in Chapter 19, the Polish logician Alfred Tarski proposed a 'theory' of truth, which he suggested captured the idea of correspondence, while characterising the unique role of truth as the foundation of logical discourse. His unspoken starting point was the account of reference proposed by Frege, in which truth features both as the aim of discourse, and as the semantic value of successful utterances. Tarski therefore called his theory the 'semantic theory of truth' – the theory which shows the role of truth in semantic interpretation.

To what conditions, Tarski asked, should a 'definition' of truth conform? What would lead us to accept it as a definition of *truth*? First, it should assign truth-conditions to each sentence of our language. Secondly, it should derive those truth-conditions from the semantic values of the parts of a sentence (so satisfying Frege's requirement, that the semantic value of complex expressions is determined by the semantic values of their elements). Thirdly, it should meet what he called a 'condition of adequacy', namely, that every instance of the following 'convention':

(T) *s* is true if and only if *p*

should come out true, where the letter *s* is replaced by the name of a sentence, and the letter *p* by the sentence itself. One instance of the schema is this:

(S) 'Snow is white' is true if and only if snow is white.

(Tarski in fact said 'true-in-English', since he believed, for reasons to which I return in Chapter 26, that truth could only be defined for each language taken on its own, and moreover that it must be defined not in that language but in another, which he called the 'meta-language'. For the sake of simplicity I am ignoring this complication, which is irrelevant to our present interests.)

Why does convention T state a condition of adequacy? The answer is simple. Its instances, such as the sentence S, express exactly the idea of correspondence: they relate a sentence to the fact that it is used to express, by first naming the sentence and then using it. We know *a priori* that the sentence 'snow is white', used according to the rules of English, identifies the *very state of affairs*, whatever it may be, that makes the sentence 'snow is white' *true*. A theory that entailed every instance of (T) would therefore capture all that can be captured of the idea of correspondence: all that can be captured in *language*. (And, as we have seen, what can be captured in language is probably not sufficient to settle what is at stake between correspondence and coherence as *criteria* of truth.)

Tarski builds on his theory in surprising and suggestive ways. For our present purpose, however, the most important result of his argument is that it gives respectability to a kind of 'minimalism' about truth. Instead of searching for profound metaphysical theories, Tarski simply returns us to the indisputable *platitudes* about truth, and asks what would be necessary, in order to provide an adequate theory of *them*. Perhaps we should not ask more of a theory of truth.

Quine, inspired by Tarski, therefore considers the predicate 'true', not as describing the metaphysical status of a sentence, but simply as what he calls a 'predicate of disquotation'. By using it, we pass from words quoted to words used: and *that*, indeed, is its function. Having named the sentence 'snow is white', and then appended to it the truth-predicate, I thereby *use* the sentence; I make it my own, so that it takes its place beside the other sentences of my discourse, as a fragment of my theory of the world.

Return now to the first section: you can envisage a philosopher who regarded the platitudes that I listed there (along with that uttered in the next paragraph on Aristotle's behalf) as containing the whole truth about truth: or at least all that we need to think about truth, in order to use the concept successfully, as the predicate of disquotation. So here is a new philosophical project: to give a theory of truth, so conceived, making the minimal metaphysical assumptions. As Tarski discovered, it is by no means easy: indeed, he came to think that it was impossible, and that theories of truth could only be devised for artificial languages, and then always at the expense of constructing another language in which to discuss them. Still, maybe he wasn't right about that. So many recent philosophers have thought, at least; with the result that a wholly new approach to the problem of truth has become standard in the literature.

Minimalist theories could be embraced equally by defenders of correspondence and by defenders of coherence: and each will protest, no doubt, that his theory *is* the minimalist theory: the least that needs to be said (and also the most that can be coherently said) to give an account of the concept. But the ground has now shifted. Maybe there is nothing to choose between correspondence and coherence; maybe these are just rival descriptions of the same idea – the idea contained in convention T. If there is a dispute, perhaps it concerns two conceptions of *reality*. It is to this topic – and in particular to the distinction between appearance and reality – that I now turn.

10

Appearance and Reality

The distinction between appearance and reality can be locally described: sometimes things appear as they are, sometimes not; sometimes appearances are deceptive, sometimes not. But the sceptical argument, combined with the desire for a comprehensive view of things, has led philosophers to attempt *global* theories of the distinction, so as to give an anchor against sceptical doubt, while using that doubt to subvert the complacency of common sense.

It is useful to return again to Descartes, specifically to the argument about the piece of wax, which occurs at the end of the second Meditation. The wax has shape, colour, size and smell – qualities that I perceive through the senses. When I approach it to the fire, I find that the wax changes in all those respects. Yet the wax itself remains. It follows, Descartes thought, that 'sensible qualities' are not of the nature of the things that possess them: they are accidents – respects in which a substance may change while remaining the *same* substance. What, then, is the essence of physical objects? The only properties that the wax seems to have essentially are extension in space, together with a capacity to exist in a variety of spatial forms. To put it another way: material substance is essentially extended, and its extensions can be 'modified' in various ways. Extension and its modes are the sum of physical reality. (This result was confirmed for Descartes by the science of geometry, which convinced him that we have a clear and distinct idea of space – and therefore can gain knowledge of its essence by reasoning alone.)

This argument was a precursor, both of Locke's theory of primary and secondary qualities, and of rationalist approaches to the distinction between appearance and reality. The rationalist tends to argue that the appearance of the world is a poor guide to the world's reality. The *reality* of the world is presented to reason, through arguments that resolve the vagueness and inconsistency of sensory perception. The world as it appears to the senses may therefore be entirely misleading. At one part of his life Leibniz, for example, was tempted by the idea that the real world of 'monads' exists neither in space nor in time, and exhibits no causal relations of the kind that we observe. He tried to show that the *apparent* world could be fully explained in terms of this underlying reality, even though the apparent world is one of space, time, objects and causality.

As a rule, philosophers have had little difficulty in discrediting the senses. The 'argument from illusion', which I shall consider in more detail in Chapter 23, is one of the platitudes of philosophy. It emphasises the wholly believable point that, because things do not always appear as they are, we need some criterion for distinguishing true from illusory perceptions. It was one of Berkeley's contentions that no such criterion exists. Now it is in general hard for an empiricist to come to terms with the argument from illusion. For his fundamental premise is that knowledge comes to us only through experience; hence if there is no criterion for the reliability of experience, we cannot be sure that we have any grasp of reality. Locke tried to show how a conception of physical reality *emerges* from our sensory experience, through certain operations that enable us to develop a rich and fruitful distinction between appearance and reality. When distinguishing primary and secondary qualities, he did not mean that the first were objectively real, the second merely subjective. The distinction was part of an elaborate *theory* of the world, as the underlying cause of our sensory experiences. In terms of this theory, Locke hoped to arrive at a conception of physical reality that would enable us to distinguish true from false perceptions. Berkeley's argument are slap-dash, and depend upon large-scale misunderstandings of Locke's position; but they are also challenging. For they emphasise the difficulty in finding grounds for the distinction between appearance and reality, when the only grounds for anything are, according to empiricism, appearances. Either we must reconstruct the distinction *within* appearances (as Hume does); or we must abandon it altogether (as Berkeley would have liked to do, had he seen how).

1. The Radical Assault

Berkeley's intention was to show that appearances ('ideas') are consistent. Contradictions and absurdities arise only when we suppose that there is a 'material' reality, which our ideas represent to our perception. Rationalists have tended to the opposite view: that the appearances delivered through the senses are inconsistent, and that the search for the really real is part of a battle to overcome the incoherence of our common-sense beliefs. This rationalist project is as old as Parmenides, and still continues. In order to grasp its radical character, it is worth taking a look at the last of the great idealists, F.H. Bradley, whose *Appearance and Reality* marshalled a vast array of arguments, many of indifferent quality, in order to show that the common-sense world is not, and could not be, real.

(a) Primary and Secondary Qualities

Bradley shares Berkeley's view that secondary qualities are mere appearances, and cannot possibly be attributed to any reality independent of the perceiver. He also thinks that this is what Locke believed – though modern commentators disagree. Bradley goes on to argue that primary qualities are in the same boat as secondary. Primary qualities are such properties as extension, solidity and mass, which belong to an object by virtue of its occupation of space, and which are attributed to it not only by our common-sense view, but also by any plausible physics. Bradley argues that such qualities are not attributed to an object, save in so far as they are *perceived* in it. We have no other source of information concerning primary qualities, save through their relation to experience. If a quality is 'non-existent for us except in one relation,' he adds, then 'for us to assert its reality outside of that relation is more than unwarranted' (p. 13). He goes on to argue that, in any case, we cannot conceive of primary qualities except as attached to secondary qualities, whose status as appearances cannot be doubted. I cannot conceive of an extended thing except as having a certain feel, look, or whatever. By the same principle, therefore, I cannot detach the primary from the secondary qualities, so as to assert the possibility of their existing independently of this relation.

The problem goes deeper. There is something confused, Bradley thinks, in the very idea of attributing qualities to things. Always we end up attributing those qualities to the 'thing-as-perceived' and therefore to ourselves, the perceivers. At the same time, we acknowledge that we ourselves do not possess them. When I observe something square and red, it is not *I* who am square or red. Yet there is nothing else that is presented to me, which could be the real bearer of these properties.

(b) Substantive and Adjective

That last argument, slippery though it is, is part of a general scepticism about the very distinction upon which so much traditional philosophy has been based: the distinction, enshrined in the subject-predicate sentence, between a thing and its properties – between 'substantive and adjective' as Bradley puts it. Bradley's scepticism is similar to Locke's scepticism about the Aristotelian idea of substance, as Locke understood it. If a substance is defined as the *bearer* of properties, then what exactly is it? Any attempt to identify it must involve attributing properties to it. But in that case I merely identify the properties, and not the thing in which they inhere. And if I strip away the properties, so as to reach the substratum that underlies them, I embark on a journey which has no goal: I seek the core of the onion, with all its layers removed. At best the distinction between substantive and adjective makes sense from our point of view: but then it is not conceivable except in relation to that point of view, and certainly cannot be attributed to an independent reality.

(c) Quality and Relation

Someone might reply that at least *qualities* exist: that redness exists, even if we are not entitled to attribute it to an independent entity. But can we even say this? Bradley argues that we cannot. For qualities are all relative to the point of view from which they are attributed. I cannot conceive of qualities, except through the idea of relation. Even if we try to derive qualities by separating them from the relations in which they are embedded, this merely introduces another relation – that of separation – which attaches qualities all the more firmly to the relations from which we were trying to extract them. But then the question arises: What is a relation? A relation seems to join qualities to each other: but how, if not by some other relation – a 'joining' – which is as much in need of explanation as the terms it relates? By this kind of vertiginous abstract argument (reminiscent of the classical discussions of universals), Bradley proceeds to disestablish the whole idea of quality and relation:

> But how the relation can stand to the qualities is, on the other side, unintelligible. If it is nothing to the qualities, then they are not related at

all; and, if so, as we say, they have ceased to be qualities, and their relation is a nonentity. But if it is to be something to them, then clearly we now shall require a *new* connecting relation. (p. 27)

That kind of argument is scarcely intelligible outside the context of Bradley's enterprise. But it enables us to obtain some grasp of his method. By taking the ordinary features of our common-sense world-view, and redescribing them in terms of philosophical abstractions (quality, relation, substantive, etc.), he is able to generate the impression that a contradiction inheres in thoughts which seem to us so familiar that we are rarely tempted to question them. Bradley's critics (G.E. Moore and Russell, for instance) would argue that the fault lies, not with our common-sense view, but with the gratuitous and unexplained abstractions that are used to describe it. But, tempting though this response may be, it does not help us to find out exactly *what* has gone wrong in Bradley's argument.

(d) The Scientific Picture

Bradley goes on to extend his scepticism to the core concepts of physical science: to space and time, substance (thing) and causality. All these, he believes, are so riddled with contradictions as to fall victim to the first assault of philosophy. At best we can regard them as parts of the appearance of the world. In themselves, however, they have no reality. As soon as you attribute reality to them, you find yourself locked in contradiction. In future chapters, I shall return to space, time and causality, in order to explore the arguments that idealists and their fellow-travellers have used in order to reject these features of our world. If you can undermine these concepts too, you will effectively deprive science of any authority in adjudicating the question of the really real. Without space, time and causality there is no such thing as physics: and if these concepts merely describe the structure of appearance, then physics too can never advance beyond appearances: it can never describe the world as it really is.

Descartes would not have been surprised by this result: for already the demon had awoken him to the philosophical fragility of physical things. But Descartes would certainly have been surprised by Bradley's next argument, which is directed against the very thing upon which the Cartesian world-view is founded: the self, as the substantial centre of its universe. What Bradley says, in effect, is that we have no conceivable grounds for asserting that the appearances which we encounter can be attributed to an observing self: no grounds for assuming self-identity through time, and no possible conception of what a self might be, over and above the appearances that seem to inhere in it.

2. The Common-sense Rejoinder

Bradley, and his contemporary McTaggart, sparked off a reaction among English philosophers, led by G.E. Moore, and culminating in the ordinary language philosophy of J.L. Austin and Strawson. These philosophers tried to build a revised conception of the common-sense world-view, by obstinately refusing even to *begin* the argument that its opponents believed was required in its favour. Moore insisted that at least some physical things are real – *really* real – because he had two hands, and they were physical objects. He was more certain of this fact than he ever could be of the sophistical arguments designed to refute it. And he was inclined to go further, and say that he never *could* be as certain of the validity of those arguments as he was of the fact that he had two hands, or that there was a window in the wall before him. (Once he said that in an unfamiliar lecture hall, and pointed assertively to a *trompe l'oeil* window that had been painted on the wall.)

This looks like a petulant refusal to play the game. But as Wittgenstein pointed out in *On Certainty*, it is more than that. Moore is saying that the concepts employed in philosophical argument are real concepts. If they are real then they must get their sense from their *application*: they must be anchored in reality. And they can be anchored in reality only if there are occasions where doubts about their application come to an end: occasions where we can say, *here* it is right to say 'hand', and so on. We can, of course, make a distinction between appearance and reality. But this is a distinction *within* our common-sense world-view, and cannot be used to reject that world-view entirely without undermining its own *raison d'être*.

This approach to the problem is capable of considerable refinement. But it is worth giving some of the simple arguments in its favour. The first is that we do readily distinguish appearance from reality in ordinary discourse, and are seldom deceived when the two conflict. Most of the cases of 'illusion' studied by philosophers (the round coin that looks elliptical, the stick that looks bent in water, the sweet orange that tastes bitter to the fevered patient, and so on), are *not* cases where we make mistakes, or cases where we are at a loss to discover reality. Our common-sense world-view is not simply a jumble of appearances, ordered for administrative convenience. It is a shared and public *theory*, designed to explain and predict the way things appear. We come to *conclusions* about the world on the basis of experience, and form a picture of its reality in accordance with the everyday need for consistency and explanation. The 'contradictions' explored by Bradley are *already* ironed out in the formation of our common-sense view, which distinguishes with remarkable ease between illusory and veridical appearances, and builds into its conception of the world an embryonic theory of our own epistemological limitations. Only by distorting the nature of appearance, and the concepts commonly used to describe it, does Bradley generate the array of paradoxes with which he seeks to dismay us.

Furthermore, the defender of common sense might argue, the Bradleian arguments threaten to undermine the very distinction upon which they depend for their credibility: the distinction between appearance and reality. When I describe the way something appears, I am describing an appearance: but it is the appearance *of* something. There is, in the very use of this term, the implication that appearances are related to an underlying reality, which they may represent more or less accurately. Appearance is, as Bradley said, a relational idea: but the relation is not to the observer; rather, it is to the thing observed. At the end of Bradley's radical arguments, however, that relation has been entirely severed. We are no longer entitled to assume, either the reality of the thing observed, or even the idea of it. The appearance is not an appearance *of* anything at all. It is 'mere appearance', which contains no hint of any reality outside itself. In what sense, therefore, are we entitled to describe it even as an appearance? In losing all reference to the object, we have lost the very contrast on which the concept of an appearance depends.

3. The Rationalist Project

Bradley would not be too disconcerted by that rejoinder. He would argue that it is too dependent for its plausibility on the assumption that our language, and the concepts deployed in it, are 'in order' as they are. And this assumption is the very one that he questions. Like many rationalists before him, Bradley would insist that we have an ability to reason which is not trapped within the contingent conceptions of our everyday use of words, but precisely transcends them, and attempts to view the world as it is in itself, freed from our self-created contradictions.

Having disposed of appearance, therefore, Bradley feels free to search for the true reality. Clearly sense-perception will not reveal it; nor will ordinary scientific reasoning. We need a method that will enable us to advance to a point of view outside that of common sense and scientific inference. The picture that it provides must be self-consistent, and it must enable us to see the world in its completeness (otherwise self-consistency cannot be guaranteed). It will therefore lead us to a conception of the whole of things, the absolute totality. Borrowing a term from Hegel (whose influence he vehemently denied) Bradley called this conception, and the reality that it presented, the Absolute.

Many of his contemporaries thought that Bradley's project was absurd, mystical, an attempt to rejoin the abandoned God of our religion, by scaling rickety ladders of argument propped against the clouds. In fact, however, the project is common to all philosophers in the rationalist tradition. Return yet again to Descartes. His argument has left him, he believes, with a realm of certainty – the self and its conscious states. Of these he is sure. Yet what *is* it that he is so sure about? What precisely *is* a self? And what kind of a world does the self inhabit? Well, the self has a conception of itself, as a mental substance; and a conception of the material world, derived from *a priori* reflection on the piece of wax. But is there any guarantee that these conceptions correspond to reality? Certainly, they belong to my *point of view* on reality – which is the point of view of a pure subject, whose experiences are indubitable to himself. But how would I show that they represent the world as it really is? Behind Descartes's subsequent arguments is the nagging suspicion that he has still not fought off the demon, and will never fight him off because he cannot obtain a view of the world as a whole, and of his own place within it, which will be such as to prove that he really does see and conceive things as they are. This search for an 'absolute conception' of reality, as Bernard Williams has called it, became the fundamental quest of rationalist philosophy. Leibniz and Spinoza both sought for a view on the world as a whole, which would be a view from outside the first-person perspective, showing the structure of reality as it is in itself, from no particular point of view. And it is surely not absurd to think that the search for that absolute perspective is integral to the entire epistemological enterprise. If we cannot obtain it, then we remain locked within our own point of view, unable either to transcend or truly to understand its limits. In which case, how *can* we assert that the way we conceive the world is the way it really is?

It was Kant's distinctive contribution to philosophy to argue that the absolute perspective is unobtainable. We do not even have a *concept* of the world as it is in itself – this is a pure 'idea of reason', a redundant by-product of our thinking which cannot lead to knowledge. The world is *our* world, and although we remain enclosed within our own perspective, the limits of that perspective are the limits of thought itself, and therefore the true limits of the knowable world. The rest is silence.

Suppose you remain wedded, however, to the rationalist enterprise. How *would* you ascend to the absolute perspective, and avail yourself of the vision that it affords you? Plato gave one answer. Descartes another. And for Descartes the crucial concept in defining the really real was the concept of God. If there is an absolute perspective, then it is God's. In order to establish the fully objective viewpoint that he needed, therefore, Descartes supposed that he must first establish the existence of God.

11

God

We have seen how the search for the really real has tempted many philosophers to look beyond this world, for a perspective that will be 'absolute' and error-free. But there is no point in aspiring to this perspective, unless one believes that there is something that resides there, and which has knowledge of the world as it really is. For it is only as a repository of knowledge (of the ultimate truth about the world) that this perspective can underpin our metaphysical convictions.

Traditional theology developed a conception of God that exactly suits him for the purpose. God is immanent within the world, but he also transcends it. His vision of reality is from no partial point of view: it is a vision of the whole world, as it is in itself, regardless of its appearance to this or that finite perception. God is all-knowing and infinite: thought is of his essence, and he is himself the object of his thinking. (God, for Aristotle, is 'thought thinking itself'.) To establish God's existence is to establish precisely that 'view from nowhere', as Thomas Nagel describes it, which provides us with absolute truth.

The subject of this chapter is God, and the arguments for his existence. However, it is worth stepping down from metaphysics for a moment, in order to discuss how such a concept could have arisen, and how it might fit into the 'naturalised' epistemology of the modern philosopher. One of the principal failings of the philosophy of religion has been its tendency to concentrate on the abstract conception of the Supreme Being (the 'God of the Philosophers') and to ignore the religious experience on which he depends (if 'depends' is the right word, which it isn't) for his earthly credentials.

1. God and gods

Modern people are frequently puzzled by the idea of God; and, for the modernist, this puzzlement *becomes* a god. (Hence the barely concealed passion of the modernist, when he addresses those questions which were once pre-empted by religion. It is this crypto-religious passion that draws people to modernism: let us at least believe in our unbelief!) In fact, however, the concept of the divine is not puzzling at all. In every pre-modern society the conception has spontaneously arisen of a supernatural world, inhabited by powers which have the same form as human powers (i.e., which are expressions of will and desire) but which are vastly superior to ours, both in their ability to get things done, and in their ability to understand the what and why of doing it. Why is this? In *The Elementary Forms of Religious Life*, the sociologist Émile Durkheim (1858–1917) gave an ingenious answer. Moral beings can exist only in communities. But a community depends upon loyalty and sacrifice, and these precious commodities do not exist simply because people associate by agreement, make contracts with one another, or have habits and customs in common. They exist because of the experience of *membership*. I cannot lay down my life for you, a stranger; but I can lay down my life for the greater thing of which we are both a part, and which forms our shared identity. The experience of membership is the core experience of society: the bond which guarantees social durability, and which also gives a point to every moral injunction.

One who is a member sees the world in a new light. All about him are events and demands whose meaning transcends their meaning for *him*. The destiny of something far greater – something, nevertheless, to which he is intimately bound – is at stake in the world. This thing is something that he loves, and that lives in him. But he is not alone in loving it: he has the support of his fellow members, and he shares with them the burden of a collective destiny. This, Durkheim suggests, is the core religious experience. And it translates at once into a conception of the sacred. Those objects, rituals and customs which provide the criterion of membership come to possess an authority that transcends the authority of any human power. It cannot be I or you who decreed them. Nor can it be my will or your will that they enact. Yet they are eminently personal: they are the 'real presence' of the thing that we love, and they address us with a moral imperative: you belong, they say, and owe the duties of belonging. But to whom are these duties owed? The answer is dictated by the question: to another being, who is like us but greater. The god steps from the experience of community, clothed already in the rights of worship. It is to him that we owe our sacrifice and our obedience. The awe that we feel in the rituals of the tribe has the god as its object. It is because he is *present* in these rituals that we must perform them correctly; and his legitimate demand to reveal himself only to those who obey him, authorises our vigilant exclusion of the heretic and the intruder.

Who does not know this experience – if not in real life, then at least in imagination? And the anthropological evidence does nothing to qualify Durkheim's view. It would not be absurd to suggest that the tie of membership is a *function* of religion in those communities fortunate enough to exist outside modernity. Before examining the God of the Philosophers, we should therefore describe the God of religion – meaning by 'religion' what it means in Latin, namely a 'binding' of people to the collective that includes them.

Some nice examples are contained in Homer: those lusty, irascible and laughter-loving Olympians, with their inexplicable interest in the human world (though not inexplicable, if you accept Durkheim's hypothesis that they are not merely part of the human world, but also produced by it. N.b. Durkheim was the son of a rabbi). What is most remarkable to a modern reader is that the Homeric gods have no passion which is not also a human passion: from resentment to anger, from love to desire, they enjoy the full fruits of our common servitude. They are as much 'overcome' by natural forces as we are, and at times as little able to resist them as a cat is able to resist a mouse.

At the same time, the power of the gods is supernatural. It is a power that defies the laws of nature: nothing that we can derive from experience, about the way things proceed in the natural world, sets limits to the actions of a god. Although subject to the passions, a god might at any moment inexplicably master them. While actively engaged in the battles of mortals, he might use some hitherto unknown force to end them. Most important of all, gods are immortal – they may come into being (since they are themselves children of other and more ancient gods), but they do not pass away. For they represent the community itself – that which is unpolluted by decay. Such is their function, if Durkheim is right: to guarantee the survival of the tribe, through every mortal danger.

The Homeric gods share certain important features with other objects of worship: first, they make demands on us which are of supreme importance. Disobey these demands, and you will be in a special kind of trouble – religious trouble, which lasts for ever. (This trouble comes from being 'cast out' from the community, which is the sole source of life and joy.) Secondly, they act 'supernaturally': however much they may choose to go along with natural laws, they retain the power to overrule them. Thirdly, they are, as Thomas Hardy says of the sun, 'brimful of interest' in the human world. Nothing escapes their attention, and everything engages their emotions. Finally, they are revealed in this world, in events which are by their very nature 'magic': an intrusion of the supernatural into the natural. From all this it follows that the places where the gods reveal themselves are sacred, and governed by mysterious interdictions. It follows, too, that we must strive to honour the gods through acts of piety, and also to win them to our mortal purposes, so far as we can.

It goes without saying that God, as the Judaeo-Christian-Muslim tradition has envisaged him, is not like that. On the other hand, his triumph over his many rivals would not have been possible, had he not answered to the needs that they served. He is not the god of a tribe, but the god of a universal community. Anybody is entitled to worship him, and to claim the benefits of worship, just so long as he has a soul to be saved. (This implies that there is a revised conception of *society* underlying the view of God that we have inherited.)

Nevertheless, he started life, so to speak, as the God of a tribe (or of twelve tribes), and his original character was heavily marked by this fact, as was the character of those tribes, so bold and indeed foolhardy in their choice of such a deity. (See Dan Jacobson: *The Story of the Stories: the Chosen People and its God*.) The God of the Philosophers was shaped (in conception, that is) by a long process of reflection on the God of Israel. Certain features seem to be essential to his divine status: notably, the possession of supernatural powers and more-than-human knowledge, together with that consuming interest in the world which is best explained by the supposition that he created it. He must also retain, if he is to perform his social function, the central feature of the object of worship: he must discriminate between members and non-members, the saved and the fallen, us and them. Other attributes of the tribal deity are, however, demeaning – notably, those attributes which seem to make God into a *part* of nature, and a subservient part, rather than the over-mastering sovereign. The war against 'graven images', which began with Moses, still rages today. And it goes hand in hand with a hostility to anthropomorphism: to the practice of attributing human characteristics (notably human passions) to God.

On the other hand, if God is to remain in communication with us, it seems impossible that our nature should be strange to him. Hence arises the belief that we are made in his image (rather than he in ours), and experience even our passions as pale reflections of some godly archetype. However, many philosophers would agree with Moses Maimonides and Spinoza that we cannot attribute passions of any kind to God: not even the passion of interest in our condition. To love God is precisely to cease one's childish demand that he return our love; it is to know that divine love cannot be expended on such trivia as us. (God's love is love of the *whole* of things, and of us only as subsumed into, and in a sense annihilated by, that whole.)

Perhaps the most important development of all came through reflection on the singularity of God. The cheerful pagan picked up his gods as he went along, adding each day, as Gibbon puts it, to the store of his protectors. Sometimes the new gods could not be assimilated within the old social forms: like Bacchus, they portended a new kind of community, with new demands, and a new experience of the sacred. (See Euripides, *The Bacchae*.) But, apart from the vague belief in Zeus or Jupiter as the 'father of the gods', there was no clear conception that any one of the immortals had absolute sovereignty over creation: often the ruling god had himself acquired his powers by usurpation, and endured under the threat of losing them in the very same way. The generosity of the pagan towards the many contenders for a place in the pantheon was of a piece with his recognition that men live in many communities, according to ancient and incompatible customs, and so can retain peaceful relations only by respecting one another's gods.

Side by side with the paganism of ancient Greece, there arose a philosophical monotheism. The position of Zeus as 'father of the gods' was gradually granted to a new entity, 'the god' (*ho theos*), who, after allowing tantalising glimpses of himself through the veils of Plato's Forms, finally steps into the centre of philosophy in Aristotle's *Metaphysics*, as the 'prime mover': the being in terms of which all that happens is to be explained. Such a being corresponds exactly to the increasingly remote and solitary God of Israel: the two ideas were made for each other, and duly fused. The impersonal 'prime mover' acquires a personality: that of the severe patriarch of The Old Testament, qualified, for the Christian, by the personality of God incarnate. Like every god, this one protects a community. But, having extinguished all competitors, he is left with an obligation to *everyone*. Maybe the Jews can claim a privileged relation to him; but they cannot claim sole rights of worship.

Nevertheless, the core religious experience, of the local community and its sacred artefacts, remains. Worship of the one God combines with an idea of 'heresy', which condemns the person who worships him wrongly, or who fails to understand his nature, or who in some similar way shows himself to be one of 'them'. There is pressure, therefore, to develop an agreed doctrine concerning God's nature and his relation to the world; and this conception must support two seemingly conflicting things: God's sole title to divinity, and the community's claim for *Lebensraum* among its competitors. It is the dynamic relation between those two requirements that brought about the modern conception of God.

2. God's Character

The supernatural attributes of every divinity are subsumed, in the new conception of God, under the idea of his 'transcendence'. He is not subject to the laws of nature, but stands above them as their creator. Perhaps there is no place for him *in* the world, even if he occasionally reveals himself there (so fulfilling his only true obligation, which is to fill the world with his presence, in case some lesser deity should find a niche in it). Although God is 'transcendent', he nevertheless has personality, in the philosophical sense. He is a free and rational agent, who can be offended and pleased by other persons. This *personal* relation to each of us replaces the old relation of the tutelary deity to the tribe that he protected. In one sense the new conception welcomes us into a new and ever-expanding community (the 'communion' of the saints); but it also threatens, as in the severer forms of Protestantism, to dissolve the experience of community entirely, leaving each individual alone in his tremulous encounter with God.

How can a transcendent being possess personality? Perhaps we can answer the question as follows: Personality is not identical with human nature but is a *form* imposed upon it. We are *stamped* with personality, which consists in capacities and attributes that may exist in non-human instances, and in varying degrees. The capacity to reason, to act freely, to form intentions and realise them, to respond to a conception of right and wrong: all these may exist in angels, divinities and corporations, maybe even in dolphins, as well as in human beings. And all admit of degree. We can therefore easily understand God as a person: he is the *supreme* person, the one who has, to their fullest extent, those capacities which are the principles of a personal existence: rationality, free will, the power to understand and act upon the good.

Membership is the greatest source of obligation. Since this obligation is transmuted in the new monotheistic religion into a personal relation with God himself, God becomes the *judge* of my conduct. He is the Supreme Person, so his rewards and punishments are also supreme. Moreover, he has the power and the will to enforce them. Nothing, therefore, can matter more to me than my relations with God.

It is part of what Chateaubriand called the 'genius' of Christianity, that it gives flesh to God's abstract personality. In the Christian religion the Word is made Flesh, as the evangelist puts it: the personal nature of God's interest in the world causes him to become a person *in* the world, and to suffer as other persons suffer (only more so). The incarnation of God in Christ has therefore been a fruitful source for the understanding of God's personality, and of the special features of our relationship with him. And it is the matter which divides Christians from Jews and Muslims, who cannot accept the *derogation* from divinity that is implied by God's trial, judgement and crucifixion.

All that is familiar, and helps us to understand how the religious experience (the experience of the sacred) can be transferred to a single Deity, and transmuted into a personal relationship with him. But there is another side to him – namely, that which is normally summarised in the idea of God as the creator of the world. From Aristotle onwards, the God of the philosophers has been presented to the public as 'the ground of Being'. He is the ultimate answer to the question 'why?', the prime mover, first cause and final explanation. He created the world and also sustains it. The major arguments for God's existence depend upon *this* feature of the Divinity, and not upon his moral personality. There is therefore a theological lacuna in the idea of God as we have inherited it. What is the connection between God's nature as the Supreme Person, and his nature as the Supreme Being? To put it another way, why does the first cause have to be a person? Why is the final explanation of the world to be obtained through Will? This question may be unanswerable, and in any case denotes a mystery. Indeed, if you think about it, this *is* the mystery of the incarnation, as Christianity envisages it.

In his *Dialogues of Natural Religion*, Hume entertainingly pulls apart the many strands in the received idea of God, and is able to create an air of absurdity in our more anthropomorphic beliefs about him. As soon as you recognise that there is no easy way of proving either that the world was created, or that if it was created it was created by a person, or that if such a person exists he is benevolent, or that if he is benevolent he is *supremely* benevolent – and so on – you will recognise the extreme difficulty of founding theology on philosophical premises, and also the danger for religion in the attempt to do so. For the process whereby God is rendered philosophically respectable also jeopardises many of the things that we spontaneously worship in him.

3. The God of the Philosophers

The Supreme Being, as described by Aquinas and his contemporaries, is not without highly specific properties, even if there is a mystery concerning his personality (or, for a Christian, his three personalities). The 'God of the philosophers', as he has come to be known, is described with remarkable consistency by scholastic logicians, church divines, Islamic, Jewish and Christian theologians, rationalist philosophers (such as Descartes and Leibniz) and modern Thomists. To put the matter simply, God is conceived by all of them as timeless, immutable, omniscient, omnipotent, and supremely good.

But what do those fearsome abstractions mean? Can we really understand them? And if we can, are we sure that they are mutually compatible? Here lies the subject-matter of 'natural theology', as it came to be known: the exploration of God's nature, and the attempt to give a consistent account of him that would not derogate from his majesty, or undermine the impulse to worship.

We can easily see that there might be problems about the concept of God, as the philosophers propose it. If God is timeless, then can he really know those truths about the world which depend upon its temporality – for example, the truth that I am writing this page *now*? If so, *when* did he know it? And if he always knew it, does it follow that I was predestined to write this page now? If so, am I predestined in all that I say and do? Then is God the real author of my evil deeds – since he both foresaw them and also created the man who would inevitably perform them? Perhaps I have free will, however; but if that is so, is God powerless to prevent my actions? In any case, what sense can we make of an omnipotent being: is the concept not inherently paradoxical? Consider the following argument: suppose God is omnipotent; it follows that he can make any possible object. What about this possible object: a creature that God cannot control? If he can make it, then he is not omnipotent, since he cannot control it; if he cannot make it, he is not omnipotent. He loses either way. (Stated thus, the argument could be taken to show merely the impossibility of a creature whom God cannot control. Nevertheless, its author, J.L. Mackie, has revised it, with a view to proving that it presents the theist with an embarrassing dilemma. See *The Miracle of Theism*, pp. 160–161.)

You can see from that random selection of questions that the concept of God is bound to be problematic; and the problems, because they touch on the destiny of the believer, have been the greatest motor to the philosophical imagination. Had we space to dwell on them, we should certainly profit from the wonderful intricacies of reasoning that the pious and the impious have spun from the most creative of all concepts (the creativity of which, for a believer, is no accident). But there is another matter that must detain us.

Descartes's refuge from the demon was the self: there, at least, he was free from doubt, and there – in his inner sanctum – he could gather the weapons that would allow him to repossess the world. Looking inwards, he found many useful things; but none more useful than the idea of God. My reflections, he argued, have convinced me not merely that I exist as a thinking thing, but also that I am prone to error, that in all my endeavours I am limited, that my knowledge is patchy at best – in short, that I am a finite and imperfect thing. But lo – how can I think of imperfection without the idea of its opposite? How can I describe myself in these terms, except by contrast with something that denies them? My sceptical argument has not only led me to a place of refuge; it has furnished that place with the greatest of all ideas – the idea of a being with all perfections, whose power, knowledge and goodness are unsurpassable. And this deduction is conducted within the inner sanctum: Descartes knows that he has the idea of God, and not even the demon can snatch it from him. (The idea of God is an 'innate idea'.) The question that remains, therefore, is whether that idea can be used to prove its own veracity.

4. Arguments for God's Existence

Before discussing Descartes's argument, we must return to the theological tradition. The God of the philosophers is infinitely powerful, infinitely wise, and infinitely good; he is also creator and judge of the finite world in which we are situated, and the final ground for the existence of everything. Is there any argument for his existence?

Certainly many arguments have been offered, most falling considerably short of the desired conclusion. Aristotle's conception of the Prime Mover inspired many of the medieval philosophers, particularly the Arabic commentators, Al-Farabi and Al-Ghazali. The Arabs in turn influenced Moses Maimonides, who influenced Aquinas, whose *Summa Theologica* attempts to give a comprehensive account of everything that is known about God. This vast work, of indescribable density and dryness, is undoubtedly one of the greatest works of philosophy. And the five arguments for God's existence with which it begins (the 'Five Ways') are all worthy of study. Consider the third:

> We find in nature things that are possible to be and not to be, since they are found to be generated, and to be corrupted, and consequently, it is possible for them to be and not to be. But it is impossible for these always to exist, for that which can not-be at some time is not. Therefore, if everything can not-be, then at one time there was nothing in existence. Now if this were true, even now there would be nothing in existence, because that which does not exist begins to exist only through something already existing.

Aquinas goes on to argue that we can therefore deduce, from the mere fact that something exists, that something exists *necessarily*. This necessarily existing thing is God. Not many philosophers are satisfied with the crucial step in the argument: 'that which can not-be, at some time is not.' What is wrong with the thought that something whose existence is contingent may nevertheless have simply existed from all eternity?

The argument is sometimes described as 'the argument from contingent being', and remains a cornerstone of orthodox Roman Catholic theology. A similar argument from contingent being, equally contentious, is this:

I exist contingently: which is to say, I might not have existed. My existence is a contingent fact; but contingent upon what? What conditions had to be fulfilled, in order that I should exist? And were those conditions also contingent? In which case, contingent on what? And so on. If anything is to be truly contingent, therefore, something must be necessary. There must be a being whose existence is contingent upon nothing, if there are to be contingent beings. This necessary being is God.

I will not discuss the 'five ways', which are the subject of a fairly illuminating book of that title, by Sir Anthony Kenny. Instead, I shall turn to Kant's later discussion, in which he plausibly argued that all proofs for the existence of God must belong to one of three varieties, which he classified as teleological, cosmological and ontological respectively. (The argument from contingent being is a version of the cosmological argument.)

(i) *The Teleological Argument:* the argument from purpose or design. (The name is derived from 'teleology', the study of goals.) This is really a family of arguments, to which Kant himself was sympathetic, to the point of providing his own rather remarkable instance. (See Chapter 29.) Beginning from a positive assessment of nature and our place within it, the argument moves to the suggestion that only an all-powerful and benevolent being could possibly have designed the world that we know. Kant's sympathy for the argument is more religious than philosophical. For, in a sense, to argue in this way is *already* to engage in worship. The very premise of the argument requires us to interpret our experience in religious terms (as a 'revelation' of the goodness of the world); while the leap to the conclusion, which far outstrips anything that reason alone could sanction, is a leap of faith. That we should jump to *this* conclusion is a *consequence* of religious belief, rather than a ground for it. For the whole plausibility of the argument depends upon a prior conception of God, and a belief that, if the world has an explanation, it is to be found in the will of a Supreme Being, with a moral purpose of his own. The argument must assume what it has to prove. This does not mean that it is viciously circular. For the theistic assumption, when combined with the teleological argument that is built on it, may render the world more intelligible than it would be otherwise. But it does mean that the teleological argument can never stand alone as a proof of God's existence.

Indeed, there is no difficulty, as both Kant and Hume saw, in refuting the argument. It involves three moves: a premise, a covert assumption, and a deduction, each of which can be separately questioned:

(a) *The premise.* The world exhibits an order that is (a) good and (b) the kind of order that manifests design. Are either of these true? We do not know. Sometimes the world seems good, harmonious, beautiful, and full of the signs of a benevolent power. But when the bombs rain down, and the death squads force the doors, this impression is apt to disappear. Even if we can find some intellectual way of accommodating the manifest fact of evil (see below), our experience of evil will not go away, and infects our conception of the world as a whole. Watching the sun rise over the Vistula from the monastery at Tyniec, touching with its golden light the bell-towers of the churches in their wooded hilltops, you are naturally moved by the beauty and order of the things you see. But behind you lies the site of Auschwitz, and the rose-tinted cloud that crowns the sun contains the night's emissions from the steel works at Nova Huta. These are things you know (or ought to know); and knowing them, the experience of order seems like a brief illusion.

Even when we escape the world of men, and look at the works of unmolested nature, the impression of design lasts only as long as our ignorance. William Paley, in his great work of natural theology written in the early years of the nineteenth century, compared the works of nature to a perfect watch, which we come across ticking in the grass at our feet. If that is what we find, how can we doubt the existence of the watchmaker? But – as Richard Dawkins (*The Blind Watchmaker*) reminds us – this impression too is an illusion. The theory of evolution shows how the appearance of design – design more intricate and wonderful than any we could ourselves encompass – may occur in things that were not designed at all, but which came slowly into being, through a process of random change and repeated disaster. For every lovely horse, a million misfits have miserably perished; and the same is true of you and me.

The theory of evolution has certainly changed the way in which the argument from design is formulated. But it does not refute the argument. There is no reason why God should not choose this way – the way of blind evolution – in order to achieve his inscrutable purpose. Moreover, there is another and more remarkable design in nature: the design of consciousness. Is it not the most wonderful of facts, that the world *knows* itself in us? How could there be consciousness of reality, if consciousness were not the cause of all? For think what a strange chance it is, that of all the infinite possibilities, just this one should come to be: the possibility which also knows itself. Surely it is a rare world that contains knowledge, a still rarer world that is, like ours, so thoroughly knowable. In our daily lives we do not notice this stupendous fact; but once we *do* notice it, we are apt to be overcome by a sense of awe. Wherever we turn the world yields to our inquiries; its order and system are through and through scrutable, and one by one its secrets are transcribed as knowledge.

There is an answer to that argument, which parallels the argument from evolution. It comes in two forms, one scientific and the other philosophical. The scientific argument makes use of what Stephen Hawking (*A Brief History of Time*) calls the 'weak anthropic principle'. This says that the world that we know exhibits just as much order as is required for men to exist in it. It is therefore not surprising that we know it, since the amount of order required by our existence is so great as to render the world entirely scrutable. However, there is no reason to suppose that the universe is ordered in its entirety in the way that we observe. All we can conclude is that the part of the universe that is accessible to us is orderly. Maybe this part is only a random fragment of some vast array of possibilities, the great majority being for ever forbidden to us, by virtue of their unruly departure from laws which just happen to obtain where we are, and which permit our own existence.

The philosophical argument derives from Kant, in the *Critique of Pure Reason*. It is not an accident, Kant argues, that the natural world is ordered, according to the ideas of space, time, causality etc. For this order is the precondition of knowledge, and in particular of the self-knowledge which is the mark of our condition. Hence we can know *a priori* that any world in which we exist will exhibit the very order that is required by science. Moreover, we can have no conception of any *other* world. No other world is even *thinkable*, since we could think it only by means of those scientific categories which it is supposed to defy. The supposition that the world might have been unknowable becomes empty. In which case, we can build no argument from the premise that our world is knowable.

In *The Critique of Judgement*, however, Kant stepped aside from that argument. There is another sense, he suggested, in which the harmony between our faculties and the world – the harmony that leads to knowledge – demands an explanation. Yet it is an explanation that we cannot give. We can merely confront this fact – in the immediacy of aesthetic experience, and in the suddenness of religious awe – and acknowledge that there is something here that points beyond the limit of our thinking.

Of course, that is not an argument. But it returns us to the premise. In the end, the mystery lies not in the nature of the world – for we have no other with which to compare it, and therefore are unable to judge whether it is remarkable, or unlikely, that it should be as it is – but in the fact that the world exists. Why *is* there anything? Why is there *something* rather than nothing? And whence did that something issue? To ask *those* questions is to leave the argument from design behind.

(b) *The assumption.* Even if we can surmount the difficulties of the premise (which some would say are more emotional than intellectual, and record a scientistic fixation with our own limited view of things), we encounter a deeper difficulty. The argument has no force without a questionable assumption, namely that the world has a cause which lies outside the world. The teleological argument covertly supposes that nature can be explained as a whole, not in terms of natural processes, but in terms of their transcendental ground. But surely, this is what has to be *proved*, and cannot be assumed throughout the argument? Indeed, this assumption is precisely the *conclusion* of the second (cosmological) argument; Kant argued, therefore, that the teleological argument leans on the cosmological, and is of no *intellectual* force without it.

(c) *The deduction.* Suppose we allow ourselves the premise, and the covert assumption that the cause of the world must be found outside the world. We are entitled to deduce no more than that the cause has those qualities which are *necessary* to produce the world as we know it. Since the world is finite and imperfect, it could have been produced by something finite and imperfect; a giant spider say, scattering its eggs in interstellar space. The teleological argument advances a *sufficient* condition for the production of our world: namely, the existence of a supremely powerful and benevolent being, motivated by inscrutable purposes. But in seeking for an explanation of contingent facts, we are not entitled to infer the existence of anything beyond what is *necessary* to explain them.

(ii) *The Cosmological Argument.* A moment's reflection will convince us that the teleological argument has failed to come to terms with the *metaphysical* problem of God's existence. It is an argument that remains locked in Homeric conceptions, seeking for the signs of divinity in the here and now. We need a stronger argument, that will take us *out* of the world of nature, so as to postulate a transcendental ground for everything within it. This is the aim of the cosmological argument, which comprehends many sophisticated and sophistical variants. In its simplest form it is the argument for a 'first cause'. Take any event in the world: either it is unexplained, or else it has a cause. The same is true of that cause, and of *its* cause. Does the series of causes extend for ever? If so, we never reach a first cause, so that the series as a whole is without an explanation. In which case all the explanations that occur in it are insufficient: none of them really explains *why* the effect occurs. If we are really to answer that question – really to find an explanation for *anything* – we must find a cause which is the explanation of *everything*.

The argument says that either the world lacks an explanation, in which case none of our attempts to understand it has a foundation and the theological hypothesis is as good as any other; or else the world has an explanation, which is the first cause. But obviously there are two glaring weaknesses in the argument: Why should we suppose that there must be an explanation of *everything*? Why cannot we simply accept the irreducible contingency of the world and all that is contained in it? And what about the first cause himself (itself)? What caused *him*?

To the second question there is only one satisfactory answer: namely that he caused himself. He is the full explanation of his own existence. It follows from his nature that he exists. And another way of saying this (according to Maimonides) is that the first cause is a *necessary* being. The argument works therefore, only if we can show that there is a being who is 'cause of himself' (*causa sui*), whose existence requires no further explanation than is contained in his own nature. This is not proved by the cosmological argument, but assumed by it. However, it *is* proved by the ontological argument (assuming the argument's validity). Hence, Kant argued, all arguments come to rest at last in *this* one, the third in his hierarchy of proofs.

(As for the first question above, this leads us into matters concerning causality and explanation which I must here postpone.)

(iii) *The Ontological Argument*. The beauty of this argument is that it supplies the deficiencies of the other two, while also proving that God has all the attributes which piety and tradition require in him. God is a necessary being, with every positive attribute – infinitely powerful, infinitely wise and infinitely good.

The argument is normally attributed to St Anselm, Archbishop of Canterbury in the eleventh century (a time when it meant something to be Archbishop of Canterbury). But it has Aristotelian and Arabian antecedents, and is, in truth, the only possible argument for the existence of God, and the only one that has not been finally disposed of. (Kant *thought* he had disposed of it; but recent philosophers, notably Norman Malcolm and Alvin Plantinga, have cast doubt on Kant's confident dismissal.) Briefly, we understand by God a being greater than which nothing can be thought. This idea clearly exists in our mind: it is the idea of a being endowed with every positive attribute and every perfection. But if the object of this idea were to exist solely in our mind, not in reality, there would be an idea of something superior to it, namely of the being that possessed not only all the perfections already conceived, but also the additional perfection of real existence. Which is contrary to hypothesis. Hence the idea of a most perfect being must correspond to reality. Existence belongs to the nature of the most perfect being: it follows from his nature that he exists. In other words, he exists necessarily and not contingently.

Once this argument is accepted, the lacunae in the other two can be filled; only one major difficulty remains, which is that of reconciling the perfection of God with the imperfection of the world: How could such a God have created the world as we know it? I return to this below.

But can we accept the argument? Kant's reply has been sympathetically received by modern philosophers, since it contains a premonition of Frege's logic. The argument assumes, Kant argued, that 'existence is a true predicate': i.e. a predicate not just grammatically, but metaphysically. In other words, when listing the properties of a thing, I am fully entitled to list existence as one of them. But this idea misrepresents the logic of existence. When I say that a green cow exists, I do not add anything in thought to the concept, I merely say that it is *instantiated*, that there *is* something in the world that corresponds to it. My description of the cow is not augmented by the claim that it exists. Suppose two farmers were to list all the 'perfections' that they require in a cow: health, stamina, abundant milk, fertility, and so on. And suppose that their lists coincide exactly, except that one farmer adds 'existence' to the list of perfections proposed by the other. Do they really differ concerning the nature of the ideal cow? Surely not. Existence adds nothing to the properties of a thing.

The point is hard to put precisely, and had to wait on the Fregean logic for its formulation. And not every philosopher has been satisfied in any case. For the Kantian reply seems to pay too much attention to the peculiarities of existence, and not enough attention to that which is common between existence and ordinary properties: namely, that they can both be 'truly attributed'. Recent philosophers have therefore continued to wrangle over the ontological argument, some believing that it is invalid, others believing that it proves something, though something less than piety requires, others still (like Plantinga) remaining content with the original conclusion (though not the argument in its original form); namely, that a necessary being exists, and is also the most perfect being and the ultimate ground of everything. I shall therefore return to the ontological argument in the next two chapters.

5. Descartes's Ladder

Descartes offers two arguments for the existence of God in the *Meditations*, versions of the cosmological and the ontological argument respectively. Briefly, they go as follows:

(i) I am an imperfect being, as is proved by the fact that I can fall into doubt and error. But I have the idea of a most perfect being. Whence came this idea? Not from me, since there must be 'as much reality (perfection) in the cause as in the effect' (a principle that Descartes regarded as self-evident). In particular, there must be as much reality in the 'formal' cause of an idea (in the true explanation of the idea) as there is contained in it 'objectively' (i.e. in the thing that the idea represents). The idea of a perfect being must therefore have a perfect cause, namely God.

(ii) The ontological argument. I have an idea of a most perfect being. I clearly and distinctly perceive that such a being must contain all perfections, and hence reality in every degree. Hence this idea contains existence. God's essence is to exist. Of no other thing, Descartes adds, can this be said.

With those two arguments Descartes ascends at last to the divine perspective. The world contains a supreme being who has full knowledge of all that it contains. Can I share in that knowledge, or am I for ever cut off from it? The answer is suggested by the demon himself. Since God is supremely good, he is no deceiver; it therefore follows, Descartes argues, that those faculties which we have innately will, when used according to their true and God-given nature, yield truth rather than error. The existence of God underpins and guarantees my claims to knowledge. I can share in the divine perspective, and so transcend my point of view. Thereby I come to know the world as it truly is.

But is this really possible? Consider the world as I perceive it: it contains colours, smells and sounds – qualities that are perceivable only by beings with sensory organs. These secondary qualities are no part of the primary constitution of matter, and play no part in the scientific world-view. Surely God does not know the world in this way, nor does he have the experiences on which secondary qualities depend. His conception of the world must therefore be very unlike mine. Is it really possible that I can come to share in it, by refining away the pollutions that are due to my finite nature? Yes, said Descartes (and following him Leibniz and Spinoza). For reason is a faculty that I share with God, and reason views the world as God views it, from no point of view. Reason surveys the whole of things; and this alone tells me what the world is really like.

One of Kant's great achievements was to show that such a conception is mistaken. We can rely on our reasoning powers, Kant argued, only so long as they are applied from our own point of view, which is that of limited and experience-bound creatures who are part of nature. If we emancipate reason from the constraints of experience, it falls into self-induced illusions – and misleads us the most, precisely when it aspires to that 'view from nowhere' which God alone possesses, and which we can never comprehend.

6. The Problem of Evil

Our discussion has touched on another problem in the philosophy of religion: the problem of evil. It troubled Plato greatly, that the poets attributed to the gods such imperfect and criminal natures. Being a philosopher rather than an anthropologist, he found himself disinclined to credit these tawdry beings. Their imperfections seemed to cast them in the role of creatures rather than creators of the world. The Supreme Being of the Judaeo-Christian-Muslim tradition has been cleansed of any such faults, precisely in order to fit him for the role of creator. (Though he too started life with some fairly delinquent tendencies – see Exodus 4: 24, 25, in which God fails in an attempt to murder Moses.) But in that case, why did he make such a bad job of it? Or did he make a good job which we merely *see* as bad?

Very briefly, there are the following answers to the problem of evil, none of them entirely satisfactory:

(i) God made the best of all possible worlds, to use Leibniz's famous (indeed notorious) idiom. Among the many goods that he bestowed on us is the good of freedom. He could not have intended a world without freedom, for freedom is a necessary part of moral goodness. It is also the precondition of love, and God intended to love the world and to be loved in return. Freedom, however, cannot exist without the possibility of abuse. Evil entered the world through mankind's abuse of his freedom. Nothing else is evil; only this.

(ii) That which appears evil from our finite perspective is not evil at all, but either good in itself, or else part of some totality in which its apparent evil is subsumed by a greater good. Just as the joy of the last movement of the Ninth Symphony would be empty rhetoric (and therefore not joy at all) without the striving and mourning that precede it, so is the joy of our redemption – the supreme blessedness – not possible without the fall that made it necessary. *Felix culpa.* Alternatively, evil is a necessary part of the beauty of the world, in just the way that catastrophe is a necessary part of the beauty of a tragedy. (See Plotinus, *Enneads*, II, 3, 18.)

(iii) Evil is a nothing, a negation, and has no intrinsic reality. We fear it, but only because not being is repugnant to being. Properly understood, however, we cannot be harmed by evil. To understand the world is to understand that we are absolutely safe – even in suffering and death, which take nothing from us, since they are nothing themselves.

The last position is more mystical, and has recommended itself not only to believers, but to unbelievers too (e.g. to Rilke in the *Duino Elegies*). I shall return to these questions in Chapter 30. Clearly they are real questions, and we cannot live as philosophers without some attitude towards them.

12

Being

For many philosophers Being is the true subject-matter of metaphysics. From Parmenides to Heidegger there has been a constant return to this mysterious idea, and to those queasy 'being' emotions, which lead to drink or metaphysics. What exactly do philosophers have in mind by Being, and what *are* the problems associated with this idea? It is useful to survey a few of the classic discussions. They are not discussions of the same thing; but they have important elements in common.

1. Aristotle

In *Metaphysics* IV, I, Aristotle refers to being *qua* being (as his Greek is usually 'translated') suggesting that this might be the ultimate subject-matter of philosophy. What does the word *qua* mean? (See Lucky's monologue in *Waiting for Godot*, and Jose Bernadete, *Metaphysics*, ch. 1.) Aristotle's discussion is so dense that, as David Wiggins puts it, a matchbox-full weighs as much as a universe. But we can extract the following argument:

We say of many things that they *are*: of objects and events, of qualities and relations, of processes and possibilities. Is there any *single* thing that we mean in these many applications of the term 'being'? When we say of Socrates that he is a man, that he is snub-nosed, that he is a philosopher, that he is the teacher of Plato, that he is talking to Xanthippe in the kitchen, we surely cannot mean *one* thing by the word 'is'? Aristotle recognises a primary sense of 'is', displayed in the first of those predications: to say that Socrates is a man is to say *what* he is. It is to characterise him as a *substance*, and not merely as having this or that property which he might have lacked. Substance is one of

Aristotle's categories; others include quality, quantity, relation, action, passion, place, time and so on. In understanding the categories we understand the deep structure of reality. And in understanding substance we understand the primary – or focal – meaning of the verb 'to be'. This is the meaning that we *have* to grasp, if we are to understand the other applications of the term. The study of being *qua* being is therefore the study of substance: in the first place of individual substances like Socrates. (Predicates, for example, have being only in so far as they are predicated *of* substances.) Substances are the things that we have to identify as being, if we are to attribute being to anything. But the study of individual substances leads us further, Aristotle argues: to that paradigm of substance on which all other 'beings' depend (alternatively, of which all other beings are 'predicated'), namely God himself. (*Metaphysics* VI.)

Aristotle's argument is complex, and involves many excursions. Nevertheless, it suggests that the concept of being is the first subject-matter of metaphysics, and that we shall have no satisfactory metaphysics if we cannot find the 'ground' of being: we need to discover what makes being intelligible to us, what ultimately exists, and what explains why there should *be* anything. The answer to all three questions is God.

From Aristotle's theories we may depart in many ways. We may embark towards the Sparse Theory of Being, as defined by the 'analytical' philosophers. Or we may prefer the Encumbered Theory of Being, characteristic of Hegel and his followers; or even the Anxious Theory of Being, espoused by Heidegger. In each of these theories there lies a god-shaped hole, which we can fill only if we can depart in another direction: towards the theory of *necessary* being, bequeathed to us by Aquinas.

2. Analytical Philosophy: the Sparse Theory of Being

We have already seen this theory at work, but it is worth recapitulating it, in the new context established by our discussion of God. The standard analytical approach is to argue that there is no such concept as that of being – if you mean by concept what is expressed by a predicate. For 'existence is not a true predicate', to repeat the Kantian slogan. More precisely, existence is a quantifier: to say that there *is* a golden mountain is to say that the concepts *golden* and *mountain* are instantiated by some object. (More simply, it is to say that there is an x, such that x is golden and a mountain.) All questions about being are reducible to questions about quantifiers. Over what classes of entity may our quantifiers range?

However, the problems of ontology remain. How do we prove that a certain kind of thing exists? One suggestion is that of Quine, who invites us to see our language as a kind of theory, the aim of which is to bring order to experience by generating reliable predictions. This theory obliges us to 'quantify over' certain objects, and so to bind the variables attached to predicates: as when we say that there *are* elephants. We may sometimes use quantifiers, however, when their use is not necessary for the best theory of experience. (For instance, we may quantify over fictional entities, saying there is a Prince of Denmark, who could not make up his mind. But we do not strictly have to do this, in order to describe the world as we know it.) It is only if we *have* to quantify over some predicate, in the course of constructing our theory, that we can say that the theory commits us to the existence of things of that kind.

This means that existence is 'theory-relative'. We can say that elephants or witches or gods exist, relative to the theory that requires them. We then have the further question, whether the theory is true, to which Quine, as we saw, is disposed to give a pragmatist interpretation. Quine is quite radical about this, defending what he calls 'ontological relativity'. He is not inclined to specify what exists, except in the context of a theory. And the theory that is useful for one purpose, may not be useful for another. The question of being – what really exists and why – dissolves into that of the relative utility of various 'conceptual schemes'.

Much of analytical philosophy has consisted in exercises of ontological slum clearance; demolishing the crowded tenements where gods and spirits breed. The result can be depressing, especially when the cheerful old streets with their abundance of life and mystery are replaced by regimented barracks, of Corbusian design, such as you find in Quine. We have the sense that those old gods and spirits, however dubious their ontological credentials, had something to say to us that we need to hear: something that cannot be conveyed in the clipped bureaucratic language of the new town-planners. Perhaps this is because there is more to the question of being – or at least other questions of being – than analytical philosophers have been prepared to countenance.

Even within the analytical tradition, there has been a growing rebellion against the view that quantification and its logic say all that is to be said about existence. The old Aristotelian idea, of a distinction between fundamental and derivative kinds of being, has been revived and discussed in a variety of forms – notably by Strawson (in *Individuals*) and Wiggins (in *Sameness and Substance*). In both these difficult but rewarding books, a connection is made between being and time, and priority given to those things that exist in time and can be re-identified through time. This reverses the ontological order of the Platonists, who accord fundamental being only to that which is outside space and time – changeless, eternal and knowable to reason alone. It is precisely the connection with time, modern philosophers suggest, that gives to being its foundations. For it is in the reference to spatio-temporal particulars that language is rooted. On the other hand, it is also the connection with time that lends to being much of its mystery. How can something be *now*, and be the same thing as what was *then* and as what will be *tomorrow*? How can I identify these spatio-temporal particulars through change? Reflection on those questions has given further content to Frege's view that existence and identity are conceptually connected. It has also, in Wiggins, led to the Aristotelian conclusion, that only certain ways of enduring through time confer *substantial* unity and identity. This conclusion involves a systematic repudiation of 'ontological relativity'. There really *is* a distinction, independent of our 'conceptual scheme', between substantial items, like cats and dogs and people, and factitious items, like heaps, orchestras and snowballs. How can a modern philosopher arrive at such a conclusion? The answer lies in the concept of identity.

3. No Entity without Identity

In using the quantifier to convey existence we rely on identity. 'Socrates exists' becomes $(\exists x)(x = \text{Socrates})$. In translating definite descriptions into predicate logic we also need identity. These facts point to a deep metaphysical truth, summed up in a slogan of Quine's: 'no entity without identity'. The discovery of this truth, or rather of its importance for any logically-based metaphysics, is often credited to Frege, who, in section 62 of *Foundations of Arithmetic*, writes thus: 'if we are to use the symbol *a* to signify an object, we must have a criterion for deciding in all cases whether *b* is the same as *a*, even if it is not always in our power to apply the criterion'. Frege's way of putting the point in fact runs together two distinct ideas: no object without identity, and no *naming* of an object without a *criterion* of identity. The relation between those ideas is extremely difficult to define.

Recall Frege's theory of the subject-predicate sentence. In the sentence 'Jack is angry' the singular term 'Jack' refers to an object, Jack. I understand the term by knowing *which* object it refers to. And it is plausible to suggest that I therefore need a procedure whereby to identify Jack, to distinguish him from others, and to settle the question whether this or that object is the same as him. Maybe such a procedure deserves to be called a 'criterion of identity'. At least, we can surely agree that, if it is true that Jack is angry, there must be something in the world that *is* (identical with) Jack. The concept of identity has been surreptitiously invoked, in the very act of reference.

But what about the general term 'is angry'? Frege says that this refers to a concept, and also that it determines a *function*. This is confusing, so let us stick to functions, which are what do the work in Frege's semantic theory. Frege says that functions lie 'deep in the nature of things', though he himself is astonished by this, since they are peculiarly incomplete (or, as he says, 'unsaturated') entities, which determine one object (a truth-value) only when supplied with another. There are therefore real problems about the identity of functions. Is the function referred to by 'is angry' the same as the function referred to by 'is furious'? How could we know? It would not be sufficient that all and only angry things were furious. For consider 'has a heart' and 'has a kidney' – two predicates which apply to exactly the same things, but which surely refer to different functions. And if we say that 'is angry' and 'is furious' refer to the same function if and only if they *mean* the same, then not only do we need a criterion for sameness of meaning (something which Quine, for example, would say is unobtainable, or obtainable only by fiat); but we have now effectively abolished the distinction between sense and reference, by making sameness of reference depend upon sameness of sense.

I shall return to this problem in Chapter 25, when discussing the foundations of mathematics. But it nicely illustrates the point of Quine's slogan. Frege's objects are welcome members of our ontology, since they come already furnished with their identity-papers. Functions, however, stand helpless at the frontier, awaiting the documents that will never arrive. A function may lie 'deep in the nature of things'; but there seems to be no clear answer to the question '*which* thing is it?' The temptation is to send it back to the slums.

But what exactly *is* identity? This too is a puzzling question, which inspired Leibniz (who was the first *really* to perceive that existence and identity are deeply connected) to his most interesting thoughts.

Grammatically speaking, identity is a relation; on the other hand, it has only one term. Everything is identical with itself, and *only* with itself. So why call it a *relation*? The answer is that it can be treated as such from the point of view of logic, and can be effectively treated in no other way. But *what* relation is identity? Philosophers agree on the following four characteristics:

(i) Identity is symmetrical. If $a = b$, then $b = a$. Generalising:

$(x)(y)(x = y \supset. y = x)$

(ii) Identity is reflexive: everything is identical with itself:

$(x)(x = x)$

(iii) Identity is transitive: if $a = b$, and $b = c$, then $a = c$. Generalising:

$(x)(y)(z)(x = y \;\&\; y = z \supset. x = z)$

Conditions (i), (ii) and (iii) are satisfied by every 'equivalence relation', such as 'is the same height as', 'is coeval with', 'is congruent with'.

(iv) In addition identity satisfies Leibniz's law, which says that, if a is the same as b, then everything true of a is true of b (and, by (i) above, everything true of b is true of a):

$(x)(y)(F)(x = y \supset. F(x) \equiv F(y))$

(The symbol $\equiv$, read as 'if and only if', is defined truth-functionally, and is equivalent to $p \supset q \;\&.\; q \supset p$. The proposition $p \equiv q$ is true if p and q are both true or both false; otherwise it is false.) In stating the law, we find ourselves obliged to quantify over predicates: i.e. to move into what is called 'second-order' logic. For many (Quine and Goodman, for example) this already involves an intolerable excursion into the slums.

Is Leibniz's law true? Suppose John is thinking tenderly of Mary, and Mary is the person who, unknown to John, ate his beloved cat. Is John thinking tenderly of the person who ate his cat? If we answer no, then do we imply that there is something true of Mary – namely, that John is thinking tenderly of her – which is not true of the person who ate John's cat, even though that person and Mary are identical? Most philosophers would answer 'no' to that question, thus saving Leibniz's law, but at the cost of requiring a theory of contexts like 'John is thinking of . . .' ('intensional' contexts). I return to this topic in the next chapter.

(v) There is another, and yet more controversial principle, invoked by Leibniz in his account of identity. This is the so-called Identity of Indiscernibles, which is the converse of Leibniz's law, saying that if a and b have all their properties in common, they are one and the same:

$(x)(y)(F)((F)x \equiv F(y) \supset. x = y)$

The attraction of this principle is clear. For it enables us to give necessary and sufficient conditions for identity. From Leibniz's law and the Identity of Indiscernibles, conditions (i), (ii) and (iii) follow by logic. But is the Identity of Indiscernibles *true*?

Imagine two steel balls, which have all their properties in common. Could they not still be two? Well yes, if they are spatially or temporally separate. But then they would not have *all* their properties in common, since they would differ in their spatio-temporal characteristics. Imagine, therefore, a world which contained nothing except two steel balls, slowly rotating around each other at a constant speed and distance. What are the spatial properties of ball *a* at any moment? The answer is something like this: ball *a* is at three feet from ball *b*, moving in a figure of eight relative to *b* at two miles an hour. But exactly the *same* could be said of *b*, only here the spatio-temporal properties of *b* are defined relative to *a*. Since, however, there is no difference between *a* and *b*, they now have *all* their properties in common. And yet they are two.

The example is controversial, for reasons that I consider in Chapter 24. But it serves to raise two fundamental questions: (i) To what extent can we eliminate individual objects from our ontology, replacing them, for example, with collections of 'property instances'? (ii) To what extent is our concept of identity bound up with space and time? Those are the deep metaphysical questions that modern philosophy inherited from Leibniz. Although Leibniz accepted the Identity of Indiscernibles, he did not want to eliminate individual substances from his ontology, or to 'reduce' them to their properties. He wanted rather to establish a rational ontology of individuals, in accordance with the 'Principle of Sufficient Reason', which holds that there is a sufficient reason for the existence of everything, however contingent. Only on the assumption of the Identity of Indiscernibles, he thought, could such an ontology be devised, since the reason for each thing could only be found in the list of its properties. If there could be indiscernible objects, then one and the same reason would explain the existence of one such object, of two such objects, of three, and so on . . . and therefore of none.

It is the second question, however, that has attracted most attention from recent philosophers, and served to add substance to the sparse theory of being. I mean that literally; for it is precisely substance, in something like Aristotle's sense, that has crept back into philosophy as a result of David Wiggins's meditations on identity. His initial target is the concept of 'relative identity', defended by Geach (among others). According to Geach, identity is relative to a 'sortal concept': a concept which answers the question 'Same *what*?' If Jack is the same as Henry, then this is presumably because Jack is the same *man* as Henry. According to Geach, this opens the possibility that *a* should be the same *F* as *b*, but not the same *G*. (Thus Christian doctrine maintains that Christ is the same substance as God the Father, but not the same person.) This radical view can be shown to violate both Leibniz's law and transitivity, which is enough to show that relative identity is not a species of identity. Nevertheless, Wiggins argues, there is a grain of truth in Geach's argument. Whenever *a* is the same as *b*, there must be a sortal concept, under which *a* and *b* both fall, which defines their conditions of identity. The question 'Same what?' is always both pertinent and metaphysically fundamental. In the case of Jack and Henry, it is true, we have an embarrassing abundance of appropriate sortals: man, human being, animal, person. And there is a real question as to which of those is basic. But that takes us precisely back to Aristotle's theory of being, which seeks for the *ultimate* constituents of reality, those things, and those kinds of thing, which we must identify if we are to identify anything.

The individual substances in our everyday ontology are all in space and time. So they must be, say Strawson, Wiggins and others, if we are to identify them, individuate them, and reidentify them as the same again. But not every sortal under which entities are so identified is a true substance-sortal. For there is a distinction between those which divide the world arbitrarily, or according to our passing interests, and those which divide the world according to its intrinsic order. If I rebuild my car, it is an arbitrary matter whether I decide to call my car the *same* as the one I owned last week; or at least, the decision will be dictated by legal and fiscal convenience, and could be made differently by different people. The sortal term 'car' serves to count things, and offers answers to the question 'same what?'; but it sorts things relative to a human interest, and touches only superficially on the nature of things. The sortal term 'horse' is quite different. It is not for me to decide whether this horse in the stable before me is George, or whether it is the same horse as the one who stood here yesterday. A horse's history is determined by the laws of equine nature, and without reference to human interests. By identifying George under the sortal 'horse', we not only avail ourselves of a real criterion of identity; we also say what George fundamentally *is*.

All that is highly controversial, and we shall have cause to return to it. But it illustrates the important observation that the sparse theory of being, having tripped over the concept of identity, finds itself deep in the metaphysical mud.

4. Necessary Being

The ontological argument was dismissed by Kant, with the dictum that existence is not a true predicate. Russell believed he could clinch the matter, by showing the absurdities that followed from the opposite assumption. If existence is a predicate, he suggested, we could construct an argument like this one:
(1) Donkeys exist.
(2) Eeyore is a donkey.
Therefore (3) Eeyore exists.
(Russell compares this argument to 'Men are numerous; Socrates is a man; therefore Socrates is numerous.' But is it really comparable?) Russell's counter-instance is in fact question-begging. Someone who thought that existence is a predicate would read (1) as: *Some* donkeys exist. From which, given (2), (3) does not follow. Those who think that existence is a predicate say that it is true of *some* things that they exist, false of *others*. This was Meinong's position. Does it force us to distinguish being from existence, and to say that there *are* things which do not exist? Maybe it does. But, if we can do so without falling into contradiction, and while expressing all that we know to be true, what prevents us?

The point is reinforced by a consideration of fictions. Surely we all know who Eeyore is; we know a great deal about him, and can fruitfully compare him to our friends and acquaintances. We have an interest in the question whether he exists. This is a question *about* Eeyore; and we can quite cheerfully answer in the negative, without denying the many things that we know to be true of our favourite donkey. We can account for fictions far more easily on the assumption that existence is a predicate, which may be affirmed or denied of an object, than by trying to translate the name 'Eeyore' into some definite description, and then to translate all the sentences containing it into the cumbrous formulae of Russell.

If existence is a predicate, it is far more easy to accept the suggestion that some things exist necessarily. There may be things whose existence is part of their essence: whose complete conception *contains* existence. Traditionally only one entity has been thought of in this way, namely God. This is because we have a conception of God which enables us to understand *how* his existence follows from his nature. If God is the greatest being, and if existence is a property, it must be that God exists. For by taking away existence you diminish a thing in every respect relevant to its goodness and power. Some philosophers (e.g. Norman Malcolm) have defended the ontological argument by arguing that, while existence is not a predicate, *necessary* existence *is* one. Malcolm's idea is this:

In saying that something exists necessarily, I am saying that it is dependent for its existence on nothing outside itself. This is, surely, a real perfection. By adding necessary existence to the list of God's properties, we vastly add to God's goodness and power. Moreover, Malcolm argues, this is precisely what St Anselm had in mind in his original argument.

This amendment makes questionable assumptions about necessity. It also has the unfortunate result that you cannot deduce from the premise that something has necessary existence that it actually exists. At best you can conclude that *if* it exists, then it exists necessarily. Consider the following concept: a bird, with red plumage, six-foot wingspan, speaking perfect Latin, competent on the alto-saxophone, and with necessary existence. That *seems* like a consistent concept. But *is* there such a bird?

Plantinga argues more plausibly (*The Nature of Necessity*, pp. 214–215), that we can provide a perfectly coherent concept of a being with 'maximal greatness', and that this concept satisfies the requirements of the ontological argument. We can therefore construct a proof that necessarily such a being exists. I shall return to this proof in the next chapter. As with all such arguments, we should ask whether this 'greatest being' really does correspond to our conception of God. In particular, how does he (or it) stand in relation to the created world? Do contingent beings depend upon him? Did he cause them? Is he interested in them?

Traditional versions of the argument were phrased so as to link God to the world, through the idea of causation. A necessary being, argued Maimonides, is one that is 'caused by itself', in just the way that a contingent being is caused by others. The very same thing that exists necessarily is therefore the thing to which we are led, when seeking the ultimate explanation of the world that we know. As to quite what is meant by 'cause' in this argument, that too is a topic for another chapter. But the God of Maimonides has a precious attribute that Plantinga's God may lack: he overcomes the contingency of the world. Since all contingent beings depend upon him, and he exists by necessity, the riddle of existence is solved. There is nothing which merely exists, unexplained and inexplicable; nothing whose existence is a 'brute fact'. And that goes for you and me. The search for God's purpose in creating us can therefore begin.

There is another worry about necessary existence. Are we sure that one and only one thing can possess this feature? What about numbers, for instance? If the number two exists, it is hard to conceive how it could exist *contingently*. Is there a possible world in which there is no number two (but all the other numbers), or no numbers at all? The supposition hardly makes sense. But the number two purchases its necessary existence at the expense of its causal power. With no ability to affect the world, numbers are 'causally inert'. Might not the same be true of Plantinga's God? At any rate, he is not alone in that realm of necessary existence: infinitely many things exist there beside him: numbers, universals, possible worlds, and many others. God becomes king of the ontological slums.

(That last problem was noticed by the medieval theologian Duns Scotus, who argued that abstractions like mathematical objects do not exist independently, but only in God's thought, and that they owe their eternal nature precisely to the eternity of the being who thinks them.)

5. Hegel: the Encumbered Theory of Being

One of the most influential accounts of being in the whole history of philosophy is that given by Hegel, in his *Logic*, a work which prompted Russell to remark that, the worse your logic, the more interesting its results. It is very important to have some understanding of Hegel's ideas, which recur in many guises throughout the subsequent literature. And the first thing to understand is that Hegel did not mean 'logic' in the modern sense – according to which logic is the study of inference and argument. He meant the abstract study of the thing that the Greeks called *logos*, which means, variously: word, description, explanation, reasoning, account. Hegel's *Logic* is an *a priori* study of thinking, describing, applying concepts. And the first of the concepts that it studies is the one on which all others depend – the concept of being (usually written, for no good reason, Being in the English translations).

The most controversial assumption has already been made, in the description of being as a *concept*. But another controversial assumption soon eclipses it. We do not, according to Hegel, compare our concepts with an independent reality; rather, concepts *unfold* and thereby generate the reality that is described by them. Hegel is an advocate of the coherence theory of truth: he holds that it is impossible to measure thought against something that is not thought. Reality is determined by our ways of conceiving it. To understand reality is to understand our conceptions, and vice versa. The structure of the world just *is* the structure of thought, and all that exists is spirit (*Geist*), seen now in one way, as 'objective spirit', now in another, as 'subjective spirit'.

Hegel believes that concept-application exhibits a peculiar triadic structure, which he calls the 'dialectic'. All thought involves the application of a concept, and the first 'version' (or 'moment') of any concept is *abstract*. In trying to grasp reality, I begin by applying some abstract conception to it: such as 'object' or 'thing'. I then acquire a more 'determinate' grasp, by understanding the inadequacies of this abstract conception; and so arrive at a more 'determinate' idea. But this determinate idea wars with the abstract idea, with which it is, in a certain sense, in contradiction. ('Every determination is a negation', says Hegel, quoting from Spinoza.) Out of this conflict a new concept is born, one which is 'truer' than the first, both in making finer discriminations, and in presenting a more complete picture of reality. Hegel puts the idea in the following way:
(i) First moment: a concept is applied; but it is abstract, 'immediate' and indeterminate. (Immediate = roughly that it stems directly from thought, without any further conceptions.)
(ii) Second moment: the abstract concept can yield knowledge only by being 'mediated' by rival conceptions, hence becoming determinate.
(iii) The conflict between the abstract and the determinate conception is resolved by an intellectual 'transcendence' (*Aufhebung*), to a 'truer' conception that embodies both.

It is easy to see how concept-application may proceed from abstract to determinate, in progressive stages: as when I understand an object first as a thing in space, then as a living thing, then as an animal, and then as a cat. But what is meant by the idea that the various stages are reached through a conflict (even a contradiction)? Hegel's thought is roughly this: concepts are by nature universal and hence abstract. Yet their *application* is always to an instance, a particular. However there is nothing outside concepts which could introduce the element of particularity – we have no access to a preconceptual reality. Concepts must in some way *apply themselves*: they must contain within themselves whatever is necessary to identify the *particular* to which they are applied. Hence the abstract, universal element in every concept must be counterbalanced by a concrete, particularising element: a vector, if you like, tending *against* abstraction, and therefore against the concept in its abstract form. The clash of the two is what leads to the idea of a concrete reality, which both *is* cat and yet is *not* cat, since it is not identical with the universal.

So what does Hegel say about existence? Our Fregean conception of existence is like that of an anchor: the existential quantifier ties our concepts to reality. It is not itself a concept, but an operation which attaches concepts to their instances. This is what leads to the 'sparse theory of being' – the sense that there is nothing interesting to be said about being, apart from the theory of the quantifier. But there is no place for such an anchor in Hegel's philosophy. There is no way of fixing our concepts to a pre-conceptual reality. Thought actively creates the reality to which it refers, through the very process of applying itself. Being too is involved in this process: it is the supremely abstract concept, the concept under which everything falls. Hence being is the starting point of 'logic'.

The concept of being provides an illustration of Hegel's dialectic. As initially conceived being is entirely abstract: it is 'indeterminate immediacy' as Hegel expresses it. I can understand this idea without the aid of any others (it is 'immediate'), but that is only because it is entirely indeterminate: it applies to everything, and so says nothing in particular about anything. (Always, in Hegel, there is the interesting thought that we purchase immediacy at the expense of determinacy, and so certainty at the expense of content. The more certain our knowledge, the less we know.) It follows that, in predicating being, we say nothing about *what* is. To say that there is being is therefore to say nothing. Hegel thinks of this as a contradiction: we have applied not only the concept of being, but also that of nothing or not-being, which was lying concealed, so to speak, within being, and eager to wage war against it. Not-being determines or 'limits' being, and compels it to 'pass over' into the next concept in the dialectical chain: that of determinate being. Determinate being is the kind of being that genuine particulars have. A table, for instance, exists; but there is a limit to its existence: there are places where it is not, and when we apply the concept *table*, we divide the world into things that are tables, and things that are not. All this is comprehended in the thought that tables have determinate being, in which both being and not-being are comprehended and transcended. Hegel uses the German word *dasein* to denote this idea. ('*Dasein*' means to exist, but signifies, etymologically, 'being there': this, for Hegel, captures the 'determinate' element in our idea of existence.)

An interesting complexity here enters Hegel's theory. The relation between subject and object, inner and outer, self and world, is, he believes, entirely mysterious, unless we recognise that we are dealing with a division within thought (or spirit) itself. Thought can point in either of two directions: it can express itself in objects, realising its potential through the construction of an objective order. Or it can reflect on itself, studying not the object but the 'appearance' of spirit to itself. Hence there are two kinds of 'logic': objective logic, which studies the application of the concept to the object; and subjective logic, which studies the concept as it is in thought. The one studies 'being-in-itself' (or the concept as *being*); the other studies 'being-for-itself' (or the concept as *concept*). The distinction between 'being-in-itself', and 'being-for-itself', became immensely important in the history of philosophy: it was pillaged and mangled by Marx, and resurrected in something like its original meaning by Heidegger and Sartre. The interest of the distinction lies partly in the suggestion that self-conscious beings are not merely different in kind from non-self-conscious beings, but actually have another kind of *being*. They *exist in another way* from rocks and stones and trees. This thought promises a metaphysical underpinning to the view that we are *in* nature, but also 'apart from' it.

Hegel recognises the importance of time in the philosophy of existence, arguing that this idea arises out of the dialectical opposition between being and determinate being, which can be resolved only through temporal ways of thinking. We give sense to the idea that one and the same thing both is and is not, by postulating its existence at one time, but not at another. Through time we may discriminate among entities; hence we can count them, and distinguish them. Time provides us also with the concept of 'becoming' (the next stage in the dialectic), through which we understand the being of organisms. Organisms are entities in a constant state of becoming, which yet remain the same.

The theory proceeds in this way, through successive 'unfoldings' of the 'moments' of being. It is hard to understand in detail, and full of amazing and outrageous arguments. But the central notion is not without interest. Hegel is saying that we can understand the whole world through one concept – the concept of being. But in understanding this concept, we generate its successive 'determinations'. We then discover that there are many different ways of being, and that our world divides according to the divisions of the concept. The being of a concrete thing is not the same as the being of an abstraction; the being of a reidentifiable particular is not the same as the being of things outside time; the being of an organism is not the same as the being of an inorganic lump; the being of a self-conscious subject is not the same as the being of a mere organism. And so on. All these are interesting suggestions. But they all depend on a contentious premise: namely, that existence (being) is a concept like any other. Being does not anchor our conceptual scheme in reality, but *generates* that conceptual scheme from its own instantiation. The result, as Schopenhauer scornfully remarked, is an ontological proof of the existence of everything. And while this proof, too, has the benign result of removing the contingency from our world, the costs in credibility are greater still than those incurred by the followers of St Anselm.

6. Heidegger and the Anxious Theory of Being

Before leaving this topic it is worth glancing at one of the most notorious of all works of philosophy: Heidegger's *Being and Time*. Heidegger's purpose in this work is unclear, as is the method that he uses to accomplish it. Nevertheless, it is possible to give a few guidelines, which will point to the questions that trouble him.

Heidegger was a phenomenologist, a pupil of Husserl, and deeply influenced by his teacher. He was also influenced in his language by Hegel, and in his argument (or lack of it) by meditations on the pre-Socratics. Almost nothing else seems to have influenced him; so far as his writings reveal, Frege, Russell and Wittgenstein might never have existed.

Phenomenology sets itself up as the study of appearances. But Heidegger takes it to be a study, and indeed the only true study, of 'the things themselves'. This is because appearances are 'phenomena' which means (according to Heidegger's reading of Greek) 'things that show themselves'. Phenomenology therefore studies the revelation of a thing in appearance, and not the appearance itself. (Most of Heidegger's arguments are not arguments at all, but fragments of amateur etymology, usually focused on a Greek word. Much of Central European philosophy between the wars was like this. Compare, for example, Voegelin, Rahner and Patočka.)

It is in the context of phenomenology, so conceived, that Heidegger poses what he calls the 'question of being'. This question has 'ontological priority' over all other questions: which is to say, not merely that other questions must wait on it for an answer, but that *we too* depend on that answer. My *existence* is at stake in the question. And I find the answer only by *existing* in another way.

Like Hegel, Heidegger recognises many ways of being: but he adds some interesting new categories to those forged by Hegel. There is 'being-in-the-world', and 'being-for-self'; but there is also 'being-with-others', and 'being-towards-death'. Indeed, the argument of the book involves a journey through successive forms of being, showing that each presents a problem that is resolved in the next. (Echoes of the dialectic.)

So what, then, is the question? Clearly, it is not the question discussed by Frege. Heidegger is not concerned about the meaning of 'being', in the sense of the semantic role of that term. But he *is* concerned about the meaning of being, in the sense of the significance of our being in the world. The question of being can be formulated variously; but in essence it is this: Why am I here? That sounds far too simple, so Heidegger drops the word 'I' and replaces it with *Dasein*, which he contrasts with *Sein*, mere being. *Dasein* is an 'entity for which Being is an issue'. In other words, the entity which has not only being, but also the question of being: in other words, a self-conscious entity, or 'I'. We cannot translate *Dasein* as existence since Heidegger introduces the term *Existenz* to denote a further layer in the concept of being. Existenz is 'the kind of being towards which Dasein can comport itself and always does comport itself' (p. 32).

Heidegger is quite clear that temporality is part of *Dasein*, which locates itself in time, and seeks for the meaning of its existence in time. (Dasein has 'historicity'.) He also argues that the essence of Dasein is its *Existenz*, which would sound like an ontological argument for the existence of the self, if it were not the case that *Existenz* does not clearly mean existence (or anything else). At any event, we know this about Dasein: that being is an issue for it. It is in the world, and stands to the world in a problematic relation.

Part of the problem is summarised in the term *Geworfenheit*, or 'thrownness': things in the world are 'thrown' into it, without explanation. I see them in this way, and also react towards them accordingly. I too am 'thrown'. This makes Heidegger's problem seem like the problem of contingent being. How can I come to terms with contingent being? But coming to terms for Heidegger does not involve the proof of a necessary being. On the contrary, his bleak and godless world is one in which each of us must come to terms with his own contingency without the assurance of any firm point outside our own predicament. We must find a meaning in contingency itself: only then will the problem of being be solved for us. This problem, and the solution to it, are *existential*. That is, they concern our way of being in the world. It is only by *being* in a certain way that we solve the problem of being. But we then find the meaning of being, not in a theory or an argument, but in a way of being through which being ceases to be an issue. The answer to the problem comes when it ceases to be a problem.

The details of Heidegger's account lie beyond the scope of this chapter. In outline, however, he develops the following conception:

(i) A pure theory of things. Heidegger adopts from a long tradition of Kantian thought the distinction between person and thing (though he does not use that language). And he develops a theory of things, as 'ready-to-hand' and to be used. The theory is a phenomenological one: that is, it describes how things are understood by those to whom they appear.

(ii) A theory of persons. A person is an other for *Dasein*. More simply: self-conscious existence recognises others of its kind, and distinguishes them from things.

(iii) A theory of personal relations. These can be authentic or inauthentic. They are inauthentic when I use them to conceal the question of being, and so to flee from assuming responsibility for what I am. Inauthenticity comes when I allow others to direct my life: when I surrender myself to 'them'. (See Chapter 30.)

(iv) A theory of anxiety. *Angst* follows the recognition of my contingency and of my gratuitous separation from the world. I 'fall' into anxiety, and one source of inauthentic existence is the attempt to rescue myself by surrendering responsibility to others. (The false community.)

(v) The true overcoming of anxiety is through the attitude of 'care' (*Sorge*), in which I answer for myself and for the world as it appears to me. This is possible only if I move into a new phase of being which Heidegger calls 'being-towards-death'. I then accept death as a limit, and also transcend it.

It is not difficult to see in those ideas (which I have sketched very contentiously) some of the phenomena of religion as I described it: the sense of loss and solitude outside the community; the refusal of the inauthentic community; the desire nevertheless to overcome the solitude of the outcast; the anxiety of contingent being, and the steps to overcome it through an acceptance of death. All these ideas belong to the natural history of religion. The problem of being as Heidegger presents it seems very like the problem to which religion offers an answer. In the last chapter I shall touch on this matter again.

13

Necessity and the *a priori*

I have assumed an intuitive understanding of the distinction between the necessary and the contingent, and between the *a priori* and the *a posteriori*. It is time to explore these distinctions more thoroughly, and to relate them to a third, which has often been associated with them: that between the analytic and the synthetic. (The latter terms are due to Kant, whose introduction to the *Critique of Pure Reason* contains what is still the best statement of the problems that I shall be discussing.)

As normally considered, these distinctions are among propositions – with the usual caveat that, where some philosophers speak of propositions, others speak of beliefs, judgements, thoughts or sentences. A necessary proposition is one that is necessarily true – i.e. whose negation is impossible. A contingent proposition is one that, if true, might yet have been false. An *a priori* proposition is one whose truth can be proved by reasoning alone, and without the benefit of experience. An *a posteriori* proposition is one whose truth must be established by experience. Finally analytic propositions are those whose truth value is determined purely by the concepts involved in them. If they are true, this is because the words used to express them guarantee their truth: any other assumption being incompatible with the *meaning* of those words. (All other propositions are synthetic.)

Kant did not explain the last distinction in that way. He defined an analytic judgement as one in which the concept of the predicate is included in that of the subject. (Aquinas had defined a 'self-evident' truth in the same way.) Thus when I say that all bodies are spatial, I have expressed an analytic truth, since the concept of a body *includes* that of spatiality. Modern philosophers reject Kant's account for two reasons: first, it applies only to subject-predicate sentences; secondly it employs a metaphor that is difficult to translate into literal terms: the metaphor of 'containment'. However, as philosophers have discovered, it is extremely difficult to find a better version of Kant's idea. 'True by definition' is of little help, since most terms do not have an agreed definition. 'True by virtue of the meanings of words' is equally hollow, since the phrase 'by virtue of' stands in need of a commentary, and no one has provided it. We must therefore begin from examples. Here are some:

All bachelors are unmarried.
Nothing is both red and not red.
Anger is an emotion.
The supreme ruler is sovereign.

The assumption is that, if you understood the meanings of the terms in any of those sentences, then you would see that the sentences are true, without having to discover any further facts about the world.

1. The Empiricist Position

Although Kant introduced the terms 'analytic' and 'synthetic', he did not entirely introduce the distinction that they denote. Hume spoke instead of a division between the 'relations of ideas' on the one hand, and 'matters of fact' on the other. He believed that all judgements dealt with one or other of these, and that there is no further category to which judgements could belong. This thesis is sometimes referred to as 'Hume's Fork', and is tantamount to the view that our three distinctions are really one distinction, described in three different ways. For what Hume meant by the 'relations of ideas' is what Kant meant, when he referred to a concept's 'containing' another. When I say that bachelors are unmarried, my judgement is made true simply by the relation between the *ideas*: *bachelor* and *unmarried*. When I say that bachelors are unhappy, then what I say, if true, states a matter of fact. Matters of fact can be known, Hume argued, only through experience. And all propositions stating matters of fact might have been false. Necessity belongs only to those judgements whose truth is guaranteed by the relations of ideas.

Hume's position is characteristic of empiricism, and was inherited by the verificationists. According to empiricism, necessity can be explained only if we can also explain our *knowledge* of it. How can I, a finite and experience-bound creature, claim to know that some proposition is *necessarily* true? Surely, only if I can establish its truth *a priori*. There cannot be an empirical proof that *p* is necessarily true (as opposed to just true). With this much Kant was in agreement. But, adds the empiricist, I can establish the truth of a proposition *a priori* only if its truth is grounded in information that I already possess through understanding it. What could this information be? Surely the only information that I derive from the mere fact of understanding a proposition, lies in the concepts that compose it (the 'ideas'). Concepts are the meanings of words. Hence it is from my knowledge of the meanings of words that I derive the truth of *a priori* propositions. In short, necessary truths are true by virtue of the meanings of the words used to express them. And the class of necessary truths is identical with the class of *a priori* truths.

This position is implied in the verificationist slogan, that necessary truths are 'merely verbal'. The empiricist theory disposes of the mystery involved in the idea of necessity. *We* are the creators of necessity, and a truth becomes necessary only by becoming empty. What seem to be the deepest truths about the world are really no more than the conventions that we adopt in order to describe it. In explaining necessity in such a way, the empiricist also shows how we can *know* necessities.

The theory has the consequence, however, that all our claims to *a priori* knowledge are either false or trivial. If we know *a priori* that *p* is true, then this is because *p* is made true by our own linguistic stipulations, and by nothing else. Empiricists have always welcomed this result, since it draws a sharp limit to our philosophical pretensions. Indeed, it goes hand in hand with the claim that metaphysics – which could be true only if it is *a priori* – must consist of thoughts that are either trivial or false.

2. Kant

Kant was not satisfied, however. Not only does the empiricist position rule out any interesting results in metaphysics. It is also unable to explain a host of *a priori* truths which are manifestly 'synthetic'. Kant used this term to denote the fact (as he saw it) that, whereas analytic judgements merely break down a concept into its constituents, other judgements effect a 'synthesis', advancing in the predicate to an idea that is not implicitly 'contained' in the subject. We do not need to accept that particular theory, however, in order to find a use for the expression that Kant made famous: 'synthetic *a priori*' knowledge, meaning knowledge of truths which are *a priori*, and yet not derived merely from the meanings of the words used to express them. How, Kant asked, is synthetic *a priori* knowledge possible? If it is not, then there is no such thing as metaphysics.

But it *is* possible, he argued. For *mathematics* is possible, and mathematics is synthetic *a priori*. There is no way of proving mathematical truths by analysis of the meanings of mathematical symbols. Yet all rational beings are able to grasp these truths, to prove them *a priori*, and to acknowledge that, if true, they are necessarily true. Whether Kant was right about this has been one of the principal contentions of philosophy since his day: and it is a topic to which I return in Chapter 25. He was certainly right in his claim that no philosopher had come anywhere near *proving* that mathematical truths are analytic, and that Hume's complacent assumption that they too speak of the 'relations of ideas' was entirely unfounded.

Whatever we say about mathematics, Kant was convinced that there are also synthetic *a priori* truths of metaphysics, and that the task of philosophy is to discover them. If the task is so difficult, it is because we lack a theory of the synthetic *a priori*: we lack any clear conception of how synthetic *a priori* judgements are *possible*, and whence their truth derives. In the *Critique of Pure Reason* he set out to provide that theory, arguing that metaphysical truths tell us how the world must be if we are to understand it. We establish them by studying our intellectual capacities. From this study we conclude that only certain forms of experience, and only certain kinds of reality, are intelligible. This is not an empirical study. It involves us in reflecting *a priori* on the understanding, so as to draw the limits of what can be known. (It is not that we say: this is as far as *human* knowledge extends; rather, knowledge can go no further.) A moment's thought reveals that Kant's theory does not really explain how synthetic *a priori* knowledge is possible. For what of the theory itself? If it is to uphold the *a priori* status of the truths that it establishes, it too must be known *a priori*. We have an explanation of the synthetic *a priori* nature of metaphysics, only on the assumption that Kant's theory in turn is synthetic *a priori*. If on the other hand Kant's theory is really analytic (based on an analysis of what we *mean* by 'knowledge', 'understanding' etc.) then so are the conclusions that derive from it.

Kant argued that there are two criteria of *a priori* truth: strict universality and necessity. We could never make a strictly universal judgement (concerning what is always and everywhere true) if we could not be confident of its necessity. And we can be confident of necessity only where we have an *a priori* proof. He also thought that we constantly encounter these signs in our ordinary thinking:

scientific laws, for example, which we assert as universally valid. Science would be without foundations if we could not find some synthetic *a priori* principle in which to ground it. Synthetic *a priori* truth is not a luxury of metaphysics, but a component of everyday reality.

One problem, however, is that there is very little agreement as to which propositions *are* synthetic *a priori*. Those chosen by Kant are so tied up with his own contentious system that few philosophers are persuaded by them. Here are some instances of propositions that have been held to be *a priori* yet not analytic:

(1) Nothing can be red and green at the same time.
(2) God exists. (One of Kant's objections to the ontological argument was that, if valid, it would make God's existence into an *analytic* truth.)
(3) Every event has a cause.
(4) Truth is correspondence with the facts.

You will not be surprised to learn that examples like (1) are far more often discussed in modern philosophy than examples like (2). And of course they are interesting. It is implausible to argue that, when we teach a child the term 'red', we also teach him that red means 'not green'. It is not the *meaning* of the terms 'red' and 'green' that tell us that nothing can be red and green all over, but the qualities themselves. It is a truth about the *nature* of red, and the nature of green, that they cannot jointly inhere in a single surface. (But what is meant, here, by 'nature'?)

3. Kripke

Kant agrees with the empiricists, that the distinction between the *a priori* and the *a posteriori* coincides with that between the necessary and the contingent. But he disagrees with them about the analytic/synthetic distinction, which he regards as a distinction of quite another kind. Hence, he believes, there can be synthetic *a priori* truths, which would also be necessary truths. Saul Kripke goes a stage further, arguing that we have not two distinctions but three. For let us consider them in the abstract: the necessary/contingent distinction is metaphysical, a distinction between two ways of being. The *a priori/a posteriori* distinction is epistemological, between two routes to knowledge. The analytic/synthetic distinction is semantic, between two ways in which the truth-value of a sentence is determined. Why should we suppose, therefore, that these are three names for one distinction, or that the distinctions have anything in common? Just as Kant had been prepared to countenance synthetic *a priori* truth, Kripke shows how we might accommodate the 'contingent *a priori*' and the 'necessary *a posteriori*', as well as other equally surprising things.

Naming and Necessity is a fascinating work, and one that has changed the course of modern philosophy. Since I cannot hope to summarise all its arguments, I shall concentrate on the central two or three. The first concerns identity, and exists in a variety of forms – including a formal proof in modal logic, presented by Ruth Barcan Marcus. (On modal logic, see the final section of this chapter.) Everything, it seems, is not merely identical with itself, but necessarily identical with itself. Moreover, nothing is identical with anything else. Consider, then, the identity statement '$a = b$'; it is true only when 'a' and 'b' refer to the same object; otherwise false. But if 'a' refers to the same object as 'b', namely the object a, how could it be false that $a = b$? Surely, the statement merely identifies an object (namely a) with itself (namely b); and a statement which does *that* is not merely true, but necessarily true. There is no possible situation in which a is not identical with itself.

The argument can be rephrased in terms of Leibniz's Law, which says that, if a is identical with b, then everything true of a is true of b. One of the things true of a is that it is necessarily identical with a. So this must be true of b also.

Kripke connects these observations with the questions concerning identity and reference that I have already touched on. Frege (and many philosophers after him) felt that terms which refer to objects (the genuine *names* of a language) are connected with a procedure for *identifying* those objects. It is by virtue of this procedure (which is the *sense* of the name) that we can make statements of identity. But what *are* the genuine names of a language – the words which, so to speak, enable us to step down from language to the world of existing things? Russell believed that a 'logically proper name' has a meaning *only* if there is an object denoted by it: the object denoted *is* the meaning. Many apparent names in our language therefore turn out to be disguised definite descriptions, whose bearers must be guillotined from our ontology like the unfortunate king of France.

Although Russell was led into difficulties by his strange idea that the referent of a name *is* its meaning, he was surely right to suggest that the point of names is that they *refer*. On any account, this must be the fundamental fact about the meaning of a name. One suggestion is this: the sense of the name '*a*' is *that* it refers to *a*. Its role in language is fully defined by its connection with its object. This, roughly, is the line taken by Kripke, and it enables him to give a new force to Frege's thesis, that names have a special role in statements of identification and identity. Some philosophers, notably J.R. Searle, have assumed that the sense of a name is given by a description, or a bundle of descriptions. If this were so, Kripke argues, it would follow that Aristotle might not have been Aristotle. For that man – namely Aristotle – might not have satisfied the description that is contained in his name (whatever the description is). The fact is, however, that we do not understand names in this way at all. If we did, then we should be hard pressed to know whether we all meant the same by any name in our language. To understand a name is simply to know *which* object it refers to. The name could not refer to any *other* object without changing its sense, and so ceasing to be the same name. Kripke expresses this point by saying that names are 'rigid designators'. They refer to the same object in all possible situations (or all 'possible worlds').

Frege worried about such identity statements as 'the morning star is identical with the evening star', or 'Phosphorus is Hesperus'. These are true but *a posteriori*: we discover their truth through observation, by discovering that the two 'singular terms' refer to the same thing. Frege was inclined to say that they are *contingently* true identity statements. But no identity statement, Kripke insists, is contingently true. If 'Hesperus' and 'Phosphorus' are names, then they are rigid designators, which pick out the same object in all possible situations. If it is *ever* true that Hesperus is identical with Phosphorus, it is *always* and *necessarily* true. Things are more complex with 'the morning star' and 'the evening star', since these phrases can be read in two ways. It is obviously possible that the star which is visible last in the morning should *not* be identical with that visible first at night. But it is not possible that *this* star, the morning star, should not be identical with *this*, the evening star.

So where does that leave us? Kripke suggests that we are dealing with truths that are necessary and *a posteriori*. He gives other examples, corresponding to the traditional idea that certain things, and certain kinds of thing, have 'real essences' (in Locke's phrase), which they cannot lose without ceasing to exist. Thus if Moggins is a cat, she is *essentially* a cat: she couldn't cease to be a cat without ceasing to be. Yet it is an *a posteriori* conjecture that this creature, whom I have called Moggins, is a cat. My researches into the matter are far from complete, and one day someone may well surprise me with the information that Moggins is a marmoset or a bandersnatch.

Kripke also tries to prove that there are contingent *a priori* truths: true propositions which might have been false, but whose truth we know *a priori*. One example concerns the standard metre rod in Paris. Surely it is *a priori* obvious that the metre rod is a metre long. But this rod might not have been a metre long: it could have been the very rod that it is, and yet have expanded or contracted. This aspect of Kripke's argument is, however, rather more controversial. Can I really know *a priori* that the metre rod in Paris is a metre long, when its actual length seems to depend upon factors – temperature, pressure, atmospheric conditions, etc. – which are known only *a posteriori*? But perhaps we do not need Kripke's example. For do we not already know of a truth that is contingent and *a priori*, namely the truth expressed by the proposition that I exist? Descartes established his existence by *a priori* (and valid) argument; yet he might not have existed. If this example too is controversial, it is because it contains the word 'I'. Surely, it might be said, I have not specified a proposition by uttering 'I exist': for I have not yet said *who* exists. An *a priori* proof that I exist is not an *a priori* proof of anything that could be common knowledge. I need to deduce from the thought that I exist that therefore Roger Scruton exists. And that does not follow. (I may be mistaken in thinking I am Roger Scruton; maybe there is no such person.) Puzzles can be multiplied here, and I shall return to some of them in Chapter 31. But we need not linger over them, since the real force of Kripke's argument does not depend upon the claim that there is a contingent *a priori*, but on the metaphysically revolutionary claim that there is a necessary *a posteriori*.

This latter claim has overthrown many established ideas, and many intellectual complacencies. But it raises awkward questions of its own. In particular, just what *do* we mean by *a priori*, necessary, and the rest? Before turning to these questions, it is worth glancing at a fourth position, which holds not that there is one distinction involved in our discussion, nor that there are two, nor that there are, as Kripke maintains, three; but rather that there are none.

4. Quine

Quine's 'Two Dogmas of Empiricism' had the kind of impact in its day as Kripke's *Naming and Necessity* has had in ours. His primary target was the empiricist view of necessity, as 'truth by virtue of

meaning'. But his arguments tell equally against the other two positions. Briefly, Quine's attack goes as follows:
(1) Analyticity. What do we mean by this? To say that the sentence 'All bachelors are unmarried' is analytically true, is to say that it is made true by the meaning of the words that compose it. What words? Well 'bachelor', for a start. This surely *means* unmarried man. So the sentence reduces to 'All unmarried men are unmarried' which is a logical truth. However, why are we so sure that 'bachelor' and 'unmarried man' are synonymous? What is our criterion of synonymy? The answer is that 'All bachelors are unmarried' is *analytic*. In which case our definition of analyticity goes in a circle.
(2) Necessity. What does this mean? What is my criterion of necessity? One suggestion is that a necessary truth is one that we are prepared to affirm, whatever the course of experience. But, Quine argues, we can affirm the truth of any sentence in our total system, in the face of whatever experience, just so long as we are prepared to make adjustments elsewhere. No sentence is tested against experience on its own: I can test the truth of 'Moggins is a cat' only by testing the truth of 'Moggins is a mammal', 'Moggins is a predator', and so on. And since experience is proof of nothing until *described*, I can always adjust the evidence, in order to hold on to my sacrosanct ideas. Of course, someone will reply that I cannot hold on to the sentence 'Moggins is a cat' in the face of all the evidence that she is a bandersnatch, without changing the meaning of what I say: 'cat' will become synonymous with 'bandersnatch'. But then I must rely on the notion of meaning and synonymy, in order to make the distinction between those sentences that are *really* necessary, and those which are not. And that, in turn, will require us to lean on the notion of analyticity. We are beginning to go round in a circle again.

Similar considerations apply to the *a priori*. We can define analyticity, necessity, *a priori*, and synonymy, but only in terms of one another. They form what Quine calls a 'circle of intensional terms', whose utility remains in doubt until we can give some clear and independent criterion for their application. (I explain the term 'intensional' below.) Since, Quine argues, we do not need these terms, we ought to throw them out altogether. All we can say, and all we need to say, is that our language forms a single system or 'conceptual scheme', which faces 'the tribunal of experience' as a whole. We have no need to refer to meanings – ghostly metaphysical entities whose conditions of identity can never be defined. For the dimension of sense is not required in order to relate our language to the world: reference alone suffices. And when we consider our conceptual scheme as a whole, we see that there is no distinction that we could possibly make between those items in it that are necessarily true, and those that are merely contingent. The only distinction that we could make (and the only one that we need to make) is between sentences that we are reluctant to give up in the face of recalcitrant experience, and sentences which can be jettisoned without compunction. This distinction is a distinction of attitude; it does not describe two 'ways of being' in the world.

Needless to say, all that is highly controversial. (See, for example, the attempt at a reply by Grice and Strawson, 'In Defence of a Dogma', and the full army of Quinean arguments marshalled by Gilbert Harman in his articles in the *Review of Metaphysics*, 1969.) The problem is that, if you accept Quine's conclusion, you find yourself drawn to nominalism, pragmatism, and a highly scientistic worldview. Yet, if you look closely at his arguments, you will find those very positions built into the premises, and so protected from their much-deserved interrogation. Even if there is no non-circular account of necessity for example, this justifies our rejection of the idea only if we accept Quine's view that nothing is defined until it is defined in the terms that a radical nominalist would countenance. But try giving a non-circular definition of 'thing', say, or 'experience'. You can advance a little way, but your definitions become steadily more obscure until the point where you save yourself from the labyrinth, and go back to where you started.

5. Necessity and Possible Worlds

Remember Frege's theory, according to which the reference of a singular term is an object, and that of a sentence a truth-value. Instead of 'reference' logicians often use the term 'extension' (borrowing a word from the nineteenth-century logician, Sir William Hamilton). This does not express quite the same idea as is expressed by Frege's term 'reference'; but it captures what we need for logic. The extension of a name is an object, of a predicate a class, of a sentence a truth-value. Extensional logic is the logic that issues from the assumption that the extension of a complex term (such as a sentence) is fully determined by the extension of its parts – in the way that the truth value of p & q is completely determined by the truth value of its components, and the truth value of 'Mary is angry' is fully determined by the fact that 'Mary' refers to Mary, and 'angry' to the class of angry things. In extensional logic, the law of substitution is universally valid. This says that terms with the same extension can be substituted for each other without changing the extension of the whole. (Leibniz's law is a special case.) Modern logic was built on the assumption that the law of substitution is obeyed (the assumption of extensionality). And where it is not obeyed, the writ of logic ceases to run – at least, that is what radical thinkers like Quine are inclined to believe.

The idea of necessity cannot be introduced into our language, Quine thinks, without violating the principle of extensionality. For consider the sentence:

(1) Necessarily the winner of the Derby won the Derby.

That would seem to be true. Yet substitute 'George' for 'the winner of the Derby' – since it was, after all, George who won – and the result is:

(2) Necessarily George won the Derby.

But that is surely false: for George might not have won. (Otherwise who would be a bookmaker?)

Quine expresses the point by saying that the term 'necessarily' creates a context that is 'referentially opaque'. Others have spoken instead of 'intensional' contexts. All of Quine's 'intensional terms' ('analytic', 'necessary', '*a priori*', 'synonymous', etc.) create such contexts, in which the law of substitution fails. Hence his word for them. And hence his hostility to the thing that they describe. There is simply no place for intensional contexts in logic, he believes; for they destroy the transparency of language, and cloud its relation to the world.

Two responses have been made to that argument, both germane to our discussion. The first distinguishes two kinds of necessity: necessity *de dicto*, and necessity *de re*. A necessity of the first kind holds by virtue of the words used to express it (i.e. by virtue of 'what is said'). Thus 'Bachelors are unmarried' is a necessity *de dicto*: it holds on account of our choice of words. In describing someone as a bachelor, we thereby imply that he is unmarried. But consider John, a bachelor. Is it necessarily true that he is unmarried? Clearly not: he might marry at any time. He is 'necessarily unmarried' only *as described*; in reality he is not necessarily unmarried at all. A *de re* necessity would be one like this: Necessarily, Hesperus is Phosphorus. For it is not merely *as described* that the evening star is identical with itself. This necessity lies in the nature of things.

It was one of the assumptions of Hume's empiricism that all necessities are *de dicto*: i.e. they are artefacts of language. We create necessities by our use of words; they point to no special 'way of being', of the kind explored by the metaphysicians. Quine's assumption that the concept of necessity leads inevitably to intensional contexts is of a piece with his acceptance of the empiricist view, that *if* there is necessity, it must derive from our use of words. But look back at propositions (1) and (2) above. If we take 'The winner of the Derby' in (1) to be a 'rigid designator', and ask ourselves what it stands for, the answer is evident: it stands for George. In which case the sentence is clearly false. George does not have the property of necessarily winning the Derby: nor does anything else. But on this reading, (1) and (2) *are* equivalent in truth-value. In other words, if you read them as expressions of a necessity *de re*, it is not at all clear that they involve an intensional context. In a *de re* necessity the element of necessity is, as it were, absorbed into the predicate, and ceases to cast its shadow over the sentence as a whole. If Moggins is necessarily a cat, then she is necessarily a cat however identified: the thing hanging upside-down from the curtain is necessarily a cat.

The second response is equally interesting. When logicians began to pay attention to the concepts of necessity and possibility ('modal' concepts), they developed languages in which these ideas could feature as 'modal operators'. It is clear that we have some understanding of necessity and possibility. For instance, we can define them in terms of each other. 'Necessarily *p*' is equivalent to 'It is not possible that not-*p*'. Moreover, we know that 'Necessarily *p*' implies '*p*'; that '*p*' implies '*p* is possible'; that 'Necessarily *p* and necessarily *q*' implies 'Necessarily *p* and *q*'. And so on. Building on these intuitions, elaborate systems of modal logic were developed by the American logician C.I. Lewis (himself a pragmatist, like Quine). The problem then arose, how to interpret those systems. What should we assign to the modal operator 'Necessarily' as its extension or *value*? Since it creates an intensional context (for certainly it does not follow from 'Necessarily *p*' that 'Necessarily *q*', just because '*p*' and '*q*' have the same truth-value), we cannot assign to it any kind of truth-function. What then *should* we assign to it? Or does the whole enterprise of interpretation break down, as Quine suggests? If it does, Quine is right to be suspicious of modal logic.

The answer to the problem is suggested by the modal operators themselves. For 'possibly' and 'necessarily' are interdefinable in exactly the way that the quantifiers 'some' and 'all' are interdefinable. 'Possibly *p*' is true if and only if 'It is not necessary that not-*p*' is true, etc. So perhaps we should construe the modal operators as quantifiers, which refer not to the propositions that are attached to them, but to the possible situations in which those propositions are true. This would lead to a coherent interpretation of modal logic, provided we can show that the normal requirements of a semantic theory – finitude, consistency, completeness, and so on – are satisfied by the resulting interpretation. This was shown by Kripke, in a brilliant mathematical paper written at the age of thirteen, and published in the *Journal of Symbolic Logic* for 1963. To say that *p* is necessarily true, he suggested, is to say (with a few refinements) that *p* is true in all possible worlds. We can now convert our seemingly intensional modal sentences into perfectly extensional sentences, in which 'Necessarily' ceases to be a sentence-forming operator on sentences, and becomes instead a quantifier, ranging over possible worlds. And we can construct from the result a theory which shows how the truth-values of modal sentences are determined by the references ('extensions') of their parts. All we need to assume is the existence of possible worlds, among which the actual world is one.

With this there re-entered into philosophy a notion that had first been discussed by Leibniz. We can re-phrase everything that we wish to say about necessity and possibility by use of the idea of a possible world, and the result will be 'transparent' in the logician's sense. To say that Moggins is necessarily a cat, is to say that, in every world where Moggins exists, Moggins is a cat. (Though she is not black in every world where she exists.) (I have made a contentious assumption here: namely, that Moggins herself exists in other worlds. David Lewis argues that we should speak instead of the 'counterpart' of Moggins in each possible world. For our purpose this dispute may be set aside.)

The apparatus of possible worlds has been used by philosophers for many purposes, and has had considerable impact on logic and metaphysics. It has also enabled Plantinga to revive the ontological argument in a novel form. He gives two versions of the proof, the simpler of which reads like this:

Consider the property of maximal excellence, which includes omniscience, omnipotence and moral perfection. It is possible that something should be maximally excellent: i.e., there is a possible world in which a maximally excellent thing exists. Now consider the property of 'unsurpassable greatness', which is 'maximal excellence in every possible world'. This property, Plantinga argues, is instantiated either in every possible world or in none. So that if there is *any* world which contains an unsurpassably great being, then every world contains him. But there *is* a possible world in which unsurpassable greatness is exemplified: hence an unsurpassably great being exists, and exists necessarily (in all possible worlds).

The argument makes a key assumption in modal logic, namely that what is necessary or impossible does not vary from world to world. It is arguable that this assumption can be made only under certain conditions, and that these conditions are implicitly denied by Plantinga's manner of argument. But the really important premise is that maximal greatness, and therefore unsurpassable greatness, are instantiated in a possible world. *Is* this a real possibility? Leibniz said that the ontological argument does not prove God's existence. It proves, rather, that God's existence is either necessary or impossible (impossible, because there just *could not be* such a being as God). J.L. Mackie strenuously attempts to show that Plantinga's argument does not really surmount the hurdle that Leibniz places before it: that Plantinga has, and can have, no proof that the concept of unsurpassable greatness is possibly instantiated. (*The Miracle of Theism*, a useful book by a man who spent much of his life lecturing God on His non-existence, and who is now being lectured in his turn.)

The argument takes us into technical regions, and we must allow it to vanish over the horizon. But I recommend it to those who wish to see how important the concept of a possible world has become, and also how controversial. Indeed, it is *inherently* controversial. A possible world is a non-actual world, and how can such a thing exist?

Most philosophers accept that the reference to possible worlds is merely a 'manner of speaking', a convenient way of addressing issues which could be explained in other and more cumbersome ways. Not so David Lewis, who in *Counterfactuals*, ch. 5, and *The Plurality of Worlds*, has vigorously maintained that possible worlds really exist. Of course, he adds, only one of them exists *actually*; but the word 'actually' should be understood like 'here' and 'now': it merely picks out the particular place (the particular possible world) where *we* happen to be. Only monstrous egoism could lead us to conclude that no other world *exists*.

Paradoxically, Lewis's ontological extravagance stems precisely from the sparse theory of being. If quantification is the measure of existence, then everything exists over which we have to quantify. We know that there are modal truths (we could not make sense of the world without them); and we know that modal truths can be understood only if we quantify over possible worlds. Therefore possible worlds exist. (But if that is so, do they exist necessarily, as numbers do? Many of their defenders are inclined to say yes. But the actual world is also a possible world: does *it* have necessary existence? And if so, does everything contained in it have necessary existence also?)

Rather than sink into this quagmire, let us return to *de re* necessity. For, if the Kripkean approach to the matters I have been discussing is right, this idea lies embedded in our ordinary thinking, as well as being fundamental to natural science. There are real necessities, which lie in the nature of things, and are not merely projected into the world by language. Empiricists find it very difficult to accept this, partly because they believe, with reason, that we could never discover such necessities, and are therefore unable to prove our title to believe in them. The argument originated with Hume, and it is to his thoughts on causality that we must now turn.

14

Cause

The topic of natural necessity leads us to the problem of causation. This became a problem with Hume's devastating, though not entirely original, account of it – an account that his contemporaries understood as thoroughly sceptical. (Modern commentators, notably Stroud and Pears, tend not to agree with that interpretation.) It is worth noting that the term 'cause' has not always meant what it means for Hume and his successors. Traditionally (i.e. in those Latin authors inspired by Aristotle, which means just about everyone up to Leibniz and Spinoza) the word 'cause' signified any valid explanation. There are as many types of cause as there are ways of explaining things: Aristotle distinguished four of them. If an explanation is any satisfactory answer to the question 'why?' we certainly ought to acknowledge that explanations may not be all alike. For we should distinguish different senses of the question 'why?'. There is the 'why?' that seeks the end: 'Why did you do that?' 'In order to . . .' There is the 'why?' that seeks a reason, whether justifying ('because it was the right thing to do',) or explanatory ('because I wanted to'). And there is the 'why?' that seeks what Aristotle would have called the 'efficient cause' – the kind of cause that we most usually mean by the term: 'Because she pushed me.' When the judge asks me why I put poison in my wife's tea, he will not be satisfied by my saying 'Because electrical impulses from my brain caused my hand to reach for the bottle of arsenic and tip it into the waiting teacup' – although that may be a true answer to the question 'why?' construed as a request for the 'efficient' cause.

Philosophers therefore distinguish various kinds of explanation: teleological, rational, and mechanical, often reserving the term 'causal' for the last of those. A teleological explanation explains an event in terms of its end (i.e. in terms of something that comes afterwards). A 'causal' explanation, as now understood, explains an event in terms of events that *produce* (and therefore precede) it. Just where ordinary reasons for action fit into the picture is a problem that I shall here postpone.

1. Hume's Problem

Consider the statement: She fell because he pushed her. This refers to a 'causal connection'. Between what does the connection hold, and what kind of a connection is it?

(i) The terms of the relation. Some recent philosophers (e.g. Donald Davidson) argue that causal relations hold between events. Others prefer facts or states of affairs. Does it matter which we choose? Is there anything at stake, other than words, in the decision to say that the cause of her fall was the event of his pushing her, or the fact that he pushed her? In favour of events we find the following arguments: that events are individuals, that they can be precisely located in time and space, that they are unambiguously part of the physical world, and finally that the sentences that refer to them do not contain 'intensional contexts' of the kind discussed in the last chapter. (This means that different descriptions of the same event can be substituted without changing the truth-value of the sentence.) None of those, it is sometimes argued, are clearly true of facts, which, as we have seen (Chapter 9), are metaphysically problematic in a peculiar way of their own. For example, there is no agreement as to the identity conditions of facts. If it is a fact that the cat is on the mat, then it is also a fact that the mat is under the cat; are those one fact or two? And what about the fact that a small feline is on a square of woven wool? How would we decide to count the facts? And are we really sure that facts belong to the world rather than our way of describing it? All those difficulties are already familiar from criticisms raised against the correspondence theory of truth, and they are clearly serious.

In favour of facts, however, we find the following arguments: explanation involves giving a reason for things, which means giving true propositions, and it is facts, not events, that correspond to true propositions. A causal connection is one which makes one thing more probable, given another, and relations of probability hold between facts, not events. (This is made much of by D.H. Mellor, in his *Matters of Metaphysics*.)

For the purpose of this chapter, there is no need to make any final decision between those two positions. But it is important to bear them in mind, since the difficulties which they each encounter will re-emerge in our discussion. For convenience, I shall refer only to events; but nothing that I say will specifically favour the view that it is events and not facts that enter into causal relations.

(ii) The relation. The statement I have chosen is a 'singular causal statement'. That is, it relates one event to its cause, and says nothing about any other event. What makes this statement true? The common-sense answer is that she *had* to fall, given that he had pushed her. Causation involves necessity. Otherwise, the apparent connection between the two events is merely 'accidental'.

Hume was not happy with this. For how could we possibly observe such a necessary connection? All that we observe is one event, and then another. Not only do we never observe necessary connections, Hume argued; it is difficult to see how we could even acquire a legitimate 'idea' of them. For there seems to be nothing in our experience from which such an idea could derive. (Hume does in fact try to *explain* the idea of necessary connection, but his explanation merely confirms his view, that we have no grounds for affirming that things in the world are necessarily connected.)

Hume's view is prompted in part by his empiricist account of necessity, according to which necessities merely reflect the 'relations of ideas'. (See the previous chapter.) Necessities do not inhere in the world, but only in our ways of describing it. But there is another, and more important, reason for what he says. This is what we could call Hume's law, that there are no necessary connections over time. If one thing precedes or follows another, then the connection between them is at best *contingent*, for it is always possible that the world should end before the second event occurs. The two arguments are connected. The empiricist believes that the only form of necessity is logical necessity. A proposition is logically necessary if its negation is self-contradictory. But there can never be a contradiction involved in supposing that the world ends at a given time t: the description of the world up until t will always be consistent with the assertion that nothing happened thereafter. If that were not so, then the sequence of events until t just *could not have happened*; since something necessary to its occurrence (namely the future event which is implied by its description) was lacking. And that is absurd.

Hume's law has other applications. But its application to the concept of cause follows at once, on the assumption that causes precede their effects. Is this assumption warranted? Perhaps not. For there seem to be clear cases in which cause and effect are simultaneous: my sitting on the cushion causes the dent in it. But could there be *backwards* causation? That is a question to which I shall return.

2. Causation and Laws

Suppose we accept Hume's argument. What then *is* the connection between *a* and *b* which makes it true that *a* caused *b*? Hume's contention is that there is nothing to be observed *at the time* which distinguishes causal from accidental connections. All we witness at the time is one event, and then another which is contiguous with it (i.e. in the same approximate space). But temporal priority and contiguity do not amount to causation. So what else must be added? The only thing that we *can* add (since it is the only thing that we could observe) concerns *other* places and *other* times. We can observe that events like *a* are regularly followed by contiguous events like *b*; this constant conjunction of *a*-type and *b*-type events gives us our ground for asserting that *a* caused *b*.

So is that what we *mean* by 'cause'? Verificationists say 'yes'. Committed as they are to identifying the meaning of a sentence with the grounds for asserting it, they jump at once to the conclusion that '*a* caused *b*' just means: '*b* follows and is contiguous with *a*; and *b*-type events are regularly connected in just that way with *a*-type events'. In short, causation means regular connection. There is nothing more that we mean or can mean by the concept. Hume is sometimes taken to have said this, though in fact he was rather more subtle, believing that the idea of necessary connection remains part of the concept of cause, even if there are no grounds in nature for applying it. But the verificationist position has recommended itself to many philosophers in our time, from Russell to Donald Davidson. Even if their reasons for accepting it are various, it has assumed the status of a philosophical platitude.

But what does it mean? What exactly is implied by the idea of regular connection? Do we mean that *a*-type events are *always* connected with *b*-type events? Or that they *tend* to be so connected? Or that they have been so connected *up until now*? Hume, Russell and Davidson seem to concur in saying that we mean 'always'. It is the universal quantifier that distinguishes causal statements from statements of some lesser connection. But Kant responded to this in the following way: if the connection between *a*-type events and *b*-type events is *strictly* universal, then we could not conceivably have any empirical grounds for asserting it. Strict universality is not merely a sign of *a priori* truth. It brings with it an implication of necessity. We can make judgements of this kind only when we think that there is a necessary connection between *a* and *b*, enabling us to venture beyond the limit of our observations. Hume's theory of causation is therefore no alternative to the one that he attacked.

Empiricists resist Kant's rejoinder. There are ways of defining causal regularity, they believe, that do not lean on the idea of necessary connection. We can say what we wish to say in terms of the notion of a 'law of nature', which states fundamental regularities, from which other regularities flow. And a law of nature is not a *necessary* truth. It might have been false. Empiricists have based their belief on the contingency of laws of nature on the premise that laws cannot be established *a priori*, but only by experience. From which they have inferred that laws are contingent truths. If Kripke is right, however, such arguments are worthless. The question whether laws of nature are necessary is not settled by the proof that they are or are not *a priori*.

All this leads to two large questions: What is a law of nature? And is it true that all causal statements are relative to such laws? I shall deal with each question in turn.

3. Laws and Counterfactuals

Suppose that every time I light a fire in my fireplace an owl hoots outside. This is, let us suppose, a long-term coincidence. Until I have carried out some investigations I am not entitled to infer a causal connection between the two events. I am entitled to conclude that every time I have lit a fire an owl has hooted. This is a strange fact, but a fact none the less. But I am not entitled to infer that if I *were* to light a fire now, an owl *would* hoot outside. I have no evidence for the truth of *that* assertion, which goes beyond what I have observed. In making it, I imply just the kind of law-like connection between the two events that is the mark of causality. Suppose that I discover, however, that an owl is nesting in my chimney, and is driven out each time I light the fire, only to return undiscouraged when the smoke dies down. That would entitle me to say that if I were to light the fire now, the owl *would* hoot. For the one event, I may surmise, is the cause of the other.

This has suggested to some philosophers a criterion, whereby we may distinguish an accidental regularity from a law. The first implies the truth of a 'counterfactual', while the second does not. A counterfactual is a special kind of conditional, which says what *would* have been the case in some counterfactual state of affairs. Such conditionals are interesting for many reasons. For one thing, they are not 'truth-functional': we cannot reduce them to material conditionals, or fit them easily into our standard systems of extensional logic. They also illustrate one of the most interesting and fertile areas of human thought – which is our ability to speculate about possible situations, and to form an understanding of what they might involve. It seems that this very ability is exercised in our ordinary causal thinking. Every time we assert the existence of a causal connection, our thought reaches beyond the actual, to embrace the possible as well.

But this gives us further ammunition against the empiricists. Thought about the possible is precisely the *same* kind of thought as thought about the necessary. Causality is firmly established as a *modal* notion. And the best way to understand counterfactual conditions is through necessary connections. I can know that if something were red it would be coloured, precisely because there is a necessary connection between being red and being coloured. Am I not saying something similar, when I say that if the fire were lighted, an owl would hoot?

Thus we find that modern philosophers, when they discuss counterfactuals, often have recourse to the concepts involved in understanding necessity. David Lewis, for example, in his book *Counterfactuals*, analyses them by means of the concept of a possible world, just as you might analyse necessity. You may say, if you like, that he is defining an idea of *physical* necessity. But it is an idea of necessity none the less.

Someone might reply that the assertion of a necessary connection between the lighting of the fire and the hooting of the owl is unwarranted. It doesn't *have* to hoot, even if its hooting *is* caused by the fire. For we have not identified the law that governs these events. Laws of nature underpin causal regularities, but are not exemplified by all of them. Perhaps it is a law of nature that owls avoid smoke. But in asserting this, we have gone deeper into the nature of things. We are now approaching those absolute and exceptionless regularities which are the sole examples of natural laws.

Such a reply, however, merely concedes the point at issue. For it implies that all our causal thought is anchored in exceptionless laws. True causal statements are made true, in the last analysis, by necessary connections.

4. Causality and Determination

But here is where you all go wrong, argues Elizabeth Anscombe, in a famous paper ('Causality and Determination'). The commitment to exceptionless generalisations is no part of causal thinking at all. The Kantian and the Humean are both wrong. The first asserts that causal connections hold universally, and therefore necessarily. The second tries to replace the concept of necessity by an idea of 'mere' universality, believing that he can capture what is distinctive of causal connections (their strength, so to speak) while remaining in the realm of the contingent and the observable. But we are looking for causality in the wrong place: namely in an idea of 'determination', according to which the effect *has* to follow, once the cause has occurred. This is, in fact, no part of the concept.

Miss Anscombe considers the following example (obliquely derived from Schrödinger). A lump of radio-active matter is placed next to a Geiger counter, wired to a bomb. Things are so arranged that, if an alpha-particle strikes the counter, the bomb will go off. We know, therefore, that, when the bomb explodes, the explosion is caused by the proximity of the radio-active material. But there is no law of nature which tells us that in such circumstances the bomb will go off. The only relevant law tells us that there is a fifty per cent probability that an alpha-particle will be emitted in the space of an hour. This could be true, even though no alpha-particle is emitted, however long we wait for it. Here then is a causal connection that can be observed, despite the fact that the cause does not determine the effect.

We can go further. For surely, we can understand and influence our world only through the knowledge of causal connections. But few of us know the laws, if there are any, in which those connections are grounded. Most of our causal knowledge is contained in singular statements, of the kind considered in section 1. I see the knife cutting the bread: and in doing so I witness a causal connection. I see the horse stumble on a stone, the tree shudder in the wind, the apple fall from the branch, the child cry out with joy: all these involve causal connections, which I grasp and know without the benefit of science, and in total ignorance of the laws of nature which supposedly control them. Causal thinking can hardly be supposed to commit me to ideas which may be forever absent from my mind, despite the richest store of causal knowledge.

There are two familiar responses to this. One says: Quite right. Causal connection is something *sui generis*, which can be observed and discussed without reference to the unobserved and the counterfactual. The other says: You are moving too quickly. I do not infer *q* from my thought that *p*; but *p* may nevertheless imply *q*; and this implication may be all-important in determining what I really mean. (Perhaps I must say what I mean; but must I mean only what I say?) After all, even children distinguish things from their properties. But how many children grasp what is implied in that distinction, or realise what they really mean when making it?

5. Causation and Probability

Nevertheless, we have come across a real difficulty, to which I shall return in the next chapter. For it is hard to deny the force of the Geiger counter example. It reminds us that scientific laws do not as a rule state universal connections, but only probabilities: the probability of *a* given *b*. In the case of quantum mechanics (which claims to be the ultimate truth about the universe) these probabilities are ineliminable: they arise at the very limit of what can be known. In all our search for scientific principles, we find ourselves bound, and happy, to accept statistical laws. There is a good chance that an owl will hoot when I light the fire: and that is all I can deduce, from the hypothesis of a causal connection.

The question then occurs: Which comes first? Do judgements of chance and probability depend upon causal statements? Or do causal statements depend on judgements of probability? Some philosophers – notably Suppes and Mellor – believe that the concept of probability is fundamental, and contained *within* the idea of cause. In witnessing causal connections we are gaining insight into evidence. We are experiencing what provides reason for what, and what is likely given what. This riveting suggestion will encourage us towards two conclusions: first, that it is facts and not events which are the terms of the causal relation; secondly, that the relation we capture in the idiom of evidence (the relation between *p* and *q* when *p* is evidence for *q*) is not a relation in thought, but a relation in reality. The real is rational, as Hegel said, and the rational real.

6. Causality and Time

One final point. Why did Hume insist that causes must precede their effects? What is the connection between this component of the traditional analysis and the component of regular connection? For surely, the regular connection exists equally between effect and cause. Why must explanation always be of the *later* event in terms of the *earlier*? Or must it? Whichever of the standard approaches we adopt, the thesis of temporal priority remains isolated and unexplained by it. Suppose someone were to advance a counterfactual analysis of causation. To say that the light went on because I pressed the switch is to say that, in these circumstances, or circumstances sufficiently close to them, had I pressed the switch it *would* have gone on. Does that imply that the pressing of the switch *preceded* the light's going on? No. For it is equally true, given the causal connection, that, in these or sufficiently similar circumstances, if the light had gone on, then I would have pressed the switch. (To use David Lewis's idiom: any world with these circumstances, and in which the light goes on, is a world in which I press the switch.) So why does explanation go from past to future, and not the other way? This is a deep puzzle, for which it is difficult to find a satisfactory answer. Some philosophers argue that explanation is possible only because the event explained is *brought about* by the event that explains it. And this idea of 'bringing about' applies only from past to future.

But is this so? At least one philosopher, Michael Dummett, has doubted that the puzzle can be solved so easily. In an article entitled 'Bringing about the Past', he offers an example designed to show that 'backwards causation' may still be a live philosophical option. A certain tribe sends its braves each year on a lion hunt in some distant region. They are absent for six days, during two of which they travel, two of which they hunt and two of which they return home. During their absence, the chief dances in order to cause them to act bravely. He dances for all six days, and justifies this by saying that whenever, in the past, a chief has danced for only four days (assuming perhaps that the braves are by then no longer in need of his help) the results have been disastrous.

You might respond to this by arguing that at least the chief ought to *test* his belief that his dancing during the last two days helps to bring about the bravery of his subjects during the days before. Maybe he should dance after their return, when it is known that they have *not* acted bravely, and see if the situation changes. (Why do we think that it will not?) But of course, the chief may find some other, equally rational, explanation of the fact that, in this case, he is – say – unable to dance, or unable, by dancing, to change the braves' past behaviour. By imagining the circumstances in the right way, Dummett supposes, we can reach the conclusion that we have no better reason to reject the idea of a backwards causality than to accept it. Backwards causation becomes a live option, and one which may be reasonably incorporated into the planning of the tribe.

The argument, and the example, are highly controversial, for many reasons. If backwards causation can occur in this case, why can it not occur generally? And if it can occur generally, will we not find ourselves involved in paradoxical beliefs? (Could I bring about some event in the past which totally changes my present life – even to the extent that I am no longer interested in bringing about the past?) Many philosophers argue that forwards causation is bound up with our very concept of time, and that we should land in contradictions just as soon as we took backwards causation seriously.

The matters that I have touched on in the last two sections are intricate and hard to understand. To gain a better appreciation of them, we need to look more generally at the nature of scientific thought, which is the topic of the next chapter.

15

Science

The questions dealt with in the last chapter belong to metaphysics. They concern the nature of the world – in particular, whether the world contains real necessities. But the considerations that generate this metaphysical conundrum also produce a problem in epistemology: the problem of induction. Whenever we make a prediction, and whenever we infer a general law from its instances, we make an inductive inference: an inference from the observed to the unobserved. In mathematics and logic we reason deductively, and our inferences are valid just so long as the conclusions cannot be false while the premises are true. Deduction is not without its problems; but it is bound to lead from truth to truth. Induction, however, leads us to conclusions which are not entailed by the premises: conclusions which might be false, when the premises are true. In which case, how do we know that these conclusions *are* true? In what sense is a conclusion justified, when its falsehood is compatible with the arguments that lead to it?

The problem is sharpened by two observations.

(1) Inductive inference does not merely pass from the observed to the unobserved; it does so, as a rule, by postulating a law. I have observed the sun rise each day, year after year. I therefore conclude that the sun rises every day. And from that general law I deduce that the sun will rise tomorrow. But the law covers indefinitely many instances. So my evidence falls far short of my conclusion. What kind of validity could such reasoning aspire to?

(2) Hume's law again. There are no necessary connections across times. Whatever the truth about the world up to time t, it will always be compatible with any course of events whatsoever after t. Hume's law is able to generate ever more virulent sceptical challenges to those

who place their trust in induction. It is perhaps the most powerful negative force in philosophy since Descartes's demon.

There is a spatial version of Hume's law too. Just as we can envisage the world ending at any point in time, so that a description of the world up to that point is logically independent of a description of the world thereafter, so could we envisage each area of space as bounded, with nothing beyond it. The description of the world inside the boundary is then logically independent of the description of the world beyond. (But the case is more difficult. Such a boundary cannot run through the middle of Moggins, since half a cat implies the existence of the other half. I beg leave to set this difficulty aside.)

1. Some 'Solutions'

A variety of solutions to the problem have been attempted. Here are some of them:

(1) *J.S. Mill.* In his *System of Logic*, Book III, ch. 21, Mill suggests that our scientific thought assumes the Uniformity of Nature. We suppose that the future will be like the past, and unvisited regions like those we have observed. This assumption is so deeply embedded in our thinking that we could not begin to question it. (Maybe, if we did question it speech itself would fail. In which case, we should cease to question.) Every inductive argument is therefore really a deductive argument, with the following form:

Major premise: Nature is uniform.
Minor premise: All observed instances of *F*s have been *G*.
Conclusion: Therefore all *F*s are *G*.

There are two weaknesses in this argument. First, how do we know that the major premise is true? Because we have observed nature to be uniform up to now, and so assume it to be uniform always and everywhere? Then of course we are assuming precisely what has to be proved, namely the validity of induction. Secondly, even if we grant the major premise, the argument is still not deductively valid. For the *F*s that we have observed may be atypical. Or *G* may be a very superficial characteristic of the things that possess them, so that *F*s might easily cease to be *G*. (Cats could cease to have tails; academics could cease to be left-wing; marmite could cease to taste delicious.) So just when are we entitled to infer 'all *F*s are *G*' from 'observed *F*s are

G'? That *is* the problem of induction, and Mill's invocation of the major premise has done nothing to solve it.

Mill was not unaware of those problems, and had important things to say to which I return below. In particular, he denied that the circularity involved in the argument is vicious.

(2) *Probability*. Maybe we should abandon the paradigm of deductive argument, and look for principles of reasoning which allow premise and conclusion to be logically independent. In fact there are such principles: the axioms and rules of probability theory. Of course, these axioms and rules are controversial; but we are not without intuitions in the matter. We frequently measure the likelihood of one event, given the occurrence of others. And if we could reduce this form of argument to its fundamental assumptions, we should have a standard of inductive validity. Inductive validity will be a matter of degree: an argument will be valid *to the extent* that the conclusion is rendered probable by the premises.

There are 'subjective' and 'objective' probabilities. People *assign* probabilities to events and theories, and this is a 'subjective' fact: a fact about the people themselves. From some given assignment all kinds of conclusions follow; but the assignment may itself have no further justification. However, there is also a notion of objective probability, or likelihood. The likelihood of an event is a fact, independent of our predictions. If the concept of probability is to answer the riddle of induction, it must show how to justify beliefs about likelihood. But how is that possible? One answer is that we discover probabilities by means of 'long-run frequency'. If, in the long run, 9 out of 10 *F*s are *G*, then the probability of an *F* being *G* is 90%. Such measurements permit an objective calculus of probability, in terms of which to weigh the evidence that we possess for all our laws. The only problem, however, is that we still do not know how to justify our beliefs about 'long-run frequency'. Necessarily our observations cover only a finite sample. But the 'long run' either is infinite or at least possesses no boundary. The assumption that our sample is typical is precisely what we need to prove. The inference to the conclusion that a sample is typical just *is* induction. And the problem is: how is it justified? (As a discussion of probability this is the barest beginning: but I return to the topic in section 7.)

(3) *Popper*. For a variety of good and bad reasons it is difficult to ignore the name of Popper, who (in the 1930s, when he was a kind of fellow traveller of the positivists) made the striking suggestion that the crucial concept in scientific method is not verification but

falsification. We can never conclusively verify a scientific law; but we *can* conclusively refute it. One counter-example suffices to overthrow what no finite number of positive instances could prove. Scientific inference is not a matter of induction, but of conjecture and refutation. Scientific hypotheses last just so long as they are not refuted. That we have actively searched for a refutation, and so far failed to find it, is our best guarantee of the truth of a scientific law, and the only guarantee that we really need.

What Popper says is plausible. But it does not solve the problem of induction. The method of conjecture and refutation has served us in the past, and led to true hypotheses. But will it serve us in the future? Only if we can assume the principle of induction. Moreover, it is not simply that we accept scientific laws provisionally, while seeking for their refutation. We accept them as *true*, on the basis of the evidence. The fact remains that the evidence will always be insufficient to prove them.

Nevertheless, Popper's account of scientific method has been extremely influential. For it suggests a criterion for distinguishing genuine science from pseudo-science. Genuine science is an adventure of discovery – an active exploration of the world. Its hypotheses and theories are framed in terms that facilitate their own refutation, by defining exactly what a counter-example would be. Pseudo-science avoids refutation. Its theories are framed with a view to adjusting them in the face of evidence; its laws are vague and indeterminate, and contain escape clauses that will permit them to survive the emergency of a temporary disproof. According to Popper the theories of Freud and Marx are like this. If the work of Thomas Kuhn (*The Structure of Scientific Revolutions*) is important, however, it is largely because it questions this simple dichotomy between science and pseudo-science. Even in the most rigorous of theories, Kuhn argues, a cherished 'paradigm' is retained and protected; and a 'paradigm-shift' is a seismic event, in which the whole structure topples.

(4) *Strawson*. The problem of induction, Strawson argues, arises because inductive inferences are not deductively valid. If they *were* deductively valid, however, they could not yield any new or interesting result – they would merely lay out the implications of what we knew already. It is therefore pointless to complain that they are not deductively valid, when this fact is a necessary consequence of their role in scientific discovery. Moreover, we have no other principle of scientific inference. Any other method that we could propose for advancing from the observed to the unobserved must

either lean on induction, or else give no grounds at all for its conclusions. It is an analytic truth, in fact, that inference to the unobserved depends upon induction. Once we have seen this, we must surely give up the attempt to find a justification of induction in terms that do not presuppose it. The attempt to find such a justification is no more coherent than the attempt to find a justification for *deduction*, in terms that do not presuppose the validity of deductive argument.

Strawson's account belongs to a family of similar responses; and it is easy to sympathise with it. It *does* seem as though the search for a non-inductive justification of induction will lead us nowhere, unless to a snark or a boojum (probably the latter). Braithwaite (*Scientific Explanation*) goes further, however, and argues (following J.S. Mill) that the use of the principle of induction to justify induction is not viciously circular at all. It involves what he calls an 'effective' circularity. Induction has always worked in the past. Therefore it will work in the future. In drawing this conclusion, however, I do not accept the validity of induction as another *premise* of the argument. I *use* induction, as a rule of inference. And the rule validates itself, by generating its own reliability as a conclusion. Is such an argument circular or not? And if so, is the circle vicious?

(5) *Pragmatism and naturalised epistemology*. In contrast to the above approaches, which accept the classical statement of the problem and confront it directly, there has arisen a rival tradition, typified by C.S. Peirce, John Dewey, and W.V. Quine. According to this tradition, we should study the principle of induction in terms of its utility to the creature who deploys it. The principle will enable him to make successful predictions, and so to control his environment. Even if the world suddenly changes, so that whenever n instances of F had turned out to be G, the $n + 1$th would be not-G, you would still be better off using the principle of induction. For this would lead you to make the most useful predictions in the long run, even in these strange circumstances. (After a while, having observed n Fs, you get into the habit of expecting the next one to differ. And you will turn out to be right. But that habit just *is* induction.)

The believer in a 'naturalised' epistemology would shift the focus of the problem, from a first- to a third-person perspective. Creatures who depend on induction, he will argue, have a better chance of survival than those who use some rival principle, *however* the world turns out to be. Hence this principle will be favoured by evolution, and will drive out its competitors in the struggle for survival.

Is that a vindication of induction? Yes and no. It depends on the *way* in which the principle of induction is supposed to be useful to us. The simple answer is that it is useful because it generates *true* conclusions. But how do we know that? We are back with the problem. Indeed, if Durkheim is right, false beliefs are often more useful than true ones. In matters of religion, for instance, the reckless pursuit of truth is likely to destroy a society's chances of survival. A measured taste for falsehood may be a decided evolutionary advantage. It may even be one of the factors that enable us to face the truth. (For it is interesting to note how prone to superstition are those rationalists who scoff at the old religions, and how unable they are, in a crisis, to look reality in the face.)

2. Hempel's Paradox

The 'solution' that has won most sympathy is the one that rejects the problem. It says 'induction is induction, not deduction; moreover there is nothing wrong with it, and it is the only conceivable principle whereby to pass from the observed to the unobserved'. But maybe there *is* something wrong with it. This is the thought prompted by a famous paradox, due to Carl Hempel (one of the Austrian verificationists) and sometimes called the 'paradox of confirmation'. Take the sentence 'All ravens are black'. According to the principle of induction this general law is confirmed by its instances. But what are its instances? Black ravens of course. But are these the *only* instances of the law? Perhaps not. For 'All ravens are black' is logically equivalent to 'All non-black things are non-ravens'. Are these too instances of the law that all ravens are black? If that is so, then every time I observe a red pullover I confirm my belief that ravens are black!

Why is this a paradox? It does not involve a contradiction. But it does cast doubt on the normal ways of formulating the principle of induction. If every law is confirmed by its instances, we need to know how to recognise an instance. A red pullover is not an instance of the law that all ravens are black. For this is a law about *ravens*, and surely can be confirmed only by observing *them*. At the same time we cannot escape from the fact that 'All ravens are black' is logically equivalent to 'All non-black things are non-ravens'. And inductive reasoning would be a hopeless mess, if we could not infer, from '*p* confirms *q*' and '*q* entails *r*', that '*p* confirms *r*'. To abandon this principle would involve abandoning all hope of an inductive logic. Or would it?

3. Scientific Method

Hempel's paradox has not been laid to rest. But it has awoken philosophers to the fact that induction does not stand alone in scientific inference. It is the keystone of a methodological arch. Scientific method involves several operations, besides the gathering of evidence. Our search for confirmation goes hand in hand with a search for refutation. And the search is part of theory building. We seek explanations of the observable, not just generalisations about it. This means that our laws will be prompted by the search for causal connections.

Hence we make a distinction between those regularities that *explain* their instances – such as 'All ravens are black'; and those which, even if true, explain nothing: for example, 'Everything in my pocket is a non-raven'. Ravens form a significant class, with a common nature. Things in my pocket do not.

The philosophy of scientific method addresses such questions as the following:

(1) What are theories, and what are denoted by their terms?

(2) How do we arrive at theories, and how ought we to arrive at them?

(3) Which classes of object are the primary subject-matter of scientific laws?

(4) Are there real essences, and if so how do we know them?

(5) How does the world described by science relate to the world of appearance?

(6) What is probability? Are there real (objective) probabilities?

I shall briefly survey some of those problems.

4. Theories and Theoretical Terms

Suppose you seek to explain an event: for instance a flash of lightning. 'Why did it occur?' you ask. One answer is to give a general law. 'Lightning always occurs (or is very likely to occur) when there are storm clouds.' Most people would not be satisfied with that. It seems like an explanation only because it suggests a general connection between events. But it offers no further account of the connection. It is a general law, but not yet a law of *nature*. Suppose someone continued as follows: 'When there are storm clouds, an electrical charge accumulates within them. This charge may be suddenly earthed by the damp air between the cloud and the ground. The resulting discharge is lightning.' This is far more satisfying. For it unravels the causal chain that puzzled you. It explains *why* by showing *how*.

The example is an instance of theory-building. According to many philosophers (Popper included) theories are really 'hypothetico-deductive' systems. They are systems from whose highest principles (or axioms) chains of deduction depend, coming to rest at last in 'observation statements', which may be confirmed or refuted by experience. The passage down from theory to observation is one of deduction; the passage up from observation to theory is one of induction and 'abduction': the term coined by C.S. Peirce to denote the process of forming an hypothesis. And the two processes constrain and shape each other. It is through this dynamic of mutual constraint (which is 'dialectical', in something like Hegel's sense) that we learn to distinguish the real from the apparent order of the universe.

But what of the terms of our theories? To what do they refer? This question has proved troublesome to empiricists. For in speaking of fields, waves and sub-atomic particles, we are not referring to anything that is directly observable. Of course we observe the *effects* of these things; that is precisely what the theory *says*. But we do not observe the *things themselves*. Sometimes theoretical entities seem inherently paradoxical, like the entities studied in quantum mechanics, which are both waves and particles, and concerning which there may sometimes be no categorical truths, only probabilities.

The first response to these questions is to distinguish a theory from its models. A model (like Niels Bohr's famous model of the atom as a solar system), gives us a way of envisaging theoretical entities. A perfect model reproduces every feature that is attributed to the theoretical entity by the theory. But it will also have further properties, not mentioned (and perhaps not even mentionable) in the theory. (Thus Bohr's atom has definite shape, where there is in reality only a probabilistic distribution of forces and fields.) Theoretical entities seem paradoxical, only because we confuse them with the models that we use to visualise them. In themselves they are not paradoxical – merely unimaginable.

The second response is to ask what we mean, when we say that theoretical entities exist. Roughly there are three answers that are commonly entertained:

(1) Realist: we mean that they really exist, out there, in the world, just as you and I exist. Theories tell us what reality is like. They could not

do so, if they did not describe reality as it is. That's the whole point of them. Any other supposition is incompatible with the belief that theories really *explain* appearances.
(2) Reductionist: theoretical entities are really 'logical constructions' out of the observable entities from which they are inferred. A theory is given by the totality of its observation-statements.

Reductionism was a favourite posture of the logical positivists, and in particular of Carnap, whose single-minded attempt to reduce the world to the observations whereby we gain knowledge of it, is responsible for much of the philosophy of science. Reductionism took heart, too, from a result in formal logic known (after its discoverer) as Craig's Theorem, which shows how any theory can be replaced by another which lacks its theoretical terms. But the appeal of reductionism was short-lived. For the observation-statements of a theory are infinitely many. We can never complete the reduction of theory to observation. Moreover, our observations are 'theory-laden'. They prove what they prove only because the terms of the theory, and the existential pre-suppositions thereof, are built into the record of our observations.
(3) Instrumentalism: theories should be understood as tools for moving from one observation to another. Their sole value lies in their predictive power; and they have this power by virtue of the logical relations which take us from observation to theory and back again. We do not need to suppose that the terms of the theory refer to anything. All that matters is that the predictions come out true.

Instrumentalism, like pragmatism, to which it is closely related, looks fine so long as you don't examine it too closely. The problem is, however, that the inference from observation to prediction is possible only on the assumption that the existential claims of the theory are *true*. So we are back with the realist. Moreover, the entities postulated by a theory may be observed at some later stage – like the moons of Saturn, or bacilli in the bloodstream. It seems strange to say that we must, after observing them, assign to them a completely different metaphysical status.

5. Real Essences and Natural Kinds

But how should theories be constructed? Can we have a theory about anything (the things in my pocket)? Why do we cheerfully accept 'Ravens are black' as the first step in a theory, but not 'Non-black things are non-ravens', despite their equivalence? Mill suggested an answer. The world, he argued, contains *kinds*. It is not we who compose these kinds; they exist in the world. Their members are bound together by a common nature, and belong together regardless of how we describe them.

Thus we identify ravens as a fit subject for scientific investigation, since we recognise that they belong together as members of a 'natural kind'. The same is true of the fundamental stuffs of our universe: gold, for instance, or water. We did not create the kind *water* simply by our classifications. Contrast the kind denoted by the term 'ornament'. We choose to regard now this, now that, as an ornament; and the things themselves need have nothing in common, besides our desire to bring them under this common label, and to use them as the label suggests.

The distinction between natural and non-natural kinds has been associated (for example by Kripke and Putnam) with another – the distinction discussed sceptically by Locke, between real and nominal essence. Natural kinds have a real essence, apart from the 'nominal' essence whereby we pick out their instances. For example we pick out diamonds by means of their hardness and transparency. But diamonds are not essentially hard or transparent. Their real essence is *discovered* by scientific investigation. The true theory of diamonds tells us what they essentially are: namely carbon, the very same thing as charcoal.

This is another instance of an *a posteriori* necessary truth. Diamonds are carbon in every world where they exist: they are necessarily carbon. But a diamond might lose any of its accidental properties – its hardness, sheen and transparency – without ceasing to be what it is.

The idea of a natural kind has aroused much interest among recent philosophers, for two reasons. First, it suggests that science looks for necessary connections. There are real necessities, involved in the very structure of the world, and it is the task of science to discover them. Secondly, it enhances the authority of science, by detaching it from our observations and attaching it to an objective order. Science seems less and less like the systematic description of appearances (which is what it was for Carnap) and more and more like the patient exploration of reality, the attempt to 'divide nature at the joints', and so to replace our interest-relative concepts with concepts of natural kinds.

6. Abduction

The postulation of a natural kind is part of the process of 'abduction'. I do not arrive at the conclusion that ravens form a kind by observing a multitude of ravens: I postulate an order in nature, and understand what I observe in terms of it. My observation of this object ceases to be the perception of a black and feathered thing and becomes the perception of a *raven*. And what I observe, in studying it, are the features of ravens.

The first move in scientific inference is not the inductive 'proof' of a law, but the abduction of a theory; and abduction is impossible without the classification which defines what we observe. Our instinctive sympathy for nature prompts us to make lively and illuminating guesses at the true order of things: the species and substances from which the world is composed. Of course, we make mistakes: for a long time whales were classed as fish, fire as a substance, and glass as a crystalline solid. But, from our first shots, we advance by clear steps to a truer picture, and so amend our classifications. Only because the first shots are largely accurate, however, can the process begin.

If we now combine what was said about theory-building with these brief remarks about abduction, we arrive at the view of science as 'inference to the best explanation'. This inference begins with the very classification in which inductive reasoning is founded. The enterprise of science is to find an order in the world, which will enable us to understand *why* the world appears as it does. Now there are many – infinitely many – theories which explain any particular observation. In searching for the *best* we are searching for the theory which explains as much as possible with the fewest assumptions.

And is the best theory also *true*? The answer is simple: it *may* be false; but the considerations that lead us to adopt it are the considerations which distinguish true theories from false ones. Explanatory power and ontological economy are criteria of truth. If this is so, however, it points to a profound feature of reality: namely, that it is *ordered*, so as to be maximally intelligible to us. Nothing that happens lacks a sufficient explanation, and the explanation of one thing provides the explanation of others. The world obeys what Leibniz called the 'Principle of Sufficient Reason'. Why is that?

There are two popular answers, but they are not the only ones. The first, and simplest, answer is that God made it so. He would not have wished his creatures to live in a world that eluded their intelligence.

Hence he created an ordered world, which reveals its secrets step by step to the inquiring intellect. The second answer again invokes what Stephen Hawking (*A Brief History of Time*) has called the 'weak anthropic principle', namely: observation is possible only in a world that sustains life like ours. For that to be true, the world must exhibit causal order of a complex and systematic kind. It is therefore a law of nature that nature will be observable only where nature is also explicable. (Kant would have said something similar, although for him it would be a law of *metaphysics*, rather than a law of physics.)

7. Probability and Evidence

Hempel's paradox was taken by its author to show that we are by no means clear about the concept of evidence or confirmation. The unclarity is compounded by the notion of probability, which seems to play a major role in scientific inference, but which is as fertile of philosophical problems as any concept employed by science. Although this is an intricate and technical area, it is important to grasp some of the main issues. This section will not be the barest beginning (you have already had that); but it will be indecently exposed.

If *p* entails *q*, then we say that *p* is a sufficient condition for the truth of *q*, while *q* is a necessary condition for the truth of *p*. If *p* and *q* entail each other (are mutually deducible), then *p* is necessary and sufficient for *q* and vice versa. In the middle years of analytical philosophy it was assumed that the search for necessary and sufficient conditions is the central task of philosophy – hence the triumphant appearance of 'if and only if' (or the more barbarous 'just in case') on so many of those dreary pages. Wittgenstein changed all that, first with his idea of a family resemblance, secondly with the distinction between symptoms and criteria. If someone asks for a definition of 'game', he suggests, I cannot provide him with a set of necessary and sufficient conditions. But I could describe a family of features, groups of which may be sufficient, and no one of which is necessary for an activity to count as a game. ('Count as' betrays the underlying nominalism.) Different games resemble one another as do the members of a family; and the concept of a game is a 'family-resemblance' concept.

For a while this idea exerted considerable influence, and at least one philosopher (Renford Bambrough) used it to propound what he took to be a solution to the problem of universals (though it is no such thing). More important, however, was the second of Wittgenstein's contributions: the distinction between symptoms and criteria. Looking out of the window I see pools of water on the ground and dark clouds above the farm: symptoms of rain, but neither necessary nor sufficient conditions for its presence. A symptom is evidence that is both inconclusive, and dependent for its authority on some contingent (usually causal) connection. Reaching from the window I find my hand wet with drops of water. Further evidence of rain, but not a *symptom*. This is what rain *consists in*; falling water is a *criterion* of rain. Even so, it is not a sufficient condition, nor even a necessary condition. The water may have been tipped from an aeroplane; it could be raining, even though a wind holds the drops above the ground. Nevertheless, falling water is not just *contingently* connected with rain: it is *necessarily* evidence for rain.

What kind of a necessity is this? Wittgenstein's immediate disciples assumed it to be a *de dicto* necessity (since they did not recognise necessities *de re*). They would argue that it is 'part of the concept of rain' that falling water is evidence for its presence. Interestingly enough, the only clear example that Wittgenstein gives (in the *Blue Book*) is of a natural kind: the symptoms and criteria of a disease. So perhaps he is referring to a *de re* necessity; in which case we should argue that falling water is evidence for rain *in the nature of things*. Either way, we seem to have two ideas of evidence – contingent and necessary – neither of which can be understood in terms of the old idea of necessary and sufficient conditions.

This subject is fraught with controversy. But it is worth touching on for two reasons. First, it disconnects truth-conditions from evidence, while reconnecting them in a novel way. You could agree with Frege, that the meaning of a sentence is given by the conditions for its truth, while denying that you have a grasp of *evidence* for every sentence that you understand. For instance, I understand the sentence 'My Redeemer liveth': it is true if and only if a superhuman person exists, who loves me completely, who is empowered to purge my guilt, and who intends with my help to do so. But I do not know what evidence would lead me to believe such a thing. For certain concepts, by contrast, evidence lies in the nature of the thing denoted by them. I could not grasp the truth conditions of 'Apples are nutritious', without also understanding the evidence that would establish whether it is true. This raises again the question of meaning: Do we *really* understand a sentence when all we have grasped are its truth-conditions, abstractly conceived, and not the criteria which tell us when to use it? There is an important argument here to which I return in Chapter 19.

The second reason for taking an interest in criteria is that they show the centrality of probability in the formation of belief. Evidence for *p* consists in whatever makes *p* more probable or likely. If relations of evidence can be *necessary*, then probabilities lie deep in the nature of things. But what exactly *are* they?

There are three kinds of situation in which we talk of probabilities, and it is a matter of dispute whether we mean the same in each case:

(1) The probability of throwing two sixes with a pair of 'true' dice. Here we have a clearly defined set of alternative outcomes, and a definition of 'true' which makes it a purely *mathematical* question, how likely any particular outcome is. In cases like this we employ a 'calculus of chances' whose theorems are *a priori*. If a particular pair of dice seem to defy our mathematical expections, that merely shows they are not 'true' but biased.

(2) The probability that it will rain in London on the first day of September. Here we are dealing with frequencies which – we believe – represent real tendencies. Thus, if it has rained in London on the first day of September in 45 of the last 50 years, we might conclude that there is a 9 in 10 chance of its doing so again. In reasoning thus, we are obviously distinguishing real likelihoods from mere coincidences: such reasoning only makes sense on the assumption that reality tends in the direction represented by our figures. In other words, we are assuming that our sample fairly represents the 'long-run frequency', and that this long-run frequency is *no accident*. In the case of the weather this is a very flimsy assumption. The deep question is: When is the assumption *not* flimsy? When can we infer long-run from short-run frequencies?

There is another problem too, in that frequencies are relative to classifications. Suppose I ask the following question: What is the likelihood that Smith, an academic, will vote Conservative? The statistics tell us that in the last two elections only 16 per cent of academics voted Conservative. So we answer 16 per cent. But Smith is *also* a reader of the *Daily Telegraph* and a keen follower of foxhounds. And the statistics tell us that 90 per cent of such people vote Conservative. Classifying Smith in this new way, we answer 90 per cent. Which answer is the correct one?

(3) The probability that the Big Bang theory of the universe is true. Here we are not dealing with frequencies or samples, but with the *weight of the evidence* for a particular hypothesis. Philosophically speaking this is the most interesting case, since it is impossible to slip out of the problem as to what 'probability' means. We cannot point to

a 'long-run frequency' in reality; nor can we derive our judgement from the *a priori* calculus of chances. Probability here appears as the *basic* fact. Some therefore argue that, in such cases, probability judgements are merely 'subjective'. Others, more plausibly, say that judgements of probability, here as elsewhere, are measures of the 'degree of rational belief': such was the position advocated by Keynes, Jevons and Jeffreys, the three thinkers (all with an interest in economics) who put the concept of probability on the modern philosopher's agenda.

On such a view, there is a single idea of probability in all three cases: what distinguishes them is the kind of evidence on which the judgement of probability is based. Indeed, Keynes and Jeffreys went further, arguing that probability is essentially a *relation*: one proposition is more or less probable, *relative* to another. There is no sense in an absolute assignment of probabilities, as the example of the conservative academic shows. After all, either he *does* vote Conservative, or he doesn't. But the probability of his voting Conservative, relative to this or that piece of evidence, is a perfectly coherent idea, provided we understand that what we are measuring is not some feature of the physical world, but the degree to which it is reasonable to believe one thing, on the basis of another. There may *be* some corresponding feature of the physical world, such as a long-run frequency. But that is at best part of our *evidence* for a judgement of probability, and not what we mean by it.

Such a position is attractive, though it has been disputed – for example by J.R. Lucas, who advocates an absolute conception of probability, according to which the true probability of a proposition is the target that we approach, as we call in the evidence. The probability that Smith will vote Conservative in our example is not 16 per cent: for the evidence on which this judgement is based is far too slender. Nor is it true that there is a 90 per cent probability either, for there is more to Smith (or at any rate, a little more) than his reading of the *Telegraph* and his following the hounds. Gather together *all* the relevant facts about Smith – his deprived childhood, his hatred of Tory grandees, his contempt towards Trade Unions, his liking for pickled onions – and you will be able to tell us what the real probability of his voting Conservative will be.

If that sounds daft, it is because it is indifferent to an important distinction emphasised by recent writers (notably David Lewis): the distinction between epistemic and objective probabilities. Some of these writers (Dorothy Edgington for example) are attracted by the parallel between probability and possibility, arguing that 'probably', like 'possibly', should be treated as a modal operator. (So, if you like that kind of thing, you can begin to define objective probabilities: 'Probably *p* is true in the actual world if and only if *p* is true in a majority of physically similar worlds'. But problems enter when you try to quantify probabilities, so defined.) We have a variety of notions of possibility; in particular we distinguish between what is epistemically possible and what is physically or metaphysically possible. A proposition *p* is epistemically possible if its truth is not ruled out by our existing body of knowledge; but it may not be metaphysically possible, all the same (i.e. there may be no possible world in which it is true). It may not be *physically* possible either: it may be false in all worlds where the laws of physics hold.

Perhaps we should treat probability in the same way, and distinguish what is epistemically probable (i.e. what is probable relative to our existing body of evidence), from what is physically or metaphysically probable. There are then at least two kinds of probability: probability relative to evidence (the epistemic kind), and objective probability – which in turn may be either physical or metaphysical. The idea of objective probabilities has seemed puzzling to many people. What does it mean to talk of the objective chance that *p*? Surely, *p* is either true or not true? Those who defend objective probability tend to argue that probabilities can change over time. It was objectively probable in 1990 that the Labour Party would win the next General Election; it ceased to be probable after John Major became leader of the Conservative Party; and in 1992 the probability of a Labour win is zero: the falsehood of the proposition now lies for ever in the nature of things.

Whether or not we countenance objective probabilities, the question arises how probability is measured. In the first kind of case – the 'true' dice or coin – we have a simple and *a priori* standard: namely, the number of alternative outcomes. The probability of throwing a six with one dice is 1/6; the probability of a 'true' coin falling heads is 1/2; and so on. In the second kind of example also we have an intuitive measure. Here probability = long-run frequency. And we assume that, the longer the run, the more representative our sample.

In the third kind of case, however, measurement of probability is rather more problematic. We certainly compare epistemic probabilities, saying that it is more probable that the Conservatives will win the election than the Labour Party, for instance. But can we assign precise mathematical values to the respective probabilities? We can certainly assign a value to a particular person's degree of *confidence* in the truth of p, given q. For this can be measured by betting odds: what he is prepared to risk on the assumption that p is true. But a measure of subjective confidence is not a measure of probability. Whether epistemic or physical, probability is not simply what a particular person believes it to be. It really is *objectively* more rational to believe that the Conservatives will win. But by how much?

We can take steps towards answering that question, on the strength of certain intuitive ideas that find mathematical expression in the axioms of 'probability theory'. Suppose we rank probabilities on a continuous scale from 0 to 1. Then the probability of a certain (e.g. a necessary) truth will be 1, the probability of an impossible proposition 0; every other kind of proposition will fall somewhere between those poles.

Now surely, one or other of p and not-p must be true. It is intuitively obvious, therefore, that the probability of p is 1 minus the probability of not-p. Furthermore, it is intuitively obvious that probabilities can be summed. If p and q are logically independent, then the probability of p or q is the arithmetical sum of the probability of p and the probability of q, less the probability of their being true together. The probability of their *both* being true (the probability of p & q) will be expressed as a product rather than a sum, of individual probabilities – specifically, the probability of p multiplied by the probability of q given p. This will always be less than or equal to the probability of either being true alone. So here are three neat axioms of probability theory:

(1) $P(p) = 1 - P(\sim p)$.

(2) $P(p \vee q) = P(p) + P(q) - P(p \;\&\; q)$.

(3) $P(p \;\&\; q) = P(p) \times P(q \text{ given } p)$.

Those axioms are not independent: the first is simply a special case of the second. Any assignment of probabilities will have to satisfy the constraints implied by such axioms. They provide a crucial test of the *consistency* of someone's belief about probabilities.

This all looks straightforward, until we remember that epistemic probabilities are relative to evidence. There is no such thing as the probability of p, in this sense, but only the probability of p given e, the evidence. Even so, a set of intuitive axioms can be devised, in the same *a priori* fashion as the three that I have given.

One of the problems for a theory of epistemic probability is to account for the way in which judgements of probability *change* with changes of evidence. Suppose that I have a sack containing 100 billiard balls, 70 black and 30 white. What is the probability that any ball I take from that sack will be black? The natural answer is 0.7 – supposing that we have no other facts to take into account. But if the first ten draws from the sack produce black balls, what is the probability that the next one will be black? If we had no other information, we should say that the probability is by now pretty small: there must be a white ball due to come up. But most people would not draw that conclusion, since we already have more information. Given what has happened up to now, it is highly probable that the balls are not distributed randomly, and that the black balls are all on top. It is not *less* probable that the next ball will be black, but *more* probable.

The example illustrates the dynamic nature of judgements of probability, which endlessly feed on themselves as the evidence changes. One of the constraints on the calculus of relative probabilities, therefore, will be to show how judgements of probability respond to changes in the evidence. Two theorems capture this idea. They are both due to Bayes and are named after him. The first defines the change in probability of some hypothesis h, given new evidence e, in the following formula:

$$\frac{P(h/e)}{P(h)} = \frac{P(e/h)}{P(e)}$$

where $P(h/e)$ means the probability of h given evidence e.

The second compares the changed probability of two *rival* hypotheses, h and g, given new evidence e, in the following terms:

$$\frac{P(h/e)}{P(g/e)} = \frac{P(h)}{P(g)} \times \frac{P(e/h)}{P(e/g)}$$

In other words, the ratio of the probabilities of the two hypotheses, given the new evidence, is equal to the original ratio of their probabilities, multiplied by the ratio of the probability of the evidence given the first hypothesis, and the probability of the evidence given the second. If the new evidence is more probable on the assumption of h than on the assumption of g, then the relative probability of h increases.

Bayes's theorems are purely mathematical, as is the theorem of Bernouilli (sometimes called the Law of Large Numbers), which says that the longer the run, the more probable that it is representative. As mathematical results it is not possible to dispute these theorems. What is controversial, however, is the use that is made of them, especially when dealing with those difficult cases (the third kind of case in my classification), where probability has no basis in long-run frequency or the calculus of chance.

When the verificationists first turned their attention to the problem, it was exclusively the epistemic form of probability that interested them, since it was the relation between evidence and hypothesis that seemed to promise a 'verificationist' solution to the riddle of induction. (See, for example, Rudolf Carnap, *The Logical Foundations of Probability*.) The hope was to define a relation of 'probabilification', holding between propositions, and analagous to such logical relations as entailment and consistency, while admitting degrees. However, no one seems to have produced a plausible method for determining these degrees: for showing exactly to what extent a given piece of evidence 'probabilifies' a given hypothesis; the project has therefore been largely abandoned.

This does not mean that the epistemic concept of probability is resistant to rational analysis. People have degrees of confidence in propositions and, as the practice of betting shows, are prepared to assign precise values to certain hypotheses on the strength of the evidence available to them. Moreover, there are definite rational constraints upon this practice. Assignments of probabilities should obey the axioms of probability theory, which provide a test for their simultaneous consistency. But they should also change in response to changes in the evidence, and change in certain uniform ways. For example, if you assign a certain conditional probability to p given q, and then learn for certain that q obtains, your new value for the probability of p should be the same as your old value for the probability of p given q. This requirement that rational people should change their minds in response to new information, and change in certain recognised directions, has been the principal motive of probability theory in the Bayseian tradition.

However, it remains an open question whether the rational constraints are strong enough to bring reasonable people into agreement when faced with the same evidence. Maybe rational people, faced with new evidence, must adjust their beliefs in similar ways: but what constrains them to start with one belief rather than another? Some people seem to regard the 'Big Bang' theory of the universe as inherently unlikely, and therefore require far more evidence before they will assign to it the same probability as is assigned to it by those for whom it has an intuitive appeal. But who is to say which of the two is more rational?

8. Probability and Scientific Realism

Modern physics seems in fact to have come close to embracing a notion of objective chance: of probabilities *de re*. These probabilities seem to lie in the nature of things, and do not vary with the evidence. Such, at least, is implied by quantum mechanics, in its currently favoured interpretation.

Modern philosophy contrived to ignore quantum mechanics during the years of its discovery. Paradox is exactly what you should expect, the philosophers argued, if you probe so deeply into nature that your everyday concepts can no longer describe what you find. The fault does not lie in our concepts – which are perfectly well adapted to the use that we normally put them to. Nor does it lie in reality – which is bound to look peculiar when observed on the smallest scale. It lies merely in our desire to describe the microphysical world by means of images and ideas that belong to the world of 'medium-sized dry goods', in which we pass our days.

Through a famous example, Schrödinger showed that the paradoxical nature of quantum phenomena cannot be confined to the microphysical world, but infects the whole realm of nature. Suppose a cat is placed in a box, along with a tiny quantity of some radioactive substance. The quantum law indicates that there is a 50 per cent probability that the substance will emit an alpha-particle during the next hour. Things are so arranged that any such particle impinges on a Geiger counter which operates a hammer which breaks a flask of cyanide which kills the cat. On one interpretation (that favoured by Niels Bohr, and hence known as the 'Copenhagen Interpretation'), there is, in the absence of any observation, no determinate answer to the question whether an alpha-particle has been emitted during the hour in question. The event is, as it were, smeared over time, and becomes an 'occurrence' only with the act of measurement. The sole question is whether the total system – the decaying substance together with the apparatus for observing it – registers a particle emission at a given time. Without the apparatus of observation we can only say that the system is in a 'superposition' of two states – one with a decayed atom, and one without it. But that means, one with a dead cat and one without it. Do we therefore say that the cat is in a state of suspended animation, until someone does the favour of observing it? This is surely an intolerable suggestion; or at any rate, in Richard Healey's words, 'an unusual variant on the maxim that curiosity killed the cat' (*The Philosophy of Quantum Mechanics*, 1989).

Idealists sometimes get carried away by such thought-experiments, imagining that quantum mechanics shows physical reality to be dependent, in some way, on the mind that observes it. This is unwarranted. The crucial factor is not mind but measurement – the physical process whereby events are registered as information. Since all processes affect the things which affect them, it can easily be deduced that some measurements will be in principle impossible. For, in some cases, the process of gathering information will destroy the conditions in which measurement occurs. Hence Heisenberg's 'Uncertainty Principle', which holds that it is impossible to determine the velocity of a particle at the same time as measuring its momentum: any measurement which fixes the one quantity will destroy the conditions that enable us to fix the other. What does 'impossible' mean here? The obvious answer is '*physically* impossible'. At this level of theorising, however, physical impossibility is little different from metaphysical impossibility. The Uncertainty Principle tells us how the world is, in its most fundamental nature. The question is whether the resulting 'gaps' in our knowledge are also gaps in the physical world. The 'realist' is the one who answers 'no' to that question.

Realism takes heart from another famous thought-experiment – that of Einstein, Podolsky and Rosen – designed to restore our faith in the independent reality of the physical world. Suppose you were to split a particle into a pair, A and B, and suppose that it were a law of nature that the properties of A are systematically correlated with those of B. For example, if A has positive 'spin' (intrinsic angular momentum) along some axis, B has negative spin along that axis; and so on. The Uncertainty Principle tells us that the spin of a particle along one axis can be measured only by rendering spin along other axes indeterminate. But suppose we measure the spin of A along one axis, and that of B along another; surely we can then deduce the spin of A along the second axis too. This enables us to give empirical sense to something that Einstein believed: namely, that the properties of all entities are determinate, whether or not we can directly measure them. Each particle *has* a spin along each axis. Even if only one such spin can ever be measured by the study of that particle alone, the spin along some other axis persists as a 'hidden variable'.

When we study the Brownian motion of particles suspended in water, we observe a chaotic to-ing and fro-ing. We can formulate laws which govern this motion, without inquiring into its deeper causes: for example, laws which tell us how far particles of a given size and mass will move on average at a given temperature and viscosity.

Such laws will be statistical, like the laws of sociology. However, we can eliminate the statistical element, just as soon as we find the 'hidden variable' – in this case, the molecular motion that accounts for the observed behaviour of the particles. A theory of this motion makes available to us something finer than a statistical law: namely, an explanation, for each individual particle, of why it moves as it does. Similarly, Einstein thought, the statistical laws of quantum mechanics are not ultimate, but will be replaced at last, when we discover the hidden variables that explain the statistical regularities. There are no probabilities *in rem*; for 'God does not play dice'. Einstein's thought-experiment was designed to show how the requisite discovery could be made.

In 1964 the physicist J.S. Bell published an article that entirely transformed this debate. If there really are hidden variables, he argued, then we should be able to calculate *a priori* which distributions of a given group of properties are statistically possible, and which statistically impossible, for a sequence of particle-pairs projected from some given source. Bell derived a theorem, whose validity is independent of the truth of any actual physical theory, since it focuses instead on the rules of inference that govern all systems in which statistical regularities of the relevant kind are measured. The theorem says that there is a certain statistical inequality between the spins of a proton-pair, as measured along the three geometrical axes. This inequality holds good, whatever the factors affecting the particles at birth (when the split occurs) and during their subsequent journey, provided we make two assumptions: first, that spin along each axis is an independent variable, whether measured or 'hidden'; second that there is no 'action at a distance'. (This second assumption, known as 'locality' or 'Einstein locality', holds that there can be no passage of an influence at a speed faster than the speed of light. This is a deep physical truth – maybe even a truth of metaphysics, since its denial involves unthinkable paradoxes of time-travel.)

Consider, then, a proton-pair, A, B. Quantum theory predicts that the properties measurable in A will be correlated with those measurable in B, regardless of the distance between them. Because of this correlation, the measurements will violate Bell's inequality, which arises only on the assumption that the state of A is independent of measurements conducted on B. In which case, it seems, one or other of the two assumptions must go: 'realism' (that is, the determinacy of the hidden variable), or Einstein locality. (There is a third possibility, once espoused by Bernard d'Espagnat, which involves abandoning the basic rules of scientific method – a desperate expedient, which would leave us unsure that these rules apply to anything.)

In a remarkable experiment, the French physicist Alain Aspect has put the matter to the test. The margin of error is too great for certainty (and maybe there are philosophical grounds for thinking that it *must* be too great for certainty). Nevertheless, it seems that the predictions of quantum mechanics prevail over those of Einstein, Podolsky and Rosen: Bell's inequality is violated. The assumptions of realism and locality have been put to the test and found wanting.

What should a philosopher make of these bewildering discoveries? Should they lead him away from 'scientific realism', towards a more anthropocentric vision of the physical world? Or should he merely shrug his shoulders, and content himself with the belief that the physical world is real, but paradoxical? Here are two thoughts: first, notice that neither the 'realists' nor the 'anti-realists' in this debate reject the authority of experiment and observation. Einstein's commitment to the 'hidden variable' arises from his belief that we can actually (though indirectly) *measure* it. Even Einstein seems to concede that a quantity that is in principle *inaccessible* to observation or measurement has no part in physical reality. Einstein's realism is what Kant called 'empirical realism' – realism about the empirical world. It is not a 'transcendental realism', and arises from exactly the same view of science as the 'anti-realism' of his opponents: the view which holds that physical reality *is* what it is discovered to be, when all possible observations have been made.

Secondly, there is no contradiction involved in the quantum view of matter. If it is paradoxical, it is so only in the sense of defying our prejudices. The correlations predicted by quantum mechanics – which tells us that the observation of A will reflect the measurements performed on B, however far apart the two particles may be – do not establish that there is 'action at a distance'. Such correlations cannot be used to pass a signal faster than light: there is no such thing as a 'Bell telephone'. If there were such a thing, then reality would indeed be paradoxical. Van Fraassen has fruitfully compared the case with historical instances in which the demand for an explanation was not met but rejected:

> The Aristotelian question about the law of inertia – 'But what keeps a body moving if there are no forces impressed on it?' – was not answered but discarded by the seventeenth century. It appears to be very hard to treat such demands ruthlessly, for even Newtonians would still speak of a

vis inertiae, paying lip service to the old demand for an explanation. (*Quantum Mechanics*, Oxford 1992.)

In like manner, Van Fraassen suggests, we should recognise that the laws of quantum mechanics are the best we have by way of an explanation: to attempt to go behind them, to some 'underlying' structure, is simply to compound the mystery.

Maybe that is all that we can say: the theory is strange to us, but only because it breaks through the barriers of prejudice. It does nothing to show that the physical world is either mind-dependent, or fundamentally inaccessible to science. The world just *is* as a true theory says that it is. And Quantum Mechanics is a true theory.

9. Hume's Law Again

But this returns us to an old dispute – that between realists and nominalists concerning the nature of universals. To a nominalist it is intolerable to suppose that things are as they are by their nature, and regardless of our attitude. Or rather, say that if you will. But in doing so, you merely attribute to things the 'nature' created by your classifications. By way of reinforcing this position the nominalist Nelson Goodman devised a most ingenious application of Hume's law. This is 'Goodman's paradox'. (See his book *Fact, Fiction and Forecast*, and in particular the chapter entitled 'The new riddle of induction'.)

Consider the following predicate: 'Green if examined before 2000 AD; blue if examined thereafter'. Let us replace this cumbersome predicate by the term 'grue'. All the evidence that I have now for saying that something is green, is also evidence for saying that it is grue. So which is it? It cannot be both, since they are logically incompatible. Yet induction gives me equal evidence for both hypotheses: there is no reason to choose between them on inductive grounds. (And of course, we can invent infinitely many predicates like 'grue', all competing for instantiation in those things we describe as green.)

This is an application of Hume's law. Nothing that is true of the world now logically implies what the world will be like in the future. Just as the world could cease to exist in the year 2000 without contradicting anything that was true before that moment; so could it be instantly replaced by a world exactly like ours except that every instance of green was replaced by an instance of blue. But if that is possible, how do we know that our world is not a world of grue things, rather than green ones? How do we know that grass is not grue? Or that it *is* grue for that matter? (There is a spatial version of the paradox, corresponding to the spatial version of Hume's Law. Suppose all emeralds within the boundary of the world as so far observed are green; then we have equal evidence for saying that they are grue, where grue = green within the boundary, blue outside it.)

Some people respond as follows: 'grue' is not a real predicate but one artificially constructed. It is so defined as to change its application at an arbitrary point in time or space. Such a predicate could surely play no part in a scientific theory, which deals with the way the world is at *every* time and place.

Such a response is beside the point. Imagine someone who divided the world using predicates like 'grue' and 'bleen' (= blue before 2000, thereafter green). He thinks that I mean grue by my use of 'green'. But then suddenly, in the year 2000, I begin to apply the word 'green' to bleen things. 'What an extraordinary thing', he says. 'You mean by "green" grue until 2000, bleen thereafter! What kind of a predicate is that, with this arbitrary reference to a point in time built into it? What part could such a predicate play in a scientific theory, when the whole purpose of theory is to say how the world is at every time?'

The argument connects with another, due to Wittgenstein, concerning following a rule. In his book *Wittgenstein on Rules and Private Language*, Kripke uses such arguments in order to justify a radical scepticism about meaning. How do I know that I mean *green* and not *grue* when I use the word 'green'? Surely, you reply, I will know when the year 2000 comes, and I start applying the word 'green' to blue things. But that is not so. For *if* I apply the word 'green' to those blue things, I shall say that those things are the *same* colour as the things to which I used to apply the word, namely green (meaning grue).

This conundrum is extremely hard to solve. For Goodman it serves to establish that our classifications are *our* classifications, and that the attempt to ground them in some translinguistic reality is inherently impossible. Such distinctions that we make among our classifications concern their 'projectibility' into the future, rather than their ability to 'divide nature at the joints'.

10. The Scientific World-view

While nobody has a final answer to Goodman's paradox, few people accept the radical conclusions that Goodman himself derives from it.

Let us, then, take refuge with the crowd, and review the scientific picture of the world, as philosophers tend to present it.

The world divides into natural kinds: stuffs on the one hand, including elements and compounds, and kinds of individual on the other, including animals, and sub-atomic particles. These kinds have real essences, from which all their properties flow. It is the task of science to explore those real essences, and to derive the laws of transformation that govern them. Whatever features are mentioned in the scientific inventory of the world are primary qualities in Locke's sense: qualities that things possess in themselves, and which explain how they appear to us. Scientific theories *deduce* the appearance from the reality; but they do so only by describing the reality in terms that have little or no relation to appearance. Secondary qualities, and non-natural kinds, play no part in the description of the really real. The fact that we perceive and classify things as we do may be explained by science. But the explanation will undermine the thing explained.

But now it looks as though science has usurped the place of Descartes's God. It seems to be aspiring precisely to that 'view from everywhere' which shows the world as it is, undistorted by the quirks of human perception, organised according to its intrinsic nature, rather than by categories that express our limited perspective. Science implies that we are merely part of nature. Our perspective on the world is not sovereign, but a by-product of the evolutionary process which created us. Its authority stands always to be usurped by the imperial ambitions of scientific theory. In which case what and where *are* we? This is the question to which I now turn.

16

The Soul

The word 'soul' is no longer much in use, partly because of its religious connotations. Instead philosophers speak of the mind, of mental states, of consciousness, or of the self. None of those idioms is entirely satisfactory, so I shall use the term 'soul', in its traditional sense, to denote everything that is comprised in the inner life. We think, reason, and form beliefs: in other words we have minds. We feel emotion, sensation and desire: in other words we have passions. We act, intend and decide: in other words we have will. And we know our present mental states and decisions with a peculiar certainty and immediacy: in other words we are self-conscious. The term 'soul' refers to all those things: to mind, passion, will and self. And two central problems about the soul will concern us in this chapter: the so-called mind-body problem, concerning the place of mind in nature; and the problem of the self, concerning the nature and extent of self-knowledge.

When people say that animals have minds, but not souls, it is usually because they acknowledge the existence of animal beliefs, desires and sensations, but deny the existence of a 'self'. Religion goes further, according to the human soul a capacity to survive bodily death which is not possessed by the minds of animals. Yet are we so very different from the animals? Aristotle used the one word – *psuche* – to denote the vital principle that is possessed by every living being. Reason and self-consciousness belong to *nous*, which is the immortal part of *psuche*. But a horse, a dog, even a cabbage, has a *psuche*; if my soul, like theirs, is the principle of life in me, then death should be the end of it, and the end of all its parts and properties, self included.

Those questions are deep and troubling. The favoured modern response to them is to argue that we differ from the lower animals only in the complexity of our mental life. There are capacities and states that we have but they do not. But these capacities and states are not metaphysically distinct from theirs. Like theirs, our minds are part of nature, causally connected not only to our bodies but to the entire physical world.

Thence arises the theory which, in one form or another, has become first contender for the truth about the human soul: the theory of 'physicalism'. A hundred years ago philosophers speculated about the relation between 'mind' and 'matter', thinking of the first as some kind of airy nowhere-stuff, the second as solid, lumpy and dull. Nowadays philosophers are far less certain that they know what is meant by 'matter'. This term no longer appears in the favoured theories of physics, and is a survivor from Aristotelian science, which advocates a distinction, and also an indissoluble conceptual connection, between matter and form. It was indeed Aristotle who described the body as the matter of the human being, and the soul as his form.

The modern philosopher, having discarded 'matter' as the stuff of the universe, is hard pressed to find a substitute. Perhaps energy would be more in keeping with the spirit of modern physics. But rather than tie himself to a science that might be as soon surpassed as the one he has discarded, the modern philosopher prefers to leave the question open. It is for physics to discover what there is; but it is not for the philosopher to set *a priori* limits to discovery. In place of matter, therefore, he speaks of 'physical reality', meaning the sum total of the ontological commitments that a true physics would make. A physicalist is someone who thinks that there is nothing apart from physical reality, and that the soul therefore either does not exist, or is part of physical reality.

Of course, we still do not know what *physics* is. As it progresses, physics comes to seem more and more peculiar, and 'physical reality' less and less distinct from the observations through which we discover it. But let us be content, for the moment, with this: physics is the true theory of things in space and time, and of the causal relations between them. Physicalism is the theory that the soul is one of those things.

1. Descartes Revisited

Once again it is Descartes who set the agenda for modern philosophy, arguing for a 'real distinction' between mind and body. Precisely what his argument was for this 'real distinction' is a matter of dispute.

The key premise depends upon a technical idiom, introduced by Descartes in order to encapsulate his theory of *a priori* knowledge: 'clear and distinct' conception. That which we know *a priori* we know as necessarily true, and as part of the essence of the thing known. This is because clear and distinct ideas (ideas unmixed with others or with sense-perception and grasped by the intellect in their completeness) contain the intrinsic marks of truth, and show the world from the point of view of reason (the point of view which is also God's). In the sixth Meditation, Descartes argues as follows:

> I know that everything which I clearly and distinctly understand is capable of being created by God so as to correspond exactly with my understanding of it. Hence the fact that I can clearly and distinctly understand one thing apart from another is enough to make me certain that the two things are distinct, since they are capable of being separated, at least by God.

He adds that I have a clear and distinct idea of myself as a thinking thing: I clearly and distinctly conceive that thinking belongs to my essence. I also clearly and distinctly conceive that nothing *else* belongs to my essence: in particular, no fact about my body or any other extended thing. Likewise, I have a clear and distinct idea of body in general, and know that nothing belongs to the essence of body besides extension. In particular, thought does not belong to the essence of body. I therefore clearly and distinctly perceive that the mind is essentially distinct from the body and therefore in principle separable from it.

The details of the argument are set out in the sixth Meditation and the sixth set of Descartes's replies to objections, and many are the fallacies that have been discerned, rightly or wrongly, in this immensely subtle piece of reasoning. The real difficulty for Descartes resides, however, less in the argument for his conclusion, than in the conclusion itself. For if extension is the essence of physical reality, and thoughts do not possess it, thoughts cannot be part of the physical world. The mental and the physical belong to separate ontological realms. In which case how do they interact?

Descartes had no doubt that they *do* interact, and indeed that they exist in a thoroughgoing interconnectedness, so that 'I am not lodged in my body as a pilot in a ship'. But he could find no satisfactory explanation of this fact, and eventually fudged the issue with a notorious hypothesis concerning the mediating function of the pineal gland.

Indeed, this is the point where a physicalist might attempt to stand Descartes's argument on its head (or rather feet, to use Marx's jibe against Hegel). We know that there *is* causal interaction between mind and the physical world. Thus my thought of Jim's remark causes my anger which causes me to strike him, which causes him to fall, breaking the coffee-cup. The shards of porcelain cause his wound, which causes his pain, which causes his resentment, which in turn causes his decision to sever our relations. In that perfectly normal sequence of events, the chain of causality runs from mind to the physical world and back again, with no sign of an interruption. But suppose we could prove that only physical things can be causally related. Would we not then have a proof that the mind *is* a physical thing?

2. The First- and the Third-person Perspective

Let us wait a while for that proof, and consider instead another aspect of Descartes's argument. His view of the mind derives from a study of the first-person case. The mind is what *I* discover in myself, by looking inwards, when I cease to meditate on the 'external world'. This is the Cartesian theory that we have already explored in Chapters 4 and 5. But there is also the view of the mind, the third-person view, which sees it from 'outside', as we see the minds of others. (There is yet another possibility, of great importance in ethics: the second-person viewpoint, explored by Martin Buber in his poignant essay *I and Thou*. But let us put this to one side.) The third-person viewpoint is necessary to us, and is intimately bound up with our attempts to understand and explain one another's behaviour. We attribute beliefs and desires to animals, since their behaviour seems to be caused in something like the way that ours is caused: by information gathered from the surrounding world, and by the desire to change the world in their own favour.

Indeed, if it were not for the first-person perspective, the Cartesian theory of mind would be entirely without appeal. It is this perspective that nourishes the idea that what I *really* am, I am for myself alone. Surely, nothing like that is true of a dog or horse. Their sketchy souls exist because of *our* perspective: it is we who train them to shuffle up to the boundary of the human world, and stare at us from that place of half-exclusion.

So what is it about the first-person perspective that offers such an obstacle to physicalism? There seems to be little agreement. The following thoughts, however, are fairly common:

(1) I know my own mental states immediately and incorrigibly. Yours I know only through the study of your words and behaviour. This suggests that there is some *fact* about my mental state, which is revealed only to me. Moreover, it is an important fact, perhaps the *most* important fact, since it provides sufficient proof of what I am thinking or feeling.

(2) If we enumerate the physical truths about someone, including the truths about his brain, nervous system and behaviour, there is still something left out: namely 'what it is like' to be him. Moreover, if we say everything that can be said about a mental state from the third-person perspective, still there is a further truth about it, namely, 'what it is like' to be in that state. This argument, much abridged from Tom Nagel (see 'What is it like to be a bat?' in *Mortal Questions*) may even be applied to animals. There is something that it is like to be a *bat*, and something that it is like to experience its pleasures and pains. This suggests the existence of 'purely subjective' facts about the mental. Such facts could play no role in a physical theory, since they are inseparable from the very point of view that physical theory endeavours to transcend: the point of view of the subject (which, in the case of language-users, is that of the first person).

(3) Mental states possess irreducible phenomenological qualities, 'raw feels' or *qualia*, and this is something that we observe in our own case by introspection, but which others are forever barred from perceiving.

Sometimes the first and third ideas are run together; sometimes the crucial fact that is supposedly overlooked by the physicalist is called 'consciousness', the 'inner' aspect, or the 'subjective' reality of the mental. In a striking variant of Descartes's argument for the 'real distinction', Kripke (*Naming and Necessity*) argues that no physical process could be identical with a pain, since it is a necessary feature of pain that it is felt painfully, and no physical process could have such an essential property.

What do we make of all those arguments? That is the subject for a whole book, an outline for which is given in the Study Guide to this chapter. Meanwhile here are two thoughts: it does not follow from the fact that I know my own mental states immediately, and yours only through the study of behaviour etc., that there is, in my case, something else that I know. That would be true, only if there were another basis for my first-person knowledge, another 'way of finding out'. But I don't *find out* at all: first-person knowledge has no basis.

Suppose that there is a 'purely subjective' aspect to our mental states. Imagine a society of beings exactly like us, except that, in their case, there *is* no subjective aspect. Their language functions as ours does, and of course there is nothing that we can observe in their physical make-up or behaviour that distinguishes them from us. They even speak as we do, and say such things as 'you don't know what it's like, to have a pain like this'. Their philosophers wrestle with the mind-body problem, and most of them are even Cartesians. Is that an incoherent suggestion? If not, then maybe that is precisely the case we are in.

3. Intentionality

Before examining physicalism in detail, we must become acquainted with the second major objection to it. So far, I have spoken of soul, mind and mental state, and assumed an intuitive understanding of those ideas. In fact, however, do we really know what we mean by these terms? What makes a state into a *mental* state?

There are roughly three answers to that question:

(1) A mental state is one on which there is a 'first-person point of view', or which has a 'subjective' aspect. This I have already explored.

(2) A mental state is one with a certain role in the understanding and explanation of behaviour. To this I shall shortly return.

(3) A mental state is a state with 'intentionality'. This is the suggestion that I shall now consider, since it has been associated with a peculiar and interesting objection to physicalism.

Modern discussions of intentionality (note the spelling: with a *t* not an *s*), begin with Brentano, a late-nineteenth-century Austrian philosopher, who taught Husserl, and who some consider to be the true founder of phenomenology. In his *Psychology from an Empirical Standpoint*, Brentano raises precisely the question that philosophers generally ignore, namely, what makes a state or condition (he says 'phenomenon') *mental*? He proposes a criterion of the mental, a property that all and only mental states have:

> every mental phenomenon is characterised by what the scholastics of the Middle Ages called the intentional (and also mental) inexistence of an object, and what we could call . . . the reference to a content, or direction upon an object (by which we are not to understand a reality in this case), or an immanent objectivity.

Brentano's language is obscure and hesitant. But here, roughly, is what he had in mind. Whenever John is afraid, he is afraid of something; whenever he thinks, he thinks about something; whenever he believes, he believes propositions about something; whenever he is angry he is angry about something; whenever he has an experience, it is an experience of something. And in all those cases the 'something' – which we might call the object of the mental state – may exist only *in* his thinking, and have no independent reality. (That is what Brentano means by 'in-existence'.) This feature, called intentionality, from Latin *intendere*, to aim, was mentioned in Chapter 4. Brentano believes that it is a feature of *all* and *only* mental 'phenomena'. Every mental state is directed towards an inexistent object. This means that, when referring to mental states, we are obliged to employ special idioms. If it is true that John is afraid, then John is afraid of something – a mouse, say. Take the sentence 'John is afraid of a mouse'. This looks as though it describes a relation, as in 'John trod on a mouse'. But from 'John trod on a mouse' you can infer 'There is a mouse, upon which John trod'. In a real relation, both terms exist. But from 'John is afraid of a mouse' you cannot infer 'There is a mouse, of which John is afraid'. His fear would be just as real, even if there *were* no mouse.

There is a long story to be told here, and I shall return to it in Chapter 18. But to cut the story short, it has seemed to many philosophers, including Brentano himself, that this peculiar grammar involved in reference to the mental is precisely what distinguishes the mental from the physical. All physical relations are real relations; no physical state of affairs can have this special kind of relation to an 'inexistent' object: for in what, so to speak, would the object exist? It needs a mind to harbour it; and that is precisely what no physical reality can contain.

This has struck many philosophers (notably Roderick Chisholm) as an obstacle to physicalism. For if we are to capture mental reality in physical terms, then we must describe *physical* things by means of intentional language. And this we cannot do. There is just no sense in the idea of a physical 'relation', one term of which does not exist. Such quasi-relations make sense only because there is some act of 'reference' or 'aiming' which may, so to speak, go astray. And this is what distinguishes the mental realm from the physical.

There are several reasons for discontent with Brentano's position: (1) It is not clear that *all* mental phenomena exhibit intentionality.

Cannot there be mental states which are not 'of' or 'about' anything at all? (Think of sensations, nameless fears.)
(2) It is not clear what the property of 'intentionality' really *is*. If it is supposed to be a grammatical property, then perhaps it is simply intensionality (with an *s*), a property that belongs equally to certain modal contexts. And if it *is* a grammatical property, what does it tell us about the thing that it describes?
(3) It is question-begging to suppose that physical processes cannot exhibit intentionality. If mental states *are* physical, then some physical states *do* have intentionality. The only problem is 'how?'. In answer to this certain philosophers, notably Dennett, have developed theories of 'intentional systems' which will enable us to envisage exactly how a machine (a computer, say) could exhibit intentionality.

For those and a variety of other reasons, modern philosophers tend not to regard intentionality as an insuperable obstacle to physicalism. Indeed, Davidson regards it as an argument in *favour* of physicalism (or something like physicalism) for reasons that I touch on below. The study of intentionality has nevertheless become central to the philosophy of mind, since it challenges those who believe, for whatever reason, that the mind is a kind of computer.

4. A Note on the Unconscious

This is perhaps an appropriate place to return to the topic that was mentioned at the close of Chapter 4. Can there be unconscious mental states? And if so, do conscious and unconscious states belong to the same thing? Such questions are both difficult and – following Freud's highly imaginative use of the concept of the unconscious in psychotherapy – of considerable urgency. There are those who dismiss the idea of unconscious mentality as a metaphor, or even as nonsense. Others, however, see no difficulty in the concept, on the grounds that the mental can be defined without reference to consciousness, by means of the feature that I have just described – the feature of intentionality. In effect, we have at least two ideas in view, when we refer to states as mental: first consciousness, in the sense of the 'first-person perspective', and secondly intentionality – the 'direction on to an object' that we observe in emotion, belief and desire. It is interesting, indeed, that the examples usually given of unconscious mental states are always characterised in terms of their intentionality: unconscious desires, thoughts, and emotions – whose unconscious character is frequently explained in terms of the pain involved in confessing to them. There seems to be something inherently paradoxical in the suggestion that a mental state that has no intentionality (toothache, for example) could equally exist in an unconscious form.

But there is a difficulty here. To what are the unconscious states to be ascribed? Philosophers in the Cartesian tradition rely on consciousness not merely as the defining feature of the mental, but also as the procedure whereby we attribute mental states to a single subject. The entity which *has* the mental states is the very same entity that is conscious of them as its own. The mind is a 'centre of consciousness', and to possess a mental state is *ipso facto* to be conscious of it. The very least that can be said is that, if there are unconscious mental states, then mental states must be states of something other than the Cartesian subject. Of what then?

One suggestion is the organism. Mental states are really states of an active animal and part of its sensitive response to the environment. We define them as mental on account of their specific role in the formation of behaviour – the role which is exemplified by intentionality. It so happens that some of these states can also come before the consciousness of the organism. Others remain hidden from it.

But that suggestion, when examined, turns out to be highly implausible. For it comes to seem like a total accident if mental states are conscious, and consciousness itself appears to ride on the back of the animal like an irresponsible jockey, who might at any moment fall off, and who has no special role to play, other than reporting on states which he observes with no true authority. The peculiar intimacy that exists between conscious states and the consciousness that shines through them seems to vanish.

Furthermore, those who speak of the 'unconscious' often refer to an unconscious mind – as though unconscious mental states had some *other* bearer. The mind acquires its unconscious *Doppelgänger*, which inhabits the organism in the *same* way (whatever that is) that it is inhabited by the conscious mind. But that too seems unsatisfactory: for how do these two minds 'communicate'?

One solution to these difficulties is to admit the existence of the unconscious, but as a *special case* of the conscious mind. Unconscious mental states exist when (a) a person behaves *as though* he thought *x* or desired *y*, and (b) just such a thought or desire can be brought into consciousness (say by psychoanalysis). If (a) is satisfied without (b), we may be reluctant to speak of a *mental* state, and refer instead to some kind of pathological condition of the organism – as in hysteria, hypnotism, and the like. Such a view still gives precedence to consciousness, as the defining criterion of the mental.

Much of the difficulty here is caused by the failure to distinguish accurately between consciousness and self-consciousness. Animals have mental states, but presumably do not attribute them to themselves. The horse feels pain, but does not entertain the thought: *I am in pain*. Do we want to say that the mental states of animals are unconscious? Surely not. When referring to an unconscious mental state we are referring to a special condition of a *self*-conscious creature: one who can attribute thoughts and feelings to himself, and normally does so with first-person authority, but who, for some reason, is not attributing to himself *this* thought or feeling, which he clearly has. How is this possible? We could answer that question only if we had a theory of self-consciousness: that is, a theory of a highly specific mental capacity, which a creature might lack, while still possessing mental states. (See the discussion in Chapter 31.)

However unsatisfactory, the topic of the unconscious serves to remind us of what is perhaps the crucial difficulty in the philosophy of mind. Mental states are properties of something: but of *what*? Of consciousness? (And what is that?) Of a 'mind'? Of the body? Of the brain? Of the whole organism? Failure to answer those questions vitiates many of the current contenders for the truth about the mental, including physicalism in its more common forms.

5. Varieties of Physicalism

Physicalisms are many – almost as many as physicalists. But the following broad divisions encompass most of what is now propounded by those who think that the mind is part of the physical world:

(i) Behaviourism

Two theories go by this name: one, popular among hard-headed psychologists earlier in this century, holds that, since mental states are irreducibly 'inner' and subjective, they are not suitable subjects for scientific investigation. If there is to be a science of psychology, therefore, it must be a science of behaviour. The other, more philosophical, says that the mind just *is* behaviour (taken in conjunction with the bodily preconditions of behaviour). The severest form of behaviourism is reductionist, and argues that mental states are logical constructions out of behaviour, in something like the way that the average man is a logical construction out of men. Like all such reductionist theories it finds itself unable to show how statements about the mind might be replaced by statements about behaviour.

In a famous book, *The Concept of Mind*, Gilbert Ryle defended a kind of sophisticated behaviourism in the following way. The Cartesian, he suggested, has been misled by the superficial grammar of our mental language, into thinking that there is some entity – the mind or soul – to which it refers. This 'ghost in the machine', which has haunted philosophical discourse down the centuries, is in fact a grammatical illusion (the result of a 'category mistake'). Mental terms like 'thought', 'feeling', do not refer to objects; they are like adverbs, describing complex ways of behaving. To say that John dug the garden thoughtfully, is not to say that his digging the garden was accompanied by some peculiar inner event; rather, it is to describe the way in which he dug. It is like saying that he dug the garden slowly. Once we understand the 'deep grammar' of reference to the mental, we shall escape from the Cartesian illusion, and see that we are not two things – a machine, and a ghost who drives it – but one thing, vastly more complex and interesting than either.

(ii) The 'Identity' Theory

For a long time philosophers thought that it was quite absurd to say that mental states are really physical, since the sense of our mental predicates is so radically different from the sense of any physical description. How could we assert identity between pain and a process in the brain, when we learned and applied the word 'pain' in a way that we could never have learned and applied the term 'brain process'? At a certain point, led by the Australian philosopher J.J.C. Smart, there was a rebellion against this line of argument. Surely, Smart protested, the terms 'pain' and 'c-fibre stimulation' (for example) can differ in sense, and yet have the same *reference*? For a while this protest was bound up with a defence of the view that the identity in question is a 'contingent' one: a view that, since Kripke, philosophers are reluctant to endorse. All that Smart meant, however, was that the identity between pain and c-fibre stimulation is *a posteriori*.

A vast literature has grown from discussions of the identity theory. Perhaps the two most cogent objections to it (apart from objections to physicalism in general) are these:

(a) Pain is a state; it exists only where there is a whole *mind* in which it inheres. You cannot assert identities between the instances of proper-

ties, unless you are in a position to assert identity between the entities that possess them. (Imagine the situation where a c-fibre, which has been extracted from the nervous system of a human being, is being stimulated in a laboratory. Do you say that there is pain going on in that test-tube? In which case *whose* pain? Mrs Gradgrind's?) All the hard work, of identifying the *kind of thing* that can be in pain, and identifying the physical object with which it is supposed to be identical, has yet to be done. Until that work has been done, the assertion of identity between mental states and physical states is a sham: the best you can prove is that the two are causally connected.
(b) Suppose I come across a creature who looks and behaves like me, who responds to my overtures of affection and takes up the kind of role in my life that is characteristic of my friends. Tormented by doubts, or in the temporary grip of mad scientist disease, I open up his head, to observe the seat of his mental powers. To my astonishment I find nothing – or perhaps only a dead kitten and a ball of string. I replace the ceiling of his skull and lapse into appalled meditation. My friend reacts to my ignoring him. Do I now say that he is not *reproaching* me; that he is not *hurt* by my behaviour, that he is utterly mistaken when he says, with every appearance of understanding the language, 'I am sad that you have changed'? All this is well set out by Mary Shelley, though the moral of her story is seldom drawn. Frankenstein's monster tells us that the pattern of behaviour, and the link between environment and response, are what matter to us, not the mechanism that creates them. The mechanism might be quite mysterious, and in any case different from person to person.

(iii) Functionalism

This leads to another kind of physicalism. The relation between Baron Frankenstein and his monster is like the relation between two quite different bits of hardware, each identically programmed. The soul is the software, in terms of which the connections between input and output are established. The brain is the hardware on which this programme has been impressed. The theory of functionalism holds that what makes a mental state into the particular state that it is (pain, anger, the belief that *p*) is its role in linking environmental input to behavioural output. Do not be deterred by the ugly jargon. The functionalist is saying: stop looking for the mental in the depths, and attend to the surface. You see the web of connections between an organism's environment and its behaviour; you observe its sensitivities and predilections. And, like all your observations, these are animated by a theory of what you see, a theory which tells you that this goes with that, this results from that. Functional connections are just what we notice, when we relate to the organisms that surround us. And it is not surprising, therefore, if we have developed a language with which to describe those connections – the language of the mental.

Functionalism allows for the possibility that the mind of one creature might be instantiated in processes that do not occur in another, even though the other also has a mind. The silicon creatures of science fiction might be as susceptible to pain, thought and grief as we are.

(iv) Others

There are other possible physicalisms, and I cannot hope to list them all. One theory, however, deserves special mention, if only because of its surprising treatment of the objection from intentionality. This is the theory proposed by Davidson in his paper 'Mental Events'. Like all of Davidson's papers, this forms part of a system whose apparent lacunae are filled elsewhere – a seamy web of commitments none of which is more plausible than the whole. In particular, Davidson starts from the Humean premise that, if it is ever true that *a* causes *b*, this is because there is a universal and exceptionless law, to the effect that *a*-type events cause *b*-type events. Since it is true that mental events cause and are caused by physical events, there must be universal laws connecting the two kinds of event. However, mental concepts are fatally debarred from featuring in such laws, on account of the intentional idioms by which they are identified. As Brentano argued (see section 3 above), there is no reference to the mental that does not employ 'intentional' idioms, and these have no place in a physical science. In describing events by means of mental concepts, therefore, we use descriptions that can be employed only in singular causal statements, and never in causal laws. Since there must be such laws, however, there must be other descriptions of the mental events (of those very *same* events) which relate them in a law-like way to their causes and effects. Mental events are mental only *as described*. But we know that there is another description, which identifies their place in causal laws, and therefore in the scheme of things. This is the description for which the physicalist is searching.

Davidson calls his position 'anomalous monism'. 'Anomalous' means 'unlaw-like'; 'monism' means that there are not two kinds of thing in the world, mental and physical, but one kind of thing, which can be conceptualised in two different ways. (There is an antecedent for this way of thinking in Spinoza.)

6. Emergent Properties and Supervenience

When a painter applies paint to a canvas, he creates a physical object, by purely physical means. This object is composed of areas and lines of paint, arranged on a two-dimensional surface. When we look at the painting, we see those areas and lines of paint, and also the surface which contains them. But that is not all we see. We also see a face, that looks out at us with smiling eyes. Is the face a property of the canvas, over and above the blobs of paint in which we see it?

When you play a melody on the piano, you strike the keys successively, and each produces a sound. This is a purely physical process, resulting in a succession of tones, one after the other. You hear that succession; but it is not all that you hear. You also hear movement: the melody begins on C, moves upwards to A, and down again to the G below. Is the movement a property of the sequence of sounds, over and above the tones in which we hear it?

It is hard to give a straight answer to those questions. In one sense the face is a property of the canvas, over and above the blobs of paint; for you can observe the latter, without observing the former, and vice versa. And in some sense the face really is there: someone who does not see it is suffering from a deficiency, whether visual or intellectual. On the other hand, there is a sense in which the face is not an additional property of the canvas. For as soon as the blobs of paint are there, so is the face. Nothing more needs to be added, in order to generate the face – and if nothing more needs to be added, the face is surely nothing more. Moreover, every process which produces just *these* blobs of paint, arranged in just *this* way, will produce just this face – regardless of whether the artist himself is aware of the fact. (There could be a machine for producing Mona Lisas, whose instructions mentioned only the distribution of colours on a two-dimensional graph.)

One response to these cases is to say that the face is an *emergent* property of the canvas. It emerges from the disposition of coloured patches, in the sense that any object with this disposition of coloured patches will display this very same face. But it is nothing 'over and above' the patches, since nothing else needs to be added to the canvas in order for the face to appear. Such emergent properties are also 'supervenient' on the properties from which they emerge. That is to say, any change in the face necessitates some change in the underlying physical arrangement. And the same arrangement will generate the same face. There would be something absurd in saying that canvas A is physically indistinguishable from canvas B, even though a face can be seen in A but not in B; there must be *some* physical difference between them, even if we may have to look hard to find what it is.

The example is controversial, as we shall see in Chapter 24. But it helps us to understand certain broadly sketched positions in the philosophy of mind, which argue that mental properties are 'emergent' properties of physical systems. Maybe functionalism can be dressed up in this way: certainly Davidson's anomalous monism implies that mental characteristics are emergent. The idea is this: we build up a physical system from scratch, tying together the various circuits and transistors, and linking them to the movable limbs and the sensory detectors. For simple systems, we feel no temptation to describe them in mental terms, for we are comparing them always with things more simple than themselves. But there comes a point where the natural comparison is with things more complex: with animals say, or even human beings. We could reach this point simply by adding to the 'repertoire' of the physical system. Only physical parts and properties need change as we do so. But suddenly the mental 'emerges' from the physical, just as the face emerges from those final brush-strokes. The mental is something over and above the physical, in that it requires wholly new concepts to describe and relate to it (and maybe to *perceive* it, too). But it is nothing over and above the physical, in that when the physical base is in place, the mental also is there, and without further additions.

This position will become more intelligible after Chapter 17. But I hope that the reader will see how appealing it is.

7. The Self

You can go on refining the physicalist position, so as to accommodate all that Mary Shelley tried to teach us. As the position gains in sophistication, however, it begins to awaken us to an enormous gulf in the world of organisms: the gulf between us and the rest. We have capacities that we do not attribute to animals, and which utterly transform all the ways in which we superficially resemble them. Two in particular deserve a commentary: rationality and self-consciousness.

(i) *Rationality*. We attribute beliefs and desires to animals, and already there is a vast controversy as to whether this is not a kind of metaphor. For animals have no language for their beliefs, no way of specifying to the satisfaction of an observer that it is *this* they believe, not *that*. Why do I say that a terrier believes there is a rat in that hole? How, short of his uttering some proposition, can I distinguish his believing *that* from his believing there is one of those grey things in the hole, or one of those victims that he is licensed to kill, or one of those smells that he can sink his teeth into and taste blood? The choice seems *arbitrary*. Yet surely dogs, like people, gather information from their environment, and respond to it appropriately. It seems that we need to distinguish this process of information gathering, from the more sophisticated process that goes by the name of rational belief.

We do not merely have beliefs; we express them, draw conclusions from them, reason in their favour. Our concepts are as finely divided as the words which express them; and our thought processes are emancipated from the present moment in ways that have no parallel in the animal kingdom. (Thus we can speculate about possibilities and impossibilities, about purely imaginary episodes, about theories that defy appearances, and about appearances at other places and other times.) It is no exaggeration to say that we inhabit another *world* from the world of animals, and are in a certain measure freed from their servitude to present desire. In the next two chapters I shall say something more about this. In any case, such thoughts compose a large part of what has been meant by soul. It is through this constant aspiring beyond the moment, that we attain access to eternal truths and immutable realities. How could we do that, if we were merely devices for processing information?

Some philosophers (notably Searle and J.R. Lucas) have made much of this, since it seems to point to something deeply implausible in the computer-inspired models of the human mind. A computer can certainly respond to information inputs with information outputs; it could even be wired up to a robot that uses that information as the basis for a plan of action. But could a computer know that the information is *true*? Could it ever attain the peculiar perspective of the rational being, who is able to compare his beliefs with reality, and to cast judgement on his own predicament? Our use of concepts like truth seems precisely to separate us from the world of our experience, in a way that casts doubt on the thesis that we are subservient parts of the natural order.

(ii) *Self-consciousness*. The so-called mind-body problem might never have arisen, had it not been for *self*-consciousness. A description of the world of nature could be extended to include the animals, without automatically creating a metaphysical problem. Of course, the presence in nature of creatures that gain and respond to information about their environment is a great challenge to science. But it is not, as such, a challenge to philosophy. The problems begin, however, when we acknowledge the presence in nature of a first-person point of view, which may be that of God (the 'I am that I am' that spoke from the fire), or merely that of the philosopher (the '*cogito ergo sum*' that speaks from the fireside). Is it not here that the mystery lies? In Thomas Nagel's words, 'where in the world am I?'

Nagel emphasises the problem in the following way (*The View from Nowhere*). Suppose I possess a complete description of the physical world, according to the true scientific theory of all that is contained in it. In this description are identified not only the animals, but also the people which the world contains. One of these people is called Roger Scruton, and has all the attributes that I have. Still, there is a fact that is not mentioned in the description, namely, the fact that this person is *me*. Similarly, there is the fact that this pain is my pain, this joy my joy. My ability to situate myself in my world is of a piece with my first-person view of things; and what is revealed to me in that first-person view does not feature in the scientific inventory of the really real.

We have emerged on to one of those vertiginous ledges in philosophical speculation from which the hysterical take wing to their destruction. Modern philosophers tend to respond by turning in another direction. Look, they say, you may have a point, that the scientific inventory does not tell me which of these people is me. But nor does it tell me which moment is now, or which place here. Words like 'I', 'now' and 'here' (known usually as 'indexicals') serve to locate the speaker in his world. They do not record some additional facts about that world, but merely attach our sentences to it, by identifying the place at which they are spoken.

Is that an adequate response? To that difficult question I shall return in Chapter 31. For the moment it is best to move on.

8. The Unity of the Soul

Those worries were familiar to Kant, who argued roughly as follows: our self-consciousness presents us with a peculiar idea of unity. I know immediately that this pain, and this thought, belong to one thing, namely me. What I am, I do not know, except that I am one thing, a bearer of mental predicates. Kant called this knowledge of unity the 'transcendental unity of apperception' ('apperception' being Leibniz's word for self-consciousness). The unity is 'transcendental' in the sense that I can never derive it from some more basic premise. I cannot be in the position of someone who says, this pain has such and such a property, this thought such and such a property, therefore they belong to one and the same thing. The very fact that I can identify the pain and the thought presupposes that I can identify them as mine: as states of one thing.

Many philosophers are tempted to draw the conclusion that the soul therefore has a peculiar kind of unity and indivisibility. There is no such thing as half a soul. And this fuels the rationalist belief that the soul is a very special kind of entity, a substance that is strictly apart from the physical world. For the contents of the physical world are rendered divisible by their spatiality. But, Kant argued, the inference from the transcendental unity to a substantial unity involves a fallacy. All that the first unity tells me is that, by virtue of my ability to describe my own mental states, I can identify each of them as mine. It does not follow from this that I am a substance (rather than, say a property of something else – of my brain say). Indeed nothing follows at all, concerning what or where or how I am: only *that* I am.

Nevertheless, there is another reason, Kant believed, for doubting that I am straightforwardly part of the physical world, and this reason also derives from the thing that sets us apart: the fact that we are rational beings. The rational being is not merely self-conscious, he is also free. Indeed these two ideas are, in the end, one idea, described from two different perspectives: the theoretical and the practical. It is by exploring the notion of freedom, therefore, that we discover what is meant by those who say that the soul is not part of nature.

17

Freedom

There are two sources of the metaphysical conundrum of human existence. The first is consciousness; the second is freedom. We make choices, and carry them out; we praise and blame one another for our decisions; we deliberate about the future, and make up our minds. In all these commonplace events, we assume that we are free to do more than one thing; what we *actually* do is our choice, and our responsibility. Is this assumption ever justified? If not, what remains of morality?

1. Causality and Determination

The idea that we live in a law-governed universe, in which each event follows according to immutable causal laws, evokes the spectre of determinism. This is the belief that everything is determined to happen just as it does, that things 'could not have been otherwise'.

Much labour has been spent on defining determinism. Here is one suggestion: Given the sum total of true causal laws, and a complete description of the universe at any one time, then a complete description of the universe at any other time may be deduced. In which case, the way the world is at any future time is fully determined by the way the world is now. Given how things are now, nothing hereafter could happen otherwise. This goes for my actions too. What I shall do at any future moment is the inexorable consequence of factors over which I have no control: factors which existed indeed even before I was born.

So defined, determinism is almost certainly false. The fundamental laws of the universe do not enable us to deduce the future from the past. They say only that certain events are probable, given certain others. Quantum phenomena are not simply isolated events of no general significance: they permeate reality, and imbue it with a systematic uncertainty. Hence the plausibility of Anscombe's argument about the Geiger counter. To say that every event has a cause is one thing; to say that every event is determined by its cause is quite another thing, and something that physics has refuted. If we say that 'determined' means 'probabilistically determined', that merely concedes the point.

Nevertheless, a milder form of determinism seems plausible. This says that the explanation for each event resides in the conditions which cause it, and that chains of causation stretch back indefinitely. Hence everything is now in place, that would explain the whole of things hereafter. In what sense, therefore, do our future decisions *change* the course of events, rather than merely confirm it by being part of it?

Does this mild form of determinism damage our belief in human freedom? Some would say not, since freedom requires only that the outcome of human choice be unpredictable. That could be true, even if every event, including human action, belongs to the one chain of causes.

2. God's Foreknowledge

If the question concerns prediction, however, there is absolutely no reason to invoke causality in order to recognise the threat to human freedom. Even in the absence of deterministic laws, the truth about the future might be knowable. Presumably, if God exists, God *does* know it. In which case, how can *I* be blamed for an outcome that was known before I was born? This was the way in which the problem of free will impinged on the thinking of medieval philosophers: How is human freedom compatible with God's foreknowledge? (Leibniz wrestled with this problem, with interesting results.) There is even a version of the conundrum that makes no reference to prediction at all, but simply reflects on the idea of truth itself. Aristotle considers the statement that there will be a battle at sea tomorrow. Does this statement have a truth-value *now*? If it does, it must be the same as the truth-value of the statement, uttered tomorrow, that there is a sea-battle today. If it is true tomorrow that there is a sea-battle it is true today that there *will* be a sea-battle. Statements about the future, if true, are true *now*. In which case can the future really be changed?

(One suggestion, considered by Aristotle himself, is to deny that statements about the future have truth-values. But this would mean that prediction is literally impossible, whether or not determinism is true.)

3. Could I Have Done Otherwise?

Freedom, Hobbes suggests, means choice. Do I have any choice in the present circumstances? Given that I do A, might I have done B? To say that I do A freely, is to say that I could have done otherwise, had I chosen. (See G. E. Moore in his *Ethics*.) As long as the outcome of a situation depends upon my choice, then I am free. Thus I do not freely injure you when I fall on top of you, having been pushed off balance: for there is nothing I can do to avoid your injuries. Whatever choices I make in the matter, it is too late to save you, once I have begun to fall: and I did not choose to fall. Lawyers discuss cases of this kind, and the idea of 'avoidability' plays a large part in allocating legal blame. If you could have avoided the consequences for which you are blamed, then you freely produced them. (See C. L. Stevenson, 'Ethics and Avoidability', in P. Schilpp, ed., *The Philosophy of G. E. Moore*.)

This approach has its merits. But it slips out of the question too easily. I did *x* freely, it says, if and only if I could have done something else, had I chosen. But *could* I have so chosen? And if so, could I have chosen *freely*? The question about the action has become a question about the choice that leads to it, and we seem to be launched on a regress. At some point we must simply accept that I make a choice, without choosing to do so. This just *is* my choice. But then, could I have made another choice? Or was I determined to make the choice that I did?

Such questions are familiar to us. For we do not blame people automatically for what they choose to do. We try to understand the motives behind their choice, so as to see the choice as part of the sequence of events that leads to it. And we frequently come to the conclusion, discovering the misfortunes that led to a person's present character, that it is 'not his fault' when he does what he does. He was *made* that way, by circumstances that he could not control. Having spelled out the misfortunes that twisted his character, and descanted upon his own deformity, Richard III declares: 'I am determined to prove a villain'. That is, I have chosen, resolved to do it; and I am *also* the victim of (determined by) my circumstances.

4. Freedom and Character

Such thoughts suggest that there is, after all, some kind of conflict between our belief in free will, and our knowledge of causes. The more we know of the circumstances, the less we are inclined to say that someone is a free agent. *Tout comprendre c'est tout pardonner.*

But now we encounter a paradox, hinted at in the great speech of Richard III. Suppose there is some completely *uncaused* choice: a choice that erupts into the stream of events without a cause. Is this what we mean by a free choice? Surely, such a choice would be as surprising to the person who makes it as it is to everyone else. It would be a wholly gratuitous event, and one for which he could be neither praised nor blamed, since it comes to him unbidden, and stems from nothing in *him*. Freedom then reduces to mere inexplicability. In which case it is difficult to see that freedom is something that we should either covet or judge.

Indeed, if we look at our ordinary notion of a free act, we recognise that, because we praise and blame people for what they do freely, we are eager to ascertain that their actions are truly *theirs*. We refer their free actions to their characters – saying that *he* caused this, or *she* caused that. The free act issues from the deliberations of the agent, which must therefore also be part of its cause. Freedom implies causation, rather than the absence of causation.

5. Compatibilism

We are now in a position to define some of the views that modern philosophers have taken.

(i) *The concept of freedom is incoherent.* This would be the view of someone who argued, first that freedom seems to be incompatible with causation (since we deny that a person is free whenever we can trace the causality of his action to some point beyond his own decision-making); and secondly (for the reason just given) that freedom implies causation. The concept of freedom would then be contradictory.

(ii) *Incompatibilism.* Freedom is incompatible (a) with determination or (b) with causality. Many accept that it is incompatible with determination (e.g. Anscombe), but deny that it is incompatible with causality.

(iii) *Compatibilism.* Freedom is compatible with (a) causation, or even

(b) determination. There is simply no contradiction between the two ideas.

Compatibilism has a respectable history; and admits of two varieties. There are those, like Spinoza, who provide a new concept of freedom, and those like Hume, who make do with the old one. Spinoza argued as follows: everything happens by necessity. If we describe as 'free' only those actions that originate in the agent himself, then only God is free. But we can also make a distinction between events that we understand, and events that we do not. To the extent that we have an 'adequate idea' of the causes of our actions, to that extent are we free, in the only sense that could conceivably apply to us. This freedom is a matter of degree, and might be defined as 'the consciousness of necessity'.

Hume's approach is altogether gentler, and has won considerable sympathy. Hume was a determinist. He believed not merely that every event has a cause, but also that causes determine their effects, since they are joined to them by universal laws. (He believed this despite his scepticism about 'real necessities'.) But, he argued, there is nothing in our idea of freedom that denies determinism. The idea of freedom arises when we attribute the consequences of an action to the agent, by way of praise or blame. There is nothing in this idea that either affirms or denies determination, and its grounds are unaffected by the advance of science.

6. Responsibility

Modern Humeans have been particularly impressed by the idea of responsibility, and the associated concept of an 'excuse'. (See J. L. Austin, 'A Plea for Excuses' in his *Philosophical Papers*.) It is of the greatest importance in human life to assign responsibility for actions, so as to reward the good and punish the delinquent. We could never succeed in managing or adjudicating the web of human relations, if we lost sight of this idea, and the greatest labour of the law courts is to ascertain, in both civil and criminal cases, just who is liable for what. If we study moral and legal argument, we discover that the concept of responsibility is quite different from that of choice – and is certainly far removed from 'metaphysical freedom'. A person may do something deliberately, choosing what he does out of rational considerations, and yet not be liable for the consequences. Suppose I am driving a lorry whose brakes have inexplicably failed; by steering to the right I can turn the lorry towards a frail old man; by steering to the left I can turn it towards a group of children; by doing nothing I allow the lorry to crash into a crowded pub. Am I responsible for the death of that frail old man, just because it is *his* life that I decide to jeopardise?

Equally, a person can be held responsible for results that he did not choose to bring about, and which lay outside his deliberations. Suppose it is your job to check the brakes of every lorry that leaves the factory, and you wave one of them past uninspected. Are you not partly to blame for the death of that frail old man? This is a case of negligence, of great importance in the civil law. But there is criminal negligence too. A reckless disregard for the lives and property of other people may lead to criminal charges. In imposing penalties, the law merely follows our moral intuitions.

Finally, in all cases of liability, there is the possibility of an excuse. I did it deliberately, that is true; but I was acting under duress, in a state of mental confusion, or after a heavy bout of drinking. I was negligent, that is true; but my wife had just left me, and I was unable to concentrate, or the rules were unclear and I had been given no time to study them. Excuses remove blame, by showing that it was no fault of *mine* that set the catastrophe in motion.

The concept of responsibility is vital to us, and has its own logic. We tend to agree in our judgements of blame, and to recognise the same factors as reducing or compounding it. In what, then, are such concepts based? And do they depend on an idea of metaphysical freedom?

7. Persons and Animals

Some light is cast on the matter by the distinction between persons and animals. Maybe we *are* animals, but animals of a special kind, whose actions are judged in moral terms. When you discover the violated chicken-run and the headless corpses, you are naturally distressed. But are you also *resentful* of the fox? Do you wish him to stand trial, to be judged, to suffer condign punishment? Surely not. You may stay up all night with a shotgun, or ring the master of foxhounds. But your purpose is not to *punish* the fox; it is simply to be rid of him. He did what he did by instinct, and without *mens rea*. He had no duty to respect your chickens, nor did they have a right to life and limb that could be enforced against him. Animals have neither duties nor rights, and it is not merely sentimental, but absurd, to treat them as though such moral ideas applied to them. If you try to apply such ideas to animals, the result is not merely confusion, but a radical failure to relate to them at all. You will achieve nothing that you want, and nothing that they want either.

The case is quite different with humans. Discovering that it was my neighbour Alfred who broke into the run on a chicken-biting spree, I am naturally indignant. I reproach him with the damage, try to persuade him of the error of his ways, of his duty to respect my property and the well-being of innocent creatures who had done him no harm. It is fairly likely that these procedures will not succeed, if Alfred is a hardened chicken-biter. But I can sue him too, and also bring criminal charges on grounds of cruelty. One way or another, my disposition to judge him, to blame him and to express my anger, will have an effect. He will be less disposed to do it again. And, as he mends his ways, so does my resentment dwindle. We may one day be the best of friends, and those headless chickens no more than a memory.

The case illustrates Strawson's argument in 'Freedom and Resentment'. There are, Strawson suggests, two different attitudes that we may adopt towards persons, which he calls 'objective' and 'reactive', but which we might more usefully describe as 'scientific' and 'interpersonal'. I can relate to persons as I do to other natural objects, by studying the laws of motion that govern them, and trying to adjust their behaviour accordingly. Or I may react to them as one person to another, giving them reasons, feeling resentment at the wrongs I suffer from them, and anger at their injustices. In normal circumstances, the interpersonal attitudes are self-sustaining. They are effective in bringing about a mutual accommodation, and a common understanding. Resentment is rewarded by apology, and apology by forgiveness. The give and take of reason comes to rest at last in concord.

Underlying such interpersonal attitudes is the assumption of rationality. A person is a rational being, with rights and duties to which we may appeal. He is able to understand and act on reasons; and his projects can be altered and amended in the light of argument. It is because of this that I feel resentment when he harms me, and gratitude when he acts for my good. Those reactions are reasonable; they are also *effective*. My resentment persuades him to change his ways, while my gratitude encourages him to retain them.

In certain cases, however, resentment and gratitude, praise and blame, reason and casuistry, are ineffective. Alfred silently bears my reproaches, and the next night is biting again. He is in the grip of an obsession that lies outside the reach of interpersonal response. I now waver in my view of him. I can no longer relate to him as I do to other persons: interpersonal emotions are no longer profitable; nor does he acknowledge their sway. Eventually I am driven to take up a more scientific outlook, to seek for the causes of his condition, and to attempt a cure. If I lock him up, it is not by way of punishment, but in order to treat him for an illness. Alfred has fallen out of the web of personal relations, and become a thing. Something like this may happen in mental illness.

And here lies the conflict, Strawson ingeniously suggests, between freedom and causality. It is not a conflict *in rem*; it lies rather in the attitudes of the observer. Personal relations cause us to ignore the deeper causes of one another's actions, and to respond to others through our spontaneous ideas of responsibility and right. Interpersonal emotion gives us a far more effective handle on the human world than we could ever obtain through a science of human behaviour. But there comes a point when the interpersonal approach ceases to be rewarding. It is then that we begin the search for causes. In consequence, we demote the other from person to thing.

The conflict, therefore, is not between actions that are free and actions that are caused: our science of human nature applies indifferently to both and denies the reality of the contrast. The conflict is between attitudes that require us to overlook causality, and attitudes that require us to attend to it, and to define what we see in terms of it.

8. The Kantian Position

I have defined Strawson's position in terms that he does not use, in order to move on to its original. This is the position given by Kant, which can fairly claim to be the deepest of all philosophical answers to the problem of freedom.

We know that we are free, Kant argues, because we are bound by the moral law. We are self-commanded by reason to do what we ought and to avoid what we ought not. Such commands would not make sense, if we could not freely decide to obey them; for that which we do by nature cannot also be a duty. There is no place for freedom in the world of nature, whose ruling principle is the law of causality. It seems, therefore, that I am both part of nature – since I am an animal, subject to the passions, and prompted to act from all kinds of non-rational motives – and apart from it, since I am the originator of my actions, which stem from reason, and express my free obedience to a transcendental law.

How do I reconcile those two ideas? Kant's view is that they cannot be reconciled, but only transcended. They offer complete descriptions of the world from rival viewpoints: the viewpoints of understanding and practical reason.

The understanding gives us the following picture of the human condition: we are organisms, part of the natural world, and bound by the world's causality. Like other animals, we are influenced by our passions and desires, which cause and explain our behaviour. We are objects among others; our desires have no authority in the struggle for survival, but can be overridden and impeded by the desires of others, and of the animals that compete with us for the earth's resources.

Practical reason gives us another picture. We are persons, with rights, duties and moral values. As such we stand outside nature, judging both nature and ourselves according to a higher law. Reason does not tell us only what to do, so as to fulfil our desires; but also what we *must* do, whether we desire it or not. It is addressed directly to the agent, with a 'categorical imperative'. He cannot deny this imperative but can at best merely avoid it. We are not objects among others, but subjects, with a first-person perspective on the world that calls us to account for it, and which reminds us constantly of our apartness from the natural order. Hence persons are not just things, and cannot be treated as things without defying the law of reason. They are to be treated as ends in themselves and not as means only: they have rights, and their rights are inviolable. But they have duties too, and it is through the reciprocal web of right and duty that persons relate to one another, and so aspire towards the 'kingdom of ends' which is their ideal community.

Kant's picture contains many elements that we might question. But it is confirmed by our pre-philosophical intuitions. For example, we make a distinction between a bodily movement and an action: between my arm's rising and my raising my arm. What is the difference? Kant has an answer. The one is a natural process, understood through the causes that produce it. The other is an expression of the rational being, to be understood in terms of his reasons for action. Likewise, we make a distinction between causes and reasons. The first explain the movement in terms of natural laws (e.g. 'my arm went up because the synapses fired'), the second give an account of the action, in terms which may also justify it. ('I raised my arm in order to warn him, and this I had to do because . . .') Again, we make a distinction between creatures like us, with rights and duties, whom we cannot treat as we will, and who have claims on our consideration, and the rest of nature. (There are problems, of course, about animals: but the unclarity of *their* case serves only to remind us of the clarity of ours.) And this distinction is illuminated, not only by the contrast between person and thing, but also by the more metaphysical contrast, between subject and object. The self-conscious subject is not just a part of nature, swept along by forces that he does not control. He is a judge of nature, who does not merely do things, but also has the *question* what to do. He *situates* himself in his world, and is not merely 'contained' by it. (This is part of what Heidegger meant by the 'question of Being'.)

As we proceed in this way, distinguishing persons from things, and our attitudes to persons from our attitudes to things, it becomes steadily more plausible to suppose that this *is* the distinction that we have in mind, when we consider the place of freedom in nature. Moreover, we can now see why the concept of responsibility is so important. For it links the world of nature to the world of persons; it tells us that this or that part of the natural world expresses the rational self of this or that person. This is the beginning and the end of judgement.

But how are the two views to be reconciled? Kant, as I said, did not believe that they *could* be reconciled. The contradiction between them is, however, a contradiction only for the understanding, and not for practical reason, and could therefore be transcended in the moral life. (This is the true inspiration for Hegel's theory of the dialectic.) Kant's approach is far too drastic; so let us see if we can replace it with something a little closer to common sense.

18

The Human World

Return for a moment to the discussion of Chapter 15. I mentioned the argument that mental states are distinguished by their 'intentionality': the property of 'reference' to an 'inexistent object'. Just what property is this? Frege's theory of sentence-structure depends upon the principle of extensionality, which says that the reference (extension) of a complex term is determined by the reference of its parts. Frege recognised that certain contexts seem to violate this principle (see Chapter 6). In the context 'Mary believes that . . .', we cannot reliably substitute terms with the same reference. From the fact that Mary believes Lady Blackstone to be a woman it does not follow that Mary believes the master of Birkbeck College to be a woman, since Mary may not know that Lady Blackstone is the master of Birkbeck. (Since when have ladies been masters?) For Frege this failure of extensionality is only apparent, since terms in contexts like 'Mary believes that . . .' have a new and 'oblique' reference. Frege's suggestion is that terms here refer to their normal *sense*. You can substitute words with the same sense, he argues, without changing the reference (truth-value) of the whole. That is the Fregean theory of intensionality, and it is now rejected.

The context 'Mary believes that . . .' is more often regarded as genuinely non-extensional, in other words, as 'intensional'. Interestingly the context is completed by a term which identifies an intentional object: namely, the object of Mary's belief. Whenever we refer to intentional states, in fact, we end up with intensional idioms. From 'John fears the masked man', it does not follow that 'John fears his father', even though the masked man is his father. From 'Mary is jealous of Alev', it does not follow that 'Mary is jealous of the poorest girl in Turkey': and so on. The intensional context contains a name or description, which is also the 'description under which' the object of belief, fear or jealousy is conceived by the subject. This 'description under which' identifies the 'intentional object' of a mental state. By a strange fluke, therefore, intentionality is a kind of intensionality: the kind that arises when mental items are described. (Or is it a fluke? Surely, it cannot be.)

That is not *exactly* right, but it will suffice. Intentionality reflects the fact that mental states are also conceptualisations of the world; in order to describe them therefore we must identify the 'descriptions under which' their 'world' is presented. A study of the mental realm is also a study of the intentional realm – the world as we conceive it in experience. Husserl called this intentional realm the *Lebenswelt* or 'world of life'; others, following Wilfrid Sellars, have written of the 'manifest', as opposed to the 'scientific' image. I shall refer to 'the human world'. And one of the tasks of philosophy, I shall maintain, is to describe and explore this human world – maybe even to vindicate it against the world of science.

1. Varieties of Intentional State

Intentionality means that words like 'of' or 'about' are needed to describe the mental. These terms denote 'quasi-relations'. But do they always denote the same quasi-relation? The question is extremely hard to answer. As soon as we penetrate below surface grammar, we discover a host of different ways in which conceptions enter consciousness. Here are some of them:

(i) *Perception.* When I see an object it *looks like* something: a cow, say. But maybe it is not a cow. On the other hand, that is how it appears. What do we mean by a perceptual appearance? Is it a kind of belief? I shall return to this question in Chapter 23.

(ii) *Belief.* Beliefs can be true or false. What makes them true is the world, which may or may not be as our beliefs represent it. Moreover, each belief involves a conception of its object. Some conceptions portray the world in terms of our experience rather than its real constitution. For instance, my belief that this book is red applies a concept – *red* – whose content must be given through perception. Such a belief is a belief about *appearances*.

(iii) *Desire.* When I want a glass of beer, there is an 'object' of desire, which I conceive under what Anscombe (*Intention*) calls a 'desirability characterisation'. I think of the beer as cool, refreshing and vaguely permissible. The object of desire lies in the future, and may disappoint my expectations; it may even fail to materialise at all.

(iv) *Emotion*. An emotion is founded on a belief: for instance, if I fear something, it is because I believe that it threatens me. Hence emotions share the intentionality of beliefs: they may be true or false, just as beliefs are true or false, depending upon whether the world matches the conception. But they have a second dimension of intentionality: for they also involve desires. If I fear the masked man, I desire to get away from him. My fear has the intentionality of a desire: it involves a conception of the desirability of some future state (of maskedmanlessness).

(v) *Thought*. When I think of a landscape, I do not necessarily believe anything: the landscape may be entirely imaginary. Yet still I 'entertain thoughts', which could be true or false, and whose content may represent or misrepresent the world.

(vi) *Imagination*. When reading a fictional story, or looking at a picture, I conceive an imaginary world, and allow my thoughts and feelings to roam in it. This too involves me in conceptions, which by their nature are freed from the demand that they be true. (Though maybe they should be true to life.) There are deep questions here to which I shall return.

The list could be extended. But it serves to show that the appearance of the world is constructed in many ways, according to the requirements of many human faculties. All the mental states in my list would be described by a certain kind of philosopher (Fodor, for instance) as 'mental representations', since they 'represent' the world as being thus and so, rather as a picture does. But the term 'representation' adds little that is not already suggested by Brentano's 'reference'. (In aesthetics, philosophers often try to understand representation in *terms* of reference.) The core idea is that of *truth*. The world can *falsify* our mental states. But sometimes we do not care. In imagination, for example, we expect exactly that. Our imaginative conceptions are nevertheless precious to us, and enter into our vision of reality.

2. The World of Science

This returns me to the matters discussed in Chapter 15. Science, which begins from the attempt to explain appearances, rapidly begins to replace them. Seventeenth-century scientific thinkers like Gassendi and Boyle were quick to notice this, and introduced the distinction between primary and secondary qualities which was then made famous by Locke. When explaining how things appear, we describe how they (probably) are. And we then attribute to reality only *some* of the qualities that are naïvely perceived in it: the 'primary' qualities. The other qualities are defined in terms of the sensory experiences involved in detecting them. To say that an object is red, is to say that it appears red in normal conditions to normal observers. (Who are *they*? Well, those who, in normal conditions, see red things as red! There are lots of circles here, but the consensus is that none of them is vicious.) Does this mean that nothing really *is* red? Berkeley thought so, but then Berkeley was like that. It is more sensible to say that things really *are* red, but that the classification 'red' is experience-relative, with no place in the really real (the world as it is in itself, from no point of view).

The argument can be generalised. Concepts that we use from day to day reflect the world *as we encounter it*. They can be used in true judgements, as well as false ones. The proof that redness is a secondary quality does not show that judgements about redness are all false. On the contrary: it enables us to understand how true and false are here distinguished. Similarly, 'folk psychology' (the implicit theory of the human soul contained in our everyday mental conceptions) describes only the appearance of the mental world, and not the underlying mechanism. But this would invalidate our everyday judgements about the mind only if the true theory of the phenomena turned out to *contradict* our everyday beliefs.

But there's the rub. How do we know in advance that our everyday conceptions do not involve errors? How can we be sure that science will not sweep them all away, offering in exchange only its cold abstractions, a skeleton on which nothing human hangs?

This worry is not just philosophical; it is also spiritual. The meaning of the world is enshrined in conceptions that science does not endorse: conceptions like beauty, goodness and the soul which grow in the thin top-soil of human discourse. This top-soil is quickly eroded when the flora are cleared from it, and nothing ever grows thereafter. You can see the process at work in the matter of sex. Human sexuality has usually been understood through ideas of love and belonging. An enchanted grove of literary ideas and images protected those conceptions, and man and woman lived within it happily – or at any rate, with an unhappiness that they could manage and control. The sexologist clears all this tangled undergrowth away, to reveal the scientific truth of things: the animal organs, the unmoralised impulses, and the tingling sensations that figure in those grim reports on the behaviour of American humanoids. The *meaning* of the experience plays no part in the scientific description. Since science has absolute sovereignty over what is true, the meaning comes to be viewed as a fiction. People briefly try to reinvent it, sometimes even hoping to do a better job. Failing, however, they lapse into a state of cynical hedonism, scoffing at the fogeys who believe there is more to sex than biology.

Scientific exploration of the depth of things may therefore render the surface unintelligible – or at least intelligible only slowly and painfully, and with a hesitancy that undermines the needs of human action. As agents we belong to the surface of the world, and apply to it the classifications which inform and permit our actions. We cannot replace these conceptions with anything better than themselves, for they have evolved precisely under the pressure of human circumstance, and in answer to human needs: in particular our need for meaning.

3. Non-natural Kinds

Our classifications divide the world according to our interests. Science ministers to an interest in the nature of things, and therefore employs classifications whose purpose is to reveal how things are. Concepts of natural kinds 'divide nature at the joints', and hence feature in our scientific laws. But we may be more interested in the relation of objects to ourselves than in their causality and construction. We seek not only for the causes of events, but also for their meaning – even when they have no meaning. For example, we group the stars into constellations according to fictions of our own, and in doing so we commit astronomical outrage. For the astronomer our concept of a 'constellation' displays nothing but the superstitious emotion of those who first devised it. For the astrologer it conveys the deepest insight into the mystery of things. And modern people do not doubt who is right, even when they consult the horoscope.

But there are legitimate classifications which cut across the divisions recognised by science. A good example is that of ornamental marbles. The purpose of this classification is to group stones in accordance with their aesthetic similarity. An ornamental marble can be polished; it has a grain, a colour, a depth and a surface translucency full of sacred connotations. The classification includes onyx, porphyry and marble itself; scientifically speaking, it is therefore utter nonsense. For onyx is an oxide, porphyry a silicate, while limestone – a chemical allotrope of marble – is expressly excluded from the class. A science of stones must aim to replace all such classifications – whose subservience to human purpose diminishes their explanatory power – with the natural-kind concepts that would finally explain things (including the appearance of things).

A science of stones would classify marble and limestone together, as different crystalline forms of calcium carbonate, generated by the decomposition under pressure of once living things. Such a science would find no single explanation for the fact that the appearance and utility of marble approximates so closely to the appearance and utility of onyx and porphyry. Hence it would contain no classification corresponding to 'ornamental marble'.

Something similar could be true of folk psychology. In Chapter 16 I asked what distinguishes a mental from a non-mental state. One currently popular answer is: 'explanatory role'. A mental state has a particular role in the explanation of behaviour. (Functionalists take this line.) But our ordinary explanations of things are superficial: they express rough and ready theories which permit us to respond to events with the rapidity required by survival, but with no more knowledge of their real constitution than that implies. The 'theory' that tables and chairs are solid, for instance, is extremely useful to us; but it was long ago discarded by the true science of wood. Likewise, our concepts of belief, desire, emotion and intention – useful though they may be in classifying behaviour and focusing our response to it – may be as superficial as the concept of an ornamental marble. Maybe the true theory of behaviour will make no use of *those* concepts, but will cut across the barriers between them in something like the way that the concept of calcium carbonate straddles the divide between limestone and marble. The true theory of the mind will then not be a theory of belief, intention and the rest; those ideas will have been eliminated. The theory will not analyse our mental concepts, but replace them. (The lesson of intentionality, Davidson suggests, is that it *must* replace them.)

At the same time, however, science may provide no substitute for the concepts which order and direct our everyday experience. A sculptor armed with the theories of chemistry, geology and crystallography, but without the concept of an ornamental marble, will not have the immediate sense of similarity in use which enables his less erudite colleague spontaneously to envisage a decorative masterpiece. Their very perceptions will be different, since the erudite sculptor will be without the concept under which the stones should be perceived by someone who is seeing them as material for *sculpture*. Likewise, the 'true science' of the mind will provide no substitute for our ordinary mental concepts. For these concepts bring order to our mutual communications. If the fundamental facts about John are, for me, his biological constitution, his scientific essence, his neurological organisation, then I shall find it difficult to respond to him with affection, anger, love, contempt or grief. So described, he becomes mysterious to me, since those classifications do not capture the intentional object of my interpersonal attitudes.

4. Intentional Understanding

We may therefore contrast two modes of understanding: science, which aims to explain appearances, and 'intentional understanding' which aims to interpret them – i.e. to describe, criticise and justify the human world. Intentional understanding studies the world in terms of the concepts through which we experience and act on it – the concepts which define the intentional objects of our states of mind. The Kantian philosopher Wilhelm Dilthey (1833–1911) coined the term *Verstehen* to refer to this kind of understanding – and the term has entered sociological usage through the writings of Max Weber. Intentional understanding engages directly with the world as we perceive it; it aims not so much to explain things, as to make us at home with them.

The concepts of our intentional understanding are not easy to analyse. They are embedded in feeling and activity, and difficult to bring into focus. Nevertheless, there are genuine, objective truths about the human world, which we discover through philosophical analysis. Some call this analysis 'phenomenology'; others refer to the 'analysis of concepts': but, in this area at least, where we are studying the structure of appearance, the two methods are one and the same. The attempt to deepen our intentional understanding involves exploring the public language of appearance, through which we understand the world as a sphere of action and an object of response.

Here, then, is how we should express the Kantian theory of freedom: people may be conceptualised in two ways, as elements in nature, or as the objects of interpersonal attitudes. The first way employs the concept human being (a natural kind); it divides our actions at the joints of explanation, and derives our behaviour from a biological science of man. The second way employs the concept person, which is not the concept of a natural kind, but *sui generis*. Through this concept, and the associated notions of freedom, responsibility, reason for action, right, duty, and justice, we gain the description under which a human being is seen, by those who respond to him as a person. Our response is locked into the web of interpersonal feeling. Each of us demands justification of the other, and the resulting give and take of reasons is the root of social harmony. The concepts that permit this give and take are not just useful to us; they are indispensable. And chief among them is the concept of freedom.

The scientific attempt to explore the 'depth' of human things is accompanied by a singular danger. For it threatens to destroy our response to the surface. Yet it is on the surface that we live and act: it is there that we are created, as complex appearances sustained by the social interaction which we, as appearances, also create. It is in this thin top-soil that the seeds of human happiness are sown, and the reckless desire to scrape it away – a desire which has inspired all those 'sciences of man', from Marx and Freud to sociobiology – deprives us of our consolation. Philosophy is important, therefore, as an exercise in conceptual ecology. It is a last-ditch attempt to 'save the appearances'.

Intentional understanding clarifies and endorses the 'thin' description of our world. It gives sense to functional kinds and secondary qualities. It upholds those more elusive classifications which form the background to personal life: classifications relative to emotions (the fearful, the lovable, the disgusting) and to aesthetic interest (the ornamental, the serene, the harmonious); it gives sense to our interpersonal attitudes, by reinforcing the concepts on which they depend for their intentionality; and it explores the meaning of the world, in moral and religious experience.

5. Enchantment and Error

But surely, we cannot assume that our intentional understanding is all right as it is? We may discover contradictions in our common concepts (the concept of freedom for example); we may discover, at the heart of our ordinary world-view and the folk psychology that goes with it, certain myths, fantasies and delusions. If these are swept away by science, it is surely because they deserve to be.

To criticise a belief is not the same as to criticise the concepts contained in it. Consider secondary qualities. Philosophy tells us that our beliefs about these may be true, even though the concepts used to express them are systematically misleading. Conversely, concepts which are rooted in reality sometimes feature in false beliefs about it: as in primitive beliefs about the stars. Philosophy can tell us whether our concepts are in order; but not whether our beliefs are true. But it can resist those phoney arguments against our beliefs, which derive merely from a consideration of concepts. For instance, Marxism criticises as 'mere ideology' many of the beliefs that are woven into the fabric of social life. It says that beliefs about justice have no authority, since they are adopted merely because they legitimise 'bourgeois' power. It may well be that the *concept* of justice contributes to this function: for the concept is interest-relative, and political order is the interest which it serves. But it does not follow that our beliefs about justice are *also* merely functional, with no claim to a truth of their own. (The case parallels that of secondary qualities.) It is *true* that theft is an injustice, even though the concept of justice derives its sense from the social relations that are sustained by it.

On the other hand, there are plenty of false beliefs about the human world, and our loss of those false beliefs often accompanies a loss of confidence in the concepts employed in them. This too may be part of our spiritual crisis. The reader of Homer's *Iliad* will be impressed by the lack of distance between the Greek hero and his world. The world as the hero perceived it was already shaped by the needs of action; it reflected back to him the intentionality of his own emotions, and he could read his decisions off from the reality, without ever pausing for the reason why. However, the Greek hero's world is not merely conceptualised as 'ready for action'; it is also an enchanted world, one that is understood in supernatural terms. The theology of Homer is 'thin', rooted in the top-soil, inhabiting the same surface as the Homeric hero himself. This explains its utility; but it is a utility that is bound up with its falsehood. (So much for pragmatism.)

Contrast the world of Proust, in which the hero, if that is the word for him, finds himself constrained on every side by modern science. He knows the falsehood of all those superstitions, and knows that if there is meaning in the world it is because he himself creates it. He no longer reads his will in nature; instead he tries to *discover* his will, by transcribing nature into himself. The human world becomes a world recollected; its meaning resides not in the great deeds and challenges, but in remembered perfumes and the distant echoes of an enchantment that vanished with childhood. There is meaning in this world, but it is a meaning that falls fatally short of any real decision to act.

The contrast is important. The human world flourishes best when refreshed by springs of falsehood. The concepts that flower in the Homeric jungle satisfy the hero's lust for action, because they fit the world to his intentions, and extend intentionality far into the surrounding unknown. The parched truthfulness of Proust's oasis offers little encouragement to the hero. The merely passive meaning that he finds in things is divided by only a hair's breadth from the final meaninglessness of the world of science.

Those considerations help to explain why so many philosophers are suspicious of the human world, and of the concepts that are used to describe it. We are in the vicinity of too much agreeable error, too much charming allegory and myth. Nevertheless, we should not concede the point. Science has authority only because it explains how things appear. It can therefore never have more authority than the appearances that it strives to replace. The human world is the full system of those appearances; and even if it is riddled with error, the same is true of science. Nor do we free ourselves from error by abandoning our intentional idioms: for they alone can describe appearances, and therefore they alone can provide the final proof of science.

6. Intention

There is another and greater (because morally greater) error involved in capitulating to the scientific world-view. This is shown by a mental concept that I have so far not discussed: the concept of intention.

Animals have desires, but they do not 'make up their minds' to act. When they act out of desire, the desire is sufficient cause of their action. If they hesitate, it is because some other desire has intervened and plunged them into conflict. Persons do not only have desires; they also have intentions. That is, they take decisions, form plans of action, and carry them out. I can intend to do what I do not want to do; I can also want to do what I do not intend. The distinctive mark of intentional action is that it is done for a reason, and the agent himself is able to offer this reason in answer to the question 'why?' (See Anscombe, *Intention*.) Our capacity for intentional action is therefore bound up with our ability to give and receive reasons for action. When I 'make up my mind', as opposed to merely acting out of impulse, I am involved more centrally in my action, by the chain of reasons that link the action to my self. (This distinction between intention and impulse underlies much of the Kantian philosophy of freedom.)

Not every way of conceptualising the world can offer reasons for action. Take love, for example. When John loves Mary, he loves her for the particular person that she is – irreplaceable and inexhaustible. Her being Mary conditions the intentionality of his feelings, and provides him with reasons for all that he does for her. It is because she is Mary that he sends her flowers, courts her, desires her and marries her. But this idea of a 'particular person', whose dominant feature is precisely that she is who she is, is not countenanced by science. A science of human behaviour would do its best to rid the human world of such a description, so as to find the real cause of John's infatuation. He loves Mary because of her smell, or because of some other feature that she shares with Jane, Rosemary or Inez. Even if the quality that draws him to Mary is *not* shared by any other woman, it is still the quality, not the individual, that draws him. Science recognises no such thing as an 'individual essence'. The descriptions under which Mary is perceived by John make essential reference to her name, and to the individuality that is captured in it. The descriptions are not false. But they cannot feature in the scientific world-view. Science, if given exclusive sovereignty over truth, threatens the aims of love.

Intention lies at the interface of thought and action. It links the two through a conception of the world and of the agent's place within it. Not every conception can feature in a reason for action; indeed, rational agency constrains the construction of the human world. If the world is to be *our* world, then it must invite and facilitate our projects. It must be possible for us to apply conceptions that define the ends of our endeavours: such conceptions (e.g. the lovable, likeable, beautiful and agreeable), govern also our practical reasoning.

What exactly *are* reasons for action? The problem is doubly difficult. First, the question 'why?' can be interpreted in at least two ways: as asking for an explanation, and as asking for a justification. Which of those is meant by a 'reason for action'? Secondly, reasons given from the third-person viewpoint may seem very different from those which would be given (either to himself or to another) by the agent himself. Which is the *real* reason for the action? Matters are further confounded by the suggestion that there might be unconscious reasons for action. Does this make sense? And if so, what are the limits to unconscious reasons, and how do we discover them?

I do not propose to explore those questions, which belong to the philosophy of mind. But here is *one* thought, related to the theme of the present chapter. The reasons that *I* have for my action must justify it in my eyes, whether or not they also explain it. But the justification that *I* have may not coincide with the justification that *you* would give; and *your* justification may be unavailable to *me*. Consider the anthropologist who studies a war-dance, in which I, a member of the tribe, participate. They are right to dance, the anthropologist argues, since it fortifies the spirits and engenders solidarity in a time of crisis. But this is not *my* reason for dancing. I dance in obedience to the god of war. If I were to think of my dance as the anthropologist thinks of it, I should rapidly lose interest in dancing. (That is why decent anthropologists keep quiet about their discoveries, until they are out of earshot of those who might be corrupted by them.)

The example shows how the first-person perspective may be built into a particular conception of the human world, and how the conception may be threatened by a third-person point of view. The anthropologist's theory undermines the *Lebenswelt* of the dancer, and so breaks the connection between thought and action. Likewise Durkheim's sociology, which justifies religion, also undermines it.

7. The Will

The distinction between *my* perspective on my actions and the *observer's* perspective, may be reproduced in myself. I may adopt now one perspective, now the other. I may look on the future in the light of the question 'What shall I do?', and seek the reasons that help me to make up my mind. Having made up my mind, I then decide. 'I *will* do it', I say, and admit no doubt or denial of the thought. Alternatively, I may look on myself 'from outside'; I *predict* my actions, but without actually deciding on them: 'No doubt I *shall* go home, that is just the kind of character I am.' I may doubt the truth of this prediction; and I justify it by giving reasons for believing it, not reasons for bringing it about.

The distinction between deciding and predicting has been much discussed, as has the distinction between the first- and the third-person view of action. These distinctions help us to see what philosophers have meant by 'will'. As a rational agent I do not merely speculate about the future: I deliberate, decide, and *lay claim* on future events, as part of myself. I have immediate knowledge of the contents of my decisions, and thereby project my self-consciousness outwards into the world, and forwards into the future. Only beings with a self-conscious viewpoint are capable of this. Kant went so far as to suggest that the true mystery of the self is encompassed only through practical reason. It is in the exercise of the will, and in the moral life that stems from it, that the self is 'realised' in the empirical world. Through will, the transcendental becomes incarnate, and 'dwells among us'.

Schopenhauer, who was inspired by Kant's philosophy, accorded the sovereign place in his ontology to will. The idealists, he believed, had fully established their case, that the empirical world as we know it is a system of appearances (he called them 'representations', or *Vorstellungen*). Our concepts – space, time and causality – apply to this realm of pure appearances, and organise it into the familiar material world. But is there anything *behind* appearances? Is there any 'thing-in-itself'? Yes, Schopenhauer answered. There is the thing that I know immediately and without concepts, which appears not as a 'representation' but as the 'I' itself – namely will. This is the 'thing-in-itself', the really real behind appearances, the entity that escapes the categories of space, time and causality, and which bears its nature absolutely. The will is free; it is known immediately (i.e. without concepts); and our thoughts about it are justified through practical rather than theoretical reasons. The vain attempt of philosophy, to seize the reality behind appearances, is neither fruitful nor necessary. For we have only to look inwards, and we find what we were looking for, in the workings of the will itself.

Schopenhauer's system is not to everyone's liking. It should be said, however, that he is the only great philosopher to have recognised the connection between self-understanding and the understanding of music. For this reason alone you should read *The World As Will and Representation*. And there is another reason too: it is written with surpassing ease and elegance – always serious, but never dull. It is largely thanks to Kant and Schopenhauer that the will has been identified by modern philosophers with rational agency: the conscious process of decision-making, with its background of reasoning and argument. There are other conceptions of the will, however, both more atavistic, and less attached to an incipient morality. Nietzsche, for example, wrote of will as the capacity for self-assertion, implying that its origin is not rational, but lies in the instinctive desire to seize my own advantage, and to flourish as a living organism. More recent discussions have emphasised the pre-rational phenomenon of self-movement: the ability to *try*, whose foundations lie deep in the animal psyche. Thus Brian O'Shaughnessy, in a book that is as interesting as it is difficult, has argued for a view of will as a kind of non-rational force, 'spirit in motion', which is something other than mere desire, and other than practical reason, but which is manifest in phenomena like 'efforts of will', where trying and succeeding are the paramount moments. (*The Will.*) O'Shaughnessy's discussion is distinguished by its immense range and subtlety; if it has served to restore the topic of the will to a central place in philosophy, it is also not quite the topic that Schopenhauer knew.

But it is time to move on. The theory of intention has brought us to the most important case of it – the act of *meaning* something. Do we here mean by 'mean' exactly what I have meant in referring to the meaning of the world?

19

Meaning

In Chapter 6, I gave a résumé of Frege's theory of language, in terms of which to re-cast some of the central questions of philosophy. The theory develops a systematic connection between language and truth, arguing for the following conclusions:

(1) The unit of meaning is the sentence, which expresses a complete thought.

(2) A sentence has a truth-value; and truth is the 'preferred value' of every meaningful sentence.

(3) The thought expressed by a sentence is given by its truth-conditions: the conditions which determine when it is true.

These conclusions about sentences feed back into a theory about the parts of sentences. Each meaningful component has both a sense and a reference; the sense determines the reference, and just as the sense of the sentence is determined by the sense of its parts, so is the truth-value determined by the referents of individual terms. This theory enables Frege to explain why we understand indefinitely many sentences on the basis of a small vocabulary. It also prompts him to account for the separate operations which terms encompass: naming, predicating, and quantifying.

But does such a theory tell us what meaning *is*? Certainly it enables us to *interpret* language, and to give rules for deriving the meaning of a complex term from the meaning of its parts. But does it distinguish signs which *mean* something, from signs which merely 'bear an interpretation'? Suppose I came across a pattern of pebbles on the sea-shore, forming the letters of the sentence 'God is dead'. Do these pebbles *mean* that God is dead? Surely, only if someone placed them there with such a purpose. An accidental pattern is not yet a meaningful sentence. You often find patterns in nature, in which a finite set of elements is arranged according to systematic 'rules'. (The markings on a zebra, or the flecks of foam above a barrier reef.) An ingenious person could divide these patterns into 'words', assign sense and reference to each of them, and truth-conditions to the whole 'sentences' in which they occur. But that does not show them to be meaningful, even if it were to lead to a semantic theory which made surprising sense of them. The assumption that wherever we can interpret, there is meaning, is popular with horoscopists, alchemists, witches and structuralist literary critics. But there is something wrong with an assumption that is common to *so* many kinds of charlatan.

1. Meaning and Assertion

The missing element, it might be suggested, is that of *use*. Meaning occurs when signs are used to say something. If the devil had arranged those pebbles, then there really *would* be a message contained in them. So how do we understand the use of signs?

One suggestion is that signs become meaningful when used to assert something. Frege believed that assertion is an all-important part of language. He also argued that it is something over-and-above the sense or reference of the thing asserted. In an important sense, assertion is not part of the meaning of a sentence. This is because the same sentence can occur, with one and the same meaning, both asserted and unasserted. Consider the following argument:

(1) *p*
(2) *p* implies *q*; therefore
(3) *q*.

Someone who asserts (1) and (2), will also be led, if he is rational, to assert (3). But the proposition *p*, which is asserted in (1), is not asserted in (2); just as *q*, which is not asserted in (2), is asserted in (3). But the meaning of neither sentence changes during the course of the argument. Only if '*p*' means the same in (1) and (2) is the argument valid. Assertion, therefore, is not a semantic property of the sentence; so what is it? The obvious suggestion is that assertion is an action, which the sentence is used to perform.

But is it the only action that sentences can be used to perform? Clearly not. I can assert that *p*; but I can also question whether *p*; order that *p* be made true; wonder whether *p*; entertain the thought that *p*; suppose that *p*; and so on. All these 'speech-acts' have something in common – namely, their 'content', which is the proposition *p*. But they differ in what they intend, and in how they are understood by the listener.

Those thoughts suggest a way forward. Suppose we could find a paradigm speech situation. We might then study it in order to see exactly what the speaker is trying to *do*, and how the proposition (the content) is involved in his doing it. We would then begin to understand what is involved when someone does not merely *say* '*p*', but also *means* it. This is the approach taken by Grice in a famous paper ('Meaning', 1957).

2. Grice's Theory

Grice begins by criticising C.L. Stevenson, a philosopher who worked in the shadow of logical positivism, and who was one of those useful providers of theories that are clear enough to be obviously wrong. Stevenson advocated a 'causal theory of meaning': words acquire meaning through the states of mind that cause them, and the states of mind that they cause. I believe that *p*; this causes me to utter '*p*', and this in turn generates the belief that *p* in you. Habit-forming associations ensure that '*p*' is generally uttered with just these causes and effects.

Empiricists have often come up with such accounts of meaning, relating words to 'ideas' by habit, association or some other 'adhesive' force. What is wrong with those theories? The obvious retort is that they overlook the rule-governed nature of language, and replace rule and law with mechanism and habit. But this is not Grice's retort. His concern is not with the rules that define a language, but with the *act* of meaning. We often mean something by gestures that have no rules to govern them: by waving, say. 'Speaker's meaning' is the fact in which all language is ultimately rooted. If there are rules of language, they are rules which individual speakers apply in the act of *meaning* something.

Stevenson's analysis will not do, for many reasons: thinking of Mary caused me to speak of cabbages, which prompted your thoughts of last week's cricket match. As readers of *Ulysses* know, this is the stuff of mental life; so why did I mean *cabbages* when I used that word? Conversely, my thinking that the vicar is coming to tea causes me to put the kettle on which causes you to think that the vicar is coming to tea. But surely my putting on the kettle was not an act of *meaning* that the vicar is coming to tea?

Let us return to our intuitions about meaning. First, meaning is something that people *do*. (There is also 'natural meaning', as Grice calls it, as when clouds mean rain, but this is clearly not the same phenomenon.) Moreover, meaning, in its primary occurrence, is a three-term relation. John means something by his gestures to Mary. For meaning is an attempt to communicate; soliloquies are deviant cases, to be explained only when we have grasped the central relation. Thirdly, meaning is an intentional act: to mean something is to do something with a particular intention. And that intention is directed towards the other person. Fourthly, the other person has to understand the gesture, and this is part of what I intend in making it.

So here is a suggestion: John means that *p* by his action, if he performs it with the intention to induce in Mary the belief that *p*. This will not do, Grice argues, since it fails to distinguish those cases where someone intends another to believe something but expressly *refrains* from meaning it. I want you to realise that your wife is unfaithful, and so scatter the evidence across your path. But I don't want you to accuse me of saying it or meaning it; so I hide my intention. The example suggests what is needed: namely, a second-order intention. I must intend that the other recognise my intention. So here is Grice's analysis of 'John means that *p* to Mary by doing *x*':

John does *x*

(1) with the intention that Mary believe that *p*;

(2) with the intention that Mary recognise intention (1)

(3) with the intention that Mary's belief that *p* comes about through the recognition of his intention (1).

To put it more succinctly: John does something with the intention that Mary believe that *p* by recognising that such is his intention. Meaning is a matter of second-order intention: of intentionally drawing attention to one's own intention. So says Grice, and many people have agreed with him.

A small industry developed during the sixties and seventies, devoted to the task of finding counter-examples to Grice's analysis, and to proposing further refinements designed to avoid them. (Thus, you can imagine circumstances which prompt you to add another layer of intention: the intention that the second-order intention be recognised. Maybe you could climb this ladder for a good many rungs before unambiguously *meaning* something.) There is also something wrong with the idea that meaning is always a matter of inducing a belief. Nor does Mary have to *acquire* the belief that *p* in order to understand John's meaning. She could understand what he says, and not believe a word of it. But what is understanding? Surely, it is grasping the *meaning* of something. In other words, John means that *p* only if he intends his audience to *grasp his meaning* by recognising his intention: the analysis is beginning to run in a circle, and a particularly involuted circle too, involving intensional idioms: the kind of idioms that a theory of meaning is supposed to explain.

Another objection has been mooted by Searle and others. To say that John intends to do something is to imply something about his beliefs. You can intend to do only what you believe to be possible. (I cannot now intend to jump to the moon, even though I may desire to do so.) Moreover, it is not any kind of possibility that is involved. It is possible, in one sense, that I should knock out Mike Tyson in a clean fight – there is a possible world in which just that happens. But in a real sense it is not *within my powers*; hence I cannot intend to do it unless seriously self-deluded.

Likewise second-order intentions can exist only in those circumstances in which people can seriously believe in the possibility of accomplishing them. Under what circumstances *can* I believe that you will be able to extract from my words precisely the intention that you believe that *p* for all those minute variations in the value of *p* that we encounter in our daily acts of communication? Clearly we need conventions, rules, systems with which to settle what can be expected and what assumed. That is, we need a system of meaningful signs – in short, a language. But it is not surprising if we can mean something by the use of language; for sentences in a language already have meaning, for the reasons Frege set before us.

3. Speech Acts

Some philosophers believe, on the strength of that last consideration, that Grice's attempt to reduce meaning to *speaker's meaning* is therefore bound to be a failure. The analysis will always depend on a theory of sentence meaning, which will in turn depend upon a theory of language. It is only when language is in place, that speakers can form the second-order intentions required by Grice. Far from giving a theory of meaning, Grice supposes that we can obtain one from another source.

But *is* this an objection? It is important to distinguish vicious circles from harmless ones. If you can give no account of *a* except in terms of *b*, and no account of *b* except in terms of *a*, then you are in trouble. (Or are you? It depends on what you mean by 'account'.) However, even if a *full* specification of what is involved in *a* involves a full specification of what is involved in *b*, and *vice versa*, this may not matter. For perhaps we can identify what we have in mind by *a* and *b* in some other way. In our case, we have done just that, specifying speaker's meaning through intention, and sentence meaning through semantic rules. And we are now in a position to see that the two phenomena are mutually dependent. It is semantic rules that give to speaker's meaning its full range and potential; but it is second-order intentions that bring language alive, and convert those formal rules into rules of *meaning*. Such is the suggestion made by Searle. And it is backed by an ingenious argument of David Lewis's (*Convention*). Rule-guided activities arise in a community, Lewis argues, only when its members begin to convey second-order intentions of Grice's kind. Convention and second-order intention arise together in the genesis of social life.

Those thoughts are applied by Searle to the theory of speech-acts, which had first been adumbrated by J.L. Austin, in a series of posthumously published lectures (*How to Do Things with Words*). Although assertion is a primary use of language, it is not the only use. Questions, commands, warnings and promises all involve acts of the kind analysed by Grice. Moreover, Searle argues (*Speech Acts*), they all exploit the same body of semantic rules, since every speech act has a propositional content (roughly, the Fregean thought) that is now asserted, now questioned, now suggested, now promised. The two parts of the concept of meaning therefore come together in the theory of speech-acts.

The details of Searle's theory need not concern us: they are in any case highly controversial, and certainly wrong in many respects. The important point to grasp is that, by a judicious combination of Grice and Frege, we can overcome many of the objections to both philosophers. In asserting *p* I am not intending to induce the belief that *p*. I am intending the audience to recognise *my* belief that *p*, at least in the normal case, when I am speaking sincerely. (Insincerity is a secondary case, possible only when the web of mutual confidence has been established.) In warning I intend you to recognise a danger; in promising I intend you to trust my intention; and so on. In each case I intend you to recognise something by recognising also that this is my intention. Such a second-order intention is possible because you, like me, have grasped the semantic rules that identify a proposition, which is the common subject-matter of our thought.

Speech-act theory is an interesting branch of the philosophy of language. But it occupied the centre of the stage only briefly, and only for so long as philosophers failed to attend to the notion that it takes for granted: the notion of a semantic rule. Just what is involved in understanding the rules of a language? Surely, the real questions about meaning will remain unanswered, until we produce a theory of meaning that will show how we comprehend a language, so as to use it to speak about the world.

4. What is a Theory of Meaning?

Thus began a debate that has all but saturated the channels of philosophical communication in recent years, and which no student of the subject can ignore. In another famous paper ('Truth and Meaning', 1967) Donald Davidson initiated the programme for which he is renowned. The aim of this programme was to show that truth is the central concept in the theory of meaning. What are we asking for, when we seek a theory of meaning for a language? The study of artificial languages suggests that we seek to *assign* meanings to all the sentences of the language – including those that we have never previously encountered. How is that done?

Suppose we had a theory that assigned meanings directly: which told us, for each sentence *s*, that *s* means *p*. In order to understand such a theory we should have to know what kinds of things could be substituted for '*p*' in the formula '*s* means *p*'. But this formula contains an intensional context: we cannot substitute sentences for '*p*' on the grounds of their shared truth-value, since that would generate absurdities, as in the case of 'Mary believes that *p*'. It does not follow from the fact that Mary believes that *p*, and the fact that *p* and *q* have the same truth-value, that Mary believes that *q*. Likewise, we must surely deny that a sentence means exactly the same as every sentence that shares its truth-value. To think otherwise is to confuse sense and reference. Indeed, it is plausible to assume that we could substitute for '*p*' in our formula only those sentences with the same *meaning* as '*p*'. But then we should need a criterion for assigning meanings to each of our sentences. And that is precisely what we are looking for. A theory of meaning that uses the notion of synonymy or sameness of meaning will therefore never get off the ground.

We must therefore look for an extensional theory of meaning; a theory that permits substitution of terms with identical reference. What would such a theory be? We can answer that question, Davidson suggests, by asking another. What is the criterion of *success* for such a theory? Such a criterion would show us that we have assigned to every sentence the meaning that it actually has. Consider then, some predicate T, which is predicated of sentences, and which tells us of each sentence to which it is applied that it is correctly used in these or those conditions. To each sentence *s* there will be a formula, saying that *s* is T if and only if *p*, where '*p*' states the conditions under which it is right to use *s*. If we could find a theory that generated, for every sentence of the langauge, a formula of this form:

(1) *s* is T if and only if *p*

where '*p*' states exactly what must be the case for *s* to be used correctly, then we should have a theory of meaning. But what makes such a theory successful? We need an *a priori* guarantee that *s* is true only when '*p*' is, and *vice versa*. The sentence '*p*' must therefore give the truth conditions of *s*. We can be sure of this only if '*s*' is the *name* of the sentence '*p*'. The predicate 'T' will then naturally be read as 'true', and (1) will become Tarski's condition of adequacy for a theory of truth. A theory of truth is *adequate*, Tarski says, only if it generates all sentences of the form:

(2) 'Snow is white' is true if and only if snow is white

i.e. only if it pairs each sentence with the condition that is stated by it. (That is what we mean by 'correspondence': see the discussion in Chapter 9.)

Tarski developed theories of truth for artificial languages. He tried to show how you could assign values (referents) to the vocabulary of a language, so as to develop systematic theories which determine truth-values for the completed sentences in terms of the values of their parts. In the formula (1), the left-hand side names or identifies a sentence, while the right-hand side states its truth-conditions, in terms of the semantic properties of its parts. The theory obeys the principle of finitude: that is, indefinitely many truth-conditions are built up from the finite assignment of values to the basic vocabulary. But the theory is never at a loss; for every well-formed formula (every sentence) it specifies a truth-condition. The theory of truth says all that we need a theory of meaning to say. (It also bears out Frege's idea, that the sense of a sentence is given by its truth-conditions.)

For reasons that I shall discuss in Chapter 27, Tarski was sceptical about the possibility of extending his approach to natural languages. But Davidson brushes this scepticism aside, arguing that a theory of truth is simply the best that we can achieve by way of a theory of meaning. Such a theory tells us how we understand the sentences of our language; it generates truth-conditions for each sentence, regardless of its novelty; it is extensional, and therefore depends upon no idea of sameness of meaning; and it passes the most important test of all – it gives as the meaning of each sentence precisely what *that* sentence says.

It is important to see what Davidson is *not* arguing. He is not arguing that he has produced a theory of meaning for English or for any other natural language. He is merely saying what such a theory would be like, if we could find one. He is not arguing that the concept of meaning is redundant, or replaceable by that of truth; he is simply saying that a theory will give us all that we hope for, and all that we need, just so long as it satisfies the formal criteria laid down for a theory of truth, and in particular the condition of adequacy. Finally, he is not arguing that a theory will consist merely of trivial-seeming formulae like (2) above – formulae that simply pair each sentence with itself, by first naming it, and then using it. He is saying that the theory will be adequate so long as it *entails* all formulae like (2). In other words, we know that '*p*' states the truth-conditions of some sentence *s*, so long as it is equivalent to the very sentence that '*s*' stands for.

(A word of caution: all this is difficult to state. For a theory of truth, Tarski argued, cannot be expressed in the same language as the one under investigation (the 'object' language), but only in another language (the 'meta-language'). If we do, nevertheless, try to use the same language, we must be very clear that sometimes we *name* sentences, sometimes we *use* them. In the above account '*s*' is a name; likewise " 'snow is white' " is a name, whereas 'snow is white' is the sentence named by it. The distinction can be made, as I have just made it, by first naming a name, and then naming a sentence. But inverted commas tend to get out of hand. See the dialogue between Alice and the White Knight in *Through the Looking Glass*.)

What is the upshot of all this? The most important points are these:

(1) The concept of truth has again come to occupy the central place in the theory of language.

(2) The meaning of any sentence is given by its truth-conditions.

(3) To understand a sentence is to know the conditions under which it is true.

5. Truth and Assertibility

The last of those points has been seized upon by Michael Dummett, in his extended commentaries on Frege, so as to raise a new set of questions that go to the heart not only of the philosophy of language, but also of metaphysics itself. What is it, to 'know the truth-conditions' of a sentence? Until we have answered this question, we shall not have shown how the sentence is understood. It is not enough to give a *statement* of truth-conditions: for how do we understand that statement?

The difficulty is serious, Dummett thinks, since there is nothing in the concept of truth, as discussed by philosophers like Davidson, which ensures that we can really have a grasp of it. There are many sentences in our language whose truth-value we could never know, because their truth-conditions transcend our epistemological capacities. Thus certain sentences are 'undecidable' – for instance, 'A city will never be built on this spot' – while others reach beyond any evidence that a finite being could collect for them: universal laws ('All stars contain helium'), sentences about necessity and possibility, and so on. In stating truth-conditions for these sentences we seem to describe states of affairs that reach beyond our ability to know them. If the concept of truth plays the central role in a theory of meaning, therefore, such a theory may prove unable to explain how we understand our language.

Dummett's arguments for this position rest on two plausible considerations, one derived from Frege, the other from Wittgenstein. First, the sense of a sentence is what we understand in understanding it (Frege): therefore sense must be 'within our grasp'. Secondly understanding must be 'manifest' in linguistic practice, if language is to be learned and taught (Wittgenstein). Both considerations point to the same conclusion: namely, that we must tie the meaning of a sentence to the circumstances that justify our use of it. If we cannot do this, then we cannot 'know *when*' to utter the sentence; nor can we know of another person (whom we are teaching, say) that he utters the sentence in circumstances which justify its use.

There are a variety of conclusions that are drawn from such arguments. Some philosophers argue that truth is less fundamental than 'assertibility'. We teach language by laying down rules which tell a speaker when he 'has the right' to use a sentence. These rules are bound by his epistemological limitations, and exclude any reference to unknowable 'truth-conditions'. Alternatively, a philosopher might argue that the concept of truth remains fundamental, but must be revised so as to 'fit' our limitations. We have no grasp of a 'verification-transcendent' concept of truth; but nor do we need such a concept. We could make do with another idea, according to which truth just *means* 'assertibility'.

Either way, the upshot is 'anti-realism'. This is the term coined by

Dummett to express his dissatisfaction with realism. A realist theory of meaning holds that our language may represent a transcendent reality, lying beyond the information that we may obtain about it. If our language *were* like that, Dummett argues, then we could not understand it. Although language has a built-in reference to a 'world beyond', it is a world whose metaphysical contours derive from our own epistemological capacities. What we can *mean* is determined by what we can *know*.

Put thus, the theory seems uncontentious. But it is not normally put thus. Indeed, for obscurity and self-involvement, the literature of 'anti-realism' has no competitors outside the camp of 'deconstruction'. Sometimes anti-realism looks like verificationism, telling us that the meaning of a sentence is given by our procedure for confirming it. But verificationism, being honest, has exposed itself to too many objections. The concept of 'assertibility' is therefore constantly refined by anti-realists, in order to repudiate its disreputable relations. Sometimes anti-realists admit that ordinary language contains 'realist' assumptions. They therefore propose to revise ordinary language; in particular, to revise the logic around which it is constructed. Thus we accept the law of the excluded middle, which says that there is no 'third-way' between *p* and not-*p*: if *p* is not true then not-*p* is true. But if 'true' means 'assertible', it is doubtful that we can make such an assumption. If we are not justified in asserting *p*, it does not follow that we are therefore justified in asserting not-*p*. Or, if it does follow, that is because we are defining assertibility in terms of truth, and covertly smuggling back the assumptions that are buried in our ordinary language. It seems likely, therefore, that an honest anti-realist will be committed to a 'deviant logic'.

The debate goes on, fighting its way across the field of logic. In the end, however, anti-realism seems to be identical to what Kant called 'transcendental idealism'. According to Kant, we have the idea of a transcendental reality: of a world 'as it is in itself', seen from no point of view. But this idea is no more than a shadow cast by thought: we can never really grasp it, and the task of philosophy is to free us from its illusory power. Reality as we know it is known from our point of view, which is the point of view of 'possible experience'. Our thoughts gain their content from the experiences that give grounds for them. The world as we know it is the empirical world; but the empirical world is real and objective, unlike the transcendental world, which is merely ideal. Transcendental idealism is therefore an empirical realism. The concept of truth makes sense, so long as it is used to relate our thought to empirical reality. Any other use of it will lead to paradox and contradiction.

Kant's transcendental realism is a metaphysical theory, founded in arguments concerning the nature and limits of human knowledge. One of the major objections to Dummett's 'anti-realism' is that it advances to vast and vague metaphysical conclusions from premises that are purely *semantic*. No considerations are adduced for the theory, apart from a few far from clear intuitions concerning the understanding of language. Semantic anti-realism therefore covertly assumes precisely what it ought to be proving: namely, that the nature of our world is dictated by the nature of our language, and that there is no point of view outside language from which to make sense of the really real. It is surely far more plausible to suggest that metaphysics comes first, and that the nature and limits of what can be said depend upon the nature and limits of the knowable world.

Similar to Kant's transcendental idealism, and probably indistinguishable from it, is the position advocated by Hilary Putnam, and called by him 'internal realism'. This holds that we cannot attain a point of view outside our system of concepts, so as to compare them with a non-conceptualised reality. Nevertheless, as long as we remain within our conceptual system, we are entitled to deploy a realist idea of truth, and to compare our beliefs and sentences one by one with the independent reality. There is no assumption that this reality is fully knowable, or that our methods can completely encompass it. But to suppose that there is a real world which entirely transcends our capacity to know it is to make a nonsensical suggestion. Such 'external realism' is empty in just the way that 'transcendental realism' was empty for Kant.

6. The External Viewpoint

The same Hilary Putnam has, in recent writings, cast doubt on the whole possibility of semantic analysis, as this has been conceived by Frege and his successors. According to Frege, we understand language by grasping the senses of words and sentences; sense in turn 'determines' reference, so enabling us to advance from language to the world (to objects, functions and truth-values). This picture of language corresponds to a picture of the human mind. Meaning and thinking go on 'in the head': a thought is the sense of a sentence, and to have a thought is to entertain a mental item, composed of other mental items, each corresponding to the sense of some word or phrase in the public language. Hence I can know what I mean, and what I am thinking, without knowing whether my thought is true, or whether there is anything in the world to which it refers. This picture has been called 'psychological individualism' (by Tyler Burge), since it implies that the individual mind is somehow complete in itself, and contains within itself all that is necessary to form thoughts about the world.

In fact, Putnam argues, the content of a thought may depend upon circumstances 'outside' the mind of the thinker. Frege notwithstanding, it is often the case that reference determines sense, and not vice versa. Indeed, we have already come across examples: the proper names and natural kind terms discussed by Kripke in his theory of the 'rigid designator' (see Chapter 13). Consider the term 'gold'. We introduce this into our language by means of certain paradigm instances – pieces of a yellow, malleable metallic substance, whose observable properties we quickly learn to recognise. But we do not *define* gold through these observable properties. In understanding these examples as paradigms of gold, we do not close our minds to the possibility that there may be samples of yellow malleable metallic substances which are not gold, or to the possibility that there may be samples of gold that are not yellow, malleable or metallic. 'Gold' is introduced as referring to *this stuff*, whatever it should turn out to be. The sense of the term is given by the fact that it has *this* reference. Scientific investigation is required in order to determine this reference; only then will we have a clear idea of what we *mean* by 'gold'. The scientific theory of gold, couched in terms of its atomic number, atomic weight and place in the periodic table of the elements, fixes the 'real essence' of gold, by telling us what *has* to be true of some piece of matter, if it is to be gold. It therefore determines the sense of the term 'gold': what we really mean when we use this word. It is precisely because reference determines sense, that there are necessary *a posteriori* truths about natural kinds.

Putnam proposes the following thought-experiment. Suppose there exists a planet, exactly like ours in every respect, and populated by language-using people who are also like us. But suppose that there is one tiny difference between that planet ('twin earth') and ours: namely, that there the transparent, thirst-quenching liquid that fills the lakes and rivers and sustains the life of plants and animals is not H_2O but some other compound, XYZ. The inhabitants of twin earth call this liquid 'water' and use it and describe it in every way as we do. What goes on 'in their heads' when they think of or refer to water is exactly what goes on in *our* heads. But *their* thoughts are not the same as ours: for their thoughts are about XYZ, while ours are about *water*, hydrogen hydroxide. This fact is reflected in the difference of meaning between the term 'water' used by them, and the same term used by us.

The conclusion that Putnam draws from his thought-experiment is that thinking is not simply 'in the head': to think is to stand in a complex relation to things outside us, and those things play an active role in providing the contents of our thoughts. The argument has been taken further, by Tyler Burge and others. Hitherto we have described the workings of language without reference to the context of use, as though words were attached to the world by stipulation, through arbitrary rules. But words do not come into use in that way. They are *taught* to us, by others who observe our circumstances from outside. I see the child staring at a horse, and say 'horse'; but I am already assuming not merely that the horse is there, but that he *sees* the horse – i.e. that the horse causes in him a particular perceptual experience. This causal link between the world and the observer is built into the language. 'Horse' comes to mean a certain kind of thing, which acts on an observer in the way that horses do. A link between the speaker and the world is established in the very meaning of the word.

Likewise with proper names. In teaching the name 'Mary' I must assume a link between Mary's presence and the child's act of recognition. Mary herself plays an immovable part in the teaching of her name, and is, in consequence, a real component in the thoughts that refer to her. What makes my thought a thought of *Mary*, rather than Elizabeth, is not just what 'goes on in my head', but the chain of influence which connects my present thinking to the woman herself. (Some philosophers have gone on to produce a 'causal theory of names': a word is a name, they argue, if its use is prompted in the right kind of way by the object named. The suggestion is highly controversial; after all, we give names to fictions, to unborn children, and to remote historical figures. The most that could be shown is that our *paradigm* names – the names through which a child learns the 'game of the name' – are causally linked to their objects.)

Those suggestions are highly controversial. But they point to a serious lacuna in the traditional theories of meaning. We need to ask how our words are attached to the world, and also, what kind of a world it must be, if *words* can attach to it. To speak of rules and conventions is all very well; but how are rules understood, and what must be assumed in teaching them? Maybe the real empirical world enters into our thoughts more intrusively than Cartesians would like: and maybe, if we were to give a complete theory of meaning, we should have to give a complete theory of the world.

7. Realism and Rule-following

The most famous argument of Wittgenstein, apart from that against the private language, concerns teaching and following a rule. The argument, developed in *Philosophical Investigations*, *Remarks on the Foundations of Mathematics*, and elsewhere, has had an enormous influence, even though there is little agreement as to what it implies.

If language is an instrument of communication, then it must be founded on rules, enabling speakers to recuperate the content of one another's speech. What is it, to 'follow a rule'? Suppose I give someone the rule: 'add 2'. He then proceeds to apply this rule to the natural numbers, writing down 2, 4, 6, 8 and so on. On reaching 1000 he goes on 1004, 1008, 1012. I protest that he is not going on in the same way. No, he replies, this *is* the same way. How do we resolve such a dispute?

We might be tempted by the following response: 'add 2' just means going on 1002, 1004 and so on. That is what you understand in grasping its meaning. But what do the words 'and so on' mean? Surely you have not listed *all* the applications of the rule: there are infinitely many. Well, you say, I just mean 'the same again *ad infinitum*'. But then we are back where we started. What counts as the *same*? Our recalcitrant rule-follower may retort: 'I understand now; I must go on like this: 1002, 1004, 1006', and you nod enthusiastically. But he reaches 2000, and proceeds 2004, 2008, 2012. As you can see, the possibilities of deviance are infinite.

Naïvely, we should say that, if someone has understood the rule, then he *must* go on as we do. But what does this 'must' refer to? Nothing that our deviant rule follower has done up to any given point compels him to go on in one way rather than another. (Remember Hume's law?) Whatever he has done in the past, it is only a contingent fact that he goes on as he does. When we say that he *must* go on in our way, given that he understands the rule, we mean only that this is what we *count* as understanding it. But then the matter again comes to rest in our decision. This is just what we do.

You may have noticed a similarity between the argument and Goodman's paradox concerning 'green' and 'grue'. Like Goodman's paradox, the rule-following argument can be strengthened so as to yield alarming conclusions. So far it has tended in a nominalist direction: the meaning of our rules is given by our practice. It is what we do that is the arbiter of what we mean, and the classifications that we use are dictated by our own decisions. But return to the example. What we mean by the rule 'add 2' is determined by the *fact* that we go on 1002, 1004, etc. Did we know in advance that we would go on in this way? Maybe; but maybe not. Since the rule has infinitely many applications we cannot know *all* the applications in advance. How do we know *now* what we shall do at some point that we have never reached? If our only test is to carry out the calculation, then we do not know what we mean – since the meaning of the rule is given by all its applications, including those that we have never made.

In his book *Wittgenstein on Rules and Private Language*, Kripke gives the argument in another form, so as to emphasise its sceptical force. Imagine a mathematical operation called 'quus' defined as follows: x quus $y = x$ plus y, provided x and y are both less than 50; otherwise x quus $y = 57$. There is nothing inconsistent in that mathematical function, and you could easily imagine a use for it. Suppose now that Jim is good at arithmetic, but has never dealt with numbers higher than 50. All his additions have involved the numbers up to 50; none has involved any higher number. How do we know that Jim means plus, when he says 'x plus y', and not quus? He cannot mean both, since they are incompatible. So which? The evidence is compatible with either hypothesis. Indeed, there are infinitely many functions that might be meant by 'plus' on Jim's lips, all of them incompatible with the others. Since we cannot determine whether Jim means plus or quus, we cannot know what he means. Worse, we cannot know what *we* mean either. However far we go with our arithmetic, there will always be calculations that we have not performed. And how do we know in advance that, when we perform them, the answer will come out as x plus y, rather than x quus y, for some new version of 'quus'?

Kripke draws the parallel (which should by now be obvious) with 'green' and 'grue'. And he goes on to espouse, first a radical scepticism about meaning, and secondly a kind of 'anti-realist' response to scepticism. Neither position has recommended itself to his readers. But his trenchant statement of Wittgenstein's argument has awoken all of them from their 'dogmatic slumbers', just as Hume's argument about causation (the original of all these sceptical paradoxes) awoke Kant from his.

Wittgenstein himself was concerned to shift philosophy in an anthropological direction. The ultimate facts, he suggests, are not to be found where philosophers normally search for them. They do not concern necessities, but deep contingencies about the human condition: about our customs, practices and 'forms of life'. It is these that we must take for granted as 'the given'. And it is to these that we should return, in order to answer the ultimate questions.

8. Other Kinds of Meaning

I have dealt with questions that are discussed by modern philosophers; but they are not the only questions about 'meaning'. We refer to the meaning of music, the meaning of ritual, and the meaning of life. It would be rash to assume that meaning is here the same phenomenon, or that it should be understood through the same devices.

Nevertheless, writers have been tempted to extend the concepts that we use in analysing language to other systems of 'signs', begging many questions in the process. For it is undeniable that communication occurs in many ways, and that there are 'systems' of symbols which are not systems of words. Here are some examples:

(a) Codes: such as semaphore, flags, badges, uniform markings.

(b) Conventional dress and behaviour: black clothes and slow solemn gestures at a funeral; handshakes, nods and small-talk during introductions; rituals and ceremonies.

(c) Pictures and images, which show us real or imaginary worlds: advertisements, logos, heraldic devices.

(d) Representation in art, whether through figurative painting, verbal description, or some other medium.

(e) Expression in art, as when a piece of music expresses an emotion, or a play conveys a vision of life. Expression is what is left over, when representation has been subtracted. You could give a complete account of the world described in *Faust* – a list of characters, situations, things said and done – and yet not exhaust the meaning of the play. A vision of the human condition is expressed by it: and it is the task of the critic (no easy task) to tell us what it is.

(f) Figures of speech. There are ways of meaning things in language which are not accommodated by a theory of literal meaning. Metaphors, for example, introduce a wholly new dimension of meaning. A metaphor does not describe a connection so much as create it. It may be full of contradictions, and yet mean all the more on account of it:

I have no spur
To prick the sides of my intent, but only
Vaulting ambition, which o'erleaps itself
And falls on the other side

The concentration of meaning in that tumble of equestrian images goes far beyond the sense and reference of the terms.
(g) The meaning of a landscape, as displayed in Wordsworth's *Prelude*. You may think this is not an example of communication. But why not say, rather, that the poet uses the landscape in order to communicate with himself? He *endows* it with a meaning.

Could we envisage a theory of symbols that would accommodate all those varieties (and perhaps more as well)? In his *Elementary Course of Linguistics* (1902) the Swiss linguist Ferdinand de Saussure made passing reference to a 'general science of signs', of which linguistics would be a special case. He called this general science 'semiology'. The name was later replaced by 'semiotics' (a term first used by C.S. Peirce), as people began to renounce the search for a general method. When a writer now refers to the 'semiotics of music', there is a veiled assumption that the meaning of music arises in a way comparable to the meaning of a sentence.

Saussure was also the founder of 'structuralism' in linguistics. He argued that the structural relations within a sentence are the true vehicles of meaning, rather than words taken in isolation. Saussure's way of expressing this point is so much inferior to Frege's, that he is rarely studied by modern philosophers. However, at a certain point in the history of French literary criticism, Saussure became a cult figure, and the jargon of Saussurian linguistics entered the rhetoric of the Parisian Left. Those involved in the matter (Roland Barthes, for example) were largely ignorant of Frege, Russell and Tarski. But the 'general science of signs' promised by Saussure lent itself to their radical political agenda. Their hope was to find a method for 'decoding' the artefacts and conventions of bourgeois society, so as to expose their meaning, and endorse the prevailing intellectual contempt for them. Hence arose the peculiar aberration called 'structuralist criticism', whose pseudo-scientific jargon and radical message is taken one stage further by its successor – the 'deconstructionist criticism' of Derrida. (See Chapter 30.)

Of course there are interesting analogies between language and other kinds of sign. The question is whether those analogies can be used to give a generalisable theory of meaning, of the kind that the structuralists wished for. The two analogies that interested them were these: first, all human behaviour can be seen as expressive. It reveals thoughts, feelings and intentions, not all of which would be spontaneously acknowledged by the agent. Secondly the modes of human expression seem frequently to have a language-like structure. And since meaning in language is generated through structure, it may be that the same is true of meaning generally.

According to the Saussurian model, a sentence is a 'system' composed of 'syntagms'. A syntagm can be defined as a set of terms that may replace one another without destroying the system – without rendering the sentence 'unacceptable' to speakers of the language. For example, in the sentence 'John loves Mary', 'loves' may be replaced by 'hates' or 'eats', but not by 'but', 'thinks that' or 'swims'. Now consider an example from Barthes: the menu. (See his *Eléments de sémiologie*.) A person might order the following: *oeufs bénédictine*, followed by steak and chips, followed by rum baba. That is an 'acceptable system'; but in our society the same menu in reverse would not be 'acceptable'. Furthermore, each dish belongs to a 'syntagmatic unity': it can be replaced by some dishes but not others. Steak and chips may be replaced by ham salad but not by a glass of Sauternes – for that would be unacceptable. (And, as in the case of language, there are conventions at work here, which we might wish to elevate to the status of 'rules of grammar'.) What follows?

Consider Barthes's actual interpretation. Steak and chips is supposed to 'mean' (according to an essay in *Mythologies*) 'Frenchness'. Suppose that the 'meaning' of *oeufs bénédictine* is 'Catholicism', and that of rum baba 'sensuality': what now is the meaning of the whole system? Does it mean that French Catholicism is compatible with sensuality? Or that being French is more important than being Catholic? Or being sensual a fundamental part of both? There is no way of telling, since while the system has a kind of syntax, it is connected with no *semantic* structure. There is no way of deriving the meaning of the whole from the meaning of its parts. In all the works of structuralist critics you will find the same defect: that the analogy with syntax remains a mere analogy, since it is divorced from semantic rules. Moreover, when the kind of meaning that is intended belongs to the broad category of 'expression', this absence of semantic rules is inevitable: a point to which I return in Chapter 29.

Although modern philosophers ignore structuralism, and for good reasons, they do not dismiss 'semiotics' entirely. Several have tried to generalise the theory of meaning from language to art. Most notable among them is Nelson Goodman, who has attempted to explain both representation and expression in terms of a general theory of reference (*Languages of Art*). His task is facilitated by his nominalism, and by his refusal to tie the notion of reference down to the Fregean theory that first gave sense to it. Nevertheless, his argument has been influential, if not for its persuasive power, at least for what it promises.

I shall return to this topic in Chapter 24. For the moment, however, I should counsel scepticism. There is no doubt that the meanings we find in art, literature and music, in ritual and ceremony, in religion and life, are not one thing but many. There is no doubt, too, that these meanings contain the most precious of our values. But these facts ought to warn us against extending theories of language into areas where language may trespass only in the guise of metaphor.

20

Morality

Ethics is so difficult and so various a subject, that it is usually treated as a distinct branch of philosophy. However, it is possible to understand a philosopher's metaphysical position only by grasping the place of morality within it (even if it *has* no place); and the questions of ethics cannot be answered, while leaving the rest of philosophy intact. Before returning to epistemology and metaphysics, therefore, I propose to take a detour through morality. In common with other modern philosophers, I believe that the topic has for a long time been misunderstood.

The misunderstanding derives from two sources: an impetuous belief that philosophical questions are solved through the analysis of language; and an inherited moral idiocy which has plagued English-speaking philosophy since Bentham. Neither of these defects has been *wholly* deleterious. For each has produced clarity, in an area where clarity threatens vested interests. Nevertheless, both have impeded the kind of circumspect view of the subject that is necessary if its significance is to be understood.

But what is the subject? Where in the world is morality? Is it a matter (as modern philosophers have assumed) of a certain kind of *judgement* – maybe even of certain 'evaluative' *words*? Is it a matter, as Kant argued, of action and practical reason? Is it a matter of emotion, sympathy, motive, as Hume supposed? Or is it, as Aristotle suggested, a matter of character and moral education? The simple answer is that all those things are relevant. But one of the greatest mistakes is that made by modern philosophers, in assuming that moral judgement is in some way primary, merely because it can be discussed in the terms that modern philosophers prefer. However, the precedent is there; I shall therefore begin from the idea of moral judgement, in the hope of soon transcending it.

1. The Logic of Moral Discourse

Modern philosophy launched itself into the domain of ethics with a work by G.E. Moore – *Principia Ethica*, published a few years before Russell's 'On Denoting'. The book was made famous by certain members of the Bloomsbury circle (Keynes, Clive Bell, Virginia Woolf, and Russell), who claimed that its advocacy of personal relations and aesthetic sensibility had decisively changed their minds. (It was not so much that their minds were changed, as that the book coincided with their predilections – that is the usual way books have an influence.) In retrospect the moral message of *Principia Ethica* is no more persuasive than the life-style that it briefly helped to sustain. Of far greater interest is the argument with which the book begins:

(a) The Naturalistic Fallacy

Moore thought that he discerned a fallacy in many of the traditional ethical systems: the fallacy of identifying goodness with some property other than itself. Many philosophers have tried to define the term 'good', by listing the 'natural' properties which good things possess. For example 'utilitarians' define goodness in terms of happiness: that is good which promotes the greatest happiness of the greatest number. By a 'natural' property Moore meant a property that was straightforwardly part of the natural world, as happiness is: a property whose causes and effects are discoverable by the standard methods employed in understanding nature. All such 'naturalistic' definitions of the term 'good' commit a fallacy, he argued. For suppose it were true that 'good' meant 'promoting happiness'. Then the statement 'Whatever promotes happiness is good' would be a tautology, equivalent to 'whatever is good is good', or 'whatever promotes happiness promotes happiness'. But in that case it would be absurd to ask 'Is the promotion of happiness good?' This would be like asking 'Are good things good?' But, Moore went on, it is *never* absurd to ask, of *any* natural property, 'Is it good?' It is always an *open question* whether some natural property is good. And that is tantamount to saying that 'good' can never be defined in terms of natural properties.

This 'open question' argument is clearly invalid. A question remains open just so long as our ignorance permits, regardless of whether the thing questioned is a necessary truth or even a 'tautology'. Consider mathematics. It is perfectly sensible to ask whether a Euclidean figure with three sides is a figure whose internal angles add up to 180 degrees. This is something you have to prove, by arguments that are doubtless no longer taught in school. But the two concepts are necessarily coextensive, and you could define the one in terms of the other. Likewise, it may need a long philosophical argument to establish that the concept of moral goodness is bound up with that of happiness. In which case it would remain an open question whether promoting happiness is good, even though it is a necessary truth that goodness involves the promotion of happiness, and even though this is what 'good' really *means*.

Moore drew an interesting conclusion from his phoney argument. Goodness, he argued, is a property of whatever possesses it. (Otherwise it would never be *true* to say of something, that it is good.) But it is not identical with any natural property. Nor is it definable. Therefore goodness is a 'simple' (undefinable), non-natural property. It is a property, but one whose metaphysical status sets it apart from nature.

(b) Emotivism

That is not the conclusion drawn by Moore's immediate successors. There was a widespread conviction that Moore had found the proof of something called 'Hume's law' (which is not the law that I have referred to under that label). In the *Treatise*, Hume remarked upon the ease with which moral philosophers slip from what *is* the case, to what *ought* to be the case. But, he added, this transition from 'is' to 'ought', so easily made, is far from easily explained. Later commentators took Hume as authority for the existence of a 'gap' between 'is' and 'ought', arguing that there is no logical proof of the second from the first, however complex and informative the premises.

This, then, is the explanation of the 'naturalistic' fallacy. Since statements about what is good entail statements about what we ought to do, 'good' must be in the same boat as 'ought', and statements about the good must be separated from statements about what *is* by an impassable logical barrier. By a rapid association of ideas philosophers connected both 'Hume's law' and the naturalistic fallacy with a supposed ontological divide between 'fact' and 'value'. Hard-headed sociologists like Max Weber make such a distinction, arguing that scientific sociology must deal with facts alone, and is therefore 'value-free'. Many honest readers imagined they had a clear and distinct idea of Weber's meaning, and heartily endorsed the view that science must be factual (or 'positive'), making no reference to the 'values' of this or that community of observers.

In short, bad arguments, speculative distinctions, and scientistic prejudice, combined to persuade philosophers that there is a distinction between fact and value, between 'factual' description and moral judgement, between 'is' and 'ought' – and that these are all *one* distinction marked out by the peculiar 'logic of moral discourse', which unfits it for the description of the world. Moral discourse, philosophers argued, is essentially 'evaluative': the word 'good' does not refer, as Moore supposed, to a 'non-natural' property; it refers to no property at all. Moral language functions in another way from descriptive language, and words like 'good' have an 'evaluative meaning'.

The first developed theory of evaluative meaning was emotivism, usually discussed in the form advanced by C.L. Stevenson. One purpose behind Stevenson's 'causal' theory of meaning (see last chapter) was to enable him to introduce an idea of evaluative meaning on a par with descriptive meaning. For Stevenson a sentence is a causal link between states of mind: my belief that *p* causes me to utter '*p*' which causes you to believe that *p*. But suppose there are sentences which are linked in this way to attitudes rather than beliefs. My attitude causes me to utter '*q*' which causes you to sympathise. We now have a new kind of meaning, and a new role for language in communication. Moral judgements express attitudes, rather than beliefs. Hence they are neither true nor false, since attitudes are never determined by the beliefs on which they are founded. Each person has his own 'values', which are identified through his attitudes.

The theory is obviously wrong, because the causal theory of meaning is obviously wrong. But could it be re-stated, in terms of a Gricean theory of speech acts, say? Perhaps it could. Perhaps we could make sense of the idea that some sentences standardly express beliefs, and others (moral judgements among them) standardly express attitudes. Is such a suggestion plausible? The main argument in favour rests on the 'action-guiding force' of moral judgements. The sign that someone truly accepts a moral judgement lies in his actions, rather than his perceptions and theories of the world. Attitudes are *fundamental*, and the test of sincere conviction. If moral judgements expressed beliefs, however, the test of sincerity (it is suggested) would be quite different. For actions and beliefs are logically independent. From the fact that John believes *p* it never follows, for any x, that he does x. *What* he does depends upon his desires, emotions, motives – in short on his attitudes. (The counter-suggestion, that there are in fact 'motivating beliefs', was not seriously considered until recently.)

(c) Prescriptivism

The issue becomes sharper when we move to 'prescriptivism', a theory associated with R.M. Hare, but in fact derived from Kant. Like emotivism, prescriptivism entered the modern world encumbered by theories of meaning that would now be widely rejected. But it has a logical ingenuity which makes it a more rewarding subject of study. The starting point is the 'action-guiding force' of moral judgements; and the fundamental idea is that such action-guiding force is typified by commands; hence we can understand moral judgements by assimilating them to imperatives. When I say 'Close the door!' I give an order. You *accept* this imperative by closing the door. Any reasons given in favour of it are reasons for action, rather than reasons for believing something. (Those facts explain why we do not describe imperatives as true or false. Their success does not consist in fitting to the world; it consists in making the world fit to them. Imperatives are distinguished by what Anscombe calls their 'direction of fit' with reality.)

Imperatives have a logic. From the command 'Take all the boxes to the station!' there follows the command 'Take this box to the station'. Of course the phrase 'follows from' is peculiar here, precisely because we do not describe imperatives as true or false. Nevertheless, we can explain the relation easily enough, in terms of consistency. One imperative entails another if you cannot consistently accept the first and reject the second. Since actions can be inconsistent with one another, and since imperatives are accepted in action, we can give perfect sense to the idea of a logic of imperatives.

We should now distinguish a special class of imperatives – those which are universalisable. When I say 'Take this box to the station' I have made essential reference to this particular box, and I have addressed my imperative to you. But suppose I eliminate such reference to particulars, and specify only *types* or *kinds*: 'anybody in good health is to take boxes marked urgent to their rightful destination'. The imperative is now 'universalised', and is addressed indifferently to the world at large. It is impossible to effect this universalisation without losing the imperative mood: the sentence has shifted to the indicative. Another way of framing it would be to use the word 'ought': Everyone in good health ought to take . . . As Kant noticed, however, the indicative grammar of sentences involving 'ought' is only a surface phenomenon. Scratch a little, and you will soon reveal the imperative force beneath it.

Universalisable imperatives are characteristic of 'evaluative' discourse, according to Hare. Moral judgements are those judgements that are prescriptive, universalisable, and overriding – i.e. whose force cannot be set aside because of some competing 'evaluation'. With a minimum of theoretical baggage, therefore, Hare arrives at the following picture: moral judgements are imperatives; they have a logic, which is the logic of imperatives; they can therefore feature in rational arguments, and are not merely the blunt expressions of emotion that Stevenson and others had described. Indeed, they form the major premises of our practical reasoning, and stand at the apex of any fully reasoned answer to the question what to do.

Furthermore, there really is a gap between 'is' and 'ought'. Actions do not *follow from* beliefs: no action is ever inconsistent with a belief. Hence it is never inconsistent to accept a premise involving 'is' (i.e. to acquire a belief) and to dissent from a conclusion involving 'ought' (i.e. to reject a course of action).

Finally, prescriptivism seems to explain the 'supervenience' of moral features. If John is good, but Henry evil, there must be some *other* difference between John and Henry, which explains this moral discrepancy. (Compare the supervenience of pictorial properties, discussed in Chapter 16, section 5.) This supervenience of moral on non-moral characteristics is a direct result of the universalisibility of moral judgements, which can be sincerely made only by someone who has a universal principle in mind: a principle of the form 'everyone with such and such features is good'.

Hare connects his account with Austin's remarks about speech acts, and also, in later writings, with a theory of practical reason. This theory is designed to answer the objection that, on Hare's view, moral judgements are really 'subjective', since they express the personal decisions of the one who makes them. His view is, roughly, that moral judgements are not to be viewed as subjective merely because they involve *commitment*. On the contrary, moral commitment is a commitment of a rational kind. It imposes on the subject an obligation to look always further than his immediate decisions, to the principles from which they flow. An open-ended project of rational criticism is contained in the very idea of morality, and even if the answer to a moral question is hard to find, this does not imply that there is no such thing as finding it. I find it when I discover principles that may be recommended to any rational being, regardless of his desires.

Hare attacks something that he calls 'descriptivism' (and which includes Moore's 'naturalism') – the view that moral predicates describe properties of the thing to which they are applied, and that moral judgements are therefore descriptions. (On this view there would be no 'gap' between 'is' and 'ought', since every 'ought' would really be an 'is'.) By denying the 'action-guiding force' of moral judgements, Hare argues, the descriptivist lets us off the hook. He hopes to guarantee the objectivity of morals, by tying down the crucial terms ('good', 'ought', 'right', and 'wrong') to the natural world. But all he succeeds in doing is to rid those terms of their moral force, so enabling us to pose as moral experts, while engaging in a life of crime. If moral judgements express beliefs, then they no longer constrain our actions. If that is the price of 'objectivity', it is too high a price to pay.

Neither prescriptivism nor emotivism has much of a following today. Here are some of the difficulties:

(i) Weakness of will. If Hare is right, then I cannot believe that an action is wrong, without intending to avoid it. Yet often I give way to temptation, and do what I know to be wrong. How is this possible? Hare tries to explain such cases (see the chapter on 'Backsliding' in *Freedom and Reason*) but without success. Indeed, weakness of will is a problem for many philosophies of morality and action.

It is probably not a problem for emotivism, however, since there is no contradiction in supposing that I have an *attitude* against some action which I nevertheless perform.

(ii) Dilemmas. Agamemnon must sacrifice his daughter if he is to lead his army to Troy: so the goddess has decreed. His duty to his country and his troops demands that he kill Iphigenia; his duty as a father demands the opposite. Yet he cannot obey both those commands. Does his sacrifice of Iphigenia mean, therefore, that he has simply rejected the moral judgement that it is wrong to kill your children? Hare must say yes; but that is tantamount to denying that dilemmas are possible. Surely Agamemnon is marked for ever by the knowledge of his crime, and this fact is expressed in his remorse and penitence thereafter. (Again, this is more a difficulty for prescriptivism than emotivism: see Bernard Williams, 'Dilemmas', in *Problems of the Self*.)

(iii) The *real* 'logic of moral discourse'. We describe moral judgements as true and false; they play a role in inference that is exactly like the role of 'descriptive' judgements. In particular, they feature in conditionals: 'If John is a good man, he does not deserve to suffer'. Someone could sincerely assert that sentence, even though he had no pro-attitude to John. Yet the antecedent of the conditional is supposed to commit the speaker, by virtue of its meaning, to just such a pro-attitude. How is that possible?

This objection tells against both emotivism and prescriptivism. Hare makes feeble attempts to answer it, but it remains at large.

(iv) The real meaning of 'good'. An emotivist or prescriptivist would tend to put 'ought' before 'good', and to define the second through the first: a good action is one that you ought to do, a good man one whom you ought to imitate, and so on. The consensus is that this entirely misrepresents the sense of 'good', which was far better described by Aristotle in book I of the *Nicomachean Ethics*. 'Good' does not make sense as an isolated adjective. Always we need to ask 'good as a what?' or 'good for what?' In 'good man', 'good farmer', 'good horse', the term is used 'attributively', and its sense is completed by the noun with which it is conjoined. Every such noun, Aristotle argued, defines an *ergon*, or characteristic activity, and this is what fixes the sense of 'good'. The characteristic activity of a farmer is agriculture; the good farmer is the one who farms well – i.e. successfully. The *ergon* of a knife is its function as a cutting device: the good knife is one that cuts well. And so on. We could understand the idea of a good man if we knew the *ergon* of man. This is what Aristotle set out to discover.

All that is so neat and so plausible, that it is hard to find a suitable response to it. There *are* responses, however, which are not without force.

(v) Against emotivism: the dreadful naïveté of it all! The assumption that we know what attitudes are, and have some *a priori* revelation that they are distinct from beliefs; the failure to see that, whatever they are, attitudes have intentionality, and that intentionality constrains a mental state in a rational direction; the philistine description of the moral life, and the failure to attend to the multiplicity and finesse of our real moral judgements, which hardly ever use terms like 'good' or 'ought', but which roam freely among the concepts that describe our human world, finding now here and now there the justification for our ways of life and action. At any rate, emotivism needs a few lessons in sophistication before we can listen to its message.

(vi) The shared assumption that there are no relations of entailment between belief and action. Is this not a dogma? To prove it you would have to go deep into the philosophy of mind, so as to show that there are no 'motivating beliefs'. But emotivists and prescriptivists display a donnish reluctance to engage in such a messy business. Indeed, there is a growing disposition among moral philosophers today, to argue that the question of motivating beliefs is at the heart of ethics. It would be sufficient vindication of morality, if we could show that there are beliefs which it is rational to have, and which are in themselves a sufficient motive to action. (Maybe the belief that I am responsible for your future is like that.)

(d) Moral Realism

Faced with all those objections, philosophers have tended to return to the position that Moore and Hare dismissed, calling it not 'naturalism' or 'descriptivism' (terms too polluted by defunct philosophy) but 'moral realism'. This expression also has its disadvantages, since it implies a connection with the debate between realism and anti-realism in the philosophy of language. But let us adopt it nevertheless. The moral realist holds that moral judgements are just like any other, and describe features of reality: *moral* reality. The world contains goodness and value, right and wrong, obligations and prohibitions, and the purpose of moral language is simply to describe these things and to reason about them.

Some philosophers have tried to fight for moral realism on the enemy's territory, producing, like Searle, alleged proofs of an 'ought' from an 'is'. Others, like Phillipa Foot, have proposed criteria for the truth of moral judgements (in her case these criteria concern 'human benefit and harm'). But perhaps the most plausible defence is to return to those thoughts about the human world that I adumbrated in Chapter 18, so as to argue that moral judgements belong to our intentional understanding, employing concepts that divide the world according to our interests, providing a store of superficial truths on which to base our attitudes and actions. Consider the concept of justice. This features prominently in moral discourse, and describes a recognisable feature of human characters and human acts. It is also used to 'evaluate' those acts and characters. To say that an act was unjust is to cast judgement on it. The description has the force of a condemnation. Hare would say that really the term 'unjust' means *two* things. It describes an act, and *also* condemns it; in principle the two components could be separated, into a 'descriptive' and a 'prescriptive' part. But why say that? How can one and the same predicate obey two such disparate rules, and not be torn asunder? Only a stubborn adherence to the prescriptivist theory could lead us to say such a thing. Why not accept that the whole division between description and evaluation is founded on a mistake?

In this way discussion has begun to focus on those concepts which situate our actions in the arena of judgement, and which give us a handle on the human world. This does not mean that the realist position is adopted by everyone. In a phenomenally overrated book (*Inventing Right and Wrong*) J.L. Mackie marshals the arguments against moral realism, picking out two for special emphasis: the argument that moral judgements play a role in guiding action; and his own 'argument from queerness' which tells us that if realism is true there are moral properties, and moral properties are a *very queer kind of thing*. Well, they are a queer kind of thing. So are secondary qualities; so are all the qualities that compose the *Lebenswelt*. But queerness is just another name for 'causal inertia'. Moral properties play no part in explaining physical reality: we perceive them in the world, but the world can be explained without referring to them. From the scientific perspective, there is no fact of being good, only the fact that certain things are seen as good. So much the worse for the scientific perspective.

2. Practical Reason

Whichever way we decide between the contending parties, there will be a question that remains, and this question is far more important than the analysis of moral judgement. I refer to the question of practical reason: do we ever have objectively binding reasons for doing one thing rather than another?

Suppose the moral realist is right, and moral judgements describe a reality independent of the observer. What reason do we then have for acting, given the truth of moral judgements? Suppose on the other hand that the prescriptivist is right, and moral judgements express our decisions, so that there is no gap between accepting a moral judgement and acting on it. The question is then: What grounds do we have for our moral judgements? But those two questions are precisely one and the *same* question: the question of the 'bridge' between thought and action. Are there true thoughts, which are also binding reasons for action? If so, what are they?

There is one uncontroversial example of valid practical inference. Suppose you want to go to Iceland, and the only way is to board a plane in Manchester. Then you have a reason to board that plane in Manchester. Maybe not a conclusive reason; you could have conflicting desires: Manchester has bad memories for you; you would rather cancel your trip than go by air. But still, you have *a* reason to board that plane; and if there is no reason to the contrary, that is what you should do. The example concerns reasoning about means. If you desire *y*, and *x* is the means to *y*, then, other things being equal, do *x*. That is one form of what Aristotle called the 'practical syllogism': by which he meant the process whereby desire and thought come together in a rational action. It is practical because the conclusion is an action; it is a form of reasoning since the action is the rational answer to the question 'What shall I do?' – an answer that is justified when the premises really do provide reasons for doing it.

Reasoning about means is uncontroversial, since it guarantees nothing that was ever in doubt. Everything depends upon the major premise, which states the goal or desire. How do you justify *that*? Hume famously said that 'reason is, and ought only to be, the slave of the passions'. He meant that it could play no other role in practical reasoning than establish the means to our ends. It could never provide the ends themselves. (He also believed that the attempt to rely on reason alone would undermine the authority of our moral sentiments.)

3. Utilitarianism

Although Hume was not a utilitarian, he did offer a statement of the 'principle of utility', which until this day has occupied a prominent place in English-speaking ethics. Various versions of utilitarianism exist. For Bentham the crucial idea is the maximisation of pleasure; for Mill the goal is happiness, the pursuit of which must, however be qualified by a respect for liberty.

In its simplest form the theory argues thus. There is one goal which all rational beings have, and whose desirability is self-evident: happiness. We do not need to justify the pursuit of this goal, which is more important to us than any other. It is, as Aristotle said, the 'final end'. Practical reason concerns the means to happiness, and is therefore entirely objective. We may assess all our actions according to their 'felicific' consequences. That action is right which promises the most happiness.

First problem: *whose* happiness? Maybe the desirability of *my* happiness is self-evident to *me*; but what reason have I to put your happiness on my agenda? Utilitarians have typically held that each person's happiness counts for one in the total score, and that we should aim to generate happiness regardless of whose it is. But why should we do that? Is this something that we want, or something that we *ought* to want, or something that it is *reasonable* to want? We are back with the question of practical reason.

Second problem: how can we estimate the amount of happiness that an act will produce? Is there a cut-off point in our reasoning? Or does the question remain open for ever? It is worth reflecting on the old communist argument, which said that it was right to 'liquidate' the kulaks, since this was the short-term cost for the long-term benefit of a socialist economy. That way, of course, you can justify anything, as Lenin, Hitler and Stalin knew.

Third problem: are we ever entitled to sacrifice one person's happiness for the sake of a greater quantity all round? You will all receive a bottle of claret, the tyrant says, provided Higgins is tortured to death. If there were enough winos around, would this justify the fate of poor Higgins? Absurd and repugnant possibilities are easy to envisage.

Fourth problem: is happiness measurable? And if not, can we really reduce all practical reasoning to reasoning about means? Do not all the real questions now become undecidable?

Such difficulties can be manufactured *ad nauseam*. Some philosophers attempt to circumvent them by espousing 'rule utilitarianism'. This tells us that our test of the rightness or wrongness of an action lies in its conformity to a moral rule (e.g. one of the ten commandments). The rule itself, however, owes its authority to the fact that general obedience to it promotes the cause of happiness. This subterfuge very quickly encounters the same difficulties as the original theory. In particular it licences what Anscombe (rightly) calls moral corruption: if I believe I should not kill the innocent because the rule against killing promotes the general happiness, I may easily be tempted into sin. Maybe, I could add to the general happiness by breaching the rule: so all I need do is kill the guy, while ensuring general obedience. What I do is right, provided nobody discovers it. This kind of consideration has led Bernard Williams to argue that, if utilitarianism were a true theory, we ought, on utilitarian grounds, to prevent people from believing it. A theory that justifies its own rejection is not a very hot contender for the truth.

4. Consequentialism

The astonishing success of utilitarianism in colonising the English temperament is due to two factors: its provision of a secular goal for morality, and its promise to reduce ethics to a mathematical calculation. (Bentham proposed a 'felicific calculus', which was the archetype of the 'decision theory' that is now accepted in economics.) The utilitarian morality is the morality of *homo economicus*: so at least thought Bentham.

A wiser economist, Adam Smith, had already exposed the flaw in this argument in *The Wealth of Nations*. Long term benefits, he argued, may arise by an 'invisible hand' from transactions that do not intend them. Maybe if we aim at them we miss them. Free transactions by profit-seeking individuals benefit society as a whole; but only if the transactions are bound by interdictions. Cheating, fraud, theft and the like must be forbidden. These 'side-constraints' (as Robert Nozick calls them) are what we mean by morality. It could be that obedience to the moral law promotes the general happiness. But it could also be that it does so only if it is not the general happiness at which we aim. Moral principles produce their beneficial effects only when regarded as absolutely binding.

It is part of the moral idiocy of the English that this idea of an absolute prohibition has been regarded as absurd or irrational. Surely, it is argued, morality is *justified* by its beneficial effects. So we should adjust and amend it in the light of those effects. If, in this or that circumstance, I can clearly apprehend an overwhelming benefit that follows from flouting a principle, or an overwhelming disaster that flows from obeying it, I ought to disobey. Principles must always be weighed against consequences if morality is to be rational, and not a blind adherence to prejudice. Moral 'absolutism' is even castigated as a crime. In this way, through the theory of 'consequentialism', the possibility of unlimited corruption has re-entered the modern conscience and produced some interesting casuistry in 'applied ethics'.

5. The Kantian Approach

Consequentialism is best understood through its principal adversary, Kant, whose system of ethics is one of the most beautiful creations that the human mind has ever devised.

Kant was a prescriptivist: he believed that moral judgements are imperatives, and also universal, not only in the formal sense discussed by Hare, but in the more substantive sense of applying universally, to all rational beings. Unlike Hare, however, Kant believed that these 'categorical imperatives' have an *a priori* foundation.

In reasoning about means, I use imperatives of the form 'if you want *x*, do *y*', in which the antecedent specifies a goal. These 'hypothetical imperatives' depend for their validity on a single principle: 'He who wills the end wills the means'. This principle, Kant argues, is analytic. Hence the validity (and also triviality) of arguments about means. However, we also reason about ends. We ask whether it is right to want this or that, regardless of the benefit that stems from it. Such reasoning is *also* capable of validity. In this case, however, validity means *objective* validity – validity for all rational beings, regardless of their desires. In an important sense the hypothetical imperative is still subjective. For it gives a reason only to someone who has the desire mentioned in its antecedent: the reason is *relative* to his interests, and of no independent force. If there are to be categorical imperatives, however, they must *abstract* from such personal concerns, and view the world from the point of view of reason alone.

Like every exercise of reason, this process of abstraction involves jettisoning the reference to 'empirical conditions' (such as an agent's needs and desires) that we know only *a posteriori*. The search for the categorical imperative takes us, therefore, into the realm of the *a priori*. We are seeking an imperative whose validity is guaranteed by reason alone, and which is not based merely on some analytic connection (as in the command 'Bring it about that *p* or bring it about that not-*p*!') So this is the extraordinary programme that Kant set himself: to discover a synthetic *a priori imperative*! If you cannot do that, then the whole edifice of morality is built on sand.

Kant's next move is extremely ingenious. He derives the content of the categorical imperative, not by the process of abstraction, but by reflecting on the *idea* of abstraction itself. The categorical imperative says 'Find your reasons for acting only after all "empirical conditions" have been discounted'. If you obey this imperative, you will be doing what reason requires: your action will be referred to reason alone, and not to your individual passions and interests. Hence it will be binding not only on you, but on any rational being. We can therefore phrase the categorical imperative in this way: 'Act only on that maxim which you can will as a law for all rational beings'. ('Maxim' means, roughly, your intention in doing what you do.) This famous principle captures the demand of reason, and also coincides with the Christian golden rule: do as you would be done by.

The demand of reason is a demand that I *respect* reason – that I allow reason the final say in my decisions. This means respecting reason not only in myself, but also in others. All rational beings have a claim to my respect, and this too is a fundamental axiom of morality. I cannot override another's reason, as though it counted for nothing. I must try to persuade him, to secure his rational consent for those projects in which we are engaged together. I cannot use him merely as an instrument for my purposes (as a means only) for, like every rational being, he stands as much in judgement over my actions as I do. In short, the categorical imperative demands that I 'act so as to treat rational beings always as ends in themselves and never as means only'.

Practical reason therefore makes us part of a common enterprise – each of us is constrained not only to respect his fellows, but to find reasons for his actions that would justify them in the eyes of others. The idea of a 'community of rational beings' lurks within our practical discourse, and constrains our moral thinking. Hence the categorical imperative leads us towards an ideal community, in which every right is respected, and every duty fulfilled. This too becomes a demand of reason. In Kant's words: 'Act so that the maxim of your action might become a law of nature in a kingdom of ends'.

The three formulae are considered to be expressions of *one* imperative, which is the *a priori* law on which all morality is founded. Since its validity derives from reason alone, all rational beings are constrained by reason to accept this law, just as they are constrained to accept the laws of logic.

But what do we mean by *accepting* the categorical imperative? Like Aristotle, Kant believes that reasons are practical only if they are 'accepted in action'. A reason for action is *my* reason for doing what I do: not just the justifying principle, but the *motive* too. Reason must have motivating force, if the categorical imperative is to have *practical* authority. Hume's view, that reason is and ought only to be the 'slave' of the passions, must therefore be rejected.

How is this possible? The answer, for Kant, lies in his philosophy of freedom, discussed in Chapter 17. The choices of a rational being are 'determined by reason'; there is a 'causality of freedom' implied in the very idea of rational choice. Morality, by presupposing freedom, shows that our freedom is real; all other motives enslave us. When acting from the hypothetical imperative I act as a part of nature, the instrument of my own desires, through which the causality of nature flows to its impersonal destiny. (This is why we describe desires as 'passions'. Indeed, Kant often uses the word 'pathological' to refer to them.) By contrast, the categorical imperative appeals directly to the self. It does not take up the cause of nature, but stands against the natural order, addressing me directly in the second-person singular: *Du*.

Like the rest of his moral philosophy, Kant's theory of freedom is not merely a speculative outgrowth of his metaphysics, but a serious attempt to make sense of our moral intuitions: something that it brilliantly succeeds in doing. The resulting picture of morality may be summarised thus:

Moral beings are free, rational and capable of self-legislation. We call them 'persons' in order to distinguish them from the rest of nature, as the bearers of rights and duties. A person must always be treated as an end (he cannot be exploited, manipulated, abused, enslaved or trampled on). He is an equal member of an ideal community, and takes up his place in that community to the extent that he obeys the moral law. Towards the rational being I owe the same respect that I owe to the moral law itself. This obligation is impartial and objective, and overrides all those arbitrary distinctions of race, creed, and custom that divide the nations each from each.

Here then is the morality of the Enlightenment, furnished with an *a priori* proof. It is small wonder that Kant's moral philosophy seemed to change the world. (Though see the remark about influence at the beginning of section 1.) The problems, however, cannot be easily set aside. For example, the concept of freedom, which Kant himself admitted to be unintelligible. And the reasoning that leads to the categorical imperative: Is it really persuasive? Or is it rather a subtle piece of rhetoric?

6. Master, Slave and Side-constraints

Hegel provides an interesting argument, phrased in terms of his 'dialectic', for the conclusion that all rational beings must at least *see* themselves as Kant describes them, and in doing so he brings to the fore an idea that is only latent in Kant – the idea of moral value, as a social category. The dialectical method enables Hegel to make free use of temporal metaphors in philosophical argument. He describes logical relations as though they were *processes*, since for him the 'unfolding' of a concept is also the growth of spirit into self-awareness. In the *Phenomenology of Spirit*, where Hegel's preoccupation is with the archaeology of consciousness (the conceptual layers, so to speak, from which consciousness and self-consciousness are constructed), the temporal idiom acquires a dramatic and poetic power that forbids translation. Therefore I shall summarise, in my own terms, the famous passage about lordship and bondage (master and slave). The next few paragraphs are adapted from my contribution to Sir Anthony Kenny's *Illustrated History of Modern Philosophy*.

Like Kant, Hegel recognised that the existence of the self brings with it a peculiar immediacy – the immediacy of Kant's Transcendental Unity of Apperception; and, like Kant, he argued that the 'immediacy' with which our mental states are presented to us can provide no clue as to their nature. It is the mere surface glow of knowledge, wholly without depth. The immediacy of the pure subject is, as Hegel would put it, undifferentiated, indeterminate and so devoid of content. (Compare Hegel's account of being, summarised in Chapter 12.)

It follows that the pure subject can gain no knowledge of what he is, and still less any knowledge of the world which he inhabits. Nevertheless, as Kant saw, his existence presupposes a *unity*, and that unity requires a principle of unity, something that holds consciousness together as one thing. Spinoza had spoken in this regard of the *conatus*, or striving, that constitutes the identity of organic beings. Hegel has recourse to a similar notion, the Aristotelian *orexis*, or appetite: the striving through which we seek to possess our world. In the initial stage of consciousness, this is what the self amounts to: the primitive 'I want' of the infant, the contumacious screeching of the fledgling in the nest.

But desire cannot exist without being desire *for* something. Desire posits its object as independent of itself. With this venture towards the object, the 'absolute simplicity' of the self is sundered. In positing the object of desire, however, spirit does not rise to self-consciousness: for it has no conception of itself as *other* than the world of objects, and free in relation to them. It has reached the stage only of animal mentality, which explores the world as an object of appetite, and which, being nothing *for* itself, is without genuine will. At this stage the object of desire is experienced only as a lack (*Mangel*), and desire itself destroys the thing desired.

Self-consciousness awaits the 'moment' of opposition. The world is not merely passively unco-operative with the demands of appetite; it may also actively *resist* them. The world then becomes genuinely *other*: it seems to remove the object of my desire, to compete for it, to seek my abolition as a rival.

The self has now 'met its match', and there follows what Hegel poetically calls the 'life and death struggle with the other', in which the self begins to know itself as will, as power, confronted by other wills and other powers. Full self-consciousness is not the immediate result of this: for the struggle arises from appetite, and the self has yet to *find* itself (to determine itself as an object of knowledge). This self-determination (*Selbstbestimmung*) comes only when the subject invests the objects of its world with meaning, distinguishing those things which are worth pursuing from those which are not. The life-and-death struggle does not generate the conception of the self *in its freedom*. On the contrary, the outcome of this struggle is the mastery of one party over the other: the one who prefers life to honour becomes slave to the one who is prepared to sacrifice his life for honour's sake.

This new 'moment' of self-consciousness is the most interesting, and Hegel's account of it was destined to exert a profound influence on nineteenth-century ethical and political philosophy. One of the parties has enslaved the other, and therefore has achieved the power to extort the other's labour. By means of this labour the master can satisfy his appetites without the expenditure of will, and so achieve leisure. With leisure, however, comes the atrophy of the will; the world ceases to be understood as a resistant object, against which the subject must act and in terms of which he must strive to define himself. Leisure collapses into lassitude; the otherness of the world becomes veiled, and the subject – whose self-definition is through the contrast with the world of objects – becomes lost in mystery. He sinks back into inertia, and his newly acquired 'freedom' turns into a kind of drunken hallucination. The self-definition of the master is fatally impaired. He can acquire no sense of the value of what he desires through observing the activities of his slave. For the slave, in his master's eyes, is merely a means; he does not appear to pursue an end of his own. On the contrary, he is absorbed into the undifferentiated mechanism of nature, and endows his petty tasks with no significance that would enable the master to envisage the value of pursuing them.

Now look at things through the eyes of the slave. Although his will is chained, it is not destroyed. He remains active towards the world, even in his submission, and while acting at the behest of a master, he nevertheless bestows his labour on objects, and realises his identity through them. The result of his labour is seen as *my* work. He makes the world in his own image, even if not for his own use. Hence he differentiates himself from its otherness, and discovers his identity through labour. His self-consciousness grows, and although he is treated as a means, he unavoidably acquires both the sense of an end to his activity, and the will to make that end his own. His inner freedom intensifies in proportion to his master's lassitude, until such a time as he rises up and enslaves the master, only himself to 'go under' in the passivity that attends the state of leisure.

Master and slave each possess a half of freedom: the one the scope to exercise it, the other the self-image to see its value. But neither has the whole, and this toing and froing of power between them is restless and unfulfilled. The dialectic of their relation awaits its resolution, which occurs only when each treats the other not as means, but as end: which is to say, when each renounces the life-and-death struggle that had enslaved them and respects the reality of the other's will. In doing so each accepts the autonomy of the other, and with it the categorical imperative that commands us to treat the other as an end and not a means only. Each man then sees himself as a subject (rather than an object), standing outside nature, bound to a community by reciprocal demands upheld by a common moral law. This law is, in Kant's words, the law of freedom. And at this 'moment' the self has acquired a conception of its active nature: it is autonomous yet law-governed, partaking of a common nature, and pursuing universal values. Self-consciousness has become *universal* self-consciousness.

Hegel's account gives us another and wider vision of the categorical imperative, as reflecting the agent's view of himself as a member of society. My obedience to the moral law goes hand in hand with my belief in my own moral worth. And this depends upon my self-esteem, which in turn reflects a conception of how I might appear to others. The suggestion that moral agency is a social artefact is connected by Hegel with a belief in its 'historicity'. Moral agency has distinct historical moments, and develops as the social order evolves. All these suggestive ideas seem to add flesh to the Kantian skeleton, and promise a more plausible and nuanced moral psychology than can be derived from the Kantian idea of duty alone.

The categorical imperative can also be understood, not merely as a principle binding the individual will, but as an instrument of negotiation and compromise between strangers, through which they can rise out of mutual enmity and confront each other as equals. It is this aspect of the categorical imperative that has most appeal for modern philosophers. For it suggests a modern morality, as a system of 'side-constraints'. The moral law does not tell us what to do: but it tells us what we cannot do in the course of pursuing our interests. Our goals must be found elsewhere, in the ceaseless flow of life and appetite. But the means to their fulfilment is constrained by the moral law. In particular, we cannot treat others as means. Which is to say, we cannot discount or override their interests; nor can we enslave, manipulate, exploit or deceive them. For their rational nature has as much right as ours.

Thus there emerges a theory of rights. On the Kantian view, persons are distinguished by the possession of duties (to obey the moral law) and rights (against those who would trespass on their nature as ends). Each individual is *sovereign* within his own sphere: his rights define inviolable interests, which can be set aside only with his rational consent. It is through respecting this sphere of sovereignty, that we treat another as an end. To take from him those things to which he has a right of ownership; to enslave him or deceive him; to appropriate his sexual favours without his consent – all these involve a grave trespass against another's sovereignty, a refusal to recognise that he is a self-legislating member of the kingdom of ends.

Modern philosophers – notably Robert Nozick, David Gauthier and Loren Lomasky – have therefore used Kantian ideas, in order to make sense of a liberal morality in an age of market forces. Their vision – which also derives, via Hayek, from Adam Smith – is of a society in which each pursues his own goods, and collective choices emerge by 'an invisible hand', like prices in a market. It is not for morality, still less for the state, to impose our goals. Indeed, the attempt to do so is bound to lead to conflict, scarcity and confusion, as well as to the loss of liberty. The role of morality is to co-ordinate our actions, by setting limits to them. Some philosophers even argue (in terms derived from game theory), that the Kantian morality should be seen as a rational strategy in a collective game. The categorical imperative would be the preferred choice of all rational players in the social game, where the interest of each is to enjoy the path of least resistance to his goal.

I shall return to those thoughts in Chapter 28. They do not satisfy everyone. Nor would they have satisfied Kant. Any attempt to provide a *social* justification of morality runs the risk of the 'free rider' – the one who pretends to play the game, in order to enjoy the fruits of it. If we cannot provide a reason for *his* acceptance of the rules, then we have not really given an objective grounding to morality. That is what Hegel perceived, and why he tried to show that the cheat and the exploiter are acting against *themselves*.

7. Humean Sympathy

The weakness of the Kantian theory lies in the claim that reason is sufficient in itself to provide a motive to action. Hegel tried to supply the missing link in the argument, by showing *how* the rational being is constrained by the categorical imperative. But the story is obscure at best, and leaves the sceptic unpersuaded.

Hume's moral philosophy was for a long time much underrated, by philosophers who saw in him only the precursor of modern emotivism. However, it was Hume who gave the best rival description of the moral motive. The morality that he justified is not far from that of Kant; but the grounds for it involve a complete rejection of the Kantian picture of rational motivation. We are motivated, Hume argued, only by our emotions ('passions'); reasoning, which may define the object of a passion, cannot provide the motive. This must arise from another source. Hence, if morality has a motivating force, there must be moral emotions, which are the true centre of the moral life.

Kant would object that this makes moral reasons into hypothetical imperatives. It involves a repudiation of our nature as self-legislating beings. For Kant, a man who acts out of emotion – even if it is one of benevolence – is not 'autonomous'. He is 'swayed by passion', and his act proceeds from somewhere outside his will. He commits what Kant called 'heteronomy' of the will. And this is the charge that Kant levelled against the British moralists in general – in particular against Locke's pupil, the third Earl of Shaftesbury, who was the true originator of the ethical doctrines defended by Hume.

For Hume, however, the question of moral motivation becomes clear, once we distinguish two kinds of passion: those based in self-interest, and those based in sympathy. The latter are generally fainter, but less fickle, than the former. We each pursue our goals, and resist those who impede us. And if this were the whole content of the human heart, life would be miserable indeed. But there are occasions when we are not in the grip of passion, when our goals recede from view, and when we contemplate the human world from a position of detached curiosity. This happens when we read a story, a tragedy or a work of history. It happens too when others set their case before us, as in a court of law, and solicit our judgement. In such cases our passions are stirred not on our own behalf, but on behalf of another. This movement of sympathy is natural to human beings, and informs all their perceptions of the social world. Moreover, it tends always in the same direction. Whatever our goals, you and I can agree once we have learned to discount them. If two parties to a dispute come before us then we shall tend to agree in our verdict, provided neither you nor I have a personal interest in the outcome. This discounting of personal interest leaves an emotional vacuum, which only sympathy can fill. And sympathy, being founded in our common nature, tends to a common conclusion.

This is the origin of morality for Hume: the disposition that we all have, to discount our interests, and reflect impartially on the world. Although the resulting passions are faint compared with our selfish desires, they are steady and durable. Moreover, they are reinforced by the agreement of others, so that, collectively, our moral sentiments provide a far stronger force than any individual passion, and lead to the kind of public constraints on conduct that are embodied in custom and law.

Hume's detailed account of the moral motive is a masterpiece of philosophical anthropology, and a testimony to his wisdom as an observer of mankind. Space forbids that we dwell on it; nor is it necessary to do so. One terminological point is worth making, however. Hume is sometimes described as a 'naturalist'. What is meant is not that he was a naturalist in G.E. Moore's sense (he was not); but that he derived the grounds of morality from a study of human nature, and from a theory of the place of man in nature. Kant would not count as a naturalist in this sense, for although he regarded morality as 'natural' to man, it is not something that we owe to 'our niggardly step-mother, nature', as he described her. On the contrary, morality sets us above and beyond nature, in a condition of judgement on the empirical world.

8. Aristotelian Ethics

The question that has constantly hovered in the back of your minds is this: 'Why should I be moral?' None of the philosophers I have considered gives a final and conclusive response to it. Maybe I could live without the categorical imperative; maybe I could take a 'free ride' on the social strategies of others; maybe I could live the life of a contented psychopath, immune to those costly bouts of sympathy from which the Humeans suffer. So why not?

The best answer was provided by Aristotle, whose *Nicomachean Ethics* is a masterpiece of observation and argument, justifying a lifetime of study. Without giving the details, here is the strategy:

(i) The practical syllogism. All reasoning about what to do proceeds from premises relating to the agent's beliefs and desires. Desire is the motive for action, and the practical syllogism is its translation into choice. Your choices are dictated by your beliefs and desires – provided you are rational.

(ii) Dispositions. Nevertheless, desires can be moulded over time. They are reinforced by indulgence and enfeebled by our trained resistance. Desires express *dispositions* (*hexeis*), and these can be educated by imitation and habit. (One 'enters the palace of reason through the courtyard of habit'.)

(iii) There are therefore *two* questions of practical reason: (a) what shall I do *now*? (the practical syllogism), and (b) what sort of dispositions should I acquire? It is the second question that is the real concern of ethics. In answering it we must give a complete account of the human condition, and what dispositions are best suited to it.

(iv) Happiness. If I ask the question 'why do *a*?', then a justifying reason may be found, by showing that *a* is the means to *b*, and that *b* is something desired. But then 'why do you want *b*?' Again, if we relate *b* to something desired we answer the question. But does the series come to an end? Is there a 'final end'? Aristotle says yes. The final end is happiness (*eudaimonia*); this is final in that it does not make sense to ask, 'why aim at happiness?' Happiness means the general condition of fulfilment or 'success'. It is absurd to ask why we should pursue it, since success or fulfilment is what every activity intends.

(v) What is happiness? Aristotle has two accounts of happiness. One is formal: happiness is the 'final end', the ultimate answer to practical questions. The other is substantial, an attempt to say what happiness *consists* in. Briefly, happiness is an activity of the soul (*psuche*) in accordance with virtue (*arete*). Virtues are dispositions which give the greatest guarantee of success, and they are of two kinds, practical and intellectual, corresponding to our two modes of rational activity.

(vi) The rational being. Rationality defines the form of life that is ours. Our lives go well, according to whether we can exercise our reason successfully (alternatively: according to whether our rational activity leads to fulfilment). Hence we all have reason to acquire the virtues. (Sometimes Aristotle speaks of rationality as the *ergon* of man: see above, section 1, on the real meaning of 'good'.)

(vii) Virtue. Like Plato, Aristotle emphasised the intellectual life as the highest and happiest, and the intellectual virtues as the most valuable. His argument is more plausible, however, if we consider the practical virtues: in particular those of prudence or practical wisdom (*phronesis*), courage, temperance and justice. We all have reason to acquire these virtues; in acquiring them we acquire emotional dispositions; and from these dispositions spring the motives of our actions. By justifying the virtues, we justify the behaviour of the virtuous man.

(viii) The doctrine of the mean. To every virtue there corresponds at least one, and usually two, vices: one of excess, the other of deficiency. The rash man has an excess of spirit, the coward a deficiency. But the brave man attains the 'mean' between them. Precisely how the mean is estimated is a matter of great controversy. Aristotle clearly does not consider the mean to be a middle point on a one-dimensional line. Rather, the mean is the course that reason recommends. The virtuous man is motivated to follow reason, to pursue the rational goal, even when fear or anger might tempt the vicious man in another direction.

(ix) The education of the emotions. Since virtues are dispositions to act from a certain motive, the acquisition of virtue is also an education of the emotions. We should not rid ourselves of fear, anger, resentment or whatever. We should train ourselves to feel the right amount of anger, towards the right person, on the right occasion, and for the right reason.

(x) The third-person viewpoint. At the moment of action it is too late to change my ways: I want to save myself, and so, being rational, I flee the scene of battle. But step back and look at the situation from outside. Consider, for example, the moral education of your child. How would you want him to turn out, for *his* sake? Surely, you would want him to be happy. He must therefore have the dispositions that are needed for success in action. He must, for example, be able to pursue the goal that reason recommends to him, despite fear or weakness. In short, he must be brave. Now apply that reasoning to yourself: *there* is the answer to the moral question.

(xi) The course that reason recommends. We have reason to pursue only things that we regard as valuable, once acquired. As social beings, we frame our conception of value in terms that apply to others as well as ourselves – if you despise John for his cowardice, you will be ashamed of cowardice in yourself. Thus arises the distinctive motive of rational beings – the fear of shame, and the love of its opposite, honour (*to kalon*). Every virtue is a disposition to pursue honour (i.e. that which casts credit on you), and to avoid shame. It is precisely this course of action that reason recommends.

If this strategy can be carried through, then the question of practical reason has found an answer. Hume is right, that reason is the slave of the passions: but only in the moment of choice, when the practical syllogism governs what I do. I can, however, take a long-term view of myself in which I set aside the question 'what shall I do?' and ask instead 'what shall I *be*?' This has an obvious answer in the third-person; and the answer applies equally to *me*. In short every rational being has a reason to cultivate the virtues, regardless of his particular desires.

The resulting morality differs in one striking respect from those of Kant, and of most modern philosophers. It does not lay down principles or laws. It says that the right thing to choose is the thing that the virtuous man would choose. But *how* he would choose depends on matters that a mere philosopher cannot foresee. The whole idea of justifying the distinction between 'right' and 'wrong', by laying down principles, has disappeared.

Nietzsche points out, in this connection, that while the fundamental contrast in Christian ethics is between good and evil (acts to be done, and acts to be avoided), the fundamental contrast in Greek ethics is between good and bad – meaning the good specimen and the bad specimen. (See *Beyond Good and Evil*.) This is largely true. Aristotle has defined the characteristic activity of man, which is reason. And he distinguishes those who are equipped to engage in this activity successfully, from those inferior specimens who are bound to fail. He is not afraid to attribute to the superior person all kinds of qualities which modern democratic man would find insufferable – including the disposition to despise his inferiors.

In a similar spirit, Nietzsche recommends his morality of the 'new man'. Like Aristotle, Nietzsche found the aim of life in 'flourishing'; excellence resides in the qualities which contribute to that aim. Nietzsche's style is of course very different from Aristotle's, being poetic, and exhortatory (as in the famous pastiche of Old Testament prophesy entitled *Thus Spake Zarasthustra*). But there are arguments concealed within his rhetoric, and they are so Aristotelian as to demand restatement as such.

Nietzsche rejects the distinction between 'good' and 'evil' because it encapsulates a theological morality inappropriate to a man without religious belief. The word 'good' has a clear sense when contrasted with 'bad'; it lacks a clear sense, however, when contrasted with 'evil'. The good specimen is the one whose power is maintained, and who therefore flourishes. The capacity to flourish resides not in the 'good will' of Kant (whom Nietzsche described as a 'catastrophic spider') nor in the universal aim of the utilitarians. ('As for happiness, only the Englishman wants that.') It is to be found in those dispositions of character which permit the exercise of will: dispositions like courage, pride and firmness. Such dispositions, which have their place, too, among the Aristotelian virtues, constitute self-mastery. They also permit the mastery of others, and prevent the great 'badness' of self-abasement. One does not arrive at these dispositions by killing the passions – on the contrary the passions are an inextricable part of the virtuous character. The Nietzschean man is able to 'will his own desire as a law unto himself'.

Like Aristotle, Nietzsche did not draw back from the consequences of his anti-theological stance. Since the aim of the good life is excellence, the moral philosopher must lay before us the ideal of human excellence. Moral development requires the refining away of what is common, herd-like, 'all too human'. Hence this ideal lies, of its nature, outside the reach of the common man. Moreover the ideal may be (Aristotle), or even ought to be (Nietzsche), repulsive to those whose weakness of spirit deprives them of sympathy for anything that is not more feeble than themselves. Aristotle called this ideal creature the 'great-souled man' (*megalopsuchos*); Nietzsche called it the '*Übermensch*' ('Superman'). In each case pride, self-confidence, disdain for the trivial and the ineffectual, together with a lofty cheerfulness of outlook and a desire always to dominate and never to be beholden were regarded as essential attributes of the self-fulfilled man. It is easy to scoff at this picture, but in each case strong arguments are presented for the view that there is no coherent view of human nature (other than a theological one) which does not have some such ideal of excellence as its corollary.

The essence of the 'new man' whom Nietzsche thus announced to the world was 'joyful wisdom': the ability to make choices with the whole self, and so not to be at variance with the motives of one's act. The aim is success, not just for this or that desire but for the will which underlies them. This success is essentially the success of the individual. There is no place in Nietzsche's picture of the ideal man for pity: pity is nothing more than a morbid fascination with failure. It is the great weakener of the will, and forms the bond between slaves, which perpetuates their slavery. Nietzsche's principal complaint against Christianity was that it had elevated this morbid feeling into a single criterion of virtue; thus it had prepared the way for the 'slave' morality which, being founded in pity, must inevitably reject what is dominant and strong.

9. The Moral World

Maybe none of the attempts to justify morality is finally satisfying, and the last is even somewhat hair-raising in its remorseless emphasis on success. Maybe our 'reasoning about ends' rests, at last, on the unjustified and unjustifiable bedrock of human sympathy, as Hume maintained. Nevertheless, you can see from the above arguments, that much can be said in answer to the moral sceptic and the moral relativist. (Though the definition of 'moral relativism' is a tough question that I must leave to one side.) It should not be surprising to find that moral categories are central to a description of the human world, and form the core of our 'social intentionality'. The attitudes that we have to one another (the 'interpersonal attitudes' discussed in Chapter 17), are founded on moral conceptions. The human world is ordered through concepts of right, obligation, and justice; shame, guilt, pride and honour; virtue and vice. One of the tasks of ethics is to explore these conceptions, and to show their place in practical reason.

Consider justice, for example. How should we understand this idea? Some philosophers believe that justice is a property of actions; others that it is a property of people (i.e. a virtue); others that it is a property of a state of affairs; others that it is all three. It matters greatly which we say. There are those, for example, who say that justice means respecting people's rights (Nozick), or giving to each person his due (Aristotle). And for such thinkers there is no knowing what kind of social order, or distribution of goods, would emerge, when people treat each other justly. There are others, however – Rawls is the most important – who think of justice as a criterion for the distribution of social goods. To act justly is to pursue a particular social order. Some (though not Rawls) believe that you may override individual rights, on behalf of social justice. The dispute here is deep and difficult; and it is the major *moral* dispute between the defenders and the opponents of socialism.

I shall return to these matters in Chapter 28; meanwhile it is necessary to move on to another question about the moral life – the question posed by Kierkegaard and Heidegger – of our 'being in the world'. How should we comport ourselves, in the face of our finitude, contingency, and dependence? To answer this question is to know the meaning of life.

21

Life, Death and Identity

It is clear, from the discussions in the last few chapters, that rationality is a critical concept for the understanding of human beings. It is not easy to define rationality, however; nor is it clear that rationality is *the* distinguishing feature of the kind to which we belong, more important than any *other* feature that might serve to distinguish us. There are human beings without reason, and reason may be manifest in things that are not human – in angels and gods, maybe in other animals, maybe in machines programmed to match our human powers.

Nevertheless, Aristotle's hunch – that we are essentially rational – has been shared by many later thinkers, including Aquinas and Kant. (We are referring here to *de re* rather than *de dicto* necessity: see Chapter 13.) It provides us with the only conceivable ground, short of religious revelation, for our treatment of human beings as separate from the rest of the animal kingdom. Aristotle's hunch is this: we are distinguished from the lower animals by our mental life. If we survey this mental life, and enumerate all the ways in which it transcends the capacity of apes, dogs and bears, we find that these many ways are in fact one way, and reveal different facets of a single ontological divide: between reasoning and non-reasoning beings. Here are some of the distinctions:

(a) Animals have desires, but they do not make choices. (Aristotle emphasises this in his ethical writings.) When we train an animal, we do so by inducing new desires. But *we* can choose to do what we do not want, and want to do what we do not choose. Kant made much of this distinction; for he saw it as the foundation of morality.

(b) Animals have consciousness but no self-consciousness. They do not have those peculiar first-personal thoughts, which seem to generate the mystery of our condition; they *have* their mental states, but do not *ascribe* them.

(c) Animals have beliefs and desires; but their beliefs and desires concern present objects: perceived dangers, immediate needs, and so on. They do not make judgements about the past and future; nor do they engage in long-term planning. (See Jonathan Bennett: *Rationality*.) Squirrels store food for the winter; but they are guided by instinct, rather than a rational plan. (To put it another way: if this is a project, it is one that the squirrel *cannot change*.) Animals *remember* things, and in that way retain beliefs about the past: but about the past as it affects the present. As Schopenhauer argues (*The World as Will and Representation*, vol. II, ch. V), the recollection of animals is confined to what they perceive. They remember only what is prompted by the present experience; they do not 'read the past', but 'live in a world of perception'.

(d) Animals relate to one another, but not as persons. They growl and feint, until their territories are certain; but they recognise no right of property, no sovereignty, no duty to give way. They do not criticise one another, nor do they engage in the give and take of practical reasoning. If a lion kills an antelope, the other antelopes have no consciousness of an injustice done to the victim, and no thoughts of revenge.

(e) In general, animals do not have rights and duties. It is not murder to kill them, nor is it a sin to take them into captivity or train them for our purposes. (What then is the foundation for our belief that we should not treat them cruelly? Kant was puzzled by this question.)

(f) Animals lack imagination: they can think about the actual, and be anxious as to what the actual implies. (What is moving in that hedge?) But they cannot speculate about the possible; still less about the impossible.

(g) Animals lack the aesthetic sense: they enjoy the world, but not as an object of disinterested contemplation.

(h) In all sorts of ways, the passions of animals are circumscribed – they feel no indignation, but only rage; they feel no remorse, but only fear of the whip; they feel neither erotic love nor true sexual desire, but only a mute attachment and a need for coupling. To a great extent their emotional limitations are explained by their intellectual limitations. They are incapable of the thoughts on which the higher feelings depend.

(i) Animals are humourless – no hyena has ever laughed – and unmusical – no bird has ever sung.

(j) Underlying all those, and many other, ways in which the animals fail to match our mental repertoire, there is the thing which, according to some philosophers, explains them all: namely, the fact that animals lack speech, and are therefore deprived of all those thoughts, feelings and attitudes which depend upon speech for their expression. This is consonant with the view of Aristotle, whose word for reason – *logos* – is also the word for speech. (An animal, Aristotle says, is *alogon*, which means both non-rational and without language.)

The above is undeniably controversial; yet it makes sense of many observations that have been made down the centuries. We suppose that there is some kind of *systematic* divide between us and the other animals; that the thing which distinguishes us is not manifest merely in our laborious reasonings in mathematics and science, but in all our thinking, all our activity, and all our emotional life. When Schopenhauer praised the innocent life of the animals, he did not mean that they are too good for sin, but that they exist *beneath* good and evil. Their joys are uncorrupted by remorse and apprehension; and while they flee from danger they do not, as we do, fear death, having no conception of their own non-existence. Even torture is less trouble for an animal, since it lacks the horrifying thought of what is being done to it by whom. And when it is argued that certain animals are like us in one of the above respects – animals like the higher apes, who seem to have a sense of humour, or dolphins, who seem to communicate their desires and act in concert – the arguments tend to imply that these animals are like us in the other respects as well. It seems impossible to mount an argument for the view that the higher apes can laugh, which does not also attribute to them reasoning powers, and maybe even language. It is an empirical question, whether apes are like this, or can be trained to be like this; but it is a philosophical question, whether the capacities that I have described belong together, or whether on the contrary they can be exemplified one by one.

Yet there is something important that we have in common with the animals, namely life itself. Return for a moment to Mrs Shelley's story. Are we really comfortable with the thought that a creature designed to imitate our behaviour, to speak our language and to manifest our thoughts, intentions and desires, has just as great a claim on us as any human being? For the sake of argument, we can allow that Frankenstein's monster is a person. (After all there are persons in law, like companies and universities, which do not really belong with *us* in the scheme of things.) But is this 'person' *alive*? If not, can we really feel comfortable in its presence? Would we be as wary of injuring it as we are wary of injuring one another? Is it not in some way eery, *unheimlich*, to be driven, like Frankenstein's monster, from the fold?

1. Life

We recognise life, and accord to it a special place, not merely in the human world, but also in the world of science. Vitalist philosophers like Henri Bergson have argued that the phenomena of life are distinct from the rest of nature, and involve another principle of organisation that cannot be explained in terms of physical processes acting under physical laws. The theory of evolution is often cited against such views, though it is interesting that Bergson regarded the theory as *confirming* them. (See his once influential *Creative Evolution.*) The debate is not always conducted at the highest level. Vitalist philosophers tend not to distinguish the controversial thesis that vital processes are not physical processes, from the far less controversial thesis that vital processes, though physical, display a new order of organisation, and obey a new order of law, irreducible to the organisation and the laws studied by ordinary macro-physics. This second thesis could well be true: just as it could be true that the laws governing large-scale physical events are not derivable from the laws of quantum mechanics.

More interesting, perhaps, is the distinction made by Aristotle, between organisms that are self-moving in search of what they need, from those which are inert, or which move only under the impulse of external forces. This corresponds to the common-sense division between animals and plants, hence Aristotle associates the distinction with another, between the appetitive and vegetative parts of the soul. However, like every distinction between natural kinds, that between animal and plant may not be as it seems. Perhaps there are self-moving plants, and static animals. Nevertheless, Aristotle is surely right that there is a special place in our world for those things that shunt themselves around it. For here is a manifest capacity for change, whose explanation is to be found within the very thing that changes. The animal is a 'law unto itself', and understood as such.

Later philosophers followed Aristotle, in trying to find the principle of self-movement upon which an animal's activity depends. Their interest was conditioned by two preoccupations: one scientific, the other metaphysical. They were looking both for an explanation of self-movement, which they read as the capacity of an individual to generate its successive states; and for a characterisation of the 'substances' that feature in our world-view. Leibniz argued that every monad (every basic substance) possesses a *vis viva*, or vital force; this explains not only its states, but also the order of their generation. Spinoza, who argued on metaphysical grounds that there is only *one* substance, nevertheless recognised 'quasi-substances', each animated by a *conatus*, which is the endeavour of a thing to persist in its own being. When we find something that 'holds itself together', not only at a time, but also over time, we are more inclined to say that this is a real individual, with a nature and identity of its own.

The concept of life, therefore, has been appropriated by philosophers in their search for the 'true individual'. The heap is an arbitrary individual; even the table is one thing only so long as our interests require it to be. But when it comes to the dog, the cat or the human being, their unity and identity seem to belong to them quite independently of the way they are classified. It is part of the nature of Moggins that she is one cat; and the criteria for counting cats are given by the theory of felinity.

I have referred to two marks of the 'true' or substantial individual: unity and identity. There is an organisation which makes it non-arbitrary that this part and that part belong to *one* thing; and also non-arbitrary that this thing at one time is the *same* as that thing at another. Life – or at least animal life – promises something that philosophers have always prized and never clearly obtained, namely criteria of unity and identity, and in particular criteria of *identity over time*. (See the discussion of identity in Chapter 12.)

Why should identity over time be such a problem? The answer lies in Hume's law. There is no necessary connection, Hume argued, between the world as described now, and the world as described yesterday or tomorrow. Each of these complete descriptions of the world is consistent with a denial of the others. But a complete description of the world as it is *now* mentions only things that exist *now*. Suppose there were qualitatively identical but *different* things existing yesterday. Then this fact would not show up in the complete description of the world as it was yesterday. The description would read the same, whether yesterday's objects were numerically the *same* as today's or not. So how can I have *now* any ground for asserting identity across time? The best I can find is a certain 'constancy and coherence' between yesterday's world and today's. But constancy and coherence do not amount to identity. Maybe, however, I can find something better: namely, an explanation, in the individual itself, of why it is thus-and-so at one time, having been thus-and-so at another.

2. Personal Identity

The problem of identity is particularly acute in the case of persons. For here it raises the further question, concerning the relation between our identity as animals and our identity as persons (or rational beings). We seem to have too many ways of identifying ourselves. I identify myself in the first person as 'I'. Others identify me as the rational being with whom they enter into personal relations. A zoologist would identify me as an *organism*. What is the relation between the self, the person and the animal? Three things, two things or one thing? And how do we determine its (or their) identity? There is a rich mine of philosophical problems here, and I shall briefly mention some of them.

It was Locke who first raised the question in its modern form. Man, he argued, is not the same concept as person. The first describes a part of the natural world; the second is a 'forensic' concept: it features in our inquiries into responsibilities and rights. A person is a 'thinking intelligent being, that has reason and reflection, and considers itself as itself, the same thinking thing in different times and places' (*Essay*, II, 27, ii). So we should not be surprised to find different criteria for the identity of men and the identity of persons. A human being remains the same just so long as the body ticks on; and he may exist long after all traces of personality have disintegrated. (Though is there a point where we might change the pronoun from 'he' or 'she' to 'it'?) We can also envisage radical changes of personality, just as we can imagine the 'incarnation' of one and the same person in two different humans. (The cobbler migrates into the body of the prince, in Locke's example.)

Locke proposed a criterion of identity, usually described as 'the continuity of consciousness'. So far as my memories link me to the past and my desires and intentions project me into the future, so far am I the same person over time. Thomas Reid famously objected that such a criterion could deliver two conflicting answers to the question of identity. The old general may remember the young officer, who remembers the boy who stole the apples, even though the boy has been forgotten by the general. So the general both is and is not identical with the boy. But the objection is not lethal, and suggests merely that we should amend Locke's approach. We should define personal identity in terms of a *chain* of interlocking memories, linking the general to all his previous activities: the old man remembers the middle-aged man who remembers the youth who remembers the child. If the chain is unbroken, then perhaps identity is secure.

More serious is the objection made by Bishop Butler, whose *Sermons* constitute one of the great works of moral philosophy (and of morality too). Suppose I have the thought of standing in this room once before. What makes this thought into a memory? Surely, the fact that I identify *myself* as standing in this room. But how do I know that this identification is correct? I must have grounds for judging that it was *I* who once stood in this room. False memory-claims are no grounds for identity; true memory-claims ('genuine' memories) are grounds for identity, but only because their truth depends on the truth of an identity-claim. The criterion, in short, is circular.

But is it *viciously* circular? It is important, as we have seen, to distinguish virtuous from vicious circles. Sometimes, by going round in a circle, you show that two concepts really *are* deeply connected: not that the one has to be applied both *before* and *after* the other (as in a vicious circle), but that they are both applied *together*. Maybe this is true of personal identity and the continuity of consciousness. (This is argued, for example, by Perry and Wiggins.) At any rate, Butler's criticism is far from conclusive.

Others – notably Bernard Williams – have argued that criteria based in the 'continuity of consciousness' presuppose some other criterion, based in *bodily* continuity. We think we can dispense with this other criterion only because we are deceived by our first-person perspective, which suggests that we do not need it. But if we do not need it in our own case, this is not because we use some *other* criterion; it is because we use no criterion at all. Our first-person privilege depends upon our possessing a concept of personal identity – which in turn depends upon our ability to identify and reidentify others. So how is *that* done?

The emphasis on bodily continuity does not solve the problem. The first set of arguments to show this is due to Shoemaker (*Self-knowledge and Self-identity*), who imagines two people, Brown and Johnson, undergoing serious brain operations which involve extraction of the brain. The two brains are replaced, but in the wrong skulls. Suppose that Brown's body awakens from the operation with Johnson's memories, and Johnson's traits of character. Should we say that the person 'in' Brown's body is now Johnson?

Such examples give sense to an old thought: namely that we are not identical with our bodies, but somehow *in* them. By the fiction of a 'brain transplant', Shoemaker is able to satisfy those who require some measure of causal continuity as part of the concept of identity through time. Our personal relations towards Johnson will now be frustrated and denied by the person 'in' Johnson's body; but they will be rewarded by the person 'in' Brown's body. By any reasonable idea of *personal* identity, therefore, this second person ought to be Johnson. But then suppose he is only *partially* like Johnson, and retains some of Brown's more unpleasant quirks and mannerisms. What do we say then? All kinds of puzzle-cases can be manufactured, in the face of which we may find ourselves wholly at a loss.

Hence there arises a well-known response to such puzzles, associated with the later Wittgenstein. Our ordinary concepts, Wittgenstein argued, make sense only against certain background assumptions. When these assumptions can no longer be relied upon, we are unable to use our concepts. We *begin* to imagine the deviant situations; but we soon find ourselves not knowing what to say about them. There just is no answer to the question 'Who is Johnson?', for the simple reason that there is no longer such a question.

Attractive though this response may be, it comes up against a difficulty, frequently emphasised by Williams. Even if there may be no question for us, as to who is Johnson, there is surely a question for *Johnson*? Imagine the following case: you are to undergo a brain transplant, and your body is to receive the brain of Henry, while Henry's body is to receive the brain that is currently yours. After the operation the 'occupant' of Henry's body is to be rewarded with all that he desires, while the 'occupant' of your body is to be horribly tortured. Surely it is rational for you to fear the outcome of this operation, and to wonder *which* of these people will be you? The first-person viewpoint keeps the question of personal identity *open*, in the face of the deviant assumptions that seem to undermine it.

A second set of arguments is associated with Derek Parfit (see especially *Reasons and Persons*). There are cases where the idea of identity over time does not apply. Consider the amoeba, that splits into two qualitatively identical parts, each an amoeba. With which of its successors is the amoeba identical? Any grounds for one answer will be grounds for the other. But the original amoeba cannot be identical with both. (Identity is a relation which everything has with one thing and one thing only, namely itself.) So it is identical with neither. Now transfer this idea to persons. Imagine that Brown's brain is 'read' into a computer which then reproduces all its cerebral information in a thousand other brains, each implanted in a Brown look-alike, while Brown's original brain and body both die. With which of these look-alikes is Brown identical? The concept of identity seems inapplicable, and in any case unimportant. Parfit imagines many such cases, in order to 'disestablish' the idea of personal identity, and to put the idea of continuity in its stead. What matters to me, he says, is not that I should be identical with some future human being, but that enough of me should survive in him. Our methods of answering questions of identity are in the last analysis arbitrary; moreover identity is not something that I should rationally desire.

Parfit draws alarming moral conclusions from his arguments. But the arguments do not apply only to persons, being descants on Hume's law: they could be duplicated for animals, plants, tables, chairs, even elementary particles. If they disestablish the concept of personal identity, they disestablish the concept of identity as such. But that just takes us to our starting point. For we *can* justify the concept of identity through time, so long as we identify the substantial individuals in our world. Persons are not *just* bits of software, to be incarnated at will. They are also living things, members of a natural kind, whose identity is fixed by their life, and who must live that life through.

3. Existence and Essence

This returns us to the theory of being. Animals are part of nature; their existence is regulated by the laws governing their respective 'natural kinds'. A tiger cannot become a lion, a mouse or a juniper tree, any more than Daphne could become a laurel. Its life is fixed in its course by its tigritude. Although a tiger changes through time, the process of change is determined by its constitution. Like the juniper tree, the tiger *becomes* what it essentially *is*, as its life unfolds. Hence questions concerning its identity are not arbitrary; something can be a tiger at one time only if its constitution determines that it is a tiger at other times. The mad thought-experiments of a Parfit here make no sense; for they abrogate the very laws which enable us to identify tigers in the first place. If people, like tigers, constituted a natural kind, we should have a real guarantee of their identity through time.

But do people constitute a natural kind? Perhaps not, if we allow gods, demons and angels to be people; certainly not if we include such artificial persons as firms, universities and churches. There are philosophers who nevertheless wish to *anchor* the concept of personal identity in a natural kind, sensible of the advantages that this would bring in the battle against the Humean sceptic. Such philosophers are apt to argue that the concept of a person derives its sense from the fact that it is *realised* in a natural kind. (This, roughly, is the position defended by Wiggins.) We know what persons are, only because we have this paradigm instance: the human being.

Nevertheless, it is clear, as Locke originally argued, that personal identity and animal identity are different ideas, and that the distinctive features of our condition – self-consciousness, freedom (or at any rate the belief in it) and interpersonal responses – are in apparent contest with the demands of animal life. Even if this contest is *only* apparent, it is still necessary to explain the appearance.

There is one kind of philosopher for whom the contest is far from merely apparent, but on the contrary *real* and absolute: the existentialist. I referred in Chapter 12 to the anxious theory of being advanced by Heidegger, according to which the being of a self-conscious creature is essentially different from the being of things or animals. Being, for us, is a response to the *question* of being. We 'take up a posture' towards our world, and thereby *bring* ourselves into being. Sartre puts the point more radically, arguing that whereas the essence of objects precedes their existence, in our case it is the other way round: 'there is at least one being whose existence comes before his essence, a being which exists before it can be defined by any conception of it. That being is man.' (*Existentialism and Humanism.*) (Interesting, however, that he uses the word 'man', and not 'person'.)

Existentialism is no longer very fashionable. Nevertheless, its leading idea is of considerable interest, even for philosophers who have rejected the Husserlian language that was originally used to express it. (For Husserl 'essences' are given by 'conceptions', and discovered mentally; an essence is never discovered by scientific experiment. Since Kripke it is very hard to endorse this way of thinking.) We could put the idea in the following way: the existence of an animal runs in the track laid down by its constitution (its essence); it is the same animal today as the animal yesterday, for the reason that all its possibilities were contained in it from the beginning, and are merely unfolding according to natural laws. The existence of the human being runs in no such track. I am what I *choose* to be, and what I am cannot truly be known until my life is over. Only after my existence, in which I have forged an identity for myself, can my 'essential' character be identified. Sometimes the existentialists express the point by saying that we are not substances. Heidegger, for example, writes that 'a person is not a thing, a substance'; '*Dasein* does not have the kind of Being which belongs to something merely present-at-hand . . . *nor does it ever have it.*' (*Being and Time*, p. 48.)

Why should we accept such a doctrine? The reason lies in the first-person perspective. When thinking of objects, I am thinking causally: in terms of explanation, prediction, and causal laws. When thinking of people I am not normally thinking in this way: certainly not when thinking of myself. When I make up my mind I am not *predicting* what I shall do, but deciding to do it. The person who, when asked what he proposes to do, merely utters a prediction, based on an assessment of his own character, is precisely *not* making up his mind. In making a decision I identify with my future self, place myself, as it were, in that future which lies in wait for me, and turn it in my own direction. I cannot do this if my nature is already given; for then my future is irremediable, and I have no choice. Freedom means that my existence is always incomplete, 'unsaturated', awaiting its reality. And this freedom is revealed in the immediacy with which, when deciding, I know what I shall do. (I do not, and could not, base my decisions on *evidence* for my future behaviour.)

Existentialists emphasise not only intention, but also intentionality, as distinctive of our condition, and further proof that, in our case, existence precedes essence. The argument is, like the last one, highly suggestive. But it is also, like the last one, riddled with metaphor. Intentionality involves a relation to the world as represented; and for the existentialist this relation is active, dominating, 'lived'. Consciousness involves the constant rearrangement of the *Lebenswelt*, as we forge our journey through it. In rearranging the objects of consciousness, we rearrange the subject too. For the subject is the hole in the centre of the intentional world: the perspective from which all this is understood and acted upon.

These thoughts are nicely summed up by José Ortega y Gasset, a Spanish existentialist who, unlike Sartre, was also a cultural and political conservative: 'the stone is given its existence: it need not fight for what it is . . . Man has to make his own existence at every single moment.' ('Man the Technician'.) The image of a 'fight' is never very far from the existentialist's mind, for reasons that I shall try to explain in Chapter 30.

Notice the contrast: man and stone. That given by Sartre just before the passage quoted is between man and paper-knife. That given by Heidegger is between *Dasein* and being-at-hand, which is the condition of things. But that is not the contrast to which *we* have been led by these questions: our problem concerns the relation of the person to the *animal*. Maybe existentialists can enjoy their radical conclusions because they overlook the missing link.

Certainly, if the peculiarity of our condition is bound up with intentionality, then we cannot describe it as the existentialists describe it. For animals have intentionality; they bark *at* things, strive *for* things, run *from* things, and in each of those cases the particle indicates an intentional (and therefore intensional) context. (Cf. George Eliot, on the 'intentional inexistence' of the barkee: 'the barking at imagined cats, though a frequent exercise of the canine mind, is yet comparatively feeble': *Felix Holt*, ch. 15.)

And if the peculiarity of our condition lies in our radical freedom, could we not use the existentialists' arguments precisely to cast that concept in doubt? If Sartre is right, then radical freedom entails that we have no pre-existing nature. But this would mean that there is no basis to our judgements of identity through time. On the other hand, responsibility, blame, praise, and intention itself, require us to affirm our identity through time. In which case, the postulate of radical freedom seems to lead us to a contradiction.

That is by no means the last word against the existentialist. Nevertheless, it is reasonable to suggest that the onus lies on him, to show why human life is so very different from the life of the animals; or else to show that our life is not the 'ground' of our identity. Neither option is attractive.

4. Death

But here lies another problem. Things that live must also die. And what survives thereafter, if the life and the person are one and the same?

Some ways of defining personal identity seem to justify the belief in personal survival; others seem to deny it. If the person is character, memory, and reason, then there is no problem in supposing that these things (or this system of things) may continue, when the body dies. Hence the view, defended by Plato, Aristotle and Spinoza, that the 'intellect' may survive the body and even (in Plato's thinking) precede it.

Against that, however, are the following powerful considerations: (a) There is an intimate relation between our mental states and bodily conditions. It is difficult to see what we could mean by ascribing emotion, for example, to a subject who had no means of bodily expression. Likewise sensation, perception, even belief, seem to be tied up with the body, with its sensitive organs, and with the behaviour that springs from them.

(b) Much of our mentality is part of our animal life: this is certainly true of our sensations. And even those states of mind that lie above and beyond the repertoire of the lower animals – erotic love for example – are rooted in bodily conditions and primitive responses that we share with them.

(c) Residual doubts about personal identity lead us to believe that human life, and the bodily condition implied by it, are necessary to the survival of the person. If the 'software' that programmes my mind may be realised after my body has died, there is no reason to think that the resulting hardware either is or could be me.

(d) Many things die, besides people: trees, dogs, fish, bacteria. In most cases, however, the idea that these things continue to exist after death is absurd. If people are tempted to believe that dogs survive in some 'happy hunting ground', it is for the same reason that they believe these things of the people they love: namely, their inability to accept bereavement. (But Gerard Manley Hopkins felt sincerely bereft of the Binsey poplars: are they immortal too? ('Binsey Poplars').)

Despite those considerations the feeling persists that my own death is not, and cannot be, the end of *me*. What is the source of this feeling? The following considerations seem to be involved:

(a) Our sense of personal survival is bound up with the first-person perspective. When my identity across time is a paramount consideration, the 'I' is at the centre of the stage. What shall *I* do? What shall *I* feel or think? But this 'I' can be projected beyond death. I can wonder what I should think or feel, in the circumstances where my body lies inert and lifeless. There is nothing incoherent in this thought.

(b) My own non-existence is inconceivable to me. I simply cannot think of a world without thinking also of my perspective upon it. And that means thinking of my own existence.

This argument is very tricky, and is open to the retort, expressed in luminous verse by Lucretius, and in reported conversation by Hume, that there is no more difficulty in conceiving my non-existence after death, than in conceiving my non-existence before birth, and no reason to be distressed by either. Why does my *entry* into the world forbid my *exit*?

There seem to be two quite different ideas of conceivability. I can conceive of a world in which I don't exist: if so, maybe such a world is possible. But I cannot conceive of myself not existing, if you mean conceive of a world, viewed from *this* first-person perspective, in which there is no *I*. But from that nothing follows about real possibilities.

(c) My personal relations, like my rational intellect, are not time-bound, and remain in crucial respects unaffected by death. My death extinguishes neither my obligation to you, nor yours to me. My will is enshrined in obligations and rights, and projected into an indefinite future. Death seems not to threaten the will since it leaves the web of right and duty unaffected.

Again the argument is difficult to assess. The best it can prove is that practical reason involves the belief in personal survival: not that this belief is true. A peculiar variant is given by Kant, who believed that practical reason presupposes immortality, since the weight of obligation, being infinite, requires an infinite time in which to be discharged.

(d) Death is difficult to encompass intellectually; it is also difficult to encompass emotionally, and this difficulty is felt more vividly from the third- and second-person perspective. At the time of bereavement, it is almost impossible to believe that the other no longer exists; there is a 'you-shaped' whole in my emotions, and I act and feel as if you were still existing, although far away and inaccessible.

This argument provides a powerful *motive* to believe in a 'life after death', but no *reason* to do so. If, however, such a life after death were *possible*, maybe we *ought* to believe in it if we can, out of respect for all that is most worthwhile in the human condition.

The arguments are not conclusive, and the discussion goes on. So let us pass to another and more urgent problem.

5. The Fear of Death

We all fear death; but is it rational to do so? What exactly are we afraid of? Nagel writes of a peculiar and disturbing feeling, quite unlike the fear of pain or suffering, which attends the thought that one day I will not exist: 'There is something that can be called the expectation of nothingness, and though the mind tends to veer away from it, it is an unmistakable experience, always startling, often frightening, and very different from the familiar recognition that your life will go on for only a limited time . . .' (*The View from Nowhere*, p. 225.) Is this any more than a queasy feeling?

Death, writes Wittgenstein in the *Tractatus*, is not part of life but its limit. He means that there is no such thing as 'living through death', so as to emerge on the other side of it. Death is not an experience *in* life, and there is no such thing as looking back on death, and assessing it from some new perspective.

Others have argued in a similar way for the conclusion that the fear of death is irrational. (Thus Lucretius and various Roman Stoics.) If, after death, I am nothing, there is literally nothing to fear. This, however, seems like sophistry. Death is also the *loss* of life and of the good things that come with life. And is it not rational to fear such a loss? Yet that too seems to miss the point. I could be threatened with the loss of all good things, and still regard this threat with equanimity – or at least, without that queasy feeling which comes from the thought that soon I shall not *exist*. Why is my non-existence so terrible? Why, indeed, is it terrible at all?

It is peculiarly difficult to get one's mind around this question. Every attempt to describe the evil of death suggests either that we fear the loss of goods (including the good of life), and so misses the distinctive feeling of 'ontological insecurity'; or else concludes that we fear non-existence *per se* – and that seems irrational. In another sense, however, it is plainly reasonable to fear death: for if we did not, we should fail to secure our own survival, and therefore threaten the success of all our projects. Hence a rational being needs the fear of death, just as he needs the capacity for nausea at foul smells, or the disposition to sleep from time to time. But does that make the fear into a *rational* fear?

What *is* a rational fear? Presumably it is rational to fear what will pain you. It is rational to fear some condition, to the extent that you would wish to get out of it, when you are in it. But again the criterion does not apply to death. If death is the end, then no one fears to escape from it, once it has arrived. When Achilles complains to Odysseus that he would rather be the meanest serf on earth, than the greatest prince in Hades, he speaks from a point beyond death – he speaks as a 'spirit' who has *survived* his encounter with death. But he justifies the fear of death only by showing that it leads to an irreversible decline in one's fortunes; not by showing that it brings one's fortunes to an end.

In response to this unanswerable riddle, it is tempting to turn the argument on its head, arguing that it is rational to fear the *absence* of death. Drawing on a famous play by the brothers Čapek, Bernard Williams ('The Makropoulos Case') has argued for the 'tedium' of immortality, pointing out that our joys are *mortal* joys, dependent upon death for their desirability. The central character of the play, who has lived through every love and joy only to rise to a frozen plateau of cynical disregard for others, displays the true character of a practical reason that has been shorn of mortal limits. (A more comic version of immortal tedium is to be found in the brilliant last chapter of Julian Barnes's *A History of the World in Ten and a Half Chapters*.)

Traditional defenders of immortality would scarcely be disturbed by Williams's argument. They would argue, with Aquinas (and the Dante of *Paradiso*), that our mortal desires are precisely what we lose in dying, so as to devote ourselves to those other and more mystical enterprises which never grow stale. The worship of God bears infinite repetition, precisely because its object too is infinite. Never does the Mass or the Sacred Service weary the true believer, or cause him to doubt the meaning of its inner message. If there is eternal life, why should that not be it?

6. Timely Death

Such thoughts do nothing to console the timorous pagan. Is he then caught between the irrational fear of death that the capacity for success demands, and the rational fear of a joyless longevity? This would be terrible indeed.

Looked at from the third-person perspective, death is not always an evil. Sometimes, indeed, it is a good. First, death may be conceived as a rightful punishment. A person's crimes may be sufficient reason for killing him: in which case, how can it be said that his death is an evil? (Think of Hitler or Stalin: not only were their deaths good in themselves; more miserable deaths would have been even better.) Secondly, death can be seen as a liberation from appalling torments, whether physical or emotional. Thirdly, and more mysteriously, death can be seen as the fitting conclusion to a life of great undertakings. The tragic hero is vindicated in death, which reflects back into his life the redeeming order of finality. We do not understand this; yet we feel it, and our feeling is every bit as real as the queasiness with which we contemplate our own extinction. Why should not our reflections come to rest in this more satisfying perspective, rather than dwelling on the nameless fear that gets us nowhere?

For ancient thinkers death could be vindicated in another way. Return for a moment to Aristotle's discussion of virtue. The courageous man acquires a disposition to pursue what is honourable in the face of danger. Honour is what he wants, more than he wants to flee. And it is rational to acquire this disposition, since it is 'a part of happiness': without courage one can have no guarantee of the 'success in action' which is the final end of practical reasoning. But now, consider the moment of battle. The enemy will shortly overpower me. What is it rational for me to do? For the coward, who desires to save himself, it is rational to drop his shield and run. For the courageous man, whose heart is wedded to the thought of honour, it is rational to stand, even if death is the consequence. Since the courageous man's desire springs from a disposition that all of us have reason to acquire, he is doubly reasonable. It is therefore rational to prefer honourable death to an ignominious survival. (This matter is discussed by Xanthippe and Socrates in a notorious Xanthippic dialogue: See *Phryne's Symposium*, 1158a–b.)

That is perfectly intelligible from a third-person viewpoint. We all warm to the hero, who lays down his life for his friend. Even pacifists feel this: witness the glorious tribute to self-sacrifice in Britten's *War Requiem*. And one can feel this, while deploring the 'pity' of war. But it is intelligible too from a first-person perspective. One can learn not to love death, but at least to accept it as the best outcome in a dire situation. There are circumstances in which survival is a fatal compromise of one's life, a shame from which one could not recover, a disparagement of all that one has wished for and all that one has done. Hence, according to Nietzsche, the thought of a 'timely death' may be the ground of the true (i.e. pagan) morality.

Do those thoughts justify suicide? Schopenhauer believed so; as did many of Plutarch's heroes. But it is one thing to justify acquiring those virtues which make you likely to die honourably; another thing to justify the death itself.

7. The Mystery of Death

Even if true, such thoughts do not quiet our apprehensions. Maybe nothing *can* quiet them. Maybe we should accept that the fear of death is not really a *fear*, since it is founded in no coherent thought of how we are harmed by dying. It is an *anxiety*. This anxiety, according to Heidegger, has deep foundations. For it marks the insurgence into consciousness of the thought of our contingency. Death shows us that we will not be, and therefore that we might not have been. Our existence has no ultimate foundation; it is a brute fact for which we can find no reason, since all our reasons are generated within life, and not from the point of view outside life to which we can never attain. The anxiety towards death is 'ontological'; it spreads over the face of existence itself, and undermines the 'ground of being'. What can we do to assuage it? Heidegger makes some pregnant but obscure suggestions. *Dasein*, he tells us, must assume responsiblity for its own being; and this can be done only through an ontological posture which he describes as 'being towards death': we must act out the truth of our own mortality, and never flee into fantasy or despair. We must see death as the other side of life: to look death clearly in the face is to see the meaning of life. Only then do we truly live. Maybe this is what the tragedians tell us. It is certainly one of the themes of Rilke's *Duino Elegies*. But whether a philosopher can really convey such thoughts – let alone a philosopher whose mastery of the written word advances no further than the stage reached by Heidegger – may reasonably be doubted.

22

Knowledge

In the second chapter I discussed some of the topics in the theory of knowledge. But I did not say much about knowledge itself. Lest the reader should be surprised by this, I should say that, in my view, the concept of knowledge is of no very great interest in epistemology, which concerns the justification of *belief*. If philosophers spend so much time discussing knowledge – including the mega-boring question, whether knowledge is justified true belief – it is partly because the questions of epistemology have been misleadingly phrased in terms of it. Philosophers have asked 'Do I (or we) *know* anything?', when really they sought the justification for our *beliefs*. The principal culprit is Plato, who distinguished genuine knowledge from mere opinion (*doxa*), and supposed that we could answer the questions of epistemology by inquiring into the peculiar state of mind of the one who *knows*.

There are other questions about knowledge, however, apart from those raised by the sceptic, and in this chapter I shall address some of them. For example: what is knowledge, and why do we value it? Is there a distinction between theoretical and practical knowledge? Does the person who knows that *p* have to be *certain* that *p*? And many more. The literature on this topic is vast and none too interesting.

1. What is knowledge?

Plato defines knowledge, in the *Meno*, as 'true belief with an account' (*logos*). Elsewhere in his writings he dissents from this analysis. But it is generally assumed to be the first reference in the literature to the 'traditional' theory of knowledge as 'justified true belief'. According to this theory, John knows that *p* if and only if:

(i) John believes that *p*;
(ii) It is true that *p*;
(iii) John's belief that *p* is justified.

We are interested in knowledge, according to this theory, because we are interested in the truth of our beliefs; and the search for knowledge is the search for the justifications which guarantee that truth.

One difficulty for this theory is suggested by the third condition. Surely justifications can be more or less convincing, more or less validating. What level of justification is required, before we can speak of *knowledge*? If we say that John's reasons for believing that *p* must *entail* the truth of *p*, then it is doubtful that any of us have sufficient justification for any of our beliefs, outside the areas of logic and mathematics. On this view there may be no such thing as empirical knowledge.

Plato came close to that conclusion in the *Republic*. He has even been accused of offering the following fallacious argument:

(1) Necessarily, if John knows that *p*, then *p* is true;
Therefore:
(2) If John knows that *p*, then *p* is necessarily true.

In other words only *necessary* truths (such as the truths of mathematics) can really be known. I doubt that Plato commits that fallacy (involving a misreading of what logicians call the 'scope' of the term 'necessarily'). Instead, I think he is over-impressed by the fact that it is only in mathematics, and other studies of necessity, that we really do have sufficient justification for our conclusions. He was embarking, in his own Greek way, along the road that was later to be explored by Descartes.

However, if we are to allow justification less than entailment (less than sufficient conditions) to count as knowledge, can we really have a use for the concept? If I am entitled to say that John knows that *p*, then I must know that *p* is true; I must also know that John's belief that *p* is justified. Presumably my evidence for the truth of *p*, like John's, will fall short of being sufficient. It is compatible with the falsehood of *p*. In which case how can I be certain that *p* is true? And if I am not certain, ought I not to say that I am confident that *p*, or that I believe *p* to be very probable, or some such thing, rather than that I know *p*? And ought I not to say the same of John too, since I can only know that he knows that *p*, by knowing that *p* really is true – i.e. by knowing that *I* know that *p*? So perhaps I should give up applying the concept of knowledge – saying neither that I know, nor that he knows, but talking instead only of the relative reasonableness of our respective beliefs.

Well, why not? Maybe we should adopt the position advocated by C.S. Peirce, and called by him 'fallibilism', according to which none of my beliefs is to be regarded as beyond the reach of questioning, or a matter either of certainty or knowledge. Ideas of certainty and knowledge are neither necessary nor justified: we can say all that we wish to say without them, and their introduction only generates confusion, by implying that we can reach a point in our beliefs where no further evidence will count against them.

The least that we can conclude from those considerations is that if it is ever true that someone knows anything, it does not follow that he also knows that he knows. But what should we say about the traditional theory, apart from these sceptical remarks? Maybe it is not a plausible analysis of knowledge in any case; maybe it mistakes the purpose of the concept. If so, it is hardly surprising if the theory feeds our sceptical propensities.

2. Are the Conditions Necessary?

The traditional theory gives three conditions for the existence of knowledge, which it holds to be severally necessary, and together sufficient. Nobody doubts that the second condition – the truth of the proposition known – is necessary for knowledge. But arguments have been presented against the necessity of each of the other two:
(i) Knowing without believing. Colin Radford gives examples to show that someone can know a truth, even though he does not believe it. Perhaps the most plausible case is that of the nervous examinee who, asked to write down the date of the French Revolution, panics and puts down 1789, without a shred of confidence that this is the true answer. You can imagine him completing the whole examination in such a way, giving correct answers, and telling precisely the story that the examiner is looking for, while being himself unable to say whether his answers are right or wrong. *You* know, looking on, that he is reproducing exactly what he was taught. Those history lessons were not wasted, nor was the revision that the candidate undertook during the weeks before. To put it another way: a store of historical knowledge was packed into the candidate's head; surely, it is this very store of knowledge that is now spilling out again? In other words, is it not natural to say that the candidate *knows* the date of the French

Revolution, even if, in the panic of the moment, his normal beliefs have deserted him?

The case is questionable on many grounds. Maybe the candidate doesn't exactly *believe* that the Revolution dates from 1789. But he is disposed to write down *that* date and no other. Is not this an adequate variant of our first condition, namely, the utterance of an opinion? However, maybe the example can be strengthened. For suppose he put down the *wrong* answer. Could we not still in certain circumstances say that he knew the right one? Maybe he himself will say as much when the examination is over. 'I *knew* it was 1789; what on earth possessed me to write down 1788?' And so on. (Now we are beginning to see how boring this topic is: what does it matter, that we say he knew or not?)

(ii) Knowing without justification. Although we may doubt that the examinee *believes* that the Revolution occurred in 1789, it seems that he is quite justified in writing down that date, since all the reasons for doing so have been crammed into his consciousness. He knows, because he has *learned*. (Plato was interested in knowledge precisely because it is the goal of teaching: he wanted to understand the process whereby ignorance is overcome.)

But we can imagine cases where someone knows without having learned, and without having acquired any reasons for his judgements that would tend to establish their validity. A popular case is that of the chicken sexer. It seems there are people who can feel new-born chicks and sort them at once into male and female, without knowing how or why they draw such a conclusion. And their judgements are reliably true. We are inclined, therefore, to say that such a person *knows* that the chick he has just deposited in the box marked M is a male, even though he has no justification for his belief that it is so.

This kind of case is suggestive, for a variety of reasons. For one thing, it puts in question the idea of 'justification' (*logos*) invoked by the traditional theory. Do we mean that John knows that *p* only if there *is* a justification for his belief that *p*? Or do we mean that he himself is *possessed* of that justification, in the sense of being able to give an articulate account of it? (Plato's term '*logos*' suggests the latter: the one who knows, for Plato, is the one who can *justify* his opinions.) Look at the chicken sexer from the third-person veiwpoint. Surely, it is plausible to say that he does have a justification for his belief that the chick is male – namely, that he is the kind of person who is always right in the matter of chicken-sexing. But he does not use this as a premise in a process of reasoning, which has a judgement about the chick's sex as its conclusion. Still boring, of course; but worth pondering nevertheless.

3. Are the Conditions Sufficient?

In a famous article, Gettier presented a set of cases, which seem to show that the traditional theory *must* be wrong, since even when the three conditions are fulfilled as well as they could be, we may not have a case of knowledge. The phrase 'Gettier example' was coined to denote these cases, which have become paradigms in the theory of knowledge. Here is one of them:

Smith and Jones are candidates for a job, and fall into conversation while waiting for the interview. There are no other candidates, and, during the course of their talk, Jones comes to the conclusion that Smith is so much better qualified than himself, that Smith is sure to get the job. It so happens that the conversation turns to the coins in Smith's pocket: there are twelve of them, and Smith takes them out to count them. (Twelve is his lucky number, which is why he wanted to make sure.) As a result of this conversation, Jones comes to the following conclusion: the man who gets the job will have twelve coins in his pocket. He believes this, and has the best of reasons for believing it. To his surprise, however, the job is not awarded to Smith but to Jones himself. Reaching into his pocket in his euphoria, he finds that there are precisely twelve coins in it. So it was *true* that the man who gets the job will have twelve coins in his pocket. In short, Jones has a belief that is true and fully justified. (He could not have had better reason than he had for thinking both that Smith would get the job, and that Smith had twelve coins in his pocket.) But surely he did not *know* what he thought to be true, since his conclusion was based on a judgement that was not only crucial to the matter in hand, but also entirely mistaken: the judgement that *Smith* would get the job. Jones's belief that the man who gets the job will have twelve coins in his pocket was true *accidentally*.

Gettier himself draws no conclusion from his examples, and since publishing his tiny article has retired into the obscurity whence he briefly emerged. Other philosophers, however, have taken the Gettier examples as authority for two interesting conclusions. First, we ought to cease thinking about knowledge from the first-person viewpoint, which leads us to muddle the question whether I know *p*, with the question whether I have adequate grounds for believing it. Secondly, we ought to recognise that the concept of knowledge is designed to distinguish reliable from unreliable beliefs, and is applied in order to endorse the epistemological capacities of the knower, rather than to evaluate his reasoning.

4. Reliability Theories

Various theories have been proposed along these lines, and it is worth considering two of them:

(a) *The causal theory*. This is associated with the name of Alvin Goldmann, who argues that the idea of reliability is connected with explanation and therefore with causality. In saying that John knows that *p*, I am not merely reporting the truth of his belief, but also explaining its truth, by invoking John's reliability or expertise. Hence we should analyse knowledge in the following way:
(i) John believes that *p*.
(ii) *p* is true.
(iii) The truth of *p* causes John's belief that *p*.
In other words, we rephrase the connection between (ii) and (i) as a causal rather than a rational connection. (In the next chapter I shall discuss the general interest of such causal connections in the philosophy of mind.) The idea is this: when a person has become expert in some matter, the world acts on him in such a way as to generate true judgements. He is a 'reliable cognitive machine'.

Goldmann's theory fails to account for all the cases. We have knowledge of mathematics, and knowledge of future contingents. But neither mathematical truths nor future states of affairs have a causal relation to our present mental states. The fact that 2 + 2 = 4 does not *cause* my belief that 2 + 2 = 4, since mathematical facts do not take part in causal relations. (At least, so we are inclined to believe.) If you don't like that case, then you will at least admit the other one. For it is just *impossible* that my present belief that it will rain tomorrow is caused by its raining tomorrow, even when I know that this will happen.

(b) *The 'tracking' theory*. This is the name usually given to the theory advanced at tedious length by Robert Nozick. Recall that causal relations are distinguished from accidental connections in part by the fact that they support 'counterfactual conditionals'. If it is a law that smoking causes cancer then we can infer that, were John to smoke, he would be liable to contract cancer. The element of reliability is encapsulated in the counterfactual conditional; but there can be true counterfactuals even in the absence of a causal connection. (For example, if you were to smoke 2 + 2 cigarettes, then you would smoke 4 cigarettes; but there is no *causal* connection between 2 + 2 and 4 – the connection is one of mathematical necessity.) So let us phrase the reliability condition in terms of a counterfactual: John knows that *p* only if the following is true:

(i) John believes that *p*;
(ii) *p* is true;
(iii) If *p* were not true, then John would not believe it; and
(iv) in slightly different circumstances, if *p* were still true, John would still believe it.

Those last two conditions capture the idea that John's beliefs 'track the truth'. He has the capacity to keep track of the way the world is, avoiding errors, and ascertaining truths, in such a way that we can rely on his opinion. He is a guide to reality, like a reliable textbook.

Discussion of the tracking theory has become intricate; you can easily imagine that the phrase 'slightly different circumstances' will cause problems – how different, and by what measure? But these difficulties need not concern us, since few philosophers doubt that reliability has won out over justification, as the best candidate for what is really meant by knowledge. And this enables us to return to epistemology from a new vantage point.

5. Externalism and Scepticism

If I ask myself the question 'What do I know?' I find myself immediately lost in the maze of scepticism. Even if justification is no part of knowledge, I cannot claim to know a proposition without also holding it to be true. Hence I must be able to justify my belief that it is true. And how do I do that? If, however, I step outside my own predicament, and see the claim to knowledge from an 'externalist' standpoint, the epistemological problem tends to take second place. My knowledge is like your knowledge: it consists in a system of beliefs that are not merely true, but reliably true – whose truth is guaranteed by my epistemological trustworthiness. It is surely as likely that *I* have knowledge as that you have. In claiming to know, I am not merely reaffirming my own opinion. I am counting my opinion among the judgements that spring from what is epistemologically best in me.

Of course, on this account, many of the 'traditional' ideas about knowledge must go by the board. It does not follow, for example, from the truth that I know that *p*, that I also *know* that I know that *p*. Knowledge is not 'iterative'. But that is scarcely a disadvantage. For why should knowledge possess that extraordinary property, of being always included within its own scope? Nor does it follow that I am always *certain* of the things that I know. Certainty is a distinct condition, which we achieve only through careful analysis and argument, of a kind that – when too widely or frequently applied – threatens the natural competence of the knower.

Suppose that *p* entails *q*, and I know that *p*. Does it follow that I know that *q*? To say that it *does* follow is to embrace the 'principle of closure' – i.e. the principle that knowledge is closed under the relation of logical consequence. But the externalist viewpoint gives us no grounds whatsoever for endorsing such a principle. For the principle implies that, if there is some consequence of your belief whose truth you do not know, your belief cannot possibly be knowledge. In which case, you would surely know nothing. Descartes argued that, if it is true that he is sitting by the fire with a piece of wax in his hand, it is also true that he is not being deceived by an evil demon. But he cannot *know* that he is not being deceived by an evil demon (or that he is not a brain in a vat, or whatever). Taking the 'internalist' viewpoint, he was tempted to conclude, therefore, that he could not *know* that he was sitting by the fire. But that consequence follows only if we accept the principle of closure. The externalist viewpoint shows the principle to be arbitrary: we are not forced to accept it, in order to frame a useful and cogent concept of knowledge. On the contrary, if we did accept it, we could have no such concept. So let us say, instead, that I *can* know that I am sitting by the fire – this is one of those beliefs in respect of which I am maximally reliable – even though I cannot know (without some elaborate proof which I have yet to devise) that I am not deceived by the demon.

If people are still tempted to accept the principle of closure, it is perhaps because they confuse it with another and epistemologically less damaging principle, namely, that if I know that *p*, and also *know* that *p* entails *q*, then I know that *q*. But that endorses scepticism only in the special case where I am sure of the truth of the entailment. That is precisely not the case that I am in, when meditating about the evil demon. I do *not* know whether it follows from the fact that I am sitting by the fire that I am not being deceived by an evil demon. For I do not know whether the latter hypothesis is even coherent. Maybe that is what Moore had in mind, when he claimed to be more sure that he had two hands than he could be sure of the validity of any argument to the contrary.

This is not an answer to the sceptic. On the contrary; it merely shows how insignificant is the concept of knowledge in stating or refuting sceptical arguments. However we define knowledge, we shall always have the sceptical question in the first person: *how* do I know that *p*? But this is equivalent to the question: how do I justify my *belief* that *p*?

6. The Value and the Varieties of Knowledge

Knowledge is valuable. For it enables us to rely on our beliefs: on our own beliefs and on the beliefs of others. When seeking the truth about a matter, I appeal to the person who knows. For *his* opinions are reliably connected with the truth. That is why I seek knowledge; and that is why education is so important – since education has knowledge as its goal.

But there are many varieties of knowledge: as many varieties as there are kinds of rational success. Beliefs aim at truth, and are successful when they achieve it. The one who knows is the one who reaches the target of truth dependably. But, as we saw in discussing morality, reason does not aim only at truth. There are other exercises of rationality – both practical and theoretical – each of which has its own specific target. Hence we should not be surprised to find that there are several kinds of knowledge. Here are five of them:

(i) Knowing *that*. This is the kind of knowledge that I have been discussing. Its object is a proposition, and it is our paradigm of theoretical success. Its goal is truth, and its matter is belief.

(ii) Knowing *which* and knowing *who*. You say in my hearing: 'He won the prize'. I may or may not know *which* prize you are referring to, or *who* won it. It is tempting to see this as a special case of knowing *that*. (I know which prize, when I know that he won the Hawthornden.) However, matters are not quite so simple. For there seems to be a kind of 'knowing which' involved in the understanding of language itself, and which precedes our capacity to form specific beliefs. For example, to understand a name in the language, I must know which object it refers to. This is a matter of being able to identify the object, but not of having specific thoughts about it. Gareth Evans (*Varieties of Reference*) defends what he calls Russell's principle: to grasp the sense of a singular term is to know which object it refers to. Perhaps this principle could be used to explain *why* names are rigid designators. (See Chapter 13.)

(iii) Knowing *how*. Gilbert Ryle made this idiom famous – and, following him, philosophers have often used it in order to explain what they mean by practical knowledge. It is not the knowledge of theoretical truths that tells me how to ride a bicycle. I may know how to do it, while not knowing anything much *about* bicycles. Knowing how is a matter of skill. But it is a skill rationally acquired and rationally exercised. Moreover, like other forms of knowledge, it connotes reliability. The person who knows how to do something is the person you can depend upon, when the thing has to be done. He is the person you consult for his advice and example, just as you consult the person who knows *that*.

'Knowing how' is a matter of technique. It involves mastery of the means. It can therefore be misapplied, if the end is a bad one. Hence there must be more to practical knowledge than knowing how. There must also be what Kant calls 'knowledge of ends', and what Aristotle calls 'virtue' (which he expressly contrasts with technique). This is:

(iv) Knowing *what*. A person may not know what to do or what to feel, and it is in learning what to do and what to feel that we acquire moral competence. However thoughtful and skilled I may be as a casuist, I may still not know what to do when it comes to moral choice. The person who knows what to do is the person who reliably does what is right, whether or not he possesses the skill to justify it. Involved in this, Aristotle argued, is the ability to feel what is right. The virtuous person knows what to feel, in the sense of spontaneously feeling what the situation demands – the right emotion, towards the right object and in the right degree. Moral education has such knowledge as its goal. Maybe this is what we should be teaching, when teaching the humanities.

(v) Knowing *what it's like*. 'You don't know what it's like, to suffer real fear.' The person who says this does not doubt that you know what fear is, in the sense of knowing how to recognise it in others, what are its causes, effects and expressions. Indeed, you may know all that there is to be known about fear, and still not know what it's like. For 'knowing what it's like' means *having* it. That is why Mime could not teach Siegfried what fear is like, and why Siegfried's education had to await his encounter with the sleeping Brünnhilde. Thomas Nagel erroneously assimilates this kind of knowledge to knowing *that*, arguing that there is something *that* I know, when I know what fear is like – a 'subjective' fact that cannot be listed in the scientific inventory of the world. But there is no such fact. The knowledge involved here is a matter of first-person acquaintance, and we value it because we value all the other things that flow from a person, when he is familiar with the forms and varieties of human experience.

There are other kinds of knowledge, too: for example, plain 'knowing', as when you recognise someone in the street. But we need not digress further. Suffice it to say that the reliability theory can without difficulty be extended to these cases, and justifies the intuition that, by 'knowledge', we in each case mean the same. We use this word to endorse opinions, techniques, actions, feelings and sympathies, by endorsing the capacities from which they flow. The fifth kind of knowledge is the hardest to understand: but it is this kind that forms the subject-matter of aesthetics.

23

Perception

The topic of knowledge naturally leads to that of perception, which is our principal way of obtaining knowledge about the world that we inhabit. The study of perception is a necessary part of epistemology – indeed, it is the root of the subject. It is also important in the philosophy of mind, and in metaphysics. For perception is the meeting-point of mind and world, of inner and outer – the process whereby facts are translated into consciousness, and the world provided with its mental form. To understand perception is to understand our 'being-in-the world', as the Heideggerians put it.

We can distinguish four major problems about perception, which illustrate the centrality of the concept:
(i) Are perceptual beliefs justified? If so, how?
(ii) Is perception the *foundation* of our knowledge of the 'external world'?
(iii) What *is* perception?
(iv) What must the world be like, if it is to be perceivable?
The first two are epistemological questions; the third belongs to the philosophy of mind, while the fourth is a question of metaphysics. The four questions run into one another, so that it is very difficult to answer them one by one. Indeed, there is a certain artifice in our divisions between epistemology, metaphysics, and the philosophy of mind. In the really difficult areas, such as this one, it is best to follow one's nose, regardless of boundaries.

1. The Epistemological Question

Nevertheless, the study of perception begins from a question that is clearly epistemological, and which connects directly with the argument of Descartes's *Meditations*. When perceiving through the senses we are presented with a picture of the world. How do we know that this picture is true? How can we know that the world really is as our perceptions represent it to be?

The first response to such a question is to say that the world does *not* resemble the picture painted by our senses. Perception represents the world as *we* perceive it – decked out in those 'secondary' and 'tertiary' qualities that do not feature in the scientific inventory of things. (By a 'tertiary' quality I mean a feature that is perceivable only to creatures with certain intellectual and emotional capacities: the sadness in a face, the life in a picture, the elegance of a gesture, and so on.)

Such a response misses the point. Even in describing the human world, we assume that it exists independently of those who perceive it. The rose really *is* red, regardless of how it appears to *me*. Moreover, the human world and the world of science are not two worlds, but one world under two descriptions. We are led to the theories of science by our endeavour to explain the world as it appears. These theories are about the *very same things* that occupy our everyday perceptions. If we have no guarantee that the world is as we perceive it, then we have no real knowledge of any matter of fact.

Here it is necessary to mention a notorious argument, which has been used as a ground for just about every kind of philosophical error in our century: the 'argument from illusion'. I find it extremely difficult to state the argument, since almost all the standard versions of it are dressed in language derived from the unacceptable conclusions that their proponents wish to draw. However, very roughly, here it is:

Perception has various 'sense modalities': it involves eyes, ears, touch, taste etc. Whenever I see something, it *looks* a certain way to me. It is on the basis of this 'look' that I acquire my 'perceptual beliefs'. Likewise, whenever I hear something, it *sounds* a certain way; and so on. However, looks, sounds and tastes can be deceptive. For instance, the 'look' of a straight stick in water is indistinguishable from the look of a bent stick out of water; from an angle a round table looks elliptical; when my ears are blocked things sound far away; and so on. But there is nothing in the 'illusory' look that tells me that it *is* illusory: it is indistinguishable from a 'veridical' look, obtained in other circumstances. I cannot 'read off' the properties of an object merely from the 'look' that it has when I perceive it. Indeed, there may be no object at all, even though it looks as though there were. (Mirages, rainbows, hallucinations, delusions.)

Different philosophers draw different conclusions from this argument. Broadly their conclusions divide into two, epistemological and ontological:

(i) Epistemological: the grounds for my perceptual beliefs are provided by the looks, sounds and tastes through which I acquire them. But these looks, sounds and tastes could be as they are, whether the beliefs were true or false. So how are my perceptual beliefs grounded, and how *well* grounded are they?

(ii) Ontological: when the round table looks elliptical, I am presented with an elliptical look. *Something* is elliptical; but it isn't the table, since that is round. What is this elliptical something? A *look*. But what is that? The temptation is to speak of it as an image or 'mental picture'.

Running the two conclusions together, we generate the influential concept of the 'sense-datum'. This denotes a mental item with two distinct properties: on the one hand it is or contains the information needed to ground my perceptual belief or judgement; on the other hand it is or contains a mental image, which actually *has* those properties which it tempts me to attribute to the thing perceived. G.E. Moore, Russell, and Ayer were all advocates of sense-data, and all believed that the existence of sense-data was proved, and their nature established, by the argument from illusion; though how anything could be both a piece of information and a mental image, was something that they left to the imagination of their readers. One of their motives was to identify objects of *immediate* knowledge: objects whose properties were presented immediately and incorrigibly to the mind of the perceiver, so that he could not reasonably doubt that they were exactly as they seemed to be. In this way, they supposed, you could delineate the *foundations* of perceptual beliefs, by describing the perceptual facts that could not be doubted.

Subsequent philosophers, irritated by these visitors from the ontological slums, have spent much time and effort in disestablishing the argument from illusion. Perhaps the most important assault on it is that contained in the lectures of J.L. Austin, published posthumously as *Sense and Sensibilia*. ('Sensibilium' is another of Russell's names for the sense-datum; the title is a somewhat donnish joke.) Certainly, the ontological conclusion is unwarranted. It does not follow from the fact that the table *looks* elliptical, that something else really *is* elliptical. Austin was inclined to argue that the word 'looks' indicates a hesitant judgement about reality, rather than an indubitable judgement about the mental realm. I say that something looks elliptical only in those circumstances that lead me to doubt that anything really *is* elliptical. Moreover, even if we suppose that there is a mental entity – a 'sense-datum' or whatever – that is present whenever I see something, it would surely be wrong to describe it in *these* terms, as elliptical, brown and solid. No mental entity *can* be like that. The whole argument involves a duplicitous game with the distinction between appearance and reality. It is as though we were to argue thus:

I seem to see a table,
Therefore I see a seeming table;
The thing that I seem to see is elliptical;
Therefore, there is a 'seeming table' which is elliptical.

Since there is no place in the physical world for this seeming table (which would have to occupy the very same space as the real one), it must exist in mental space. It becomes the 'sense-datum', which actually *has* all those properties that the table merely *seems* to have.

You don't have to be a philosophical genius to recognise the fallacies in such an argument, and if that is all we can say to justify the introduction of sense-data, it is clear that they had better be sent back to the slums. However, the epistemological weight of the argument is unaffected by this attack on the ontological conclusion. It still seems as though we have a problem, concerning the grounds for our perceptual beliefs. Indeed, it is the old Cartesian problem, localised to sensory experience. If our perceptual beliefs depend upon the way the world seems in perception, and if the way the world seems is compatible with the falsehood of those beliefs, how do we know that the beliefs are true?

There is much more to be said about the argument from illusion. For example, we should not accept the facile way in which, as normally stated, the argument confounds illusions (like the bent stick) with delusions (like the hallucinations of a dipsomaniac). We are not 'taken in' by the bent stick or the elliptical table-top since we recognise that these are mere 'seemings'. Nor are we taken in by rainbows and mirages, which are *veridical* perceptions, revealed precisely to the person with normal faculties. Not to see a mirage where a mirage is to be seen is to suffer an illusion. Clearly, too many things are being run together, and it would be best to start again.

2. Naïve Realism and the Representational Theory

Still, some philosophers suggest, the argument drives a wedge between perceptual experience and the physical world. It shows us that we cannot think of perception merely as a 'direct relation' between the perceiver and the thing perceived – something interposes itself between them. You may not like the connotations of 'sense-datum' or 'mental image'; but you have to accept that we do not perceive physical objects 'directly' or 'immediately', but only *via* the perception of something else – the mental content that is before our consciousness in the moment of perceiving.

Thus arises the conflict between 'naïve realism', as it is called, and the 'representational theory'. This too is a minefield, and we must try to step quickly but carefully across it. Naïve realism owes its name to the sophisticated belief that we naïvely believe in it. That is to say, we naïvely assume (the philosophers tell us) that we are in direct contact with the physical world through our perceptions: the reality of the world is 'given' to us in the act of perceiving it. There is no intermediate mental process, no image, picture or whatever, from which the information about reality is recuperated. We just open our eyes, and see what is there.

The representational theory, by contrast, argues that we perceive objects through mental representations of them. These 'mental representations' may or may not correspond to the physical reality, whose nature must therefore be recuperated by carefully weeding out the illusions, delusions and ambiguities that blight our mental careers. Sometimes the theory takes a further step, arguing that we *directly* perceive the mental representations, while perceiving the physical world only *indirectly*.

As normally stated (i.e. by its opponents), the theory of naïve realism seems daft. For it implies that perception is like glue: it sticks you to the world, and the world to you, so the world becomes part of you. That way you don't *need* to perceive it, nor can you possibly be mistaken, any more than you can be mistaken about your present sensations. The only philosopher who has come near to this theory is Berkeley, who argued that we do not perceive 'material substance' but only ideas, and ideas, being mental entities, come before our minds indubitably and immediately. Any less direct relation, Berkeley thought, makes the object of perception imperceivable and therefore paradoxical. (It doesn't make sense, he sometimes argues, to say that we *perceive* material things; alternatively, it is a contradiction to say that we do, since this makes them both objects of awareness, and therefore indubitable, and hidden by a perceptual veil, and therefore unknowable.)

Obviously, naïve realism, so defined, is in danger of collapsing first into idealism, as in Berkeley, and secondly into solipsism, as we begin to realise that our world contains nothing knowable, apart from our own 'ideas'. But does the representational theory fare any better? Surely not. For it seems to say that we perceive physical objects only by perceiving something else, namely, the idea or image that represents them. But then, how do we perceive that idea or image? Surely we shall need another idea, which represents it to consciousness, if we are to *perceive* it? But now we are embarked on an infinite regress. Wait a minute, comes the reply; I didn't say that we perceive mental representations as we perceive physical objects. On the contrary, we perceive the representations *directly*, the objects only *indirectly*. But what does that mean? Presumably this: while I can make mistakes about the physical object, I cannot make mistakes about the representation, which is, for me, immediate, incorrigible, self-intimating – part of what is 'given' to consciousness. But in that case, why say that I *perceive* it at all? Perception is a way of *finding things out*; it implies a separation between the thing perceiving and the thing perceived, and with that separation comes the possibility of error. To deny the possibility of error is to deny the separation. The mental representation is not perceived at all; it is simply *part* of me. Put it another way: the mental representation *is* the perception. In which case the contrast between direct and indirect perception collapses. We *do* perceive physical objects, and perceive them directly. (Except when we don't! I mean, except when we catch sight of them in mirrors, or observe them through their effects, as we observe a deer in the undergrowth.) And we perceive physical objects by *having* representational experiences.

This looks like a rejection of naïve realism only if naïve realism is defined as the unbelievable thing which Berkeley defended. But it is certainly not a rejection of what we all 'naïvely' believe. On the contrary, it *is* what we naïvely believe. It is normal to suppose that we perceive physical objects by having mental states, and that these mental states give correct or incorrect information about the world. The representational theory looks like a defiance of common sense only when it proposes weird descriptions of the representational character of the mental state: describing it as a sense-datum, for example, and then supposing the sense-datum to be a separately existing entity which floats before the mind of the observer, elusively part of him, but elusively out of reach. But this 'reification' of the sense-datum commits exactly the fallacy that Berkeley commits, in arguing that we *perceive* ideas, and that ideas are the true and only objects of perception. So let's move on.

3. Mental Representation

What do philosophers mean, and what ought they to mean, in referring to 'mental representation'? The phrase is part of the jargon of the modern philosophy of mind, and also of the 'cognitive science' that goes hand in hand, or at any rate hand in glove, with it. The suggestion made by the defenders of sense-data is that perceptual states are relational: on the one hand an act of awareness; on the other hand a mental image which is the *object* of awareness. The representation occurs in the second component: the image shows physical reality, rather as a picture does, and that is how we represent the world.

A moment's reflection shows this to be hopeless. For how do *pictures* represent their subject-matter? Not by resembling it – for usually they don't, and in any case resemblance is a quite different relation from representation. (Every object resembles itself; and if *a* resembles *b*, *b* resembles *a*; neither of those things are true of representation.) To cut a long story short, pictures represent their subject-matter because that is how we understand them. And we understand them by seeing their subject-matter *in* them. In other words, we understand them by having a perceptual state in which their subject matter is represented. So we have to explain 'picturing' in *terms* of mental representation, and not vice versa.

Return for a moment to Frege and the theory of reference. We saw that the world contains items – sentences being prominent among them – which can be evaluated in terms of their truth and falsehood. This remarkable property, without which there could be no such thing as meaning, is also exhibited by mental states: for example, by thoughts and beliefs. They too can be true or false, and the relation of correspondence that makes them true is the very same relation that holds between a true sentence and the facts described by it. In short, representation is typified by reference, and its ruling principle is truth. To the extent that a mental state can be true or false, to that extent is it a 'mental representation'. In which case the argument from illusion does prove one thing, namely, that perceptual experiences are mental representations.

We have already come across mental representation under another name: intentionality. Whenever we can say that a mental state is 'of' or 'about' something, we can describe it as a representation, and vice versa. The representational theory becomes therefore something quite unsurprising, namely, the theory that perceptual states are intentional, like beliefs, emotions and desires. It would be remarkable if they were not. The error of the sense-datum theory is to describe the intentional object of a mental state as an existent (rather than an 'inexistent') entity.

But here begins a vast and unfinished inquiry. Is representation possible in purely physical objects? If so, what kind of objects, how organised, and how constructed? The possibility of a cognitive science is tied up with these questions, and that is one reason why the phrase 'mental representation' is so often on the lips of modern philosophers. Before saying more about this, however, we need to turn away from the epistemological question and address ourselves more directly to the analysis of perception itself.

4. The Analysis of Perception

Let us confine our discussion for the time being to visual perception. When I say that John sees Mary, I seem to imply not only that John exists, but that Mary exists as well. The word 'see', like the word 'know', is what Ryle called a 'success word'. It does not merely describe John's mental state, but vouches for it in some way: he really does *see* her: she really is *there*. This parallels the case of knowledge, where I do not merely attribute a belief to John in saying that he knows that *p*, but also affirm that belief as *true*. The parallel with knowledge is important, as we shall see.

In saying that John sees Mary, however, I am also attributing to him a mental state of some kind. Let us call this the 'visual experience'. It is clear that John could have that experience, even if Mary had not been present – so much the argument from illusion establishes. It is also clear that, even if she had not been present, John's experience is a 'representation' to which Mary might have corresponded. But here is a peculiar thing. Even when the representation is wholly misleading, we may say that John sees Mary. For instance, he reports a shape on the far horizon, like a piece of red cloth waving in the breeze. That description gives the 'intentional' object of his experience. In fact, however, the thing on the horizon is Mary, who is very far from being a piece of red cloth. Even if John believes he is seeing a piece of red cloth, he is still seeing *Mary*. To put it another way, we do not identify the object perceived in terms of the intentional object of the perception, but independently, in terms of the 'material object' in the world. (I use the scholastic term, 'material object'; but it should not be confused with Berkeley's 'material substance'.) What makes it true that John sees Mary is that it is indeed *Mary* who stands before him. And this is so, even when he thinks that it is not Mary at all, but a fiery angel come to carry him to judgement.

Philosophers have been impressed by this, because it seems to imply that our concept of perception is tied to the external, third-person viewpoint. It also suggests that the concept belongs to the explanation of behaviour, rather than the description of appearances. The important fact is not the *truth* of John's experience, but the fact that it derives from looking at Mary, and seeking information in the place where she stands. (Compare the parallel arguments about meaning in Chapter 19, section 6.)

Two concepts therefore seem particularly important in analysing perception: causation, and information. In a famous paper, Grice gives powerful arguments for saying that John sees Mary only if Mary is causally related to John's experience. Not any causal relation will do: she must be causally related in the right way. Grice asks us to consider examples, where John's visual experience remains exactly as it now is, but the causal relation with Mary is broken, even though she is standing there before him, in the direction in which he has turned his eyes. Suppose, for example, that there is a mirror at 45 degrees in front of Mary, reflecting a Mary-look-alike who stands away to one side. Surely John is not now seeing *Mary*, but at best the Mary-look-alike.

The conclusion Grice draws is that nothing about John's visual experience will entail that he is seeing Mary: only when connected to Mary by the right chain of causes does he see her. Equally interesting is the suggestion that the experience might be wholly deviant, and still be a 'sighting' of Mary. Provided the experience comes to John from Mary, via the medium of sight, it is an experience of seeing Mary. To put it another way, the concept of perception applies just so long as John is in the business of gathering visual information from the world. No matter that the information is wholly misleading, provided it comes to him from *outside*, and not from *within*. (If, standing before Mary with his eyes open, he has an hallucination of Mary standing in front of him, he is not *seeing* her.)

So what now of the visual experience? What sort of thing is it? Some philosophers – notably David Armstrong – have argued that 'experience', as normally understood, is not necessary to perception. The crucial factor is *belief*: to perceive is to acquire beliefs through the senses, by the causal process that I have just referred to. Others content themselves with a more general idea of 'information', and leave it to the philosophy of mind to specify the varieties of information states. Part of the motive is to remove an obstacle to the idea of artificial intelligence. If perception is a matter of acquiring information from the environment, then maybe this is something that a machine could do – a Turing machine, say, or some similar device programmed with the principles of Fregean semantics. If, however, we insist that perception requires visual experience, we may have qualms about supposing that a machine could manage it.

Here begins another unfinished debate, on the topic of 'sensation and perception'. Is there anything in perception that resembles those experiences which we call sensations – i.e. the mental states that we *feel* in our bodies? We don't speak of feeling our perceptions – certainly we don't feel our visual experiences in our eyes. We can have sensations in the eyes, but they are never visual experiences. In touching things, it is true, we receive sensations: but could we not have tactile perceptions without tactile sensations? Is it not sufficient to recognise that our movements are *obstructed* by something? And what about feeling something with the end of a stick, as when I poke about in a well to retrieve the ring of Melisande?

On the other hand, perception involves sensitive organs; perceptions can be or become painful, as with a bright glow or a loud noise; they can be more or less intense. Most important, they last for a specific time, and may disappear even though the information obtained from them remains. My visual experience of Mary vanishes as soon as I turn my eyes away from her; but the information that I obtained through looking at her remains with me. Is this not proof that the experience is something over and above the information?

Besides, some perceptual information must be defined in terms of the experience involved in gathering it: information about secondary qualities. The information that something is red is to be analysed *in terms of* the experience of normal observers. It could not exist in a world where there were no visual experiences. Since secondary qualities are precisely the means whereby our senses latch on to reality, it is hard to see how there could be perception in a being which lacked sensory experience.

On the other hand, visual experiences are like beliefs, and unlike sensations, in having intentionality. Maybe we should say that they are *sui generis*, lying between the sensory and the intellectual. This would explain why they minister to two separate interests: the cognitive and the aesthetic. Visual experience gives me information, and for that I value it. But it is also a joy in itself, and this joy is inseparable from its character as *experience*.

Here then is how a modern philosopher might wish to analyse perception. He will describe it not as a mental state, but as a process in which mental states are intricately linked to the environment, receiving information, and translating it into thought and action. The process has the following components:

(i) The object, fact or state of affairs perceived. (John sees Mary; he also sees *that* she is standing there.)

(ii) The sensory organs through which information is received.

(iii) The perceptual experience, which is *sui generis*, like sensation in some respects, like thought in others. It is a 'sensory experience', but with intentionality.

(iv) The information or belief acquired through (iii), which may outlive the experience.

(v) The causal relation that leads from (i) through (ii) and (iii) to (iv).

(vi) The consequences of the information gathered – its effect on the mind and actions of the perceiver.

That last condition is not one that I have mentioned. But clearly, having strung the components of the perceptual process on a single causal chain, it would be impossible to ignore the other end of it. Perception may be input: but every input has an output, which makes it the input that it is. To understand the concept of perception we should try to see its function in a total theory of the mind. This theory belongs to 'folk psychology': it is part of the elementary endeavour to explain and understand one another, which is the foundation of our human world. But maybe it could give way, in time, to a higher science of behaviour. In that case the connections between input and output will be re-described, in terms of a theory that may make no reference to the mental, no reference to intentionality, no reference to 'visual experience' or the secondary qualities that depend on it. Perhaps we shall then have achieved what the 'cognitive scientists' are aiming at. But will we have understood perception?

5. Phenomenalism

The last section was modern almost to the point of being modernist. This resolute third-personal view of the mind has, however, much to be said for it. For one thing, it awakens us to the extent to which our concept of the mind already involves a *theory* of human action. Hence our mental concepts often turn out, on inspection, to contain a causal component. Some argue that this is true of knowledge (see previous chapter); most recognise that it is true of perception; maybe of belief; maybe even of meaning itself. Perhaps the connections between language and the world that are spelled out by a theory of reference are really *causal* connections, so that a theory of meaning is an attempt to explain what people say as an *effect* of the things they refer to. (See again, Chapter 19, section 6.)

Before such reflections seized the philosophical imagination, however, philosophers remained fixated on the first-person view of perception, and the epistemological problem concerning the 'external' world. If all our knowledge of the empirical world comes to us through the senses, of *what* is it knowledge? Sure, we can make scientific theories. But of *what* are they theories, if not of the evidence upon which we base them? Whichever way you look at it, perceptual experience occupies the centre of the stage. If we are to have knowledge of the world, it must be a perceivable world. But what kind of a world is that? What must the world be like in order to be perceivable?

Perhaps the most radical answer to that question is the one adumbrated by Berkeley, and re-expressed by the logical positivists in the following words: physical objects are logical constructions out of perceptual experiences. (Actually, they said logical constructions out of sense-data; but we should ignore this discredited phrasing.) Such is the theory of 'phenomenalism', which holds that every statement about a physical object can be translated into an equivalent statement about perceptual experience. That is all we can mean, for of nothing else do we have knowledge, and nothing else can be cited when giving the evidence for a statement about the physical world.

For a long time it was supposed that phenomenalism is simply untenable. No translation of a 'physical object statement' can be produced, since it would have to be infinitely long, in order to capture all the sensory information contained in the original. Moreover, considerations such as the private language argument persuade us that we could never identify or describe our perceptual experience, without employing concepts of a physical world. All such theories as phenomenalism have been swept away by the triumph of the view from outside, which situates mind in the explanation of behaviour, rather than in the secret inner regions.

Maybe we should not rush to conclusions, however. Does it matter that the analysis of a sentence is infinitely long, if I can have rational grounds for believing that it will only refer to one kind of thing –

namely perceptual experience? I know that the expansion of π goes on for ever; but I also know *a priori* that nothing will occur in it that is not a numeral. Hence π is a number. Similarly, the phenomenalist will say, nothing occurs in the expansion of a physical object statement which is not a statement about actual and possible experience.

And does it matter that we need physical object concepts in order to identify and describe experience, if we can be confident that *that* is what we are describing? The use of physical-object concepts may still not commit us to the view that physical objects are something over and above the experiences that lead us to refer to them.

Grice even entertains the thought that we could combine phenomenalism with the causal theory of perception – so that physical objects become 'logical constructions' out of the very experiences that they cause. This vertigo-inducing idea prompts me to move to the next topic.

24

Imagination

Towards the end of the last chapter I gave a few reasons for thinking that there are perceptual experiences, involved in seeing, hearing, feeling, tasting and smelling. The word 'experience' was meant to imply that the states in question cannot be construed in purely cognitive terms – i.e. as thoughts, beliefs or information. There is an irreducibly *sensory* component to them: a sensed character, captured by the phrase 'what it's like'. There is no such thing as 'what it's like' to believe that chlorine is a compound; but there is such a thing as what it's like to smell chlorine.

At the same time, perceptual experiences are not merely sensations, like pains and tickles. For sensations do not seem to have intentionality: they are not 'directed outwards' as perceptions are. Wittgenstein asks us to imagine the following case: certain plants have areas which, when touched, cause a sharp pain. These 'pain patches' have no other recognisable feature that distinguishes them. In such a case, Wittgenstein says (*Philosophical Investigations*, Part I, section 312), 'we should speak of pain-patches on the leaf. . . just as at present we speak of red patches'. I doubt the force of Wittgenstein's '*just as*'. For we must distinguish sensations from perceptions, and sensations do not become perceptions just by being regularly connected to a certain cause. We should not invent a 'secondary quality' which we 'perceive in' the pain patches, through our sensation of pain. We should rather say that the patches cause us to feel pain – we don't yet know why. (There is another lesson to be learned from the example, too; namely, the seeming absence of tactile secondary qualities – touch seems to deliver a purely primary-quality account of the world. See Jonathan Bennett, in *Locke, Berkeley and Hume*.) To put it another way, sensations, like pain, do not 'represent' their causes, as being thus and so. There is no such thing as a true or false pain, as there is a true or false perception.

For a long while philosophers were muddled about the distinction between sensations and perceptions. Hume uses the term 'impression' to refer indifferently to both. And this proved to be the greatest weakness in his philosophy, and the final reason for rejecting his account of mental processes. Kant changed matters, by distinguishing what he called 'intuitions' from genuine experiences. An intuition, for Kant, is an unconceptualised mental input, like a sensation. An experience comes about through the 'synthesis' of intuition and concept – that is, it arises when I bring the raw sensory material under concepts, and so endow it with a 'representational' character. Thus when I look out of my window, I see a green field, bounded by a stone wall. The concepts: *green, field, stone* and *wall*, all enter into my experience. For that is *how* I see what I see.

Not everyone accepts Kant's account of experience; for Kant does not clearly distinguish between the organisation that is intrinsic to a perceptual experience (the phenomenon of the *Gestalt*), and the concepts which we apply in and through the experience. Some contemporary philosophers refer, in this context, to the 'non-conceptual content' of an experience, meaning the order which seems to reside in the experience itself and to 'direct it outwards', even though no specific concept inhabits the perception. Nevertheless, Kant's theory offers a plausible partial account of the distinction between sensation and perception. And it gives added force to the observations made in Chapter 16, concerning the nature of the 'human world'. The concepts that inhabit our perceptions are inevitably circumscribed by our sensory powers. I cannot look from my window and see the field as a spread of photo-synthesising plant cells, bounded by slabs of calcium carbonate. Such a description corresponds to no 'way of seeing'. It employs concepts that mark no distinction among appearances. My perceptions represent the world in ways that may defy the best scientific theories. But these perceptions also authenticate the concepts that are deployed in them, and establish the reality of our human world.

When I look from my window, I do not merely see the field as green: I believe it to *be* green. My seeing is believing. And this is true of all normal perceptual experience – hesitations apart, the concepts that are applied in perceptual experience are also held *true* of the world. Suppose, however, that I direct my gaze to the far corner of the field, where the long grass is teased by wind, and strange patterns are formed as the light plays against it. It has the look of a witches' Sabbath: I see the dancing forms of the witches, and the black dog, formed of shadow, who bays in their midst. I don't believe that any such event is happening, but that is how I see it, nevertheless. Here are concepts which are deployed in perception, but which are not being applied in any judgement (as Kant would express it). This is seeing *without* believing – the kind of seeing that distinguishes men from animals, and in which imagination is the operative factor.

1. Imagination and Mental Images

The term 'imagination' is used in a variety of ways, usually to denote a mental capacity. As a technicality of modern philosophy it has at least two senses:

(a) The capacity to experience 'mental images';

(b) The capacity to engage in creative thought.

The connection between these two senses is obscure, partly because each sense is obscure in itself, and dependent upon the theory with which it is associated. Mental images occur in thinking, in dreaming, in perceiving and in remembering. They also occur when we are imagining (in the second sense of the term). Because they occur in so many different contexts, it would be quite misleading to suppose that a theory of mental images is the same thing as a theory of the imagination in the second sense, or even a necessary *part* of such a theory. (Maybe even animals have mental images: dogs, at least, seem to dream. But it strains credibility to say that they have imagination in the second sense – where imagination is the ruling principle of story-telling, painting, and creative science.)

Nevertheless, let us try to make sense of the idea of a mental image. A mental image is like a thought in the following ways:

(i) It has intentionality: it is 'of' or 'about' something. Hence a creature's capacity for mental imagery depends strictly on its cognitive powers. If a creature cannot have thoughts about the past, then it cannot have 'memory images' either. (Ask yourself the question: What makes this image, that is now before my mind, into a *memory*? It could have been exactly the same in respect of the features represented in it, and not be a memory, but a prophecy or a fiction.)

(ii) Hence images can be true or false. A true image of your friend's face is one that shows him as he is; i.e. which corresponds to the reality.

(iii) An image may stand in intellectual relations to other mental states. For example, my image of Venice may contradict your thoughts about the town; it may be compatible with the portrait given by Thomas Mann; it may imply that 'La Serenissima' is an entirely inappropriate name; and so on.

However, a mental image is not merely a thought. I can think about Venice, even produce an accurate mental description, and yet fail to be visited by any Venetian imagery. I can remember a text without having an image of it on the page; I can think my way through a musical score, without 'hearing it in my head'; and so on. Images are like perceptions: they have a component that we are inclined to call 'sensory', and which relates them to perceptual experiences. For instance, they are like perceptual experiences in the following ways:

(i) Images can be precisely dated in time; they begin at a moment, last for a while, and then cease.

(ii) They may be more or less faint or intense (like sensations): this is not a matter of detail, but of *force*.

(iii) They can be fully described only in terms of a perception. My image of Venice is a *visual* image: it is a mental state that is like the experience I have when looking down the Grand Canal towards S. Maria della Salute. My image of Beethoven's Ninth is an auditory image: it is like the experience that I have when listening to Beethoven's Ninth. The 'like' here expresses an 'irreducible analogy' – i.e. there are no 'respects' in which the two experiences resemble each other; one is simply the imagined version of the other.

(iv) There is a first-person point of view on every image, and there is also a 'what it is like' that is encompassed by that point of view. It is doubtful that there is a 'what it is like' in the case of thought. (See last chapter.)

2. Creative Imagination

Mental images occur when we dream, when we remember, and also when we *imagine* things. Sometimes we describe a person as imagining what he thinks is there but isn't – though in this sense 'imagining' means something like 'suffering from an illusion', and to 'imagine' is merely to acquire false beliefs about the world. That is not what we normally mean by the word.

In the true sense of the word, imagination means creative thought. It is not a matter of illusion; the person with a strong imagination does not suffer more false *beliefs* than his less imaginative neighbour: rather he thinks more widely, more creatively, less literally. His thought roams among possibilities and he is more ready to 'suspend' both belief and disbelief. (Coleridge wrote of the 'willing suspension of disbelief' that is involved in understanding drama.) Imaginative thoughts in this sense are not illusions about the real world, but undeceived depictions of a world that is not only unreal but also known to be so. (To be *taken in* by this world – for example, by the world of a play – is to exhibit a deficiency of imagination rather than a superabundance of it.)

Imagery may have a part to play in creative imagination, although it is neither a necessary nor a sufficient ingredient. When I imagine a dialogue between Socrates and Xanthippe, I may also envisage it, in the sense of imagining what it would be like to *see* and *hear* the encounter between them. In such a case, my imaginative thoughts are partly embodied in images. (Imagery is, indeed, an essential ingredient in 'imagining what it's like'.)

Such images differ from dream images and perceptual images in that they lie within the province of the will. It makes no sense to command someone to dream something, or to have a certain visual experience. But we can certainly command him to imagine something, and he may 'summon' or 'construct' the image without further ado, and using no method other than the direct application of his will. It is one of Wittgenstein's more interesting observations, in the *Remarks on the Philosophy of Psychology*, that mental states can be fruitfully distinguished according to whether they are subject to the will; the resulting distinction cuts across the traditional divisions between the sensory and the intellectual, between the animal and the rational, between the affective and the cognitive, and even between the 'passive' and the 'active' (as these were described, for example, by Spinoza). There are perceptions which are subject to the will (seeing the duck in the duck-rabbit figure), and also cognitive states (supposing, hypothesising); but wherever belief or sensation are involved, the will as it were withdraws. I can command you to suppose that the moon is made of rock, rather than cheese, but not to believe it; I can command you to injure your finger, though not have a pain in it; and so on.

One reason for thinking that memory and creative imagination are closely related is that both involve imagery, and in both cases the imaging process remains at least partly within the domain of the will. When I 'summon up remembrance of things past' I am *doing* something which I might have refrained from doing. I deliberately call to mind the appearance and character of past events and objects, so as to undergo, in some faint and helpless version, the experiences which were once imprinted on my senses. There is an art in this, which is not unlike the art employed in fiction; and while not everyone is able to achieve what Proust achieved, in reworking the past as though it were entirely the *product* of creative imagination, there is no doubt that 'powers of recall' and 'powers of creation' have, in this area, much in common, and speak to a single emotional need.

The voluntary nature of imaginative acts gives a clue to creative imagination. For, whether or not it involves imagery, imagination always involves the summoning or creating of mental contexts which are *not otherwise given* (as they are given, for example, in perception and the judgements that spring from it). When I stand before a horse, it involves no act of creative imagination to entertain the image of a horse – for this image is implanted by my experience, and is *no doing of mine*. Likewise, when I listen to an account of some battle, or read about it in the newspaper, my thoughts are not my own doing, and play no creative role in the unfolding of the story. In general, things perceived and things believed, in the normal course of our cognitive activity, are imprinted upon us willy-nilly, and are independent of our creative powers.

When, however, I summon the image of a horse in the absence of a horse, or invent the description of a battle which I have heard about from no other source, my image and my thought go *beyond* what is given to me, and lie within the province of my will. I can adjust both image and story, and may cancel them at will. Such inventive acts are paradigm cases of imagination. And, in so far as they involve thoughts, these thoughts are not beliefs about the actual world, but suppositions about an imaginary world.

How should we understand such thought processes? It is useful to return to Frege's theory of assertion. In the inference from *p* and *p* implies *q* to *q*, the proposition *p* occurs unasserted in the second premise, regardless of whether it is asserted in the first. Yet *p* is the *same* proposition in both premises: otherwise the inference would not be valid. It follows, Frege argued, that assertion is no part of the meaning of a sentence: a proposition does not change merely because it is or is not affirmed as true. This elementary result enables us to draw an important conclusion, namely that the content of a belief may be exactly reproduced in a thought that is *not* a belief, in which the content is merely 'entertained'. This happens all the time in inference; it is also what primarily happens in imagination.

We may therefore venture an account of at least one central

component of creative imagination: the capacity to 'imagine that *p*'. In imagining that *p* a person entertains the thought that *p*, without affirming it as true; the thought that *p* goes beyond what is given to him by his ordinary cognitive and perceptual powers; and his summoning of *p* is either an act of will, or within the province of his will (so that he could, for example, try at any moment to cancel or amend his thought). When, as may happen, the thought that *p* contains a perceptual component (as when I think how someone looks), it may be embodied in, or absorbed into, an image; and this image too is an exercise of creative imagination.

Not all imagination fits easily into this model, since not all imagination is an imagining *that* . . . Some works of imagination are pure images, without subject-matter other than the sensory forms themselves. For example, composing a melody is a work of creation: it involves putting sounds together to form a novel and interesting totality. This is a voluntary act, which goes beyond what is given in perception; but it is not an expression of a thought in Frege's sense. A melody is not a proposition; nevertheless, it is *like* a proposition, in having an intrinsic order, sense and communicative power. Such processes, which are like thoughts, but which do not involve the creation of imaginary worlds, lie in the same domain as 'imagining that', for they involve the creative transformation of experience, and therefore of the human world. This is what we instinctively feel to be true of music, abstract painting and architecture. The human world was irrevocably changed by Borromini, Bach and Braque, even if many people are unable to notice the fact. Hence we freely use the word 'imagination' of all the creative arts. Nevertheless, it is a work of theory to show that we are entitled to suppose that these various exercises of imagination involve *one* mental capacity, rather than several.

There is also what Peirce (*Collected Papers*, vol. 1, pp. 20–21) called the 'scientific imagination': imagination controlled not by the requirements of fiction, but by the goal of scientific theory. The imaginary worlds of the scientist are also physically possible, and it is no small feat to emancipate your thinking from the known reality, while keeping hold of the reins of scientific truth.

3. Imaginary Worlds

Fiction, however – whether in drama, poetry or prose, whether in figurative painting or mime – remains the prime instance of creative imagination, and one which also shows the importance of imagery in the full elaboration and understanding of imaginative thoughts. It is tempting to argue that a fiction is something like a possible world: or at least a glimpse into such a world. The work of imagination involves constructing (or, for a realist, discovering) possibilities – the purpose being, perhaps, to set the actual world in the context of its possible variants. Since our everyday thought automatically involves us in assessing possibilities and probabilities (see Chapter 15), the capacity to envisage 'possible worlds' is already implied in our day-to-day psychology. For this reason we may wish to endorse the old theory (espoused for a variety of mostly bad reasons by Hume, Kant and Hegel) that imagination is a part of ordinary thought and perception.

The suggestion is misleading in various ways. First, although we must invoke possible worlds in order to account for the meaning of modal sentences, and although modal thoughts (about possibilities, necessities and probabilities) are involved in scientific thinking, we do not have to *envisage* these possibilities, or to spell them out in narrative terms, in order to make the everyday judgements that depend on them. Conversely, when we *do* spell out the narrative of an imaginary world, we are not bound by the constraints of possibility, whether physical, metaphysical or logical. In a tragedy, Aristotle remarked, impossibilities may be countenanced, provided that they are, in the context, probable. An improbability, by contrast, however possible it might be, involves a failure of imaginative grasp. (*The Poetics.*) What is meant by 'probability' is 'truth to character'. When Fafner the giant turns into Fafner the dragon, a profound spiritual and moral truth is enacted before our eyes, even though this transformation is metaphysically impossible. (Cf. also the stories in Ovid's *Metamorphoses.*)

The creation of an imaginary world is a distinct enterprise, with a purpose all of its own. Understanding fictions involves recognising the 'fictional context' in which events, persons and objects occur bracketed not only from the realm of actuality, but also at times from the realm of possibility too. And yet, in the successful fiction, everything proceeds with its own kind of necessity: notwithstanding its deliberate unreality, it aims to be alive. Impossible worlds may also be consoling in themselves: we inhabit them with a kind of child-like freedom, and seem to breathe a clearer and purer air, as in *Through the Looking Glass* and the engravings of Escher.

The emotional response to imaginary worlds is one of the most interesting of all mental phenomena. For it seems that we can feel towards these fictitious scenes a version of the emotions that animate our real existence. We feel sympathy for the tragic character, and Aristotle assimilated this sympathy to pity and fear. Yet – because the objects of these emotions are not only unreal but known to be so – we are not motivated to act as we normally should act. We do not rush on to the stage to make common cause with the beleaguered hero. On the contrary, we relax into our emotions, and live for a while on a plane of untroubled sympathy, laughing and crying without the slightest moral or physical cost. This mental exercise is a strange one – for in what sense are we really moved by that which has, for us, no reality? And why should it be so precious to us, to exercise our sympathies in this seemingly futile way? These are among the most important questions of aesthetics.

Kendall Walton has developed, in this context, a theory of 'make-believe', arguing that our aesthetic emotions should be compared to the emotions of children at play. The drama is like a prop in a game of make-believe; and the emotions that we feel are part of the game. There is something intriguing in this idea: but it seems more like a description of the problem than an answer to it. It certainly does not settle the question of the relation between real sympathies, and the sympathies that we feel for fictions. Perhaps there is nothing further to be said, other than that the one is real, the other imaginary. Like the tears shed by Alice, the tears in the theatre are not *real* tears. Yet they are every bit as wet.

4. Fantasy and Imagination

An imaginary world is, *ex hypothesi*, not a real world. Imagination does not aim at truth, as belief aims at truth. On the contrary, it aims, in a sense, to avoid truth. And yet it is governed by the attempt to *understand its own creations*, and to bring them into fruitful relation with the world that is. We expect the work of imagination to *cast light* on its subject-matter, and on the real originals from which the subject-matter is ultimately drawn. In short, imaginative thoughts aim to be *appropriate* to reality, in the way that a tragedy is appropriate to the human condition, even in the act of idealising and dramatising its essential pathos. And though appropriateness is more nearly a moral than a semantic ideal, 'truth to life' is surely a normal part of it.

Coleridge's distinction between fancy and imagination may therefore still have a lively attraction. For we should distinguish between the kind of disciplined story-telling that illuminates reality and enables us in a novel way to come to terms with it, from the undisciplined flight into realms of fantasy. Fantasy may seem to be a step further along the path taken by imagination. In fact it is a distinct exercise of the mind, involving the creation of substitute objects for old emotions, rather than new emotions towards the familiar human world. The nature of the fantasy object is *dictated* by the passion that seeks it: as in pornography, nothing is examined, questioned or put to the test. Instead, the actual world is deleted, and replaced by another world compliant to our emotional demands. By contrast, the truly imaginative object produces and controls our response to it, and may therefore educate our passions, so as to send them back into the encounter with reality purified of their vain excesses. (Aristotle wrote in this regard of a purging or *katharsis*.)

Criticism of fiction involves an attempt to identify the kind of emotional response that is invited by it. In particular, it is of considerable importance to ask whether a work of art is really exploring the world that it creates, working through the consequences of acts, events and characters, in the light of what we know of human nature. Often, when a fiction prompts us to feel some emotion towards its characters, the situation does not really justify the feeling. We are being invited to indulge our emotions, without a sufficient cause. In such a case the danger is that we enjoy our tragical posture, without caring about its object. We learn to direct our feelings inwards, to take pleasure in ourselves as heroes of sympathy, and to ignore the real complexities of the human world. This is the vice of sentimentality; and it is the major fault of modern art.

5. Imaginative Perception

There is a particular exercise of imagination that is of vital concern to the philosopher: the kind involved not in creating an imaginary object, but in perceiving it. My image of the horse that stands before me is a straightforward perception: the horse is 'given' by the experience that I cannot help but have. My image of the horse presented in a picture is not like that at all. First, I neither believe, nor am tempted to believe, that the horse is really there. Secondly, I perceive the horse only to the extent that I am prepared to 'go along with' the lines and textures of the painting: I re-create in imagination a living creature, out of what is at best a two-dimensional outline. What I see goes beyond what is given, in just the way that a fictional thought outstrips reality. Indeed, it is the perceptual equivalent of a fictional thought. Thirdly, my experience lies within the domain of the will – a fact that is conclusively proved by such ambiguous pictures as the duck-rabbit, which I can decide at will to see, now as a duck, now as a rabbit. (It will be said that this is a special case: on the contrary, it is merely an emphatic version of the normal case. Even in the most realistic and seemingly unambiguous of Stubbs's horses I may choose to see the creature now as an 18-hand giant, now as a 15-hand ladies' horse; now as resting, now as poised for movement; and so on. It lies in the logic of the case, that what I see is only partly determined by the physical picture in which I see it. The image needs to be completed by an act of attention, and how I complete it depends on me.)

This phenomenon is sometimes described as 'seeing as' – though Richard Wollheim (*Painting as an Art*) makes a useful distinction between 'seeing as' and 'seeing in', along such lines as these: When I see *x* as *y*, *x* and *y* must belong to the same ontological category: if *x* is a particular, then so is *y*; if *x* a type then so is *y*; if *x* is a universal then so is *y*; and so on. (Consider, for example, the case in which I see a Maltese cross, now as a black cross on a white ground, now as a white cross on a black ground.) When I see *y in x*, however, no such restriction applies. I may see a face in a picture: but I may also see melancholy in the picture, or in the face; I may see pride in a monumental building, or Henry's personality in his dog or his room. The question how we should analyse this 'seeing in' is highly controversial; but there is no doubt that the phenomenon provides a paradigm for many kinds of aesthetic experience. When I hear movement in music, respond to the tone of voice in poetry, see the dignified posture in a building, something is going on which is importantly like what goes on when I see a face in a picture. We are dealing with a general phenomenon that I call 'double intentionality' (*Art and Imagination*): an experience that is directed simultaneously towards two distinct objects – the work of art (as it might be) and what is seen, heard or read in it.

There is also an important contrast between seeing *x* in *y*, and noticing a resemblance or analogy between *x* and *y*. Clearly, I can notice the resemblance between the duck-rabbit and a rabbit even when seeing it as a duck, an experience that forbids me from seeing it as a rabbit. The contrast here parallels that between metaphor and simile. In a successful metaphor one thing is embodied or incarnated in another, rather than merely likened to it. When Rilke writes: *so reiszt die Spur der Fledermaus durchs Porzellan des Abends* ('thus the track of the bat cracks the porcelain of the evening'), he does not *compare* two things, but, so to speak, fuses them together, by using the properties of the one (a crack spreading through porcelain) to describe the other (the flight of a bat through the evening sky). In metaphor a predicate is transferred from its literal use to a new situation to which it does not apply. Literally speaking, metaphors are false. This has led some philosophers to suggest that metaphors involve a new meaning of the terms used to express them. On the contrary, however, it is the *old* meaning that is deployed in the new context: that is the whole point of the transfer. (This fascinating topic occupies a growing branch of the literature: see Wittgenstein, *Philosophical Investigations*, Part II, section xi, and Donald Davidson, 'What metaphors mean', in *Inquiries into Truth and Interpretation*.) Since understanding metaphor is an integral part of all the higher forms of literary experience, it is clear that we have a clue, here, to the work of the imagination in aesthetic understanding.

6. Representation

Fictions involve the representation of imaginary worlds. We are back again with that troublesome concept encountered in the last chapter. What do we mean by 'representation'? I have already suggested that we have a paradigm case in language: here representation simply means reference, and there is no problem in understanding how works of literature can be representational – namely, by being 'about' their subject-matter in just the way that a sentence is 'about' the objects and concepts to which it refers. If there is a problem, it is that of making sense of fictions. Frege believed that all 'empty names' lack reference – or, if you prefer, they all refer to the null class, the class with no members. In which case, all fictions are about the same thing. Clearly, we do not want to say that. *King Lear* is about King Lear, and *The Castle* about the land surveyor, K. We seem to be dealing with a concealed modality. Sentences in *The Castle* are as though preceded by the phrase 'It is fictional that', analogous to 'It is possible that'. Complicated from the point of view of semantics – especially when we remember that fictional worlds may not be possible worlds (as indeed in this case); but perhaps not so difficult to grasp.

The real problem comes with the visual arts. Pictures are also representations, though of a quite special kind. They do not describe their subject matter, but depict it. Is this another relation? Or another species of the same relation? Or what? The temptation is to argue that pictures too *refer*; that they are representational just as sentences are, by virtue of their semantic structure, and the conventions which relate them to objects in the world. This is the position advocated by Goodman in *Languages of Art*, and it has much to recommend it – especially to someone like Goodman himself, whose nominalist theory of language allows him far more latitude than most philosophers are prepared to take.

The problems for the semantic theory are, however, insuperable. If pictures worked as language works, then they ought to refer by *convention*. Words are arbitrary, in that any word can be coupled with any object, provided only that a rule exists to make the connection. It is arbitrary that the word 'man' means man: 'homme', 'mensch', 'chelovek' are just as good, and none has special authority. But a picture is related to its subject matter by a natural relation. Although conventions affect our interpretation of Tintoretto's *Crucifixion*, there is an irreducible core of plain perception, available to everyone, regardless of their knowledge of the Venetian Renaissance. No convention could make this into a depiction of a rock concert or a village fête.

Secondly, sentences refer by virtue of their semantic structure. Their truth-conditions are derived from the sense of their parts, just as their truth-value is determined by the reference of their parts. That is how we understand them. Fundamental to this process is the principle of finitude: a sentence has finitely many meaningful parts, each of which has a determinate sense and reference. The same is not true of a painting. Between any two areas of a painting there is a third, which also represents something – namely the part of the subject-matter that lies between the parts depicted by the areas to either side. Paintings are not composed syntactically, and have the integrity of images, being infinitely divisible and representing infinitely many parts of the subjects which they depict. (Goodman, describing these two features as syntactic and semantic *density*, does not succeed, thereby, in giving them either a syntactic or a semantic character.)

It is therefore better to look to imaginative perception for a theory of depiction. A painting represents what you can see in it – or at least, what the painter intended you to see in it. Representation is still the same property as that exhibited by literature. For you understand it in the same way – namely, by recuperating thoughts about a fictional world. But the process of representation is fundamentally different. Indeed, paintings provide a useful case of meaning in the absence of semantic structure. Paintings *mean* by mobilising what Hume called the mind's disposition to 'spread itself upon objects'. As we shall see in Chapter 29, this is one of the most important forms of meaning, and the one whose absence is the most frequent cause for lament.

7. Imagination and Normativity

Images and metaphors may be more or less successful; stories more or less true to life; paintings more less insightful; music more or less sincere. All the works of the imagination seem to invite our criticism; for imagination is involved too in understanding them, and once our thought has been released into imaginary worlds it is bound by the laws of this new-found freedom. Imagination is a rational capacity, one which not only is peculiar to rational beings, but which also compels them to exercise their reason, to ask 'why?' of every phrase, word, and line, and to judge their appropriateness to the familiar world of reality. In the works of imagination, therefore, a peculiar form of judgement arises: we sense that, however freely imagination may roam, there is a right and a wrong way to go. In making this judgement we try to bring the imagination back to earth, to use it as an instrument of knowledge and understanding, rather than an instrument of flight. This is perhaps what Freud meant, when he described art as a passage from fantasy back to reality. It is perhaps too why Kant discerned an act of universalisable judgement – an incipient legislation – behind every aesthetic experience. At any rate, it is the origin of criticism, and the foundation for our belief that imagination is not merely a fact, but also a value.

25

Space and Time

As we have seen, we often imagine things that are impossible; indeed, if our imagination were not capable of transcending the bounds of possiblity art, religion and the self-understanding that come from them would be of little value to us. Things which are impossible but conceivable play an enormous role in our thought and feeling, and the knowledge that they do not exist has never deterred people from spending their lives in creating, debating and commiserating with them. (Here is one motive for the view that 'existence is a predicate' – see Chapter 30.)

But what about the converse: things which are possible but inconceivable? If these too must be countenanced, then we shall have to abandon the view, which has had a considerable following since Descartes, that the conceivable is a test of the possible. Just as art persuades us to accept the conceivable but impossible, science offers the possible but inconceivable: indeed, the *actual* but inconceivable! The basic items of quantum physics, which behave now like waves and now like particles, and concerning which there are only statistical truths, defy the imagination. If we say that they are nevertheless conceivable, it is because we are using possibility as a test of conceivability, rather than vice versa.

In one area, however, philosophers have been reluctant to sacrifice conceivability as a test which reality must pass. To Descartes and Spinoza, it seemed as though the validity of Euclid's geometry derived entirely from the fact that its axioms express our own 'clear and distinct' (Descartes) or 'adequate' (Spinoza) ideas. Euclid was supposed to have isolated self-evident propositions about space, whose falsehood is inconceivable; his theorems deliver the totality of *a priori* truths about geometry.

1. Euclid and Visual Geometry

The case for the *a priori* nature of Euclidean geometry was made most forcefully by Kant. His desire was first to prove that geometry delivers synthetic *a priori* knowledge, and secondly to offer a theory as to how this could be so. The idea is this: Euclid's axioms are not analytically true. There is no way of deriving their truth from the meaning of the terms used to express them. For their truth is grasped in an intuition, and this requires the *representation* of space, as in a diagram. Any intuition of space involves the endorsement of Euclidean principles: nothing that violates those principles could be envisaged by us as a *space*, in which objects can be situated, and through which they can move. Euclidean geometry is therefore *a priori* and synthetic. How is this possible? Kant's answer is that Euclidean geometry reflects the requirements of our own mental powers. If we are to represent things as *objective* in relation to us – that is, as existing and surviving independently of our experience – we must situate them in space. Space enables us, so to speak, to 'turn our back' on them, to situate them in a place where we might have been and which we might have observed, but which we do not observe. In experiencing something as objective, therefore, I represent it as spatial. In Kant's words: 'space is the form of outer sense' – outer sense being the capacity to perceive what is 'outside' me, what is objective in relation to me, what is not part of my inner realm. Space is by nature three dimensional, unitary and infinite. Those are precisely the features that are required, if space is to be the frame of reference within which we locate the objects of our perception. The spatial character of the world is imposed by our own cognitive capacities; hence we can know it *a priori*. It is only because our experience is 'organised spatially' that we have the conception of a world at all. The principles of this organisation belong to our mental equipment, and can therefore be known and studied without consulting experience – even though it is our capacity for experience that gives sense to them.

This subtle and adventurous view so impressed Fichte, Hegel and Schopenhauer, that they built it into their systems (though with the small adjustments required by self-esteem). However, nobody today is satisfied by it. The best Kant can prove by this method is that 'phenomenal space' – the space of the visual field – is Euclidean. But even this is doubtful. (Is the visual field really Euclidean? After all, parallel lines meet on the visual horizon – as they do in perspective drawings.) Even if *we* don't know what it would be like to envisage a non-Euclidean space, maybe other creatures could. It just so happens that the bits of the world that we encounter can be so quickly and usefully ordered in Euclidean terms, that we have acquired the habit of so perceiving them. Maybe evolution endowed us with a Euclidean phenomenology, our biological competitors having found their truer but more cumbersome geometry useless for those high-speed calculations required in the moment of battle. (Think of stabbing someone in Minkowski space; you're sure to miss.)

Nevertheless, Kant has a point: even if the true geometry is not Euclid's, we can know *a priori* that *objects* (i.e. entities whose existence is not dependent upon our perceiving them) are *in space*. Only if they are situated in space can we identify them through change, collect information about them when they are not observed, and give to them the full causal reality that is required by science. That is an interesting conclusion, and one defended on different grounds by Strawson in *Individuals*. But of course, it does not tell us what *kind* of space we inhabit.

Euclid's geometry is an astonishing intellectual achievement, the summary of a thousand years of patient intellectual labour, which enjoyed uninterrupted sovereignty from antiquity until Kant. But as Euclid was aware, there is one axiom of his system which is independent of the others, and which could therefore be replaced without contradicting the remainder. This, the axiom of parallels, says that parallel lines never meet (alternatively, that straight lines in a plane meet only in one point). This is certainly not made true by the definition of 'parallel', or 'straight line': it seems to be dependent on a Kantian 'intuition'. We just could not envisage things otherwise. But could they *be* otherwise?

The answer is yes. There is nothing inherently absurd in a geometry without the axiom of parallels. But why call it a geometry? For two reasons: first because it is a variation on Euclid; secondly, because it can be used to describe the very properties of physical objects that Euclid was trying to describe. Hence, a non-Euclidean geometry may be a genuine alternative to Euclid: a genuine candidate for the truth about the physical world.

2. Hilbert and Axiomatic Systems

The revolution in mathematics and logic which took place in the last century radically changed people's conception of geometry. Instead of the Kantian picture of a synthetic *a priori* science of the empirical world, geometry was conceived as an axiomatic system, the conclusions of which follow from the premises by purely logical moves, and the premises of which (the axioms) are simply uninterpreted formulae, which gain meaning only through their application to reality. An axiomatic system becomes geometry when its axioms are given a geometrical meaning – i.e. when they are used to describe spatial properties of the physical world. The many geometries – Euclidean and non-Euclidean – may be compared as uninterpreted systems. For example, we may ask whether they are internally consistent, whether the axioms are independent (i.e. not provable from each other), and so on. But all these are merely logical questions; they say nothing about which system is *true*. *That* question is empirical. It is the question whether the axioms of the theory, when interpreted in terms of physical space, actually deliver the truth. Euclidean geometry is not synthetic *a priori*: its proofs are *valid a priori*, because they are a matter of logic; but its premises are true, if they *are* true, *a posteriori*, because the world happens to conform to them. In fact they are false (though very good approximations in the small scale).

But what are the 'spatial' features of reality? What makes an axiomatic theory into a possible geometry? One of the founders of mathematical logic, D. Hilbert, who is primarily responsible for the view of axiomatic systems that I have just adumbrated, gave an answer. He showed how the fundamental notions of geometry could be defined in algebraic terms and the crucial relation of 'between' captured in three axioms. (Here, for interest, is Axiom I: 'If *a*, *b*, and *c* are points on a straight line, and *b* lies between *a* and *c*, then *b* also lies between *c* and *a*.') By such means he derived the whole of Euclidean geometry without recourse to diagrams or visual representations at all. He also showed that, while diagrams provide an interpretation of the axioms, they do not exhaust their visual possibilities. The so-called 'family of spheres' also satisfies the Euclidean axioms. In short, a sytem is a geometry provided it can be used to organise points in terms of 'betweenness' along one or more dimensions. The true geometry is the one which, as a matter of fact, describes the spatial organisation of our world.

3. Non-Euclidean Space

But how do we discover the truth about physical space? Imagine two-dimensional creatures living on a plane surface with a hump in the middle. They could discover the existence of this hump, purely by measurement of the surface. Their measuring rods have to be laid down so many more times when travelling over the hump than when travelling to either side of it. But suppose that, in the region of the hump, there is also a prevailing force, which distorts all measuring rods, making them just so long that, laid end to end, there are precisely as many rods required to surmount the hump as are required to traverse the plane to either side. The presence of the hump would no longer be detectable. In general, since measurement is a physical process, no empirical investigation of physical geometry can avoid making physical assumptions. We cannot set out to describe the geometry of our world, without assuming something about its physics: for instance, that a rigid measuring rod retains its length wherever it is placed, and at whatever speed it is travelling. Conversely no physics of the world can be developed without making assumptions about the world's geometry. Only someone who knew that there is a hump in this plane-like surface would be in a position to know that a force-field is distorting his measuring rods. Euclidean geometry was for so long accepted because it had been covertly *built into* Newtonian physics, which was in turn *built into* Euclidean geometry.

This helps us to make sense of what has happened in modern physics. First came the discovery of non-Euclidean geometries, by Riemann, Minkowski, Lobachewski and others: geometries that replace or dispense with the Euclidean axiom of parallels. Secondly came the development of n-dimensional geometries. The mathematical possibility of this is easy to grasp: a three-dimensional geometry is, in Hilbertian terms, a system of triples of numbers, x,y,z, which identify points, ordered according to the 'betweenness' relation in three dimensions. (These are the 'co-ordinates' used to define positions along three axes.) We can generalise the idea, to systems in which points are identified by n numbers, and 'between' operates over n dimensions, for any number n.

Thirdly came the recognition that you cannot describe the geometry of the world, without describing the forces that are at work in it. A physical geometry therefore involves a theory of physical objects. In our hump example, we could have spoken of a curved space, or equally of a prevailing force-field. Which we choose will depend upon the simplicity and predictive power of the theory that results.

Finally, there came the recognition that you cannot describe physical space while leaving time out of the picture. I can measure the length of something only if I know that it has not moved between the moment when I locate one end of it and the moment when I locate the other. I can also measure the length of something if I can make simultaneous observations of its extremities. I shall therefore need a criterion of simultaneity: and this criterion will be built into my geometry. Measurement requires a criterion for the simultaneity of spatially separated events. In the special theory of relativity, simultaneity is defined on the assumption that the speed of light is constant – i.e. that light signals sent across space are sent at a uniform velocity.

Now it is clear that the 'betweenness' relation orders points in time, as well as points in space. Time can be treated as a 'fourth dimension' under the axioms of a four-dimensional geometry. Indeed, this is the natural step to take, now that we have accepted that both spatial and temporal properties must be defined through the total theory of the physical world, and cannot be separately characterised. Of course, we do not, by this method, show that time is simply another dimension of *space*. As we shall see, there are features of time that make it impossible to describe it in such a way. Rather, there is a four-dimensional 'geometry', in Hilbert's sense, which delivers the truth about the physical world; three of the dimensions deliver spatial 'betweenness' and one of them delivers temporal 'betweenness'. The hypothesis may be false: perhaps we need, as some physicists argue, more than four dimensions (all kinds of numbers have been proposed, from five to seventeen and more). But which hypothesis is chosen is an empirical matter, to be tested by experiment, in the context of a total theory of matter.

When physicists speak of the 'curvature' of space, they mean roughly this. We can say that, at a certain point in space, a body will deviate from a straight line, because a force is acting on it. Or we can say that the body continues in a straight line, but that space itself is curved: i.e. the body follows a straight line which defies the axiom of parallels. We choose the latter description because it vastly simplifies the resulting physics, and delivers a theory according to which the universe is governed by a single and uniform system of laws.

Can we *envisage* any of this? Well, in a sense we can. Consider a sphere, the surface of which may be treated as a two-dimensional plane. (You can imagine being a two-dimensional creature, crawling around that surface, and trying to describe its geometry.) The two-dimensional geometry of this surface is non-Euclidean. Triangles have angles that add up to more than 180 degrees. There is more than one maximally short line between any two points. Straight lines meet more than once. And so on. And another way of putting this – easily grasped by three-dimensional creatures like us, who see the sphere from outside – is that this two-dimensional space is *curved*. Now envisage the same thing, only in three dimensions. (Well, of course, you can't. But that is precisely why geometry has to be emancipated from vision.)

4. Relative and Absolute Space

The theory of relativity is the work of Einstein, who in turn built upon the geometry of Minkowski. The idea that space is relative and not absolute is far older, and played a large part in the development of Leibniz's philosophy. The picture given by Newton was of space as an infinite container, in which objects could be situated at any point, but which had no boundaries, however far you were to travel in any direction. As Newton expressed it: 'absolute space, in its own nature, without relation to anything external, remains always similar and immovable'. (*Principia*.) Indeed, it seems absurd to suggest that space has boundaries – for what happens when you reach them? And what lies beyond? (That this is the wrong way of putting the question is apparent from the example of the sphere, whose surface is a finite space without boundaries. But it is precisely the difficulty of *envisaging* such a space in three dimensions that hampered scientific progress.) For Newton, therefore, the absolute character of space was bound up with its infinity – and Kant agreed. Space just exists, everywhere and for ever. And that is a fact, independent of any other fact – independent, for example, of the fact that there is something *occupying* space at any given point in it.

Such a view is bound to create theological worries. For it seems to imply either that space preceded creation, in which case there is something that God did not create; or else that God created not only the things *in* space, but also space itself, in which case it is just as possible that space *might not have been*. Why not therefore suppose that space has limits, beyond which there is, literally, *nothing*?

Leibniz went further. There is, he suggested, something deeply incoherent in the whole idea of absolute space. For it implies that the universe could be, as it were, moved sideways. The result would be a world different from the present one, but absolutely indistinguishable. Indeed, there would be, on this view, infinitely many indistinguishable worlds. In which case, there would be literally *nothing to choose* between them. God's choice to create *this* world, rather than any other among its indistinguishable rivals, would therefore be entirely arbitrary. Not only does this do scant justice to God's rationality and goodness; it also violates a principle which, for Leibniz, is the corner-stone of science, the Principle of Sufficient Reason. This holds that there can be no true proposition without a sufficient reason for its truth. The assumption of absolute space deprives us of a sufficient reason for the existence of anything in it. (The argument is brilliantly spelled out, and brilliantly contested, in the correspondence between Leibniz and Newton's disciple, Samuel Clarke.)

But what precisely is the alternative? Leibniz entertained the view that spatial properties are relational. The position of any object is to be given in terms of its relations to other objects. Once we have determined the spatial relations among the objects in our universe, we have fixed the position of everything within it: there are no further facts about *where* anything is. Any universe which reproduces precisely these spatial relations is the *same* universe. (This is a very modern thought, and is picked up by Hilbert in his axiomatisation of 'between'.) This solution is complicated by two further arguments. First, the notion of an 'object in space' is still obscure. If an object *takes up* space, then it must have spatial parts; and where are *they*? Is their position also to be specified in relational terms? Does not the process go on *ad infinitum*? Leibniz would say yes, emphasising that this non-finite decomposition into points is precisely what geometry analyses. But secondly, what about spatial relations? How are they to be understood? For Leibniz this was a particularly difficult question, since he was not disposed to think that relations are real: a relation is always a 'logical construction' out of the monadic predicates of the objects that are joined by it.

Kant thought of a further objection – his famous argument from 'incongruent counterparts'. Consider an object and its mirror image – a left hand and a right hand, say. Are not the spatial relations of the one exactly reproduced by the spatial relations of the other? Yet they differ in their spatial properties, since the right hand is *incongruent* with the left hand. There is no way of moving it through space so as exactly to occupy the space vacated by the left hand. A universe containing only a left hand is therefore a different universe from one containing only a right hand, even though all spatial relations in the two are identical. (Kant actually wrote of a left-hand and right-hand *glove*, having neither seen *The Beast with Five Fingers*, nor heard Brahms's left-hand transcription of the Bach Chaconne.)

How do we respond to Kant's argument? He is right in suggesting that there are features of space which are not reducible to spatial relations: namely topological features, concerning continuities and discontinuities of lines and surfaces. The orientation of an object is a topological feature: an asymmetrical object is always *turned in a certain direction*. But this is due to another fact: namely, that objects in space are not points, but are spread over many points. No point has orientation. But asymmetry brings orientation, and asymmetrical collections of points have orientation, when considered as a totality. As Wittgenstein points out (*Tractatus*, 6.36111), the problem can be reproduced in one-dimensional space. Imagine a straight line which is red at one end and green at the other. It cannot be moved through the one-dimensional space that it occupies so as to coincide with its mirror image. But it could be *turned round* through two dimensional space, so as to lie on top of its image and be congruent with it. If you could turn a right hand round in four-dimensional space, you could also lay it on top of the left hand. (Hence the eerie character of four-dimensional applause.)

There are two responses to that. First, wow! Secondly, so what? Kant's argument still shows that orientation is something over and above spatial relations. For spatial relations in *three*-dimensional space are what we are talking about. It doesn't help to be told that we can overcome incongruence by means of a fourth dimension. Give us that dimension, and we shall be able to produce a new kind of incongruence, involving *four*-dimensional objects.

All this suggests that the notion of 'occupying space' is far harder to understand than we might think. It is, as Kant rightly argued, fundamental to our conception of an objective world, that things are *in* space, and can be identified and reidentified by means of their location. But while we can describe space in terms of a Hilbertian geometry of points, the things *in* space are something more than the points which they occupy. The relation of 'being in' a place still needs to be elucidated. Perhaps we shall have to rest content with the view that it is primitive. Certainly it is not a mathematical, but a physical idea; and no purely mathematical theory will tell us the whole truth about it. Physical concepts, such as solidity, rigidity and cohesion, are necessary if we are to make sense of the occupation of places.

5. How Many Spaces are There?

Kant ran together the two questions: whether space is infinite, and whether it is absolute. He also confounded them with a third: the question whether space is unitary. The same visual ideas which persuaded him that space is a kind of medium, stretching for ever in three dimensions, prompted him to believe that there could be only one space, and that any rival candidate for the name would turn out to be simply a part of the one space, isolated at best by physical barriers, but not spatially disjoint. To put it another way: everything in space is spatially related to everything else in space. Is that true? And does it have to be true?

There is a good reason for supposing that it *is* true, namely, that theories of space are bound up with the enterprise of explanation. To find causal connections between things is automatically to situate them in the same space, since every attempt to describe the forces and fields that form the web of causal connection involves situating those forces and fields in a single space. If our universe is to be *our* universe, it must be causally connected to us, in ways that permit us to be informed about it. Anything of which we can gain information must therefore be in our space. Our world and our space are one and the same.

But do we *have* to exist in one space? Common sense says yes; but Lord Quinton says no. In a celebrated article ('Spaces and Times'), he describes what it would be like to live in two spatially unrelated worlds. Imagine that you go to sleep one night, and dream of waking in some foreign place, surrounded by strangers, who engage you in their unfamiliar life. In your dream you go to bed and fall asleep. You then wake up, back in your familiar bed at home. Each night you dream this strange dream, which gradually acquires more coherence, until it seems as though you wake into that other world at just the point that you would have reached, had you really been sleeping in it. You recount in the second world your experience in the first, and your new companions treat it as a dream. At some point you cease to be able to say that the first world is the real world, the second only a dream. All the grounds you have for attributing reality to the one, prompt you to attribute reality to the other. Unlikely though the case may seem, it is surely not impossible. But if it is possible, then it is possible that there should be two worlds which are both spatial, even though there are no spatial relations between them. (There is nowhere in the one world which is the place where the other world is.)

Do you accept that? If you do, then you must also accept that there is no problem of personal identity involved: no problem in saying that the person who sleeps in the one world is identical with the person who wakes in the other. In other words, you must attach the notion of personal identity entirely to the first-person perspective of the observer, and disregard what is happening to his body. Do you therefore say that this is one person in two bodies? The case raises many interesting questions: too many. So let us leave it behind.

6. The Mystery of Time

As I said, time can be treated, from the point of view of physical theory, as a dimension. It is structured by the betweenness relation; it even exhibits orientation. A process is incongruent with its 'mirror image' in time, just as a left hand is incongruent with a right hand. (A melody played in reverse cannot be 'moved through time' so as to display the same sequence as the melody played correctly: cf. the 'film' music in Berg's *Lulu*.) But people baulk at the suggestion that time is *just* like space, another dimension, on a par with the three that lie around us everywhere. And for good reasons. Here are some of them:

First, time has a direction (time's arrow). That is to say, it moves always from past to future, and never from future to past. This sounds clear, so long as you don't examine it too closely, hence St Augustine's famous remark: 'What then is time? If no one asks me, I know; if I wish to explain it to one who asks, I know not.' For of course it is not time that has direction, but things *in* time. Nothing ever moves backwards in time. Nothing becomes earlier than it was. (But that sounds like a mere tautology.) Here is an intuitive idea, that proves almost impossible to state coherently.

Secondly, you cannot move through time, as you can through space. You are *swept along* by it. There is no way of hurrying forward to a future point at twice the speed of your competitor; there is no lingering or dawdling by the way. The temporal order compels you to be exactly when you are at any moment, and nowhen else.

Thirdly, everything in time occupies the whole of the time during which it exists. You entirely fill one part of the temporal dimension. So too do all your contemporaries. There is no jockeying for position in time, no pushing aside of its occupants. All sit there happily (or rather unhappily) for their allotted years. Nothing in time excludes anything else.

Such features indicate that we should not speak of a 'position in time' in the same way as we would speak of a 'position in space'. Times are not locations that we can choose to occupy, or over which we can contest. They are all-embracing and inexorable.

All this has led philosophers to ask deep and agonising questions about time. First, is it necessary? Could there be a world without time? Secondly is it real? Could it be that time is in some way a creation of the thinking mind, or a way of seeing things which are in fact not temporally situated at all? Thirdly, can one and the same thing exist both in time and out of it? Could we be emancipated from the prison of time, and still be the things that we are?

Many of the arguments that have been constructed around the concept of time have their origin in antiqity. Zeno and Parmenides were famous for their proofs that time is unreal: and I shall return to their arguments in Chapter 27. Plato, and following him Plotinus, believed that the ultimate reality is timeless, and also that we participate in it, and can finally free ourselves from the temporal prison. Aristotle, in a brilliant passage of the *Physics*, anticipated just about every argument that has occurred to anyone in this area, and located as one of the sources of the mystery, the little word 'now'.

7. The Unreality of Time

Belief in the unreality of time is something of a philosophical commonplace, since it seems to make sense of the primal mystery of our existence. Schopenhauer wrote: 'A man finds himself, to his great astonishment, suddenly existing, after thousands of years of non-existence: he lives for a little while; and then, again, comes an equally long period when he must exist no more. The heart rebels against this, and feels that it cannot be true. The crudest intellect cannot speculate on such a subject without having a presentiment that Time is something ideal in nature.'

Schopenhauer accurately identifies the motive, but not the argument, for time's unreality. Aristotle comes nearer to the mark, when he says: 'One part of time has been and is not, while the other is going to be and is not yet. Yet time – both infinite time and any time you care to take – is made up of these. One would naturally suppose that what is made up of things which do not exist could have no share in reality.' Or at least, subtract from time all the bits that are not, and you are left only with 'now': the fleeting moment which vanishes, just as soon as you try to lay your hands on it.

On the other hand, is not this just a way of saying what time *is*? Are we not being misled, perhaps by the grammar of our language, into denying the reality of time, because its reality must be captured by idioms that cannot be applied to anything else? Are we not too hastily moving from the uniqueness of time, to its non-existence?

This is where the famous argument given by McTaggart comes in. In vol. II, ch. 33 of his *The Nature of Existence*, McTaggart, a Cambridge Idealist writing at the turn of the century, re-phrased Aristotle's thoughts about 'now' in the form of a paradox. On any understanding, he argued, time involves an 'order' of things in a series. Which series? He proposed three: the A series, the B series, and the C series. The first is the series past-present-future, which is implied in the use of 'now'; the second is the series earlier-simultaneous-later which is recorded in physics; the third is the *real* order of those things that we understand through temporal idioms, but which may not be temporal at all. His argument concentrates on the A series and the B series. Of the second he says that it does not really capture the idea of time at all; for time involves change, and the events arranged in the B series are, so to speak, arranged eternally. Nothing in this series changes, since each event is fixed for ever in the place allotted to it by its relation to other events. To put it in another way: the B series is a changeless sequence, ordered by a 'betweenness' relation, in which nothing changes position.

This argument is not very satisfactory; for nobody ever thought that *events* change; it is rather *objects* which change, by their participation *in* events. Nevertheless, there is a point to McTaggart's argument. The formal structure of time-relations, which is captured by the B series, does not capture what it is for something to *happen*. This idea, McTaggart suggested, can be explained only in terms of the A series. Something happens by happening *now*: that is, it becomes present, and is, as soon as present, past and irrecoverable. But the A series involves a contradiction. Every member of this series is future, present and past: and these predicates contradict one another. So nothing happens.

The obvious reply is that no event has these three predicates *simultaneously*, so there is no contradiction. This event *was* future, *is* present and *will be* past: and there is no contradiction in that. But McTaggart has a reply to this, which is best delivered in his own words:

When we say that X has been Y, we are asserting X to be Y at a moment of past time. When we say that X will be Y, we are asserting X to be Y at a moment of future time. When we say that X is Y (in the temporal sense of 'is'), we are asserting X to be Y at a moment of present time.

Thus our first statement about [the moment] *M* – that it is present, will be past and has been future – means that *M* is present at a moment of present time, past at some moment of future time, and future at some moment of past time. But every moment, like every event, is both past, present, and future. And so a similar difficulty arises. If *M* is present, there is no moment of past time at which it is past. But the moments of future time, in which it is past, are equally moments of past time, in which it cannot be past. Again, that *M* is future and will be present and past means that *M* is future at a moment of present time, and present and past at different moments of future time. In that case it cannot be present or past at any moments of past time. But all the moments of future time, in which *M* will be present or past, are equally moments of past time.

And thus again we get a contradiction, since the moments at which *M* has any one of the three determinations of the A series are also moments at which it cannot have that determination. If we try to avoid this by saying of these moments what had been previously said of *M* itself – that some moment, for example, is future, and will be present and past – then 'is' and 'will be' have the same meaning as before. Our statement, then, means that the moment in question is future at a present moment, and will be present and past at different moments of future time. This, of course, is the same difficulty over again. And so on infinitely.

McTaggart thought that this infinite regress is vicious, since it shows that we can never remove the contradiction involved in the A series except by presupposing that it already *has* been removed. He also argued that we can have no real conception of change or temporal succession (as opposed to a mere timeless series) without invoking the A series. That which is distinctive of *time* is precisely that which introduces a contradiction into the order of events. To resolve this contradiction we must suppose, in idealist fashion, that time is unreal – a reflection of our limited perspective, which does not correspond to the underlying reality.

8. Responses to the Argument

There have been many responses to this argument, and two in particular are worth considering. The first concentrates on the grammar of 'now'. 'Now' is what some philosophers call an 'indexical' expression, and what Reichenbach called a 'token-reflexive' word – meaning one that relates what is being said to the token utterance itself, i.e. to the situation of the speaker. 'Now' is like 'here' and 'I': it refers to the time at which the speaker is speaking, just as 'here' refers to his place and 'I' to the speaker. Once we understand this, we shall see that 'now' does not refer to a property of an event at all, but only to the point of view from which it is described. What is 'now' – i.e. present – from one point of view, is without contradiction 'then' – or past – from another point of view, just as a place can be here from my perspective and there from yours.

Some philosophers are not persuaded by this response. Hamlyn, for example, argues that there is no equivalent of McTaggart's contradiction involving 'here', and that this alone is sufficient to suggest that more needs to be said. Actually there is the precise equivalent of McTaggart's contradiction, since clearly no object can be simultaneously both here and there, and to say that it is here from my point of view and there from yours is just like saying that an event was future, is present and will be past. For it involves saying that it is here from here, and there from there: but just as here is both here (from my point of view) and there (from yours) so there is both here (from your point of view) and there (from mine): and nothing *can* be both here and there. But the parallel solves the paradox: to say that this is a *contradiction* is like saying that the concept of *size* involves a contradiction, just because everything that is large is also small, a large flea being a small animal, and a large elephant a small descendent of the mammoth.

Dummett, in an article purporting to be a defence of McTaggart, suggests more reasonably that, while McTaggart is culpably unaware of the logic of token-reflexives, his argument does nevertheless point to a real difficulty about time that cannot be reproduced in the theory of space. The use of token-reflexives like 'here' and 'there' is not essential to the description of space: we can locate things in space, Dummett suggests, without presupposing a point of view on it. But nothing similar is true of time; we cannot begin to identify times without first locating ourselves in the midst of them. Dummett's suggestion is difficult to follow. For the sense in which we need a point of view in order to identify times, is precisely the sense in which we need a point of view in order to identify spaces: i.e. in order to pick them out, and communicate with each other about them. Someone who knows that the battle of Hastings was earlier than the death of Queen Elizabeth I still does not know exactly *when* it was. But the person who knows that it occurred precisely nine hundred and eighty six years ago does know when it was. (And ditto for space.)

Perhaps the real point behind Dummett's response is this: that we do not only identify places as here and there; we also *travel* from here to there, and so adopt the point of view which changes there to here. This freedom of movement in space emancipates space from our present perspective, and gives sense to the idea of space as a frame within which we are situated. There is no parallel for time, and this is part of the mystery: the 'now' is all we have, and all we can have. And yet it is nothing.

Another response to McTaggart is vigorously defended by Hugh Mellor in his book *Real Time*. Mellor accepts McTaggart's strictures against the A series. It really is confused and contradictory to speak in this way, he argues, if we think that by doing so we are somehow saying what time really is. But that is because the reality of time is given by the B series. The truth-conditions of all temporal statements are to be given in terms of the 'earlier than', 'simultaneous with', and 'later than' relations. And this is true even for those temporal statements involving 'now', 'the present', 'future' and 'past'. What makes it true that Mr Major is *now* Prime Minister is that Mr Major is prime minister at a time, some part of which is simultaneous with *this* utterance. (But do we not have a similar problem about 'this'? Help!) Since it is truth-conditions which tell us what is really being said, there is no paradox: the B series is what time consists in. The A series is merely a reflection of our mobile point of view.

The responses all leave something to be desired. And, if they avoid McTaggart's conclusion, they still do not dispel the *mystery* of time: the mystery, as some have expressed it, of becoming.

9. Time and the First Person

The mystery is felt most acutely in relation to my own situation. I am in time, and time informs all my thoughts and feelings: to regret, to hope, to expect, to long for or to fear: in every posture towards the world my 'being in time' fatally qualifies my self-conception. If Heidegger is to be believed, this is the fundamental truth about my condition, and the source of my anxieties. Somehow I must come to terms with the fact that I am extended in time, and fixed to time's arrow as it flies from past to future.

Some philosophers have doubted that time as experienced is the same thing as physical time. Bergson (*Essai sur les données immédiates de la conscience*) famously distinguishes *le temps* from *la durée*, and argues that while physicists can know the first, they cannot know the second, since the character of duration is revealed only by the process of life – of *living through* the sequence of events. In living through things I acquire a knowledge of their inner order, of the way in which one thing grows from and supersedes another, and this knowledge is enshrined in my memory. Memory provides a unique overview, in which temporal order is subordinate to thought. Proust, inspired by Bergson, attempted to show that remembered time has an order that physical time cannot have – an order of meaning, in which later events cast their light upon earlier events, and are themselves in turn illuminated. The order in memory is an order of meaning rather than sequence.

Such a thought lends itself more to fictional than to philosophical exposition. The same could be said of Merleau-Ponty's theory of the 'lived present', which is another attempt to separate time as lived from time as described by the physicist. The peculiarity of self-consciousness, according to Merleau-Ponty, is that other times are brought to bear on the present, so as to 'thicken' it. The moment in which I live is one in which *was* and *will be* are actively made *present* by memory and will. I am not merely the passive victim of the arrow, as physical objects are: I *experience* time's passing, being simultaneously conscious of whence and whither the arrow flies.

It is certainly true that we can be more or less lax in our relation to time. The rational ability to assume responsibility for other times is a vital part of living well, and that means living well in the present. My present acts may be an expiation for former wrong-doing; they may also may be an invocation of future good. I do not, if I am rational, allow the stream of time to wash me unresisting with it; I am always planning for the future, and assuming responsibility for the past. This, perhaps, is what 'being in time' consists in. I know myself as process, and affirm my identity through that process: the thing which happened was *my* doing; the thing which I intend will spring from *me*. This active breasting of the current is the sum of human dignity, and if time is a mystery it is partly because we do not and cannot know how it is done. And yet we do it.

10. Process and Becoming

In contrast to those who have argued for the unreality of time, there arose during the nineteenth century a school of philosophers who were so deeply impressed by time's reality, as to wish to build it into the foundation of metaphysics. I say 'school', although it became so only in retrospect, when its various threads were drawn together by A.N. Whitehead in his vast book *Process and Reality*, and thereafter meditated and debated by a generation of devoted followers. The name 'process philosopy', commonly used to describe this school, is of obscure origin; but nobody doubts that its most important representative in recent years has been the American philosopher and theologian Charles Hartshorne, to whom is attributed also a 'process theology', of cheerful and forward-looking aspect, which American readers have found distinctly user-friendly.

Process philosophers claim among their number not only Bergson, but also Peirce, William James and Dewey, all of whom rejected the 'block universe', as James described it, in which there is no room for novelty and adventure. Although the resemblance between these philosophers is more apparent than real, it is true nevertheless that they see the concept of process as both fundamental to metaphysics, and irreducible. They are also strongly influenced in their arguments by the results of modern science, and in particular by the place accorded to time and process in modern physics and biology. Since it is not possible to summarise all the various process philosophies in a paragraph, I shall content myself with an outline of the major themes in Whitehead:

(a) Traditional philosophy fails to take time seriously. If we do take it seriously, we soon realise that no concrete entity can change (since change requires that something be both the same and not the same). Concrete entities can only be superseded.

(b) One way of not taking time seriously is to 'spatialise' it. As Bergson perceived, this is a persistent error in human thinking, which fails to grasp the way in which we are *in* time, but not *located* in it.

(c) Time can be understood only as process. Time involves 'concrescence', whereby 'eternal objects' become concrete. The fundamental entities in time are not substances, but 'occasions', which supersede one another: supersession is part of the real essence, in Locke's sense, of every occasion. In any occasion, other occasions may be 'prehended', as when an occasion is both a shooting and a wounding. But each occasion is incomplete: it cannot sustain itself in being but only fleetingly points the way to the future that is to supersede it. In a sense time consists in this incompleteness of the entities that occur in it. We

must therefore understand time through the three basic categories of supersession, prehension and incompleteness.
(d) That fearfully abstract account of temporal process is the ground for an attack on substance and quality: these categories, according to Whitehead, are merely abstractions from the flow of time, and denote no concrete realities. Indeed, the tendency to think in terms of substance and attribute (object and quality) exhibits the 'fallacy of misplaced concreteness'.
(e) We must distinguish the real from the actual. The future is merely real; the present is actual, and the past consists of an 'immortal nexus of actualities'. The process of transition, whereby the merely real is slowly colonised by the actual, is a *creative* process, which brings about an organic 'community' of actual things. All process, properly understood, is creative.
(f) Time is 'epochal'; that is to say, it is not a continuum. For if it were, each occasion would have to be constantly superseding itself. Time is occupied successively by discrete occasions, which have an inherent duration. William James wrote in this connection of the 'specious present' – by which he meant the short but genuine stretch of time, which is the minimum necessary for experience. This atomic theory of time is perhaps one of the strangest and most provocative of Whitehead's claims: but it is a theory that has found supporters among physicists as well as philosophers.

All that adds up to a distinctive and somewhat unnerving metaphysic, whose terms can be translated only with difficulty into the public discourse of modern philosophy. As Whitehead recognised, it compels us to introduce a wholly new conception of God, who can no longer be conceived as the eternal and immutable sovereign of the realm of becoming, but must himself participate in the temporal drama. However, process is creativity, and therefore already manifests the divine nature. By dint of our 'concrescence' we participate in that nature, and become 'available for all future actualities', in Hartshorne's words.

Whitehead denies the actuality of substance; but he is compelled to admit that we are persons, and that persons endure through time, embracing many occasions, and enjoying an identity and a unity which mark them out from the flow of things. This concession to common sense is, he thinks, easily accommodated within his metaphysic, though by no means every critic has been persuaded of the point. The sparse theory of being may equally lead to the conclusion that fundamental entities are essentially in time. But for most modern philosophers they are also reidentifiable and therefore durable. It is hard to reconcile that thought with a philosophy which makes supersession and incompleteness inescapable conditions of the actual. Among modern philosophers, it is perhaps only Davidson who is led in Whitehead's direction, advocating an 'ontology of events' which, if fully worked out, would surely lead to the conclusion that nothing really lasts long enough to be talked about: a conclusion which, had it been drawn by Whitehead, would have spared us not only the many hundred pages of *Process and Reality*, but also the many more hundred pages of Hartshorne.

11. Eternity

At the opposite pole to the process philosophers, but in covert sympathy with them, are those who have granted the reality of time, while according a superior reality to eternity, viewed as a condition of 'timelessness'. The greatest of them was Plato, whose vision of a realm of Forms, above all change and decay, ineffably manifest in the world of becoming, has been the inspiration for countless philosophical and theological speculations on the destiny of man. (See especially the *Phaedrus*.) What precisely is meant by 'eternity', and what is its relation to time?

We should distinguish eternity from sempiternity. Something is sempiternal if it endures forever; i.e. if there is no time at which it is not. Something is eternal, however only if it is outside time: only if temporal predicates do not truly apply to it. The traditional elucidation of this idea is through mathematics, and through the contrast between durable physical objects and numbers. It is logically possible that a lump of rock should last through the whole of time; but it is essentially *in* time, and subject to change *over* time. If the number 2 exists, then it exists at every time; but it does not exist *in* time, since it takes no part in temporal processes, nor does it change. It possesses all its properties essentially and eternally. Nothing ever *happens* to the number 2; nor does it cause anything to happen to anything else.

The ontological argument seems to imply that God, if he exists, is eternal in just that way. He possesses all his properties essentially, and exists outside time (and also outside space). He is everywhere and everywhen, but only because he is nowhere and nowhen. Plato's world of Forms is also construed on the mathematical model; and all those writers – from St Augustine and Boethius to Spinoza – who have located the destiny of man in some kind of 'redemption' from the temporal process, have been spellbound by the other-worldly majesty of mathematical truth.

Such a concept of eternity is fraught with metaphysical and theological difficulties, however. If God is really *outside* time, then how can he influence temporal processes? For instance, suppose God decides to flood the world. There is then something true of God at one time (namely, that he is flooding the world), that is not true of him at another. Furthermore, if God is related to the world (for instance, as its creator), then every change in the world will be a change in God's relational properties: he stands now in *this* relation to the created sphere, now in *that*. Yet, if God is eternal as the number 2 is eternal, no such thing could be true.

On the other hand, if God is merely *sempiternal*, like the enduring rock, he is in no way removed from the created world: on the contrary, he is a part of it. Furthermore, it is difficult to imagine a proof of God's existence which had the consequence that God is merely *durable*, rather than eternal. For such a proof would be hard to reconcile with his *necessary* existence: and without the attribute of necessary existence, God is not God. To say, with Hartshorne and the process theologians, that an unchanging God is no God, is precisely to jeopardise God's status as the Supreme Being.

Such thoughts may tempt us to believe in the unreality of time. But time's unreality does not release us from them. For it has the consequence either that all things in time – ourselves included – are unreal, and therefore that there *is* no created world; or that all created things (all contingent beings) really exist *outside* time, as God does, and merely *appear* in the world of becoming, which is a world of illusion. That second view has certainly been espoused (by oriental religions, and by Schopenhauer, among others). But it leaves us with an intractable problem: How can something that is essentially *outside* time, appear to be *in* time? Certainly, if we encounter something in time, we know that it is not the number 2; the same ought to be true of all eternal objects.

Spinoza's approach to these problems is particularly instructive. By his own version of the ontological argument, he proves that at least one substance exists, and also that at most one substance exists: substance being infinite in every positive respect. (*Ethics*, Part I.) This one substance therefore embraces everything that is, and there can be no distinction in reality between God and the natural world. Either the natural world is identical with God (the one substance), or it is 'predicated of' him as one of his 'modes'. Spinoza argues for the first of those views, and elects for the title 'God or Nature' (*Deus sive Natura*) as the correct name of the one thing that is everything.

The distinction between the creator and the created is therefore not a distinction between two entities, but a distinction between two ways of conceiving a single reality. I can conceive the divine substance now as a whole, self-dependent and all-embracing, and now as the sum of its various 'modes', unfolding each from each in a chain of dependency. The first way of conceiving substance is like the mathematician's way of conceiving a proof: studying the timeless logical connections that deliver truth upon truth from a handful of all-embracing axioms. The second way of conceiving substance is like the scientist's regimen of experiment, through which the underlying order is extracted by interrogation.

With this distinction goes another: that between eternity and time. The world can be conceived *sub specie aeternitatis* (under the aspect of eternity), as a mathematician conceives numbers and proofs; or *sub specie durationis*, as ordinary people observe the sequence of events in time. There are not two realms, the eternal and the mutable, but again two ways of conceiving the one reality. To study the world *sub specie durationis* is to study it as it is; time, therefore, is real. Nevertheless, studying the world in this way, we can never grasp the whole of it: we can never reach the sum of those necessary connections, which show how each truth contains and is contained in every other. When, as in the ontological argument, I see the world *sub specie aeternitatis*, I see that what is, must be, and that all truth is eternal truth. Then, and only then, do I have an 'adequate' idea of the world.

The appeal of Spinoza's philosophy is bound up with his monism; in Spinoza's world everything less than the whole of things becomes a 'mode' of that whole; all distinction dissolves, and individuals melt away into a vast unruffled sea of being, stretching without limit through eternity. Even if time is real, it has little real authority in the philosopher's view of things. For the grid of duration divides the one substance in ways that make no intellectual sense, and to see how things ultimately are (to acquire an 'adequate idea' of the world) we must discard it, and adopt the perspective of eternity.

There is a price to be paid for this vision, as Leibniz saw. Spinoza's philosophy lacks what the scholastics called a *principium individuationis* – a principle of individuation – whereby we might distinguish one thing from another, attribute substantial reality to the human subject, and attach our discourse to a realm of objective individuals. Suppose we were to restore that principle, to abandon monism and to accept the world as composed of a plurality of individuals, ourselves among them. What then remains of Spinoza's view of time? Could we still maintain that time is a way of ordering or conceiving individuals, and that those very same individuals could also be stripped of their temporal garments and dressed in the mantle of eternity – where eternity means not endless duration, but existence outside time? Only if some such thing is possible, can we understand the promise of eternal life. Yet how could *life* be eternal, when life is essentially a process of growth and decay? How, indeed, could any individual substance be seen in a-temporal terms, when its individuality (and therefore its essence) is bound up with its *identity through time*? Whatever can appear under the aspect of eternity, is surely not identical with *this, here, now*?

12. The Music of the Spheres

The question lies at the limit of the intelligible. But perhaps there are ways of responding to it which are not merely mystical. Here is one:

When Kant addressed the problem of space and time, in *The Critique of Pure Reason*, he suggested (following his own pre-critical arguments) that time and space are really 'forms' of sensibility: they are the frame in which we situate the world, whenever we perceive it. 'Things in themselves' are not in space and time; the spatio-temporal world is a world of appearances. As the argument of the *Critique* developed, it became increasingly clear to Kant that the thing-in-itself, conceived in such a way, could neither be an object of knowledge, nor bear an intelligible relation to the appearances that we perceive and understand. By the end of the argument, the thing-in-itself has dropped out of consideration altogether, as a mere 'noumenon'. (See Chapter 5.) To use Wittgenstein's metaphor, you can 'divide through' by the thing-in-itself, which has no place in the world-view that is defined by means of it, being no more than a shadow cast by arguments that serve to refute it.

But Kant was also dissatisfied with this conclusion. For his theory of freedom seemed to imply that we really do confront at least *one* thing-in-itself: the transcendental self whose choices are determined by reason (i.e. by 'eternally valid' grounds), and which is 'unconditioned' by the world of appearance. Indeed, Kant made a heroic attempt to defend a kind of 'individualised Spinozism', according to which I can conceive of *myself* in two incommensurate ways: as a part of nature, bound by causal laws and occupying a position in space and time; and as a member of a 'kingdom of ends', outside the empirical order, and obedient to reason alone. (Although, Kant added, the understanding stops short of this second view, which cannot be expressed in judgements, but only through the imperatives of practical reason.)

Kant's immediate followers – Fichte and Schelling – endorsed his theory of freedom, while ignoring his metaphysical scruples. A whole generation of thinkers took Kant to have proved that the transcendental self is the one reality, and that the postulate of transcendental freedom is the *ground* of any conceivable system of the world. We have access in our own case to the really real, and it is *prior to*, and therefore in essence *outside*, the conditions of sensibility, such as time, space and the categories. Yet it also 'posits itself' in the world of nature, and so becomes manifest in the temporal order. In short, one and the same individual is both in and out of time.

Nobody took this line of argument further than Schopenhauer, whose *World as Will and Representation* is a *tour de force* of synthesis, in which the philosophical richness of the Kantian philosophy is brilliantly unfolded, though with little by way of valid argument. According to Schopenhauer, the natural world is a system of 'representations', ordered by the concepts of space, time and causality. Behind these representations, and in a sense imprisoned by them, lies the thing-in-itself – the unconditioned, whose reality can be grasped by looking inwards and confronting it in me. The inner realm is a realm of immediate knowledge, in which I know without concepts, and with a certainty that I cannot otherwise obtain. It is also a realm of freedom, as Kant had demonstrated. In short, the thing-in-itself is *will*: that which expresses itself through freedom, and which is revealed to itself through practical reason, without concepts or conditions.

In order to distinguish one will from another, we need a *principium individuationis*, and this, Schopenhauer argues, is available only in the world of nature: i.e., in the sphere of 'representation', which enables us to identify objects and distinguish them, through the spatio-temporal conditions which govern their appearance. In itself, however, the will has no *principium individuationis*; nor does it exist in space and time. It is one, uniform and eternal, trapped in its temporary abode in nature only to struggle unceasingly against it. The supreme goal of the will is to return to that unconscious eternity from which it emerged: and this striving for annihilation – for the Nirvana of

Buddhism – is, Schopenhauer argues, the true secret of man's life on earth:

> Awakened to life out of the night of unconsciousness, the will finds itself as an individual in an endless and boundless world, among innumerable individuals, all striving, suffering, and erring; and, as if through a troubled dream, it hurries back to the old unconsciousness. Yet till then its desires are unlimited, its claims inexhaustible, and every satisfied desire gives birth to a new one. No possible satisfaction in the world could suffice to still its craving, set a final goal to its demands, and fill the bottomless pit of its heart. (Ch. 46.)

Precisely because it lies beyond the 'veil of appearance', the will, as thing-in-itself, is unknowable to the understanding. Any attempt to represent it in space and time will belie its inner nature, and only in our own case do we have direct acquaintance with this ultimate reality which can never be captured in words. Although the will cannot be represented, however, it can be expressed: it can give voice to itself, in those reason-guided enterprises which are free from concepts, and in which an order is achieved which is the order of inner and not of outer life. The greatest of these is music.

In a Beethoven symphony, Schopenhauer argues, we find all the human passions: yet 'only in the abstract and without any particularisation'. Grief, but not the object of grief; yearning, but not the thing yearned for; desire, without the thing desired. (To represent the objects of the passions, music would need concepts, and would therefore sacrifice its ability to glimpse behind the concept so as to confront the thing-in-itself.) Through music, therefore, we are acquainted *objectively* with that which we otherwise know only from within, as pure subjectivity.

Schopenhauer's theory of music is untenable, for a variety of reasons which need not trouble us. But it echoes an ancient Pythagorean belief, that music contains the secret of eternity: that we encounter in music 'the point of intersection of the timeless with time'. Perhaps the clearest exponent of the Pythagorean theory was Boethius who, in his *De Musica*, distinguished three kinds of music: song, instrumental music, and the music of the world, that is the source of all natural harmony. This natural music is not heard: indeed it does not *occur* in time, as human music does. It consists in the divine order of the universe – a matching of part with part and of each part with the whole, which can be fully understood by us only through the intellect. Indeed, our best way of grasping this divine harmony is through mathematics, when we study the relations between numbers, and construct those sublime proofs of geometry, in which the relations between shapes and areas are reduced to relations of number.

The music of the spheres consists, then, in an eternal and immutable order of things, revealed to the intellect in mathematics and philosophy. Yet it is precisely the *same* order, Boethius suggests, as that revealed to the senses in music. When I hear harmony, I am sensing through the ear those relations of number that resound through all eternity. Although music is a temporal process, and what I hear in music I hear spread over time, nevertheless this very time-bound experience presents me with an 'intimation of immortality': a glimpse into the eternal, and into the joy of residing there, outside space and time.

Boethius's theory is no more tenable than Schopenhauer's. Yet there is, perhaps, a core of truth in what they are each trying to say. When we hear a melody, for example, we hear a distinct movement: the melody begins at a certain point (on F, say), moves upwards rapidly, downwards more slowly, and ends at last on C. In describing the melody in these terms, I am describing what I hear: yet, in the material world of sounds, there is no such thing as movement, but only sequence. F then G; but nothing that *moves* from F to G. What I hear is movement, even though I hear only sounds, which do not move. Furthermore, I can identify this melody again: at the same pitch, or at another. I can recognise it played faster or slower; I can hear its mirror image in the imaginary space of music – as in an inverted canon. In short, the melody acquires for me the character of an individual, which recurs in various forms, always animated by its inner motion. Yet the relation of this motion to the actual sequence of events is virtually unintelligible. Although the sounds in which I hear the melody are in time and space, the melody itself seems to inhabit a space and time of its own. I encounter it in that 'ideal' space and time, even though it is not my space and time, and is incommensurate with the empirical world.

In music, therefore, we step from our time and space into an ideal time and an ideal space: time and space which are there only in the experience of music. From that ideal time it is, so to speak, a small step to eternity. Sometimes, listening to a Bach fugue, a late quartet of Beethoven, or one of those infinitely spacious themes of Bruckner, I have the thought that this very movement which I hear might have been made known to me in a single instant: that all of this is only accidentally spread out in time before me, and that it might have been made known to me in another way, as mathematics is made known to me. For the musical entity – be it melody or harmony – is only a fleeting visitor to *our* time; its individuality is already emancipated from real time. We have no difficulty, therefore, in imagining this very individual emancipated from time entirely, and yet *remaining an individual* (as Schopenhauer's will does not remain an individual). In the experience of music, therefore, we can obtain a glimpse – an intuition – of what it might be, for one and the same individual, to exist in time and in eternity.

Of course, this does not enable us to conceive how you or I might exist in eternity. But is the difficulty of conceiving this a final proof of its impossibility? Remember, we cannot, in the nature of things, conceive of a space that is three-dimensional, finite and yet unbounded. But we *can* conceive of the equivalent in two-dimensional space (the surface of the sphere). Asking someone to conceive a concrete individual (a person) as existing eternally might be a little similar. We might say: You know what it is for a melody, which exists in ideal time, to exist also in eternity. Now suppose the same of a concrete object, in *real* time. In some such way we say: You know what it is for a two-dimensional space to be finite but unbounded. Now suppose the same thing in three dimensions. And of course you can't *imagine* it!

Even if the reader does not accept any of that, it should awaken him to an important philosophical observation, namely that we shall never really make sense of time, if we do not understand the nature of mathematics. Is it *really* true that mathematical objects exist, and exist eternally? And if it *is* true, how on earth could we ever be in a position to know the laws of mathematics?

26

Mathematics

Space, Kant argued, is the 'form of outer sense', and time the 'form of inner sense'. Just as perception represents the world as spatially organised, so does all experience (including experience that is purely 'inner', like sensation) situate our world in time. Space and time are intrinsic to mental life. The first delivers *a priori* knowledge in the form of geometry, while the second delivers its own body of *a priori* truth, which is also mathematical. The 'intuition' needed to grasp geometry is matched by another intuition in the understanding of arithmetic – the operation of counting, or putting down, *first* this, and *then* that. Perhaps that operation is part of what we understand, in understanding time. Arithmetic is, if you like, the *a priori* science of counting.

Those bold and astonishing ideas are difficult to sustain. But the motive for them is one with which a modern philosopher would sympathise. Kant was trying to explain the *a priori* nature of mathematical truth, without giving way to 'Plato's temptation', which is to envisage a separate and eternal realm of mathematical objects, to which we have mysterious access through the 'ascent' of reason. On the contrary, Kant suggests, it is our nature as sensuous beings that gives us access to mathematical truths. Mathematics describes the *a priori* framework of experience – the 'form' into which experience must fit, if it is to be truly ours, and an object of our self-conscious awareness. But it describes no item *within* the empirical world. That is why, if you make the mistake that Plato makes, so as to believe that mathematics describes *objects*, you will situate those objects in a transcendental realm.

On Kant's view, mathematical propositions are *a priori* but synthetic – and in this he agrees with Plato. But his explanation of the claim removes its metaphysical force: we can accept it, without making mathematics into the paradigm of objective knowledge that it had been since antiquity.

At the same time, there is an intuitive plausibility about Plato's position. For consider the following statements:

(1) We know many arithmetical truths, and know them without a shadow of doubt.

(2) Arithmetical truths are *about* numbers.

(3) Numbers are the subject-matter of identities, and indeed identity of number is one of the primary mathematical concepts.

(4) Truth means correspondence to the facts.

It is hard to accept all those propositions and to deny that numbers are *objects*, for reasons that should by now be familiar; and it is hard to accept that numbers are objects without also accepting Plato's theory, that they are eternal, immutable and necessarily existing objects. Moreover, numbers take no part in any *change* or *process*; they are causally inert. It is therefore wholly reasonable to situate them in another, transcendental, realm.

1. The Nature of Mathematics

Mathematics has many branches, and there is plausibility in Kant's assumption, that geometry and arithmetic are distinct sciences, the one dealing with space, the other with number. However, it has been common knowledge since Descartes that geometrical forms and proofs can be represented algebraically, through a system of co-ordinates. The developments discussed in the last chapter are further proof that geometry can be studied without reference to the diagrams and figures which are normally used to teach it.

Hilbert wished to show that mathematics could be reduced to a set of axiomatic systems, each branch being distinguished from the others by its axioms. The rules of inference are common to mathematics, and involve those elementary logical steps, such as *modus ponens*, which are the property of reasoning as such. What makes a theory into a *mathematical* theory, as opposed to a piece of logic, lies in the nature of the axioms. Although these axioms are immensely various, as they must be if they are to describe space, time and measurement in all their forms, they have one distinctive feature which marks them out as mathematical, namely the reference to the algebra of numbers. Algebra is arithmetic with variables; and if we can understand arithmetic, then mathematics would no longer be a metaphysical mystery. So Plato and the Pythagoreans were right also in this: that number is the root concept of mathematics, and the final source of its metaphysical status.

Hilbert was a Platonist. This does not mean that he succumbed to what I have called Plato's temptation: he did not. He believed mathematical theories could all be reduced to axiomatic systems; but he also believed that we could never eliminate the idea of number from the axioms. We must therefore suppose that numerical expressions *stand for* objects, which have a reality independent of our calculations. This is what is usually meant by 'Platonism': the theory that mathematics describes a realm or system of *real and independently existing objects*, whose nature is known to us through proof, but which are entities over and above the proofs by which we discover them.

There are many reasons for being alarmed by Platonism. Not only are numbers very strange objects, if they are objects at all. There is a terrifyingly large quantity of them – indeed, as Cantor showed, in his theory of transfinite numbers – a 'nondenumerable infinity'. There is simply no formula or procedure which identifies the totality of the numbers, not even a formula that is applied infinitely many times. And the ease with which new numbers can be extracted from the hat of higher mathematics suggests to many people that they were never really *in* there, but are the illusory by-products of the sleights of hand that display them.

Hence arose the theory of constructivism, as an alternative to Platonism. This exists in a variety of forms, but it was given its most serious support in this century by the Dutchmen, L.E.J. Brouwer and A. Heyting. They defended a version of constructivism which they called 'intuitionism'. This is the name under which the theory is now frequently discussed (for example by Dummett, in his book of that title). The constructivist believes that we have no conception of mathematical truth apart from the idea of proof. Our mathematical theories are intellectual constructs, which reach just as far as their own scaffolding, but which never take us to another realm, or to objects that exist independently. Proof is *all* there is. Likewise for numbers; these do not exist until 'constructed', by operations which generate them in a finite number of steps. There are no numbers 'out there' awaiting discovery; all existing numbers are contained in the books and papers of the mathematicians. For to say that numbers exist is to say that there are valid proofs involving numerals.

Constructivism eliminates the vertiginous metaphysics to which Platonism seems to commit us; it also gives – or at any rate seems to give – an intelligible explanation of the *a priori* nature of mathematics.

Indeed, it is the very same explanation as the one given by Kant. Mathematical propositions are known *a priori* since we ourselves are the authors of them. We know so much about numbers, and so infallibly, because the truths about them are created by the proofs which lead to them.

However, the charm of constructivism soon wears off. As the intuitionists acknowledged, it is not possible to accept their view of numbers and to leave logic unchanged. A mathematical proposition is true only if there is a proof of it; similarly, it is false only if there is a proof of its negation. But what if there is a proof of neither? We are forced to say, either that the proposition is meaningless – which is surely intolerable, since we should not then know what we *mean* by a mathematical proposition in the absence of a proof of it – or that it is neither true nor false. In other words, we must deny the 'law of the excluded middle'. Nor is this all that we must deny. As Heyting demonstrated, we shall need an entirely new system of logic – which he called intuitionistic logic – in order to accommodate the constructivist vision of mathematical truth. This system of logic turns out to be downright *counter*-intuitive. Indeed the intuitionists anticipated, in the special field of mathematics, the difficulties that are encountered by every form of 'anti-realism'. (See Chapter 19.)

So where does this leave us? Is there any plausible explanation of the *a priori* nature of mathematics that avoids both the metaphysics of the Platonists, and the deviant logic of the constructivists?

2. Logicism

This question returns us to an earlier period of modern philosophy. The Kantian theory of mathematics as synthetic *a priori* inspired Frege to compose his *Foundations of Arithmetic*. The purpose was to show that mathematics is really *analytic* – as Hume had supposed. Frege took this to mean that mathematics could be reduced to elementary laws of logic, of the kind that must be accepted if there is to be reasoning at all. Thus was originated the 'logicist' programme – the search for a derivation of mathematics from logic, which would reduce the first to the second, leaving no residual problem about the nature of mathematical truth.

Leibniz had already tried to carry out this reduction. He produced a notorious proof of the statement that 2 + 2 = 4, which goes as follows: 2 =df. 1 + 1; 4 =df. 1 + 1 + 1 + 1. Hence 2 + 2 = 1 + 1 + 1 + 1 = 4. The proposition is proved simply by substituting definitions. '2 + 2 = 4' is 'true by definition'. But Leibniz's proof doesn't work, for there is a missing line: 2 + 2 = (1+1) + (1+1). And what entitles us to drop the brackets, and convert (1+1) + (1+1) into (1+1+1+1)? That move is precisely what the operation of addition authorises; but it is authorised by the laws of arithmetic, not by the laws of logic.

At the end of the last century, Dedekind showed that all the basic notions of arithmetic (rational, real and complex number) can be reduced to the theory of the natural numbers; Cantor then showed that the concept of one-to-one correspondence could be used to define 'equinumerosity'; and finally Peano reduced arithmetic to a set of axioms. It remained only to define the fundamental concepts involved in those axioms, for the reduction of arithmetic to be complete.

Peano's 'postulates', as they are usually called, are extremely neat:

1. 0 is a number.
2. Every number has at least one and at most one successor which is a number.
3. 0 is not the successor of any number.
4. No two numbers have the same successor.
5. Whatever is true of 0, and is also true of the successor of any number when it is true of that number, is true of all numbers.

The fifth postulate states the well-known axiom of mathematical induction, which enables us to prove theorems about all the numbers by considering only three of them. All of arithmetic can be derived from the five postulates. The logicist programme, therefore, seeks to define the three primitive terms – 'number', 'successor' and '0' – and to show that the postulates can be derived by logic from the definitions. This is what Frege and Russell set out to do, using the idea of one-to-one correspondence introduced by Cantor.

The first step is to ask what numbers are attached to. When I say that Socrates is one, and the Holy Trinity three, of what am I predicating the 'one' and the 'three'? It is immediately clear that numbers are not properties of objects: unity is not predicated of Socrates, as wisdom is. Otherwise, I could infer from the premises that Socrates is one and Plato is one that Socrates and Plato are one. The response to this is suggested by the theory of quantification. When I say that a man exists, Frege argues, I do not predicate existence of a man, but rather of the *concept* man: I say that the concept has at least one instance. (Existence is a predicate of predicates.) Similarly, numbers are predicated of concepts: to say that there are five wise men, is to say that the concept *wise man* is instantiated five times.

We can already see, therefore, how to translate adjectival expressions of number into the language of logic. This is precisely what Russell did, in his theory of definite descriptions, which shows us how to represent 'There is exactly one x such that . . . ' Generalising, we can write:

(0) 'There are no Fs' translates as $\sim (\exists x)(Fx)$

(1) 'There is exactly one F' translates as:

$(\exists x)(Fx \,\&\, (y)(Fy \supset . y = x))$

(2) 'There are exactly two Fs' translates as:

$(\exists x)(\exists y)(Fx \,\&\, Fy \,\&\, \sim (x = y) \,\&\, (z)(Fz \supset . (z = x) \vee (z = y)))$

And so on. But this does not yet tell us how to make sense of the numerical nouns: zero, one, two, etc., and does not give us the means to express mathematical laws and calculations. For that we need to define the numbers themselves.

In order to define the numbers, we must first pass from a concept to its 'extension' – i.e. to the class of things which fall under it. Every concept, Frege supposed, determines a class: for every concept F, there is the class of things which are F. The number of a concept F is the number of members of the class of things which are F. We can know that this number is the same as the number of another concept, without knowing *which* number it is. For the members of two classes can be put in one-to-one correspondence with each other. If for every member of the class of Fs there is a member of the class of Gs and vice versa, then we can say that the two classes are equinumerous, even if we do not know how many Fs there are. Our definition of 'equinumerosity' is a purely *logical* definition, which makes no use of the concept of number.

Nevertheless, we can now *construct* a definition of number by means of another logical idea: that of an 'equivalence class'. Suppose I wish to define the direction of a line in Euclidean geometry. I first of all define 'same direction': *ab* has the same direction as *cd*, if and only if *ab* and *cd* are parallel. I can then define the direction of *ab*, as the class of all lines which have the same direction as *ab*. This is an equivalence class, which fully identifies the extension of the concept: *direction of ab*. From the mathematical point of view I can say everything I want to say about the direction of *ab* by discussing the properties of this class. If I want to define direction in general, then I can say that it is the class of classes which are equi-directional.

Similarly, we can define number, as classes of equinumerous classes. The aim is to complete the definition using only *logical* concepts: concepts whose meaning and extension are determined by the elementary laws of thought. So here is how you might define zero. Consider the predicate 'not identical with itself'. It is a logical truth that this applies to nothing. Hence the class that this predicate determines necessarily has no members. It is the 'null class'. The number zero can then be defined as the number of things x, such that x is not identical with itself (alternatively, as the class of all classes of things which are not identical with themselves):

$0 = \text{df.N}x \sim(x = x)$

('Nx' means the number of x such that . . .)

Having defined zero, we then define the remaining numbers recursively, thus: 1 is the class of all classes equal in number to the null class (for it is a logical truth that there is at least one and at most one null class); 2 is the class of all classes equal in number to the class whose only members are the null class and the class whose only member is the null class. And so on. (We 'build' the numbers from the null class, while making no ontological assumptions whatsoever.) We can also define the successor relation, by means of the existential quantifier. For the number of the *F*s is one more than (i.e. successor to) the number of the *G*s if there exists an *F* such that the *rest* of the *F*s are the same number as *all* the *G*s. (Remember that 'same number as' is defined without references to number.)

The final primitive required by Peano's postulates is the concept of number itself. Frege ingeniously invokes mathematical induction (Peano's fifth postulate) in order to furnish himself with the definition. We can say that x is a natural number if it falls under every concept which zero falls under and which is such that any successor of whatever falls under it also falls under it.

That definition enables us to derive Peano's postulates, and from them the rest of arithmetic. Since Dedekind and Cantor had shown how to derive the whole of number theory from arithmetic, and since number theory *is* what is distinctive of mathematics, we have derived mathematics from logic. Q.E.D.

3. Russell's Paradox

Such, in barest outline, is Frege's theory of mathematics. At the very time that he was producing it, Russell was working along similar lines, towards the same result. However, Russell noticed a paradox which threw the enterprise into doubt. Frege's proof promiscuously relies upon his assumption that, for every predicate *F*, there is the class of things which are *F*. This is supposed to be an intuitive idea of logic, so basic as to require no further justification, like the law that everything is identical with itself. However, it introduces a concept – that of class – which is not on a par with the ordinary notions of logic. Classes constitute new entities in our ontology: they are not just bundles or aggregates, but ordered collections. So why assume that the idea of 'class membership' is a logical idea, like predication?

Worse still: from Frege's assumption a contradiction follows. Consider the predicate 'is not a member of itself'. If classes are legitimate entities, then this predicate can be applied to them. The class of tedious things is surely itself tedious: it is therefore a member of itself. The class of small things is large, and therefore not a member of itself; and so on. If every predicate determines a class, then this must be true of the predicate 'is not a member of itself'. So there must be a class of things which are not members of themselves. Is this a member of itself or not? If it is, then it is not. If it is not, then it is. Which is a straight contradiction.

There have been many responses to this paradox. But none of them succeeds in saving the logicist enterprise, since each requires us to invoke notions which are not part of the self-evident contents of logic. Here are the two most famous responses:

(i) The theory of types. This was Russell's own way round the paradox. All entities are to be arranged in a hierarchy of types, such that, while entities of the same type can belong together in a class, no entities of different type can do so. For instance, there is a class of red objects. But there is no class consisting of red objects together with the class of red objects. The class of red objects belongs to a higher type than red objects themselves. As we ascend the hierarchy the pattern is repeated. There is a class whose members are the class of red things, the class of green things, the class of blue things, and the class of yellow things. But there is no class consisting of all those together with the class of the classes of coloured things. And so on *ad infinitum*. Any attempt to construct a class whose members are of different type is ruled out. We must not introduce hybrid classes, since the result will not be a well-formed expression in our logic. Hence 'the class of classes which are not members of themselves' is not a permissible expression.

There is something intuitively appealing in this. For it echoes the constructivist's idea, that these abstract entities exist because we construct them; we must therefore take care *how* we construct them if we are not to make nonsense. Hybrid classes suffer from the same defect as the sentences 'existence exists', or 'the concept horse is a horse': they surreptitiously apply concepts to themselves, rather than to the things that fall under them. On the other hand, this intuition is at best metaphysical, and certainly not logical.

Furthermore, the foundations of arithmetic can no longer be laid down without further assumptions. We can construct the number zero from the null class by purely logical means. But consider some arbitrary number *n*. How do we know that *n* exists? We could construct it in the manner that I indicated earlier: take the null class – that is one thing; and now the class whose only member is the null class – that is another; and the class whose only member is the class whose only member is the null class – that is another. And so on. This way we could build up a class with *n* members, for any *n*, out of nothing. We can talk about all the numbers, while making no ontological assumptions whatsoever. But the resulting class will be a beastly hybrid, ruled out by Russell's theory from the very second step. So there is no way of constructing the class whose number is *n* without assuming the existence of *n* things of the same type. There must therefore be as many entities in our world as there are numbers. Otherwise mathematics cannot assume the existence of its known subject-matter. This a vast (indeed infinite) ontological assumption. Russell calls it the axiom of infinity. But it is scarcely a truth of logic: maybe not even an empirical truth. In which case the *a priori* nature of mathematics is thrown seriously in doubt.

(ii) Zermelo set theory. The response of Ernst Zermelo was simpler and neater, though in a way less intuitive. The paradox arises, he said, only for certain predicates, such as the one chosen by Russell. We should therefore conclude that there is no such set as the set of objects which are not members of themselves. Some predicates determine sets, some do not: and we can indicate this by saying that there is a set for every legitimate predicate, but not for every predicate. We construct our mathematics from the theory of sets, so defined.

But this new notion of set is very far from Russell's intuitive idea of a class. Its extension cannot be settled by the study of logical notions alone, but only by further applications of mathematical reasoning of the kind that we had hoped to explain. Numbers are reduced to sets, only because sets have become like numbers, in all the ways that make numbers hard to understand. In particular, the sign of set membership, ϵ, becomes a new mathematical primitive, whose properties are given by the theory that is built with it, and not by any excursus into the realm of logic.

4. The Primacy of Mathematics

But the tables can now be turned against the logicist. He began from the confident assumption that predication is a transparent idea, fully intelligible within the framework of Fregean logic. By means of the predicate logic he defined first class membership, and then number. But the paradox casts doubt not only on the definition of class membership, but on the very idea of predication from which it is derived. Frege, addressing the question what '*F*' stands for, is notoriously evasive. A concept, he says; but each concept determines a function. Functions are not complete entities, as objects are: they are 'unsaturated'. So what in the world corresponds to the function *F*? How are functions to be individuated? (We have already encountered this problem in Chapter 12.) In answering this question we may be tempted to move to a third level of interpretation: beyond the concept *F*, and the function *F*, there is the *extension* of the concept: the class of things that are *F*. Now at last we have a tangible piece of reality – something in the world to which the predicate '*F*' is anchored. Unfortunately, however, this elucidation is precisely what is ruled out by Russell's paradox. A contradiction follows from the assumption that every predicate determines a class. So we need a criterion, which will distinguish between legitimate and illegitimate predicates.

In fact we *have* a criterion. We need only add the sign ϵ of set-membership to our list of primitives, and then rule that there is a legitimate predicate for every set. We can define predication in terms of membership: to predicate *F* of *a* is to say that *a* is a member of the set of *F*s. We no longer refer to those strange 'unsaturated' entities, to which the concept of identity seems scarcely to apply, but to real items in the world, whose identity conditions are well understood: for sets are identical when they have the same members.

Thus we explain the fundamental operation of logic – the formation of the subject-predicate sentence – in terms of a *mathematical* operation, whose nature is defined by the axioms of set theory. Set-membership, which is the ground of mathematics, turns out to be the ground of logic too; the laws of set-membership are the laws of thought. In short, logic is not the foundation of mathematics, but a branch of it.

Things are never so simple as they seem. For although that radical line would appeal to a follower of Quine, it has a corollary that few others would wish to accept. The set of renate creatures, for instance, is identical with the set of cordate creatures, since everything with a kidney has a heart and vice versa. But logic, biology and medicine compel us to distinguish the two; and what better way than that proposed by Frege, which is to say that we have *two* properties, determining a *single* class?

We could preserve the advantages of sets over properties, if we could define properties (and not just the extensions of properties) in terms of sets. It is not impossible to do this. Here is one suggestion: property *F* is identical with property *G* if and only if, for every possible world *w*, the set of *F*s in *w* has the same members as the set of *G*s in *w*. But we have now moved into territory where no Quinean would be happy to follow us; and if the only way of defining properties involves us in quantifying over possible worlds, this could be taken as another reason for abandoning properties altogether.

5. Set Theory

The seven axioms of set theory, laid down by Zermelo and his contemporary Abraham Fraenkel, to form what is now known as Zermelo-Fraenkel set theory, are inspired by Frege, but very far from anything that would be endorsed by him. These axioms are not arbitrary, but are supposed to correspond to the most basic intellectual operation – which is that of 'holding things together in thought', as when I count things, group them, or bring them under a single predicate. The axioms are not logical truths, but part of what Russell would call an 'implicit definition' (see Chapter 7): the symbol defined being the ϵ of set-membership. Their truth is established by intuitions which lie too deep for proof, since all proof depends on them.

Consider the 'Foundation Axiom', due to Fraenkel, which says that there are no ungrounded sets: that is, no sets which contain members which contain members which contain members . . . to infinity. This captures our intuitive idea that sets are constituted by, and supervenient on, their members. Sets must 'come to earth' in their members: you have a set only when you have 'got hold of' the things contained in it. There is therefore a set of ravens, but no set of non-ravens. (If there were such a set – N for short – then it would be a non-raven. So N would contain N, which contains N, which contains N . . . *ad inf.*) Is that a solution to Hempel's paradox?

More importantly, we can now see why the Frege-Russell definition of number cannot be sustained. There is no such set as the set of all two-membered sets – i.e. the set which, for Frege, is identical with the number 2. If there were such a set – call it S – there would also be (according to the Pairing Axiom) another set T, consisting of S and Lady Thatcher. But T is a two-membered set: hence S contains T which contains S which contains T which contains . . . ad inf. Once again the Foundation Axiom is violated: the set T is an infinite ladder which never touches the ground; a ladder without a first step, like the one that was offered to Jacob in a dream.

The enterprise of defining numbers as sets of sets must therefore be abandoned; sets *themselves* become the primitives of mathematics. Furthermore, we see that perhaps Kant was right after all. The axioms of set theory are not analytic truths. There is no inconsistency involved in denying the Foundation Axiom. But this axiom is certainly not *a posteriori*; it is known *a priori*, by reflecting on the intuitive conception of membership. Moreover, the test of set theory is that we can derive the whole of arithmetic from it. But arithmetic is unlike geometry: we are reluctant to countenance 'alternative arithmetics', or to imagine possible worlds in which our laws of arithmetic do not hold. For there is no *a posteriori* subject-matter for arithmetic: no analogue of physical space, which is the *a posteriori* subject-matter of geometry. It is therefore not a *contingent* fact that our world conforms to the laws of arithmetic. In short, set theory – which is arithmetic in its final state of undress – is a body of synthetic *a priori* and necessary truth.

6. What are numbers?

So what are numbers? The answer is very hard to find. To say that they are sets is to say nothing, if we cannot say *which* sets they are. Even if we wriggle our way around the Russellian paradox (and it is only one of many such paradoxes), we are still no nearer to saying exactly what the number 3 really *is*. I could take any three-membered set, and *use* it in my calculations as the number 3. But which of the indefinitely many such sets *is* the number three?

The answer is, that there is no answer. But maybe that *is* the answer. In a striking paper ('What numbers could not be') Paul Benacerraf takes the principal legacy of logicism to be the conclusion that, whatever numbers are, they are not objects. There is no object which *is* the number three. We could choose any suitably constructed sequence of sets to serve as the items in our arithmetic. Just so long as they are ordered in the way required by counting (i.e. in the way summarised in Peano's postulates) there is literally nothing to choose between them. If the numbers were definite *objects*, with their own independent existence, there would be a specific set of objects which they are. But no such set can be found; whichever one we light upon, it will be an arbitrary choice from indefinitely many competitors.

This may not be right. But it again strikes a sympathetic chord. For it reminds us that mathematics is a practice, which gets its sense from the activities of counting, measuring, and calculating. The Platonic temptation takes us away from those activities, to a still, untroubled realm, where numbers are not so much used as contemplated, turned over and over under our intellectual gaze, the precious jewels of eternity. But mathematical knowledge is not *like* that. If it were, you could imagine someone who had a perfect intellectual acquaintance with the numbers (or at least, with a passable number of them!) but who was incapable of counting a flock of sheep. Should we not put the practice first, as Kant did, and regard the numbers as so many shadows cast by our activity – which appear to be entities only so long as we do not turn to capture them?

7. Theory and Meta-theory

The philosophy of mathematics took another step forward after Russell and Frege, with the development of meta-theories. These play an increasingly important role in the philosophy of logic, and it is worth knowing something about them.

We may define a theory as a set of axioms, together with rules of inference. A theorem is any formula which follows from the axioms by repeated application of the rules. But a theory is empty until the axioms have been provided with an interpretation. We need to assign values to the primitive terms, and to show how the values of formulae may be derived from the values of their parts. (This 'assignment of values' is part of a theory of truth for a language, as we saw in Chapter 19.) Interpretation is not something done by the theory itself, but something done by *us*, through another theory – the meta-theory. A meta-theory can tell me that a given axiom is independent of the others in a system. (Independence is proved by showing that there is an interpretation which renders the axiom false while the others are true.) A meta-theory can show that a given set of axioms is complete relative to a certain interpretation, meaning that it generates all formulae which are true under that interpretation. And it can show that the axioms are inconsistent – i.e., that there is an interpretation which generates *p* and not-*p* as a theorem. (Alternatively, an interpretation under which every well-formed formula is a theorem: see next chapter.)

We can also envisage meta-theories which will tell us that a particular formula is provable, even though we do not have a proof of it in the theory. If we can show that the axioms generate all truths of arithmetic, and that *p* is such a truth, then we have proved that *p* is provable. But we have't proved it *in* the theory. This gives some comfort to the Platonist, since it suggests that mathematical truth is distinct from any particular proof whereby we establish it. There are truths which are not yet proven: at least not *in the theory*. The final blow to the logicist programme was struck by Gödel, in his famous meta-mathematical proof that there can be no proof of the completeness of arithmetic which permits a proof of its consistency, and vice versa. We cannot know, of some system of axioms which is sufficient to generate arithmetic, that it is both complete and consistent. Hence there may be formulae of arithmetic which are *true*, but not provable. It seems to follow that no logical system, however refined, will suffice to generate the full range of mathematical truths. It follows too that we cannot treat mathematics as Hilbert had wished, merely as strings of provable formulae: the theory of 'formalism' is false.

Speculation about Gödel's theorem can be entertaining – as you can discover from Douglas Hofstadter's *Gödel, Escher, Bach*, a book which also makes some outrageous comparisons (such as those implied in the title). But it is a field in which the amateur must tread very warily. Let us therefore merely content ourselves with the thought that Platonism still has some life in it. If there can be unprovable truths of mathematics, then mathematics cannot be reduced to the proofs whereby we construct it. There is a realm of mathematical truth, whether or not we can gain access to it through our own intellectual procedures. And the extraordinary thing is that this *too* is something that we can prove. It is a remarkable fact, that we can rise above our own limitations, and project our thought into the very regions where it cannot freely wander. (But perhaps this is just one of those mathematical conundrums, which reflect the artificiality of the world of numbers? Such, at any rate, would be the retort of the constructivists.) However we approach this world, we cannot fail to be struck by a singular fact. That of which we are most certain, and which has provided philosophers since antiquity with their paradigm of knowledge, is also a realm of paradox. And the solution to each paradox seems to generate another, to the point where we feel impelled to doubt the ojective validity of the only truths that have never been doubted. What then *is* a paradox, and why is paradox to be avoided?

27

Paradox

Paradox means a proposition which is 'contrary to belief', and paradoxes are of two kinds: those that defy some familiar orthodoxy (usually without sufficient explanation); and those that begin from intuitively acceptable premises and derive from them a contradiction – something that cannot be true. It is the second kind of paradox that has been most interesting to modern philosophers, since it provides a kind of objective test of the cogency of a philosophical system: a contradiction is a *reductio ad absurdum* of the ideas that produce it.

Or is it so simple? What if reality *itself* is paradoxical? In that case, surely, we might have more sympathy for a philosophy that recorded the paradox, than for one that resolved it. This was Kant's hope, in developing his philosophy of freedom. All rational beings could be brought to see, he believed, both that they really *are* free, and that the postulate of freedom is, from the point of view of the intellect, inherently paradoxical. Others, in more mystical vein, have shown a willingness to embrace paradox in every sphere, as a sign of human limitations, and of our need for divine guidance and revelation.

Paradox has therefore had a history in Western thought that is not accurately recorded by its place in modern philosophy. Although the pre-Socratics studied paradoxes of time, space and motion, and regarded the desire to resolve them as the prime motive of metaphysics, paradox achieved its true sovereignty with the birth of the Christian religion. Christian doctrine consciously embraces paradox – the Incarnation of God, and His Crucifixion. Already in the epistles of St Paul you find these paradoxes presented in all their raw untenability as a test of faith. (The gospel is 'folly to the Gentiles, and to the Jews a stumbling block'.) The Church Father Tertullian wrote *credo quia absurdum est* – I believe because it is absurd, referring explicitly to the Crucifixion. Far from regarding this as a refutation of the Christian religion, his successors acknowledged in this utterance the true principle of religious belief.

Maybe all religions have their store of precious paradox. But paradox is not the same as mystery. It sets out to thrill and also to undermine. It is a destabilising force, and also a strange invitation to commitment. There is something in the human psyche which, faced with an unbelievable proposition, rushes forward to embrace it, to say 'yes, it *must* be so!', and to rejoice in the ruin of common sense that follows. A paradox may therefore be an act of defiance, in which the world of ordinary things is set at a distance and ridiculed.

Paradox has therefore had an important place, not only in religious thinking, but also in revolutionary politics. Rousseau, in the *Social Contract*, defiantly tells us that we must be 'forced to be free'; while his disciples, Robespierre and St Just, advocated a 'despotism of liberty' as they laughingly chopped off the heads of those who had shown themselves incapable of enjoying it. Such paradoxes abound in the literature of revolution: 'property is theft'; 'right is a bourgeois invention, so it is right to resist it'; 'repressive tolerance'; 'human will is the effect but not the cause of history; so make history!'

Maybe paradoxes in oriental religions are of another kind: not contradictions, exactly, but baffling vacuities, like the puzzles of the Zen philosopher: places where thought suddenly evaporates, and there is nothing. This is settling rather than unsettling. Paradox in our tradition sets the world against itself: the world eats itself up before our eyes; and that is the point of paradox. It seems to show the victory of thought over reality; 'I can believe anything,' it says, 'even this. Join me!'

'The thinker without a paradox,' writes Kierkegaard, 'is like a lover without feeling; a paltry mediocrity'. (Ch. III of *Philosophical Fragments*.) 'The highest pitch of every passion is to will its own downfall; and so it is also the supreme passion of reason to seek a collision, though this collision must in one way or another prove its undoing . . .' And he is prepared to go a stage further even than Tertullian: 'deepest down in the heart of piety lurks the mad caprice which knows that it has itself produced its God'. That, surely, is the origin of paradoxism: the desire to refute the world, in the agonising suspicion that it will otherwise refute you, by disproving your God.

But Kierkegaard is referring also to another kind of paradoxism: that brought into the world by Kant, and magnified by Hegel, according to which reason falls into contradiction, not through disobeying its own laws, but through obeying them too strictly. It is reason that leads us to paradox; and reason that must overcome paradox, in a transcending judgement that would be impossible, were it not for the contradiction from which it springs. This is Hegel's master-thought, and it made paradox for the first time wholly respectable. Hence the secret of Hegel's success in an era of declining faith: the paradox, he preached, lay in the nature of things, and the difficulty of believing whatever might console us lies only in the torpor of our own mental powers. 'What are young girls coming to?' asked the White Queen. 'In my day we would believe six impossible propositions before breakfast.'

1. Paradoxes of Implication

The arch paradoxist, Walt Whitman, stepped inadvertently into logical discourse, with his famous verses:

> Do I contradict myself?
> Very well then I contradict myself,
> (I am large, I contain multitudes.)

Here, in a nutshell, is the so-called paradox of strict implication. There are many paradoxes of implication, and they are interesting partly because they are paradoxes of the first kind: they do not involve a contradiction, but merely a thought that is highly counter-intuitive. You don't *have* to solve them; but your work will be harder if you don't.

Consider material implication, defined in terms of the truth-table thus:

p	$\supset$	q
T	T	T
T	F	F
F	T	T
F	T	F

This definition has the consequence that a true proposition is implied by any premise, and a false proposition implies any conclusion. Many people therefore argue that it cannot be a definition of *implication*, as normally understood. For one thing, we are always working out the consequences of false propositions – as in counterfactual reasoning – just as we are always distinguishing among true propositions between those which genuinely *follow* from our premises and those which do not.

At the same time, many logicians have argued, material implication can play exactly the role in inference that we associate with implication. In particular, it passes the crucial test, which is that of validating *modus ponens*. The proposition $(p \& (p \supset q)) \supset q$ is a truth-functional tautology: i.e. a logical truth. (But see below: Achilles II.) Such logicians have therefore been prepared to swallow the 'paradoxes'. If we don't mean *material* implication, then we don't really know *what* we mean by 'if'.

But we have another and better definition, argues C.I. Lewis: *strict* implication. The word 'if' suggests a connection between premise and conclusion, of the kind that validates an inference. In inference we hold that the premises *cannot* be true and the conclusion false. So that is how we should define 'if'. 'If p then q' means: 'It is logically impossible that p and not-q'; alternatively 'It is logically necessary that p materially implies q'. (In symbols: $\Box(p \supset q)$.)

But this definition also generates a paradox, namely, that from a logically impossible proposition everything follows (Whitman's 'multitudes'); and a logically necessary proposition follows from everything. Surely we don't want to accept that? Again, however, this is a paradox only in the *first* sense: there is no contradiction involved in accepting it; indeed, there is a considerable literature devoted to proving that we *must* accept it, and that we have no other conceivable definition of deducibility or 'entailment' that is not inherently *more* paradoxical (for example, by making entailment into a non-transitive relation, so that from p entails q and q entails r it does not follow that p entails r). You might say that it is an important logical *discovery*, that from an impossible proposition everything follows: a deep truth about impossibility, which shows exactly why we must avoid it – why paradoxes in the second sense are an ontological catastrophe. Had Hegel been apprised of this deep truth he would have been aware of just *why* his 'proofs' seemed to follow so unresistingly wherever he desired them to go. It also gives us a new test for consistency: an axiom system is inconsistent if every formula is a theorem. (See last chapter.)

Not everyone is happy with that. Dorothy Edgington has constructed arguments to show that conditionals are not to be understood in terms of their truth-conditions at all. If we try to do so, then

we end up attributing to people inconsistent beliefs. Consider the following answers to a questionnaire:
(1) The Conservative Party will win. Yes.
(2) Either the Conservative Party will win or *p* for any *p*. Yes.
(3) If the entire cabinet is involved in an horrendous scandal, the Conservative Party will win. No.
On the assumption that conditionals have truth conditions, this set of replies is mutually inconsistent. Mrs Edgington's solution is to propose that 'if' be analysed in terms of probability: that there is an assumption, in the use of this word, that the premise is the *ground* for the likelihood of the conclusion. Conditional judgements are judgements about what is the case *under a certain supposition*; and there is no way of reducing them to judgements about what is true, *simpliciter*. Hence conclusions and premises cannot be detached from each other, in the way required by an analysis in terms of truth conditions.

The area is an interesting one, but to pursue it further would distract us from the *real* paradoxes that have played such an important role in philosophy, ancient and modern.

2. Real Paradox

By real paradox, I mean paradox of the second sort: an argument that leads by rational steps to a contradiction, starting from premises that we intuitively accept. The ancient approach to paradox, exemplified by Parmenides and his pupil Zeno, was that the real paradoxes were genuine contradictions, derivable from our common-sense view of the world. They were therefore proof of the untenability of the common-sense view. They showed that reality is quite distinct from appearance, and that it is only the philosopher who can tell us what is really real. This meant that the philosopher could charge a great amount for his lessons. (See *Perictione's Parmenides*, 333 D.)

That approach survived into our century, with F.H. Bradley's arguments about appearance, and McTaggart's proof of the unreality of time. (See Chapters 10 and 25.) But it gradually lost out against the defenders of common sense, who argued that the premises from which the paradoxes seem to follow are so firmly rooted, that we should not know where we are if we rejected them. Paradoxes like those of McTaggart must be resolved, by adjusting our language and our concepts so as to retain our intuitive picture of the world. The alternative is not that envisaged by Parmenides and Bradley, in which we penetrate through the illusory veil of appearance to see the one reality. The alternative is knowing nothing at all. The ordinary world of common sense and science is the only world we can hope to know.

Nevertheless, the ancient paradoxes are still with us, and it is useful to begin with three of them.

3. The Liar

This famous paradox, which tormented many ancient philosophers, and is reputed to have caused the death of one of them (Philetas of Cos), is as hard to solve as it is easy to state. Consider the person who says 'What I am now saying is false.' Is what he says true or false? If it is true, it is false; if false, true.

How do we respond to this paradox? At first sight you might dismiss it as a trick. Surely, you might say, the sentence in question is just *deviant*: we should simply rule it out as ill-formed, or as saying nothing, or as a grammatical freak. So one suggestion has been to deny that any sentence can apply the predicate 'true' to itself. The predicate 'true' is used in the evaluation of *other* sentences. But this does not get round the paradox, since we can reconstruct it in the following form:

1. The sentence written below is false.
2. The sentence written above is true.

If 2 is true, then 1 is true; but if 1 is true, then 2 is false. So if 2 is true it is false. This and similar examples can be used to show that there is no *syntactic* solution to the liar paradox, no way of ruling out the offending sentences simply by virtue of their grammar. The paradox is a *semantic* paradox: one that derives from the meaning of the truth-predicate itself. And it can be resolved only in the context of a theory of truth.

Another response to the paradox is to deny the principle of *bivalence* – which says that every sentence is either true or false. Consider the sentence:

L: L is false.

This encapsulates the paradox in minimalist idiom. We can at once deduce that

(1) If L is true, then it is false;

(2) If L is false, then it is true.

Assuming that anything that is false is not true, and anything true not false, we can infer:

(3) If L is true, it is not true;

(4) If L is false, it is not false.

It is a rule of inference, however, that if something implies its own negation, we can infer that negation. So from (3) we can infer that L is not true, and from (4) that it is not false. So we can conclude that

(5) L is neither true nor false.

So long as there are 'truth-value gaps': i.e. gaps between truth and falsehood, there is no paradox. If there are such gaps, sentences like L are very good candidates for occupants. For they have no grounds; they purport to assert the falsehood of a sentence which does not reach beyond itself, which makes no contact with a reality that would *account* for its falsehood.

The response is appealing; but it encounters a new version of the paradox, known as the Strengthened Liar. Consider the sentence:

L′: L′ is not true.

Even if we deny the principle of bivalence, we can derive the contradiction: if L′ is true, it is not true; if it is not true, it is true. Mark Sainsbury (*Paradoxes*) envisages a sophisticated solution to this, based on the assumption that the predicate 'true' can be correctly neither affirmed nor denied of the sentence L′. But there is another general response to the paradox that we should consider, which is Tarski's.

Tarski's idea is rather like Russell's response to the paradox of class membership. Just as Russell arranged classes into a hierarchy of types, so Tarski proposes a hierarchy of languages. Truth can be defined for one language, but only in another: hence no sentence in any language can predicate truth of itself, or indeed of any other sentence in the language.

Tarski's response implies that there can be no theory of truth for a language that is not framed in a 'meta-language'. The meta-language, however, can permit us to formulate theorems about the 'object-language', of a kind that transcend the expressive capacity of the object-language itself. Tarski developed this idea in ways that anticipated Gödel's theorem. And the fruitfulness of his distinction between object-language and meta-language confirmed his intuition that truth can only be conceived in this way, in terms of a potentially infinite hierarchy of languages, each of which applies a concept of truth to the one beneath it, but none of which applies that concept to itself.

Two unwelcome consequences follow: the first is that our ordinary concept of truth, used in our language to range over its own sentences, is confused and indeed contradictory. The second is that the word 'true' becomes infinitely ambiguous, changing meaning systematically as we ascend the hierarchy of levels. This constitutes an assault on our intuitive conception of reality (which is, after all, what the concept of truth refers to), of Parmenidean proportions.

You could also imagine a Zermelo-like response to the paradox: when a use of the word 'true' produces a contradiction, rule it out. Roughly speaking, that is the approach of those who argue that Tarski's theory of truth can be generalised to cover natural languages (something that Tarski himself denied, on account of the Liar). This is not a solution to the paradox, but merely a way of confining it, so as to attend to other matters. Indeed, a solution seems to be as far from sight today as it was in the days of Philetas of Cos; rather than share his fate, we had better move on.

4. The Heap

The paradox of the Heap, known after the Greek word for heap (*soros*) as the *sorites* paradox, is as venerable as the Liar, though perhaps less deadly. We are inclined to think that, if a certain number n of grains of sand make a heap, it will not cease to be a heap merely because one grain is taken away. Hence $(n)(n$ is a heap $\supset .\ n - 1$ is a heap$)$. From this you can at once deduce, by recursive application of that formula, the conclusion that, if any number of grains of sand form a heap, so do all lower numbers, right down to zero. Conversely, if n grains are *not* a heap, neither are $n + 1$; from which, of course, you get the opposite result, that if any number of grains of sand is not a heap, then the same is true of all higher numbers, up to infinity. From those two conclusions it follows that every collection of grains is both a heap and not a heap.

That conclusion involves a contradiction: someone might reply that it does not follow, since the argument might be taken to prove merely that the predicate 'is a heap' is meaningless; in which case nothing is either a heap or not a heap. The problem with such a response, however, is that the argument can be replicated for a vast number of predicates in ordinary language, whose utility depends precisely on the fact that, like the predicate 'is a heap', they do not have precise boundaries: consider 'is red', 'is soft', 'is tall'. Indeed, just as the Liar casts doubt on the concept of truth, so does the Heap cast doubt on that of predication. It seems to imply that the very application of our ordinary predicates involves a contradiction. Perhaps we could overcome the contradiction, by inventing some new predicates without vague boundaries. For example, we could redefine red in terms of the wave length of light emitted by red things; we could give a precise ruling as to which collections of grains will be heaps, and so on. But this would have two disastrous consequences: first, that we could no longer be sure how to apply our predicates, except after laborious scientific inquiries and experiments that time and competence forbid. Secondly, that we should have to rely on the *truth* of the scientific theories employed in the definition of our new and exact predicates, even though the evidence for those theories comes from comparing them with the world as perceived – i.e. the world as described by our *intuitive* predicates, the very predicates that we are trying to reject.

The *sorites* paradoxes are many, and all are said to relate to the *vagueness* of our ordinary predicates. To say this is not to solve the paradox, but merely to describe it. How *would* we solve it? As with every paradox, we have three possible moves: to reject the premises, to reject the reasoning that leads from them, or to 'explain away' the paradox, by showing that the apparent contradiction is not a real one. It is obvious that the reasoning cannot be faulted: it is based merely on *modus ponens*, and involves no tricks or stratagems. Nor can the conclusion be accepted, for reasons that I have just hinted at. So which of the premises should we reject?

Someone might reject the idea that vague predicates have boundaries. He might say that, while there are some things which are definitely heaps, and other things which are definitely not heaps, there is an area in between the two where we cannot say either that something is or is not a heap. So we should replace the predicate 'heap' with a new and sharper predicate, which distributes those things which are neither heaps nor not heaps to either side of a clear dividing line. We sharpen the predicate 'heap' into 'newheap', and of *this* predicate it is true that something either definitely falls under it, or definitely does not. Using this new predicate, there will be a number n, for which the principle: $(n)(n$ is a newheap $\supset .\ n - 1$ is a newheap$)$, is false. This, in outline, is the 'supervaluational' account, which owes its barbarous name and its refined nature to Kit Fine. The problem is that it wriggles out of the paradox by abolishing vagueness altogether. *If* there is vagueness, then the principle $(n)(F(n) \supset F(n-1))$ seems to hold. But if there is no vagueness, how can we latch on to our predicates, and apply them spontaneously on the basis of our ordinary perceptions?

Another approach is to renounce the idea that every proposition must be absolutely true, if it is to be true at all. Maybe we should speak instead of degrees of truth, and say that, as we heap grains of sand on to the pile, it becomes *more and more* true that the collection is a heap. This certainly dissolves the paradox. Some argue, however, that the price of the solution is too great to pay. We can no longer assume, they suggest, the validity of *modus ponens*, since this is defined only for propositions with absolute truth-values; nor can we assume the tenability of classical logic, which is likewise built up on an absolute conception of truth. (By classical I mean the two-valued Fregean logic with quantification, that I introduced in Chapter 6.)

Advocates of the 'degrees of truth' solution sometimes deny those charges. Dorothy Edgington, for example, believes that all normal reasoning involving *modus ponens* must be understood in terms of probabilities. 'If p then q' should be understood as 'On the supposition that p, it is very likely that q'; but only in certain special contexts (such as arise in mathematical reasoning) can we replace 'very likely' by 'certain'. On the supposition that n grains make a heap, then it is very likely that $n - 1$ grains make a heap also: maximally likely, short of absolute certainty. But the tiny measure of uncertainty that creeps into the inference is inherited by the conclusion, and passed on down the chain. In other words it is just a little *less* certain that $n - 2$ grains make a heap, and that much less certain that $n - 3$ make a heap. Mrs Edgington compares the paradox with another, in the theory of probability, the so-called 'Lottery Paradox'. Many tickets are sold in a lottery, only one of which will win. It is therefore highly probable of any ticket, that it will not win. We can therefore construct an argument with the following premises: 'Ticket 1 will not win'. 'Ticket 2 will not win', 'Ticket 3 will not win', and so on, each of which is highly probable. But the conclusion of this argument – 'No ticket will win' is certainly false. Such a paradox is clearly only apparent, and would be solved by any theory of probabilistic inference worthy of the name. Likewise with the paradox of the heap. We should understand the principle which generates the paradox – $(n)(n \text{ is a heap} \supset . n - 1 \text{ is a heap})$ – as saying that, for any number n, on the supposition that n is a heap, then it is maximally likely that $n - 1$ is a heap also. Given enough applications of that principle, the grain of uncertainty will itself become a heap.

The argument lies beyond our scope; but it is proof of the importance of the paradox, that a solution may oblige us to acknowledge that everyday reasoning is very different in structure from the reasoning that occurs in mathematics.

5. Achilles I

Zeno's paradoxes of motion have been a source of anxiety and delight ever since he first proposed them during the fifth century BC. His original intention was to show that motion and change – maybe time itself – are contradictory ideas, and that reality is changeless. Maybe he did not succeed in showing that; but he did succeed in raising tenacious doubts about infinity. Aristotle, to whom we owe our account of Zeno's arguments, was persuaded that they must be taken very seriously; and they have been paid the same compliment by Kant, Hegel and Russell, although for different reasons. There are three of these arguments: the Stadium, the Arrow, and Achilles and the Tortoise. For the sake of brevity I deal only with the last.

A tortoise challenges Achilles to a race. Having laughed at the animal's impertinence, the hero finally consents to run, accepting the tortoise's plea that, as a beginner, he should be allowed a few paces' start. Achilles has no difficulty in reaching the place at which the tortoise began. But during the time he takes to get there, the tortoise has advanced to a new position. Achilles reaches that position too, only to find that the tortoise has advanced yet again, by a small but real amount. So Achilles must advance further, to the new position of his rival, who, however, is no longer to be found there, but just a little bit further on. And so on, *ad infinitum*. That this is the beginning of an infinite series is something that even the worst mathematician will grasp. So how on earth is Achilles ever going to close the gap between himself and the tortoise in a finite time?

As we know, there are infinite mathematical series which sum to a finite number. For example the series $\frac{1}{2}, \frac{1}{4}, \frac{1}{8} \ldots$ sums to 1. Does that not solve the problem? Does it not show that Achilles can, by running a finite length in a finite time, cover infinitely many decreasing intervals? Well, it would do so, if space were just like the continuum of numbers. But we cannot assume that. For one thing, we cannot assume that space is infinitely divisible: maybe there comes a point where we have reached the smallest possible (i.e. physically possible) unit of space: the width of a proton or a quark, say. On the other hand, that too would be a solution of the paradox, since it would suggest that Achilles does *not* have to perform an infinite task: there are only finitely many moves through space that the tortoise can make, and therefore it suffices that Achilles also make them, only in less time than the time taken by his rival.

Neither of those solutions has commanded universal assent. The second, in particular, is unsatisfactory, since it depends upon an empirical assumption that has yet to be proved. It is more acceptable to suppose that space is infinitely divisible than that it is not. And this suggests that movement through space will always involve the completion of infinitely many journeys: such is the essence of Zeno's Stadium paradox. The paradox has been generalised by J.F. Thompson, to introduce the idea of a 'mega-task': a task involving the completion of infinitely many other tasks. Life seems to abound in mega-tasks, and – short of concluding that life is impossible (which is not unlikely) – we must understand how such tasks can ever be completed. Thompson gives intriguing arguments to show that the assumption that a mega-task has been completed actually involves a *contradiction*.

6. Achilles II

The very same J.F. Thompson proposed a standard solution to the second Achilles paradox, due to Lewis Carroll, who published it in *Mind* at the end of the last century. (Whatever you think of *Mind*, it has had a distinguished history.) Achilles, being swift of foot but slow of wit, catches up with the wise tortoise who will advise him about an intellectual problem. He has been told that Paris has eloped with Helen, and also that if Paris has eloped with Helen there will be war. Is he entitled to conclude, therefore, that there will be war? Why yes, responds the tortoise, if he can proceed from p and p implies q to q. But surely that is true, says Achilles. So let us add that too to our premises, rejoins the tortoise, in which case we have the following three premises:

(1) p
(2) p implies q.
(3) (p & p implies q) implies q.

Does q now follow? Why yes, says the tortoise, if we can assume that (1), (2) and (3) together imply q. So let us add that too to our premises:

(4) (p & (p implies q) and ((p & p implies q) implies q))) implies q.

Does q now follow? Why yes, if we can assume that (1), (2), (3) and (4) together imply q. So let us add that too to our premises. And so on, *ad infinitum*. Here is an infinite task that can literally *never* be completed. Inference is impossible, and Achilles, once again defeated, retires into the slow-witted discontent from which he resolves never again to emerge.

This is not a real paradox, since it has a solution: namely, to distinguish the premises of an argument from the rules of inference. If rules of inference are translated into premises they become inert: logical truths, perhaps, but incapable of *doing* anything. *Modus ponens* is not the same as the logical truth specified in (3) above, but is *sui generis*. It is a set of instructions for *moving on* from one set of formulae to another. To put it another way, *modus ponens* is not a statement in the object-language, but one in the meta-language. And in understanding this, we have understood a profound truth about logic.

But it is a disquieting truth, too. For it suggests, like Tarski's solution to the Liar, that we cannot think in only one language. Without the capacity to rise above our own thought, and to consider it from an outside perspective, we cannot even learn how to think. What a strange thing, then, is thinking.

7. Paradoxes of Infinity

The first Achilles is one of a whole family of paradoxes surrounding the concept of infinity: a concept deployed not only in mathematics, but also in speculative theology, which cannot tolerate the thought of a finite God. The first Achilles paradox can be expressed very simply thus: how can an infinite series (for example, the series $1 + \frac{1}{2} + \frac{1}{4} + \frac{1}{8} + \ldots$) add up to a finite sum (namely 2)? Surely, if you put together infinitely many quantities, however small, the result must be infinitely large? Similar puzzles arise concerning the physical world. We are apt to think that space, and the objects contained in it, are infinitely divisible. But that seems to imply that they have infinitely many parts, each of which is of a finite, though vanishing, size; in which case, should they not be infinitely large? We believe too that time stretches infinitely in both directions: forward into the future, and backward into an infinite past. That last idea has proved particularly puzzling. For it seems to imply, as Kant argued in the *Critique of Pure Reason*, that an infinite series has come to an end: which is surely an absurdity! (Kant tried to show that it was equally absurd to suppose that only a *finite* time had elapsed up until now: so that there is a real paradox, for which a solution must be found – see the great chapter of the *Critique* called the 'Antinomies of Pure Reason'.) Wittgenstein has a similar puzzle. Suppose I were to come across a person talking to himself in a state of exhaustion, saying '. . . five, one, four, one, three – phew! I've done it!' And suppose that, on being asked *what* it is that he has done, he replies 'Recited the complete decimal expansion of *pi*, only backwards'. Would we know what to make of this? It seems that we have an idea of how to *begin* an infinite series, but not of how to *end* it. Yet, as the Achilles paradox shows, infinite series are everywhere coming to an end, and may take only a second from start to finish.

Aristotle hoped to solve the many paradoxes of infinity by distinguishing the 'potential' from the 'actual infinite'. There is something inherently absurd, he supposed, in the idea that an infinite task has been accomplished, or an infinite distance traversed; for these suggestions force us to think in terms of an actual infinity of finite things, lying complete and whole in finite space and time. But there is nothing absurd in the thought that we could begin some task that has no end: the task, for example, of repeatedly halving some spatial object. An infinite series is simply one which, however far you proceed along it, remains incomplete. And that is the concept of infinity that we deploy in mathematics.

This idea of infinity has appealed to many philosophers, Kant included: a set is infinite if, however many members you list, there are still more to come. The infinite in mathematics corresponds, therefore, to an operation which is never complete. But does that really solve the paradoxes? Unfortunately new difficulties arise, with this very idea of an infinite set. Consider the series of natural numbers, and the series of even numbers. These can be paired as follows:

1 2 3 4 5 . . .

2 4 6 8 10 . . .

Each series proceeds to infinity, and yet, intuitively, we believe that the second series has half as many members as the first. After all, for every member of the first series that is included in the second, there is one of the first that has been left out. And yet each member of the first series is 'paired off' with a unique member of the second. The two series are 'equinumerous'.

Two philosophically-minded mathematicians addressed this paradox during the last century – Richard Dedekind and Georg Cantor. Dedekind argued that there is no contradiction in the idea that the series of natural numbers should be equinumerous with a sub-set of itself. The air of paradox derives merely from the fact that we are still thinking of the two series in finite terms. Indeed, Dedekind went further, and proposed that we *define* an infinite set in precisely this way: as a set that could be put in one-to-one correspondence with a proper sub-set of itself. ('Proper' means that we do not count the set itself among its subsets.)

Using that definition, Cantor developed a complete mathematics of infinity, and was both amazed and disturbed by what he found. For it became possible to prove that there are infinite sets which are *not* equinumerous with each other. Indeed, there seem to be infinitely many infinite numbers, which can be arranged like the natural numbers, according to increasing size. He devised an ingenious proof – the famous 'diagonal argument' – to show that the set of real numbers is greater than the set of rational numbers. Here it is:

Take all the real numbers between 0 and 1. Each can be expressed by means of an infinite decimal expansion, which may terminate in an infinite series of zeroes. Thus $\frac{1}{3}$ is 0.3333 . . . ; $\frac{1}{2}$ is 0.50000 . . . ; the square root of two minus one is 0.4142 . . . ; and so on. Suppose we begin to pair off the natural numbers with this set of real numbers, choosing one natural number for every real number between 0 and 1. For example,

0 0.1029 . . .
1 0.3333 . . .
2 0.4142 . . .
3 0.5000 . . .

and so on. We can now construct a new number, by going diagonally across the infinite square of numbers on the right-hand side. The first number has 1 as its first digit; the second has 3 as its second digit, the third has 4 as its third digit, and the fourth has 0 as its fourth digit. Suppose, therefore, we replace each of those digits with another: putting 2 wherever there is a 3, and 3 wherever there is not a 3. This will differ in at least one digit from every other real number listed on the right-hand side. In other words, it will not be contained in the list of real numbers that has been paired with the natural numbers. Hence there are more real numbers between 0 and 1 (and therefore more real numbers) than there are natural numbers: the two sets are not equinumerous.

By similar arguments you can show that there are ever higher orders of infinity, which can be rationally discussed, and deployed in mathematical proofs, without ever falling into contradiction. This suggests that the old Aristotelian way of dealing with the infinite is inadequate: we can no longer treat infinite quantities in terms of tasks that are never completed; for that would compel us to say that infinite numbers are all of a piece; whereas some are larger than others. So how *do* we describe the infinite?

Cantor himself was depressed by his theories; as a profoundly religious man, he had hoped to show that we can grasp the concept of the infinite, and understand it exactly in the sense that theology requires. But the dizzying abyss that he came across, once he had unlocked the secret of the diagonal, seemed to throw all theology into confusion. Unable to think about this new kind of infinity, and unable to think about anything else, Cantor sank into a pit of melancholy from which he never again emerged. But are his results paradoxical? Surely only in the first sense, of being counter-intuitive. There is no contradiction involved in the idea of an infinite set, however surprising the results that flow from it. On the other hand, the existence of such sets was a spur to philosophy. It became imperative to explain what we *mean* by infinity in mathematics, and how we understand the proofs involving infinite quantities. This was one of the principal motives for the 'constructivist' theories of mathematics discussed in the last chapter.

8. The Role of Paradox

As such examples show, paradoxes have an important place in philosophy. It is to them that we owe our recognition that certain paths are closed to us; and it is the effort to avoid them that is the prime mover of logical argument. That is why the discovery of a paradox is always greeted with such excitement. Consider Hempel's paradox about the ravens. That has proved to be one of the most fertile discoveries in the philosophy of science: it leads us to see that scientific method cannot be summarised in the principle of induction alone. Consider Goodman's paradox about grue or Wittgenstein's isomorphous paradox about rule-following: these are the source of philosophical discontent with the naïve approach to meaning, and the spur to a wholly new conception of language, as an activity which can be understood only by an *a priori* view of our condition.

Not all paradoxes are so wonderfully illuminating: a useful exercise, as the White Queen would surely agree, is to see just how many of them you *can* accept, and how your view of the world must change in doing so. The reader should consult the important first appendix to Mark Sainsbury's well-crafted book (see Study Guide), and try to reach the end of it before going the way of poor Philetas.

28

Objective Spirit

The paradoxes discussed in the last chapter arise in the thinking of a single rational being, who follows the principles of deductive argument, to a conclusion that is against reason. If he is rational he will not accept the conclusion. But people are not always rational, and often their irrationality is motivated: they acquire a *will to believe* what is strictly unbelievable.

Irrationality is manifest also in action and desire. I may seriously pursue what I have better reason to avoid, even when aware of the reasons; and I may desire to do something that my rational nature recoils from, and which I should wish not to *have* done, once I have done it. These are paradigm cases of irrational motivation, a subject that has attracted considerable attention from recent philosophers, since it raises important questions about the nature of practical reason.

There is a third kind of irrationality, which arises when two or more people embark on a common project. Here they may settle their disagreements by a variety of ways: and only some ways of settling them are rational, by which I mean only some ways tend to further their co-operation and the common goal which requires it. Which ways are these? Here we enter the domain of political philosophy, which is the primary subject matter of the present chapter. How are individual choices to be rationally combined? And what kind of order is best for rational beings?

The problems are brought into stark prominence by the so-called 'prisoner's dilemma', which illustrates the way in which rational agents, making entirely rational choices, may nevertheless defeat their own best interests, once there is 'another player' in the game. Two partners-in-crime have been arrested and placed in separate cells. They can be convicted of robbery, if prosecuted; and could be convicted of murder, if either were to confess. Each is promised by the police that both charges against him will be dropped if he alone confesses to murder, thus condemning the other. If both confess, each will get a reduced sentence for murder. If neither confesses, both will be convicted only of robbery. Each reasons that he does better to confess, if the other confesses; and also better to confess, if the other does not. So each confesses regardless, and so helps to produce an outcome that is worse for both.

Prisoner's dilemmas arise whenever rational choice depends upon the choice of another who is rational and self-interested. And such dilemmas are not confined to 'games' with only two 'players'. (The use of these terms is natural, since the subject belongs, formally speaking, to what is known as 'game theory'.) Consider the problem posed by the dwindling stocks of fish in the ocean. Each nation reasons that the others cannot be trusted to keep to any agreement, so it is better to catch as many fish as one can, while there are still fish to catch. The outcome is worse for everyone, since soon there will be no fish. But each nation is 'imprisoned' by its own distrust, and therefore chooses to catch as much as it can.

1. The Social Contract

Such is the condition of man, argued Hobbes, in a state of nature, which is a condition of 'war of every man against every man'. In this condition, the life of man is 'solitary, poor, nasty, brutish and short', not because men behave unreasonably, but on the contrary, precisely because they behave as reasonably as they can in a situation of mutual competition and mutual distrust. The solution to this dire predicament is to 'combine' into a totality, which is the 'Leviathan' of Hobbes's title. Men must come together on terms, and regard the resulting agreement as binding. Hence they must establish a social contract. The idea of such a contract has its origins in Plato, though Hobbes derived it from the French theologian and legal theorist, Jean Bodin.

Many philosophers have based their political philosophy on the social contract. But there has been little agreement among them over its terms. Here are some versions:

(i) *Hobbes*. Having witnessed the civil war and the collapse of English society into near anarchy, Hobbes was persuaded that no contract would be of any value if it could not be enforced. Provision for enforcement must therefore be part of the *terms* of the deal. Rational beings in a 'state of nature' would contract to establish a *sovereign*, who may be a single person, or a parliament of people, but who in any case would have supreme and absolute authority to enforce the terms of the original agreement. Since the sovereign would be the creation of the contract, he could not also be party to it: he stands *above* the social contract, and can therefore disregard its terms, provided he enforces them against all others.

This is why, Hobbes thought, it was so difficult to specify the obligations of a sovereign, and comparatively easy to specify the obligations of the citizen. He was deliberately cagey about the question whether rebellion would ever be justified, partly because he believed the Cromwellian rebellion had caused such untold misery that it would be dangerous to suggest that we might legitimately step down that path again.

Hobbes regarded his sovereign as the enforcer of a system of laws, and law as necessary to a well-ordered commonwealth. Obedience to the law was the primary cost of membership, and the citizen contracted to offer this obedience in exchange for the benefits brought by civil peace. Civil disobedience, on this view, could never be justified by the original contract.

(ii) *Locke*. In his *Second Treatise of Civil Government* Locke developed a revised version of the 'social compact', as he called it. Less cynical than Hobbes, he regarded men in a state of nature as still able to recognise and obey the law of God, or 'natural law'. In entering the original contract they are not exchanging war for peace, but their natural freedom for the advantages of civilised life. 'Civil Society', which is the result of the social contract, must respect the rights that we have by natural law, which are inalienable. These rights include the right to life, limb and property, and the elementary freedoms required if consent is to be a reality.

Locke acknowledged that the contract is not explicit, but only implied in our social conduct. And he thought that its terms were accepted by a kind of 'tacit consent', whenever people chose to reside (or even travel through) a country which they were free to leave.

(iii) *Rousseau*. In the *Social Contract*, Rousseau gave his own version of a theory which, by his day, had become common intellectual property. According to Rousseau man is good by nature, and also free by nature. He becomes bad to the measure that he is unfree, and the cause of this unfreedom is institutions. These can express and enhance man's freedom only if thoroughly democratic – that is, only if every decision that they make is one in which every member has a vote. In surrendering to this arrangement, the citizen chooses to be overridden by the rest of the community whenever his choice conflicts with theirs. Hence he surrenders part of his freedom to the community. But the surrender is also an enhancement.

As in Hobbes, the contract creates a new corporate entity, with a kind of personality of its own. When considered as passive, we call it the state; when considered as active we call it the sovereign. Like any legal person it has will – what Rousseau famously called the 'General Will', which is to be distinguished from the will of all (i.e. the aggregate of the individual wills that compose society). Rousseau rejected natural rights, as Locke had defended them: the condition of society is one in which *all* rights are voluntarily surrendered to the sovereign power – which, however, loses its own right, just as soon as it strays from the path of pure democracy.

In all its forms, the social contract enshrines a fundamental liberal principle, namely, that, deep down, our obligations are self-created and self-imposed. I cannot be bound by the law, or legitimately constrained by the sovereign, if I never chose to be under the obligation to obey. Legitimacy is conferred by the citizen, and not by the sovereign, still less by the sovereign's usurping ancestors. If we cannot discover a contract to be bound by the law, then the law is not binding.

But does the social contract solve our original problem? Does it create the trust that is needed for rational social conduct, and ensure that each person, choosing rationally, will choose for the best? Well yes, if it creates the kind of person who is disposed to be bound by a contract. Imagine our two prisoners had contracted together to remain silent, and that this was sufficient motive for them to do so, come what may. But that means, in effect, that they have set their contractual obligations above any self-interested calculation. And the whole question then becomes whether it is rational to do *that*? If I know that you will not break your bond, and am a self-interested but ideally rational creature, I would *still* do better to confess. Generalised, this becomes the problem of the 'free rider': the one who takes a free ride on others' sacrifices, knowing that he can get away with it. How do we construct a society where such people will not impose on the remainder not only the burden of their misconduct, but also the sense of being cheated – from which sense, as we all know, the general temptation arises to do likewise?

2. Traditional Objections

Three important objections were offered to the idea of a social contract, other than the one that I have just referred to. The three philosophers who made them were conservatives, opposed to the dangerous liberal doctrines, as they saw them, of the revolutions that had unsettled Europe during the seventeenth and eighteenth centuries. But their purpose was philosophical rather than ideological.

(1) *Hume*. Locke's minimal condition for the contract is tacit consent. But it is a condition that can seldom if ever be met. Locke assumes that the world is full of 'vacant places' in America and elsewhere, to which the discontented subject can flee. But are these places really vacant? And is the ordinary person really free to go to them, when he can barely scrape together the resources necessary to stay where he is? In what sense is one free to choose a course of action that would bring about a certain death, or at any rate deprive one of every motive that made it possible to find a value in living?

(2) *Burke*. The social contract prejudices the interests of those who are not alive to take part in it: the dead and the unborn. Yet they too have a claim, maybe an indefinite claim, on the resources and institutions over which the living so selfishly contend. To imagine society as a contract among its living members, is to offer no rights to those who go before and after. But when we neglect those absent souls, we neglect everything that endows law with its authority, and which guarantees our own survival. We should therefore see the social order as a partnership, in which the dead and the unborn are included along with the living.

(3) *Hegel*. To speak of a contract is to suppose the existence of people who are able to communicate, agree, and recognise rights and obligations. It is already to suppose, therefore, that people have achieved some kind of social existence. (See the argument about lordship and bondage in Chapter 20.) Society cannot be founded on a contract, therefore, since contracts have no reality until society is in place. Furthermore, the theory of the social contract is intolerably naïve. It tries to construct our political obligations on the single model of consensual relations. But political life is a complex thing, with many levels of obligation. In particular we should distinguish those obligations which arise because we freely undertake them – which are, in Hegel's words, the obligations of *bürgerlische Gesellschaft* (translated 'civil society' in deference to Locke, although 'bourgeois society' would be equally accurate) – from those which come to us

regardless, such as the obligations of family life. According to Hegel, there is no coherent view of the state that does not regard our obligation towards the state as unchosen and inherited.

3. Collective Choice and the Invisible Hand

Discontent with social contract theory has led to a study of collective choice, in the search for a more plausible criterion of legitimacy. Societies exist only if there are choices which are made *on behalf* of their members. When are these choices legitimate?

One suggestion is that 'social choices', as they are called, should not merely reflect the choices of individuals, but be *derived* from them, as when the members of a society vote for a policy by a show of hands. But we must distinguish two ways in which the choices of individuals may be combined into a collective outcome:

(1) *Collective decision.* Members of a committee may vote directly on each issue confronting it. Each individual expresses *his* preferred choice about the very matter that has to be decided. The result is calculated according to some rule: for example, a policy is accepted if and only if the voting in favour is *unanimous*: in this case the individual has a veto over every matter to be decided. This gives maximum rights to the individual, and minimum likelihood that anything will actually be decided. Alternatively, there may be a rule of two-thirds majority; or a rule of absolute majority. Only when the minority decision prevails, does the procedure begin to look like a sham. For then there tends to be some kind of weighting, which gives to a certain person or group of persons the *real* right to decide. The choices of the others then play no part in the final aggregate.

These rules for aggregating individual choices might be described as a 'constitution', though, as we shall see, there is more to a political constitution than that. Although Locke and Rousseau showed a tendency to run the two ideas together, democracy is not the same procedure as contract. In a contract, members agree to be bound by a particular procedure: and they may agree at the same time to renounce any further rights to participate in how decisions are made. The social contract could justify an entirely undemocratic political *process*, even though it is, in another sense, a wholly democratic idea – since, in a contract, every participant has a veto over the terms.

(2) *The invisible hand.* Adam Smith, in a famous passage of *The Wealth of Nations*, wrote of participants in a free market, each pursuing his

individual profit, yet 'led by an invisible hand to promote an end which was no part of his intention'. The end in question, Smith argued, is the general well-being of society. You don't have to agree with Smith's defence of the market, in order to recognise here a quite different way in which individual choices are combined to produce a collective outcome. In a market each individual makes free choices, which we can imagine to be entirely rational when judged from the stand-point of self-interest. Nobody *chooses* the social outcome, which is indeed hidden from view. Yet the outcome emerges infallibly, as a result of the aggregation of a myriad individual decisions. Smith's idea was a precursor of Hegel's 'cunning of reason'. But 'invisible hand' mechanisms do not necessarily work to the benefit of the participants, as the prisoner's dilemma shows. That is why defenders of the market try to argue that there is no better way of arriving at a satisfactory distribution of resources, sometimes adding that the attempt to arrive at such a distribution by a collective *decision* is *irrational*, since this is not the kind of thing that a committee, however democratic, could decide.

The two methods of aggregation can be generalised, and lead to radically contrasting ideas of political consensus. The defender of collective decision is seeking a society which is explicitly *consented to* by its members: that is, they themselves make a choice about its institutions and material conditions. The defender of the invisible hand seeks for a society that *results from* consent, but was never explicitly consented to, since the choices of individual members range over matters which have nothing to do with the final outcome. If you think that the first kind of consent is unobtainable, as did Smith, Hegel and Hayek, then you may still look for the second kind of consent. You then have a strong motive, even if you have no reason, to believe that the invisible hand is capable of delivering something that we also *would* consent to, once it is in being.

4. Paradoxes of Social Choice

Whichever method is adopted for aggregating individual choices into a collective outcome, paradoxes arise. Here are three of them:

(1) *The paradox of democracy*. This was emphasised by Rousseau, and has the following form:

(i) If I believe in the legitimacy of democratic choice, then I believe that the policy chosen by the majority ought to be enacted.

(ii) There are two incompatible policies, A and B.

(iii) I believe that A ought to be enacted, and that B ought not to be enacted, so I vote for A.
(iv) The majority vote for B.
By (i) and (iv) I believe that B ought to be enacted; by (ii) and (iii) I believe that B ought *not* to be enacted. Hence my adherence to democratic values involves me in conflicting beliefs. (But are they *contradictory* beliefs? If not, maybe this is not a paradox of the second kind.)
(2) *Voting paradoxes.* These are of various kinds, but are typified by the following, due to Rousseau's disciple, the Marquis de Condorcet. Suppose a vote shows a majority preference for policy x over policy y, and also for y over z; it may yet show a majority choice of z over x: assuming transitivity, therefore, we have a contradiction: x is preferred to z, and z to x. This can be illustrated in the case of three citizens as follows: A prefers x to y and y to z; B prefers y to z and z to x; C prefers z to x, and x to y. In such a case there is a majority preference for x over y, for y over z, and for z over x. The paradox arises because we are dealing with 'preference orderings', where there is a choice between more than two alternatives, and we are trying to rank them. Maybe actual democratic decisions are not like that. But we know from our own electoral experiences that similar conditions do often obtain, and the winning party may reflect a minority preference when it comes to actual policies.

Condorcet was a revolutionary and an innocent; he lost his head during the Terror, having lost his mind shortly before. A woman once asked Carlyle at dinner table, 'What is the use, Mr Carlyle, of all this philosophy?' He replied as follows: 'Madam, they asked the same question of Rousseau's *Social Contract*, and the volumes of the second edition were bound in the skins of those who had dismissed the first.'
(3) *Arrow's theorem.* The voting paradoxes led to the development of a branch of applied logic known as the theory of 'social choice', which expressly concerns itself with the problems that arise when rational choices are aggregated, often to irrational outcomes. The mathematical economist K.J. Arrow made an important contribution to this subject, when he demonstrated that it is impossible to design a 'constitution', as he called it, that will generate complete and consistent rankings of alternative states of a society on the basis of the preferences of its members, whilst satisfying certain intuitive conditions. The proof defines a 'social welfare function' as a set of rules for transforming individual preferences into social outcomes. It then lays down conditions that such a function should meet: for example,

that no member should be allowed to dictate the outcome, and that the outcome should be 'Pareto optimal' – i.e. such that any change from it will make someone worse off. Arrow shows that no social welfare function will meet those conditions.

This is fairly technical stuff, and its relevance to political philosophy has been disputed. But it is part of a long series of arguments, tending to the conclusion that the aggregation of individual preferences may, after all, provide no coherent test of the legitimacy of a political decision. All such arguments are set at nought, if we attribute political decision-making to some *other* entity than the individuals who are affected by it: for example to the 'General Will' of Rousseau, or to the state that embodies it.

5. General Will, Constitution and the State

The theory of the state is complex and difficult. We can divide it up as follows:

(i) *Constitution.* A constitution is sometimes defined as the basic *law* of a state: the law upon which the validity of all other laws depends. But this does not really advance us very far, since we have yet to know what a law is; besides, there are plenty of states without laws, and, even if their constitutions are to that extent shams, we still need to know how we would explain and criticise them.

Constitutions are composed of the fundamental procedures whereby political decisions are arrived at. They may be written down, like the American constitution; but only a very naïve person would believe that they thereby become fixed, immovable and readily known. (The American Constitution has to be interpreted in the light of two hundred years of constitutional case law; it is a real question whether it means the same today as it did when written down, or whether it *ought* to mean the same, or indeed what the criterion for meaning the same might *be* in such a case.) The constitution of the United Kingdom has never been written down, and probably couldn't be, since it derives from a constantly developing body of common law, the mass of which has never been discussed by the sovereign body.

(ii) *Participatory and representative governments.* Rousseau advocated direct democracy, by which he meant a system in which every citizen participates in every political choice, by casting his vote. This worked in ancient Athens, for two reasons: ancient Athens was a small city-state (or *polis*); and the majority of its population (women, slaves and metics (migrant workers)) was disenfranchised. Direct democracy could not work in a modern society. Hence the emergence of 'representation', whereby those who vote on matters of common concern are appointed to do so by the remainder. The procedure for appointment may be democratic; in which case we have a representative democracy; but it may not be. Thus in the United Kingdom there was representation long before there was democracy, and a representative was obliged to speak for his constituents even in the days when they had little power to throw him out. (Of course, it is a moot point whether he *would* speak for them in such circumstances.) A representative is not a delegate: his constituents do not *tell* him what to say or how to vote; he is appointed on the understanding that he will follow his conscience. The only control that his constituents have is exercised at the time of elections. (Burke made much of this distinction, which is still very poorly understood.)

(iii) *The separation of powers.* Montesquieu, following and amending Locke, argued that the powers in a state are of three kinds: executive, concerned with carrying out political decisions and enforcing laws; legislative, concerned with the making of laws; and judicial, concerned with applying those laws in the individual case. He further argued that a separation of these powers is a pre-condition of individual liberty; for then they can be balanced against one another, and each of the branches of government can be made to account for its misdemeanours. The separation is never complete, though an effort was made in the US constitution to maximise it. (The result, however, has been to vest sovereignty in the Supreme Court, which is now the most important law-maker and policy-maker in the country.)

(iv) *The nature of the state.* The state can be regarded in many ways: as a kind of administrative machine, generating policy out of the choices of those who control it (this is the Marxist picture); or as a system of offices, which are distinct from those who occupy them and confer their own decision-making powers and liabilities (Aristotle); or as a corporate person, comparable to a firm or a university, which has its own will, liability and goals (Rousseau and Hegel). The Marxist picture will occupy us in Chapter 30. But it is pertinent to make some observations about the other two views.

One of the ideals of government in our tradition has been to achieve a system in which those who exercise power are always answerable for what they do with it. If any person is able to exercise absolute power, then he can effectively remove the threat of criticism, and proceed as he wills. To prevent this, power must never be granted in full, but only in small and circumscribed doses. This is achieved by the device of an office, which defines the powers of those who occupy it, while also limiting them. To go beyond those powers is to act *ultra vires*, and so to expose oneself to the penalties fixed by law. If the occupants of each office hold their position for life, they again possess an opportunity to abuse their powers. If, however, they occupy them for a period of years, or on sufferance, the office acquires a character distinct from its members, and can become an object of respect, and a source of continuity. That was Aristotle's ideal, which he held to be compatible with all the major systems of policy making: monarchy, aristocracy or democracy.

At the same time, the state must function as a unity: offices which issue conflicting decisions will destroy the state. So how are the powers to be united? This is where Rousseau spoke of a General Will, and Hegel of a corporate person. The ideal state does not, for such philosophers, have the unity of a machine, nor even that of an organism. It has the unity of a person – i.e. an entity which takes responsibility for its own decisions, and which has both rights and duties. Maybe the system of offices is a part of such a state; at any rate, this idea of unity seems more plausible in the abstract than the alternatives, since it enables us to understand how the state can be not only accepted, but also obeyed. You don't obey machines or animals; but you do obey that which has rights against you, and duties towards you, to the extent that you recognise those rights and duties.

(v) *Legitimacy*. And that means recognising their legitimacy. What makes a state legitimate? To put the question in other terms: what is the foundation of political obligation (our obligation to the state)? Here are four suggestions:

(a) *Inherited authority*. Power has been bequeathed from time immemorial, by a process which ensures that authority is preserved. Many theories of this kind have been devised, in particular by Locke's opponent Sir Robert Filmer, in order to justify some variant of hereditary monarchy. The obvious difficulty is this: whence did the authority first arise? And what guarantees that it is preserved by the transfer?

(b) *Social contract*. This is the obvious answer, though, as we have seen, it does not really get us to the point desired.

(c) *Consent*. Whether tacit or explicit, our consent to an arrangement can still provide some grounds for its legitimacy. Maybe, as Hume argued, this is the best that we can hope for. In which case a state founded on time-honoured custom, in which people freely proceed as tradition requires, is our model of a legitimate order.

That smacks of the invisible hand, of course; and it is no accident that the eighteenth-century defenders of the free market were also disposed to accept an idea of customary legitimacy.

(d) *Utilitarian theories.* According to a rival tradition, what makes a state legitimate is its ability to deliver the goods. The state is a means to the happiness of its subjects, and the more happiness it produces, the more legitimate it is. The disadvantage of such a view is that legitimacy becomes provisional: someone with a new scheme for human improvement may be entirely justified in seizing power. Indeed, the idea of legitimacy tends now to be discarded, and the state is robbed of its security.

6. Justice

Hence there has been an argument, current since Aristotle, to the effect that legitimacy rests in a negative condition. All kinds of political order can be regarded as legitimate, just so long as they pass the crucial test, whose name is justice. It is a familiar argument that utilitarianism will never pass this test, since it will authorise the destruction of the innocent for the sake of some greater good. (See Chapter 20.)

Aristotle distinguished two types of justice: or rather two applications of the concept of justice, distributive and 'commutative'. Questions of justice arise in two contexts – first, when there is some good to be distributed among those with a claim to it; secondly, where a person deserves something through his actions, whether a reward or a punishment, and where justice is the measure of what he has earned. In the first case, the concept of 'right' is all-important, in the second case the closely related concept of 'desert'. To punish someone who has not deserved punishment is surely a paradigm case of injustice; as is the distribution of property to which someone has a right among others who have no right to it.

In both cases, the question arises whether justice is a property of a state of affairs, or of some human action or relation. To put it another way, could it ever be simply *unjust that p*, regardless of how *p* came about? Socialists tend to answer yes to that question; actions are just, they say, according to whether they contribute to a just distribution. Others regard justice as a regulatory or procedural concept, which makes sense only when applied to the dealings between people, and which cannot be detached from those dealings so as to apply to a state of affairs as such. We may begin to understand this dispute by considering the theory of Rawls, and the reply offered by Nozick. These are not the last words in the matter, but they have earned their place in modern philosophy, and cannot be ignored.

Rawls's *A Theory of Justice* is a vast and intricate book, which applies every sophisticated device to the spelling out of a distributive theory of justice. Justice, for Rawls, regulates the distribution of goods throughout society. The concept of justice is the only sound device we have for justifying and rectifying distributions, and we can reach a coherent theory of justice only if we view society as a whole. We must ask which distribution of goods and benefits is the just distribution, taking everyone into account and allowing each to count only for one.

The grounds for this procedure are to be found in rational choice itself. We are all rational choosers, and we all exercise our choice in the political context. We all therefore think not only about our own advantage, but about the good society in which we would choose to live. The problem is that, in the actual situation, knowing our strengths and advantages, we are disposed to take the quickest path to success, and to override the competing claims of those who are weaker or more foolish than ourselves. We could never hope to arrive at a universally acceptable concept of justice, therefore, if we study the actual conditions of human society, since this would lead us merely to endorse and perpetuate all the unfairnesses which, by history and nature, have set us apart. We must discount our own advantages, therefore, if we are to have a measure of what is just. And this means projecting ourselves into a hypothetical situation, where we no longer know the facts that distinguish us from others. We draw a 'veil of ignorance' across social reality, and ask ourselves what we would have chosen, had we known nothing of our condition. Choosing from behind this veil can be understood too in contractual terms. We can imagine ourselves trying to establish a society with our fellows: the question what we would choose is the same as the question on what we could agree, since, in this minimal situation, nothing relevant distinguishes us. Hence the just distribution would also be the one that we should write into our social contract.

Rawls adopts a criterion of rational choice from the theory of games: that of the 'maximin strategy'. This is the strategy in which, not knowing anything about probabilities, I choose to maximise the minimum 'pay-off'. I think of the worst possible outcome for myself, and try to make sure that it is as good as could be. For that outcome might occur. In the context of the social contract this means trying to ensure that the situation of the worst off is as good as possible. On the other hand, whatever my position, I should wish to ensure maximum freedom to move out of it. I would not wish to compromise my freedom, since that would leave me without the principal source of social hope. Hence I (and you too) would choose a society in which there is maximum individual freedom, compatible with an equal freedom distributed to all. Once that principle is satisfied, we should choose to provide the best possible deal for those at the bottom. In other words, whatever inequalities are countenanced, they should have to be such that the position of the worst off is better than under any alternative distribution. (The 'difference principle'.) Finally, Rawls argues, social advantages would have to be accessible – that is, attached to positions and offices that are open to all. The principles are 'lexically ordered' as in a dictionary. Only when the first is satisfied do we appeal to the second.

All this is delivered with great subtlety and skill. But is there any reason to accept it? Is not the result too close to the favoured ideology of the East Coast intellectual to persuade us that the dice have not been loaded from the start? Among the objections made are these: first that there could be no such thing as rational choice behind the veil of ignorance. In order to secure the result that he desires, Rawls asks us to discount not only our race and sex, but also our religious values and our several 'conceptions of the good' – in short, all that makes truly rational choice conceivable. Secondly, there is no persuasive proof that people would attach the prominence to liberty that Rawls attaches to it – is not happiness a more important good, and might it not be obtained, at times, only at the cost of liberty? Thirdly, the maximin strategy is not the only one that would recommend itself to a rational being in these ideal conditions of ignorance. We sometimes play safe; but it is equally rational to take risks, and that is something that we are always doing. Fourthly, there is something contentious in the idea that we, in our empirical condition, knowing our circumstances, should be motivated by a conception which requires us so completely to set that condition aside. Rawls, like Kant, tries to derive a rational motive by stripping away all *actual* motives, leaving reason alone in the driving seat. But maybe reason cannot touch the pedals.

Those are not the objections that interest Nozick, however. His main concern is with the *goal* of Rawls's theory – and in particular with the idea that justice is a property of a distribution of goods, which can be determined regardless of the history of their production. Behind Rawls's thought-experiment lies the following picture: the sum product of a society exists because of social cooperation. The question for a theory of justice is how to *distribute* it among those who are entitled (by virtue of their membership of society) to a share. But surely, we might object, goods do not come into the world unowned: they are produced as a result of enterprise, contract, and a web of tacit agreements which already create rights of ownership. It is not possible to distribute them, without violating those rights – which would be an injustice.

Suppose, however, that the ideal distribution has been established. What guarantees that it will remain? Surely, if people are free to exchange, produce and give away their goods, the pattern will at once be disrupted. To ensure that it remains, we should have to curtail human economic freedom to such an extent that rights would again be violated.

In short, 'pattern' theories of justice are not theories of justice at all. In reality, justice is a procedural notion, and the fundamental application of it is to what Nozick calls a 'justice-preserving transfer'. (Which is a little like a truth-preserving inference in logic.) Suppose I am in just possession of a pair of shoes, and you in just possession of a pair of gloves, and we agree to exchange them. There is no coercion, and both of us desire the transfer. Surely, if we began this transaction with just possession, we also end it with just possession? If a thousand people are so desperate to hear Placido Domingo sing, that they will each pay the twenty pounds that he asks as a contribution to his fee, and if they are entirely content with the bargain, does not Placido Domingo have a right to his money? And is it not perfectly just that he ends up so much the richer? Justice-preserving transfers may therefore promote inequality in the name of justice. But any alternative would be manifestly *unjust*, Nozick argues: it would involve either forcing Domingo to sing for less than he requires, or forbidding him from making use of his talents at all.

The arguments here are intriguing, and I return to one of them below. Evidently, Nozick must be prepared to say something about what it is to be in 'just possession' of something. And this will require a theory of just original acquisition. (If something is not yours, then you do not have the right to transfer it.) At the same time, we should notice how the emphasis has shifted, away from ideas of shares and distributions to individual rights. One definition of justice, indeed, is in terms of rights. A just transaction is one in which no right is violated. (That is why punishment may be just: for if someone deserves punishment, he has no right to escape it. Hegel even says the criminal has a *right* to be punished, and you do him wrong by refusing it.) This conception of justice is a long way from anything that would recommend itself to a socialist, even if it corresponds to much that goes on in a court of law. Hence the emphasis by socialist thinkers on 'social justice': a particular application of the term, which is meant to remove us from the sphere of individual rights so as to consider the state of society as a whole.

7. Law

The rights just referred to would normally be described as natural rights. They are rights that we possess by nature – say, because we are rational beings, and all rational beings have a right to be treated as ends and not as means only. (That is the *Kantian* justification of rights: theologians might justify them in other ways.) But not all rights are natural rights: indeed, many philosophers deny that there are such things; Bentham for instance, described the doctrine of natural rights as 'nonsense on stilts'. The only creator of rights, Bentham argued, is law, and the only creator of law is the legislator. A law is a rule or convention, which assigns rights because it compels people to respect them.

Bentham was a legal 'positivist'. He believed that all law was *posited* by legislative bodies, and that none is natural. All rights, therefore, are positive rights. Hence has arisen a distinction between natural and positive law, corresponding to that between natural and positive rights. Prior to the modern period there existed a subtle and elaborate political philosophy based on natural law, from which the social contract theory was in many ways a radical departure, since it located the source of our political obligations in our own choice and agreement, and not in some God-given precept. Legal positivism was the last step in the rejection of the medieval view of things, and although still controversial, has had powerful defenders in this century – becoming near to an orthodoxy between the wars.

The question 'what is law?' is not easy. Even if we say that a law is a rule of conduct enforced by some authority, we still do not know what kind of rule we are talking about. (And, as Wittgenstein has shown, the concept of a rule creates problems of its own: see Chapter

19.) Here are some of the conditions that seem to be necessary, if a rule is to acquire the status of law:

(i) It must be definite: that is, it must permit us to determine what counts as obeying it, and what counts as transgressing it. A vague law is not a law: or if it is a law, it is a very unjust one. (Here we see how the concept of natural law constantly creeps back even into the most positivistic of theories. For there is a conception of natural justice motivating all our thinking in this area.)

(ii) It must be announced or made public. Secret laws are also not laws: or again, very unjust ones. (Soviet law contained many secret laws, however. If the citizen cannot determine whether he is on the right side of the law, he will have to be pliant to the authorities.)

(iii) It must not be retroactive. To announce *now* that someone's deed of yesterday was illegal is again to commit an injustice. Retroactive laws violate a fundamental requirement of reason: I cannot rationally obey a rule that does not yet exist.

(iv) It must belong to a legal system, regulating society as a whole. And the system must not be in irresoluble conflict with itself: there must be a procedure for resolving legal conflicts, and for harmonising valid laws.

In normal conditions law is *enforced* by the sovereign power. But this is not a necessary feature, since there are voluntary laws, like the law of nations, which countries agree to accept in the interests of mutual trust. But let us confine ourselves to law enforced by a sovereign power. What then makes such a law *valid*? This is a question that has been much discussed in recent years, since it is the matter over which positivists and naturalists are most divided. The most sophisticated recent defence of positivism is that offered by H.L.A. Hart, who argues that every system of laws owes its validity to what he calls a 'rule of recognition'. This rule tells us which regulations are to be regarded as law. For example, in the United Kingdom, a law is binding only if it has received the assent of the Queen in Parliament. An agency which attempted to enforce a law that had not received the royal assent would be acting *ultra vires*. Moreover, royal assent is *sufficient* for legal validity: nothing else needs to be done, in order to confer validity on a law.

This raises a difficulty, however. The bulk of English and American law has never been explicitly approved by Parliament or Congress, since it is 'common law': that is, law enshrined in the decisions recorded in the case books. It is a peculiar feature of common law that we do not know what it is, but only how to apply it. We assume that a given case was decided correctly, and *then* extract the 'rule' which decided it. A wrong formulation of the rule does not necessarily invalidate the original case. The doctrine of *stare decisis* (let the decision remain) tells us that, until overruled by a higher court, a case is authoritative, and we must search for the principle contained in it: the *ratio decidendi*. Hart would say that there is an implicit 'rule of recognition' here: namely, that everything accepted at common law is also endorsed by Parliament.

The study of common law has led to a sophisticated rejoinder on behalf of a kind of naturalism. Ronald Dworkin, in a series of striking papers, has shown that we cannot assimilate common-law reasoning to the model proposed even by the most sophisticated positivism. The positivist, Dworkin argues, tries to distinguish law from social convention by reference to some 'master rule'. All difficulties and indeterminacies in the law are then treated as matters of 'judicial discretion': they do not involve the discovery of genuine answers to independent legal questions, but simply the arbitrary decision of the judge. Finally, a legal obligation exists, for the positivist, only when an established rule of law imposes it. Those ideas define law as a system of command, answering to no internal constraint besides that of consistency. The view is mistaken, Dworkin argues, as are the tenets from which it derives. A 'master rule' is neither necessary nor sufficient for a system of law. It is not necessary, since law may arise, like the common law, entirely from judicial reasoning, which takes notice of judicial precedents and their 'gravitational force'. Nor is it sufficient, since a supreme legislature can make law only if there are courts to apply it, and judges in those courts must employ 'principles' of adjudication which derive their authority from no legal rule.

The existence of these principles is established by 'hard cases', in which the judge must determine the rights and liabilities of the parties, without the aid of any law which explicitly prescribes them. The adjudication of such a case is not an exercise of 'discretion', but an attempt to determine the real and independently existing rights and duties of the parties. The judge cannot think of himself as inventing those rights and duties – otherwise his decision would be an exercise in retroactive legislation. Nor can he imagine that he is exercising some 'discretion' that he does not need in the normal conduct of his profession. In a hard case law is not so much applied as discovered.

The details of Dworkin's account lie beyond the scope of this chapter. Dworkin now tends to consider law as a kind of commentary, written by many hands, on a system of constitutional rights; for he has the American legal system very much in mind. But many of his cases come from English law, which is the creator and not the creature of our constitution. It is very difficult to know, therefore, quite where his argument is tending, if not in the direction of legal naturalism.

Two vital questions must be addressed by anyone seeking to develop a comprehensive philosophy of law: What is the function of law? And what is a rule of law? Three answers suggest themselves to the first question:

(i) The function of law is to achieve some 'end state' which is desired by the sovereign power. In socialist legal systems, the law was often regarded as a means to the end of a socialist society. The problem with such a view is that means-end reasoning justifies *breaching* the law just as often as it justifies obeying it. When the citizen can claim his legal rights, he is an obstacle to the sovereign's plans, and can prevent their realisation. Which is why rights disappeared from the socialist systems. (I should say that 'socialist law' is a technical term, used to describe the peculiar systems introduced into the former Soviet Empire.)

(ii) The function of law is to 'do justice', in the varied circumstances of human conflict. This is the preferred view of the legal naturalists, and is reflected in our common law. The problem is that it does not sufficiently recognise that law is, or has become, an instrument of *policy*, and is everywhere used to impose the will of the sovereign regardless of any natural rights that may be violated.

(iii) The function of law is to resolve social conflicts, by setting intelligible boundaries to what the citizen may expect from his fellows. This is the preferred view of thinkers like Hayek, who see the law rather as they see the market, as a constantly developing response to free transactions, which endeavours to restore equilibrium when someone 'breaks the rules'.

Maybe the true answer is some combination of those. Or maybe it is something quite different.

As for the rule of law, the problem of defining this is perhaps one of the most important political issues which modern societies must address. A rule of law exists only where every act, including the acts of the sovereign and his agents, are constrained by the law. It can be guaranteed only when the victim has some redress against the sovereign who abuses his power. And it must be possible to enforce the resulting verdict. But who is to do the enforcing? Evidently the sovereign himself. For this reason, Hobbes supposed that the sovereign must be above the law: whatever he decides just *is* law, even if it violates one of his own prescriptions (since 'violation' here is just another word for 'overruling'). Others, more reasonably, emphasise the role of a separation of powers in ensuring legality. If judges are independent, they can *call* on the agents of the sovereign to enforce the law, even against the sovereign himself. But how this is done is still to some extent a mystery.

8. Freedom

Law is an instrument of coercion, and as such a restriction of human liberty. Hence the question of law and its function has been of great concern to liberals, for whom anything which restricts human freedom is *prima facie* wrong. Locke made, in this connection, a distinction between liberty and licence, arguing that the latter is not what we desire under the name of liberty, since it is in fact a threat to the liberties of others. We need law in order to forbid licence, and therefore in order to foster human liberty. Only superficially, therefore, is law a restriction of our freedom: properly understood, it is the sole means whereby freedom can be achieved.

The thought was taken up by J.S. Mill in his essay *On Liberty*. The task of jurisprudence, he argued, is to devise a system of law which maximises human liberty, by creating the greatest space for one person's liberty to grow, compatible with a like space for his neighbour. To this end he proposed his famous 'harm' principle, according to which actions should be permitted, so long as they do no harm to others. However repugnant to us morally or aesthetically, a person's behaviour should not be forbidden by the state, except in order to safeguard the well-being of his fellows.

Mill was also prepared to say *why* human freedom is a political good. Without it, he believed, there could be no progress, either in science, law or politics, all of which require the free discussion of opinion. Nor could there be progress in morality, which thrives on those 'experiments in living' from which the new possibilities of human happiness emerge.

The problem, according to Sir James FitzJames Stephen, Mill's principal critic, is that ordinary people neither thrive on such experiments, nor remain 'unharmed' by the spectacle of them. Unless we regard 'harm' as a purely medical idea, we must recognise that we are harmed by everything that threatens our happiness. And since a large part of happiness lies in the knowledge that one belongs to a secure moral order, in which shameful practices cannot be freely engaged in, we have a right to expect that our fellow citizens will be compelled to do what they should. (Stephen had been a judge in India, and had been particularly disturbed to witness the effect on ordinary decent Indians of the 'experiments in living' which their European masters had devised for their entertainment.)

Another shot at defining the liberal position involves replacing the concept of harm with that of right. On this view, an act should be permitted unless it can be shown to violate a *right* of someone else. The emphasis has now shifted: to what do we have a right, and when are our rights infringed? A philosopher like Nozick would argue that we have a reasonably clear view of rights, for the reasons stated by Kant. We have a right to the respect of others, a right not to be coerced, tricked or defrauded; a right to pursue our own way of life provided it is marked by a similar respect for others, and so on. Maybe a measure of social decency is implied in this very idea. The resulting system of law will be a system of 'side-contraints', broadly in line with the Kantian morality, but imposing no absolute obligation to pursue one goal rather than another.

The picture is certainly agreeable. But can it be made persuasive? One problem is this. We are assuming that people are rational agents, with goals of their own, able to give and demand respect from their neighbours. But it is only in the condition of society that such complex beings emerge. Maybe they will not emerge, if we allow them full freedom to emerge as they will. Maybe a degree of paternal supervision is required, to mould the raw animal material into a moral agent. And maybe that supervision will force us to curtail even the rights (or at least the abstractly defined Kantian rights) of the individual citizen. The question here is the question of 'liberal individualism', and I return to it below.

9. Property

Among the rights that have been defended by liberal thinkers, the right of property has been the most controversial. Are there any rights of *private* property? And if so, how are they justified?

The dispute here is an ancient one. Plato advocated, in the *Republic*, a kind of common ownership of goods, believing that private property is a source of conflict and discontent which is best abolished. Aristotle replied, in the *Politics*, with a subtle defence of private property as an integral part of citizenship. But for modern purposes the dispute begins with Locke, who, in the fifth chapter of his second

Treatise, produced a famous argument for the existence of a *natural* right of property. Since the right is natural, Locke argued, it lies outside the terms of the social contract, and no citizen can be assumed to have surrendered it.

Locke's major argument concerns labour, and is expressed in the following far from easy words:

> Though the earth and all inferior creatures be common to all men, yet every man has a 'property' in his own 'person'. This nobody has any right to but himself. The 'labour' of his body and the 'work' of his hands, we may say, are properly his. Whatsoever, then, he removes out of the state that Nature hath provided and left it in, he hath mixed his labour with it, and joined to it something that is his own, and thereby makes it his.

The quotation marks are a sign that Locke is feeling his way towards conceptions that he had not defined to his own satisfaction. Nevertheless, the argument from labour was destined thereafter to play a crucial role in all discussions of 'original acquisition'. It is open to an obvious objection, expressed facetiously by Nozick thus: If I pour my tomato juice into the sea, and so mix it with the ocean, do I acquire a property right in the whole ocean, or do I merely lose my tomato juice?

Sometimes Locke shifts the focus of the argument. It is not simply that I mix my labour with something, but that I add *value* to it thereby; and value of which I am the sole orginator is surely mine if it is anyone's. But *is* it anyone's? The argument could not even begin without some assurance that the thing with which I mix my labour is initially unowned. But Locke admits that it is not: for God gave us the earth to be held 'in common'. Moreover, Locke believes that I can never acquire property rights if I do not leave 'enough and as good' for others. And how can I ever be sure of doing that, given that 'others' stretch into the distant future, and there will always be someone at some distant time who is adversely affected by my acquisition?

Locke's argument attempts not merely to justify the right of private property, but also to give a procedure for answering the question *who* owns *what*? Less bold, but more subtle, Hegel (in *The Philosophy of Right*) proposes merely to justify the institution of private property, leaving the *distribution* of property to law, history and chance, over which factors the philosopher can claim no sovereignty. (Though come to think of it, Hegel claimed sovereignty over everything.) For Locke I own my labour as I own my body ('the sweat of my brow'). For Hegel bodily activity and labour are not the same. Only a *rational* being is capable of labour, since the intention to produce value is an integral part of the motive of labour. This intention is necessary to a rational being if he is to realise his potential, and become what he truly is. Moreover, it is an intention that can persist only so long as it is possible to envisage private property in the object produced. Property must be private for Hegel, since it is an expression of the *individual* self, and would lose that character if it stood in the same relation to more than one self. It is an institution through which I affirm and realise my individuality in the world. Through private property I create my home in the world, by marking out a corner of it as mine.

Both arguments are far more complex than my summary suggests; in assessing them it is always necessary to remember that a right of private property is composite. It may involve one or some or all of the following: a right to use; a right to exclude others from use; a right to sell, barter or give; a right to hold without using; a right to accumulate; and a right to destroy. A philosophical argument might justify one of those rights without justifying the others.

It is doubtful that a right of property could ever be established by *a priori* argument which made no moral assumptions; for it would be an argument from an 'is' to an 'ought', and such arguments are in chronically short supply. But this should not lead us to embrace the opposite conclusion, that therefore private property is unjustified. The arguments against private property are as speculative and metaphorical as those in favour of it. This is particularly true of Marx's attempt to 'set Hegel on his feet', by showing that the institution of private property, far from realising the individual self in the world, alienates the individual, by reducing him to an instrument. In any case, the traditional discussions overlook a far more important idea than the right of property, namely the *duty* of property. Duties, like rights, may be possessed by individuals. It is an argument for the right of private property that it is the necessary price for exacting a private *duty* of property. And we need to exact such private duties if we are to ensure that property is not misused, squandered or used to destroy the environment. Such a utilitarian argument can be taken far; all who witnessed the environmental catastrophe of communism know the price that must be paid when there is no legal person (whether individual or corporate) on whom liability for damage squarely falls.

10. Institutions

This returns me to the liberal individualism mentioned in section 8. It is useful to conclude this chapter with the Hegelian response to that position, so as to explain what is really meant by 'objective spirit'. The liberal represents every human institution as the product of human choice, and choice itself as the fount of legitimacy. According to the Hegelian, however, legitimacy stems precisely from the rational being's respect for himself and his kind as beings formed, nurtured and amplified by institutions. It is not that we have desired and chosen our institutions, for without institutions there would have been no choice to make. Nor is it that we know how to draw back from every inherited arrangement and pronounce it legitimate by some act of will – any more than we can stand back from ourselves, and ask 'Shall I, or shall I not, be this thing that I am?' The error of individualism lies in the attempt to found a vision of society on the idea of rational choice alone – on an 'abstract' notion, as Hegel put it, of practical reason, which makes no reference to history, community and the flesh. (Such is the vision of society as a collective solution to the prisoner's dilemma, from which we began.) Individualism esteems choice above everything, and regards justice as the procedure whereby each person's freedom may be reconciled with the freedom of his neighbour. The concepts of freedom and justice thereby become intertwined, as in the theory of Rawls. The modern individualist goes further, arguing with Rawls that the idea of justice must be freed from every particular 'conception of the good'. No particular scheme of values, no particular historical community, no particular custom, circumstance or prejudice, can be incorporated into the abstract statement of basic rights, which reflects only the fundamental requirement that freedom and equality are the sole excuses for government. Politics is the system whereby the atoms are maintained in their separate spheres.

But we are not like that. Our nature as rational beings emerges through the experience of membership – of family, civil society and state. We come together, not only in voluntary associations, but also naturally and inevitably. We are born, we reproduce and we die, and only through our membership of communities can these momentous episodes, in which our condition is revealed, be really accepted. And from the act of acceptance grows our sense of the meaning of life. Institutions of membership are the *sine qua non* of social life: without them, we should never acquire a conception of value; and without a conception of value we cannot justify our choices, either to others or to ourselves. Then everything is illegitimate – or rather, the distinction between the legitimate and the illegitimate can no longer be made. Perhaps this is our condition.

29

Subjective Spirit

The dispute between individualists and 'communitarians' is deep and complex. Many issues in politics, sociology and the philosophy of mind come together in the question whether society is an aggregate of individuals, or whether, on the contrary, individuals are by-products of society, who owe their nature as free rational agents to their participation in the organism that endows them with this role. Both positions are appealing, the one in emphasising the sphere of sovereignty and right that protects each person from his repulsive neighbours, the other in emphasing the mutual need and dependence that causes us to find our neighbours so repulsive in the first place.

Nobody can reasonably doubt that the rational being is by nature the member of a community – a *zoon politikon*, as Aristotle famously described him. But is there any *particular* community or kind of community of which he is by nature a member? Perhaps the most plausible form of liberalism is the non-individualist (or even anti-individualist) liberalism which accepts the view that rational choice is a social artefact, while arguing that there is no specific social order that is uniquely able to engender it.

Having acquired practical reason, therefore, we should be maximally free to conduct those social experiments which Mill recommended, in the hope that new and more consoling forms of community will emerge from them. But will they? And dare we risk experiments which threaten our little measure of peace and comfort?

The rational being finds comfort not only in communities, but also in solitude. Indeed, certain of our values are most deeply felt in solitude, and treasured all the more because they suggest that the world contains a place for the self. Aesthetic and religious experiences are of this kind. This does not mean that they are available to a purely solitary being: on the contrary, the religious experience is born from our need for membership and would be meaningless without the thought of a community of fellow-believers. Nevertheless, it is, like the aesthetic experience, enjoyed as much in solitude as in company, and seems to acquaint us with the mystery of the self, and the final goal of its otherwise absurd existence.

It is not obvious that philosophy can say much about this realm of 'subjective spirit', as the Hegelians describe it. Perhaps art and liturgy have the monopoly of coherent utterance, in regions where the thread of truth runs out. But the belief that that is so is itself a philosophical belief. Only an excursion through the field of aesthetics will tell us whether it is true.

It is a field that is difficult to enter, since we have no pre-philosophical map of it. In day-to-day thinking we refer to the mind, morality, science and God: to almost everything that figures as a philosophical problem. But the term 'aesthetic' is the invention of philosophy, and who knows whether the thing described by it is not an invention also? Perhaps the most important task in aesthetics, therefore, is to say what it is about.

Modernist philosophers tell us that aesthetics is the philosophy of art; it is therefore through the concept of art that we should define its subject-matter. But this approach soon runs into the ground. There is no definition of art that will explain why a Rembrandt portrait falls under the concept, and a rotting fish does not. There are plenty of artists, so-called, who place rotting fish on show, and proudly claim credit for the critical accolades. The concept of art has been so deformed by the romantic idea of the artist, and by the attempt by frauds and impostors to claim a distinction that they could achieve in no other way, that it has ceased to be an interesting or decidable question what belongs to it. Call anything art: for art is not a natural kind. But answer the question why we should be interested in the thing defined. Which takes us back to our starting point.

Suppose an anthropologist, engaged in the study of a remote tribe, begins to classify its activities according to the purpose which they serve. He sees people tilling the fields, planting and harvesting, and attributes this activity to the desire for food. He observes the manufacture of clothing, tools and medicines; the building of houses, and the rituals which are addressed to higher powers. But there are certain activities which seem to have no purpose. Members of the tribe take 'time off' from urgent things, in order to build weird structures, to make long wailing noises, to move in mesmerising dances, or to throw cakes of dung high into the tree-tops. Suppose that the tribesmen have a word for all these activities (though its application to each of them is disputed by at least one of their members). Suppose that, when the anthropologist asks what this word ('schmart') means, and what it is that works of schmart have in common, people rack their heads and come up with some unilluminating technicality: the point of schmart, they say, is that people have a schmaesthetic interest in it. Pressed further, they begin to claim that other tribes, at other periods of history, also take a schmaesthetic interest, though not necessarily in quite the same things. One of their philosophers eventually proclaims that the schmaesthetic is a human universal, and that it is in schmaesthetic interest that the meaning of the world is revealed. Whereupon prizes, priesthoods and professorships are conferred on those whose dung-throwing meets with the greatest applause. Do we conclude that these people have opened the anthropologist's eyes to a new branch of philosophy?

Schmart, as I said, has no apparent purpose. And if we ask our people what schmart is *for*, they are apt to tell us that it is not *for* anything, or that it has no purpose but itself. Suppose, nevertheless, that they take a great interest in schmart, and that whenever anyone is engaged in producing or performing it, others gather round to stare and applaud. Certain experts in dung-throwing become known well beyond their immediate circle, and are sought out and favoured by the multitude. Perhaps people pay to see them perform. Suppose too that people discuss the merits of various performances, vehemently maintaining that one was better than another, and looking for reasons to justify their judgement. We can conclude at least that these people *value* schmart – whether rightly or wrongly – and assign an *individual* value to the separate *works* of schmart. Suppose too that they discount altogether the judgement of one who has no direct experience of the thing he judges. The important thing, they say, is the *experience* of schmart, and it is never possible to derive by hearsay or deduction what a particular perfomance means; nor can you borrow your schmaesthetic values; you must acquire them for yourself, in honest encounter with the particular throw of dung.

As we elaborate this story, we begin to find important parallels with practices of our own. The first of these is sport. Sport, as we know it, is really two activities: that of the sportsman, and that of the observer, and the interests of the two do not always coincide. A woman playing tennis intends to beat her opponent, and does everything in her power – subject to the rules of the game – to achieve this end. The spectator may not be interested in *who* wins, or even in the winning. He is more interested in the game itself, in the skill and ingenuity of the players, and the excitement of the contest. If the woman begins to share the spectator's interest – and to concentrate on the appearance of the game, rather than the goal of winning – she is likely not only to lose, but to bring the game to a disastrous conclusion.

In other activities, however, we find the spectator and the performer more closely aligned. The actress in the theatre is not merely interested in how she appears to the spectator; she is herself the spectator of her actions. What is right, for her, is what looks and sounds right to *herself as spectator*, and, she hopes, to other spectators as well: good judgement in an actress being the ability to see her own performance through the eyes of those who merely observe it.

Dung-throwing is more like a theatrical performance than a spectator sport, for the reason that it has no *internal* aim. Of course, one could *imagine* an internal aim. The goal might be to throw the dung as high as possible: a simple form of spectator sport. But if there is no such goal, and yet the performance still attracts public attention, we find ourselves drawn to describing the practice in terms not unlike those we use for our own theatrical and artistic performances. We begin to refer to a particular dung-thrower's *style*, to the *power*, *eloquence*, or *humour* of his throws, and so on. His activity begins to seem like an elaborate act of communication between himself and the audience, in which he endeavours to attract their attention, and to adjust his behaviour so as to present the most satisfactory and most valuable appearance within his power. If the native philosopher now begins to discourse about 'schmaesthetic values', do we not have some inkling of what he means?

But the focus of attention has shifted, away from the performer to his audience. We can understand what the dung-thrower is doing, only because we have identified a certain interest that *others* have in his performance. He is ministering to that interest. In the philosophy of schmart, appreciation is prior to production. And so it is in aesthetics. There is a certain frame of mind, an attitude of contemplation, of which only rational beings are capable. Experience, for us, is not simply a matter of gaining information about the world, or forming plans to change it; it includes the contemplation of things for their own sake, without reference to our interests. Such contemplation seems peculiarly meaningful, containing an intimation of something that we can hardly put into words – maybe even an 'intimation of immortality'. If we are sometimes disposed to describe this experience in religious terms, maybe that too is no accident. If you want a simple definition of art, then art is the practice of ministering to aesthetic interest, by producing objects that are worthy of it.

1. Interests of Reason

All animals have interests. They are interested in satisfying their needs and desires, and in gathering the information required for security and well-being. A rational being employs his reason in the pursuit of these interests, and in resolving, where possible, the conflicts between them. That, according to Hume, is the full extent of reason's writ; for reason is subordinate to our interests, and has no authority to deliver any result apart from the 'relations of ideas'.

Kant argued that there are 'interests of reason': that is to say, interests that we have, purely by virtue of our rationality, and which are in no way related to our desires, needs and appetites. One of these is morality. Reason motivates us to do our duty, and all other ('empirical') interests are discounted in the process. That is what it *means* for a decision to be a moral one. The interest in doing right is not an interest of mine (i.e. of my empirical nature), but an interest of reason *in* me.

Reason also has an interest in the sensuous world. When a cow stands in a field ruminating, and turning her eyes to view the horizon, we can say that she is interested in what is going on (and in particular, in the presence of potential threats to her safety), but not that she is interested in the *view*. No animal has ever stood on a promontory and been *moved* by the prospect; no animal has ever longed for the sight of a favourite landscape. A horse may long to get out of the stable and into the field: but this longing is motivated by the sensuous interest in food and freedom.

A rational being, by contrast, takes pleasure in the mere sight of something: a sublime landscape, a beautiful animal, an intricate flower – and of course (though for Kant this was a secondary instance) in a work of art. This form of pleasure answers to no empirical interest: I satisfy no bodily appetite or need in contemplating the landscape; nor do I merely scan it for useful information. The interest, as Kant puts it, is disinterested – an interest in the landscape *for its own sake*, for the very thing that it is (or that it appears to be). This 'disinterest' is a mark of an 'interest of reason'. We cannot refer it to our empirical nature, but only to the reason that transcends empirical nature, and which searches the world for a meaning that is more authoritative and more complete than the fleeting purposes of life.

This brilliant suggestion is about the only real advance that has ever been made in the subject of aesthetics, and it will occupy us for the rest of the chapter. Before trying to understand aesthetic interest in more detail, however, we should ask what an interest of reason might be. Kant himself was a strange kind of dualist. He believed that the rational being views the world and himself in two ways: in one perspective the world appears as nature, and himself as part of it, subject to causal laws. In the other perspective the world appears as a realm of opportunity, in which the will is tested, and we ourselves are free. This second perspective is that of reason, which is always a law to itself, and which discounts empirical conditions so as to aspire to a universal standpoint, above and beyond the circumstances of the agent. Reason is motivated by the pursuit of objective validity, and takes the same view of the practical world as it does of logical and mathematical argument. An 'interest of reason' is therefore an interest in objective value.

Fichte, who was greatly influenced by Kant, disputed this account. When all is said and done, he argued, 'reason' is no more than another name for the self. The two Kantian perspectives on the world are those of the third-person viewpoint (the not-self), and the first-person viewpoint, in which the self is all. It is from this second perspective – the positing of the 'I' as the centre of its world – that morality springs. The 'I' is the source and the end of practical reason, whose prescriptions and proscriptions arise only in answer to the question 'what shall I do?' Whenever we encounter an 'interest of reason', Fichte suggested, we encounter the self: this is the only possible entity or perspective to which such an interest could be referred. The same is true of aesthetic interest. Disinterested contemplation of the world means contemplation of the world in relation to the self, and contemplation of the self as *part* of the world. (That is a long way from Fichte's language; but it is perhaps not too far from his meaning.) Such a view would explain why aesthetic experience is so gripping: we are seeking for our home in the world: not the home of the body and its appetites, but the home of the self. The endlessness of aesthetic interest reflects the fact that we can never find that home: the self is not *in* the empirical world but lies at its limit. This Fichtean thought was to resurge in Schopenhauer, and also in Wittgenstein's *Tractatus*.

2. The Aesthetic Attitude

We seem to have discovered, then, a special kind of rational interest: interest in something for its own sake, and without reference to our empirical desires. What is the value of such an interest, and what does it tell us about our condition?

Let us begin by describing its characteristics:

(i) It is an interest in the *phenomenal* world, as Kant would put it: that is, the world as it appears. The object of aesthetic interest is perceived through the senses, and the element of experience seems to be essential. You respond to the *look* of the landscape, the *sound* of the birdsong, and the *feel* of the wind against your face. The term 'aesthetic' derives from the Greek word for perception, and was first employed in its modern sense by Kant's teacher A.G. Baumgarten, in order to characterise the distinction between poetry and science. If we value poetry for its truth, then ought it not to give way, in time, to philosophy or science? Baumgarten suggested that there is a kind of *poetic* truth, which is bound up with the experience of reading and reciting: it is inseparable from the images presented, the rhythm of the lines, the sequence of the thoughts, and the sound of the words. In other words, the appreciation of poetry could not be understood in purely intellectual terms, but only as an extension of our perceptual powers.

There are real problems here: too close a relation between the meaning of poetry and the experience of it would make poetry untranslatable. At the same time we know that the translation of poetry is difficult. For the translator must preserve more than the literal meaning: there is a way of perceiving, responding to and imagining the world encapsulated in the original; and also a rhythm of thought itself, which reflects the way in which thought is *experienced*. (Cf. the 'unfolding' of a theme in rhetoric, and the barely translatable beauty of a Ciceronian paragraph.)

(ii) The object is appreciated through a perceptual experience, but also 'for its own sake'. What does this mean? Kant's reference to a 'disinterested interest' is intriguing, but scarcely more than a first shot. Every activity that is directed towards a goal is done for the sake of the goal. But what of the goal itself? Do we not pursue this for its own sake? In which case, cannot an animal be interested in something for its own sake – sexual union, for example? In Kant's sense, however, this is a profoundly *interested* interest: the animal is propelled by desire or need, and his interest vanishes when the desire is satisfied.

The simplest way of identifying the interest we have in mind is in terms of the reasons that might be given for it. If I ask someone *why* he is looking at a landscape, he may reply by referring to something that he wants from it: information, say, as to the enemy's movements. But if he replies by referring only to observable features of the *landscape*, then this is a sign that the landscape is intrinsically interesting to him, for the very thing that it is. His desire is not to change, destroy, consume or make use of it, but simply to study it, and to take pleasure in its appearance.

(iii) The search for meaning. At this point one can envisage an objection. The aesthetic interest, it will be said, always sees more in an object than its mere appearance: if it did not, then it is difficult to see why we should place such a value on it. For example, when looking at a painting, I do not see only colours, lines and shapes. I see the world that is represented by them; the drama which animates that world; the emotion that is expressed through it. In short I see a meaning. Something similar happens when I hear movement in music, and imagine the inner drama that is unfolding through the notes. Even more obviously does this happen in poetry. Someone who merely contemplated the perceived sound of the words would miss almost everything of significance. It is even true of those examples which Kantians emphasise: flowers and trees and birdsong. Those who take an aesthetic interest in flowers are pleased by their 'unity in variety' as the eighteenth-century essayists described it. This is a source of metaphor, imagery and story-telling. Writers who have tried to capture the beauty of nature have always had recourse, like Theocritus and Virgil, to story-telling, encouraging us to see in the contours of a landscape, or hear in the murmur of the breeze, some ghostly drama which is being secretly and eternally enacted there.

The reply to this is again not easy. We must distinguish two kinds of meaning: that which resides *in* an experience, and that which is obtained *through* it. The archaeologist, sifting through the sand in search of buried artefacts, is seeking information. Each experience is valuable to him, because of what it means: but what it means is something other than itself. Passing through the town where I spent my childhood, my mind is filled with memories. These are prompted by the perception of my former home; but they exist independently, and survive in my mind long after the perception has ceased. We can think of many instances of this kind, in which a thought, a belief, a feeling, a memory or an image is prompted by some experience, while existing independently. Such cases should be contrasted with the case considered in Chapter 24, of aspect-perception. When I see the dancers in Poussin's *Adoration of the Golden Calf* I am not merely prompted by the painting to think of them, or to conjure them in my mind's eye. I see them *there*, in the painting. And when I turn my eyes away I cease to see them. If I retain an image of them it is also an image of the *painting*. The meaning of this painting lies *in* the experience of it, and is not obtainable independently. Nor is the meaning a simple matter. I do not see only these dancing figures, and the scene in which they participate. I see their foolishness and frivolity; I sense the danger and the attraction of idolatry, the feebleness of the distant figure of Moses as he casts down the tablets of the law. I cast my mind back to those dreadful years of the sixties, when the world was suddenly full of middle-aged trendies acting in just this way; and a moral idea begins to pervade the aspect of the painting. The figures come before me in a new light, not as happy innocents, but as embodiments of lawlessness, and assassins of the Father.

The meaning here lies in the perception of the painting. That is why you turn to the *painting* in order to understand the meaning, so as to fall within its gravitational field. Meaning is not an 'association' or a train of images: it resides in the painting, and can be understood only through the experience of *it*. All aesthetic meaning is like that. This fact is interesting for two reasons. First, it poses a formal constraint on criticism. When a critic tells us that such and such is part of the meaning of a poem, a canvas or a piece of music, then what he says can be accepted only if we can also *experience* the work of art as he describes it. Fanciful allegories may be read into paintings in which they cannot be seen; hidden structures may be perceived in stories where they cannot be felt; a mathematical order may be discerned in music in which it cannot be heard. Clever critics who tell us of these things (like Barthes in his study of Balzac's *Sarrasine*) are wasting our time.

Secondly it raises the question why this peculiar kind of meaning, in which thought and experience are inseparable, should be of value to us. Why have we made such a special place in our lives for 'the sensuous shining of the idea', as Hegel described it? It is towards the answer to this question that all serious aesthetics tends. (The interested reader will find, however, that there is very little serious aesthetics.)

(iv) Repetition. Many of our interests, once satisfied, are dropped from life's agenda. When you have conned your law book, you set it aside. It has performed its function, which was to teach you the law. The same is true of the scientific or historical text: if you refer to it again, it is because your memory is imperfect. The interest in information is satiable; as is the interest in food. But there are interests which are by their nature insatiable, since they have no goal. Aesthetic interest is like this. People are often led to say such things as 'I have listened to the Jupiter Symphony a hundred times, and each time I find something new in it'. What they mean, however, is that they each time find something old in it: the very same experience calls to them, again and again, and still they repeat it. For there is nothing they are seeking which could bring their seeking to an end.

But this leads me to the fifth feature of aesthetic interest, and the aesthetic attitude that is built on it, which deserves a section to itself.

3. The Antinomy of Taste

An experience to which we are repeatedly recalled, which is imbued with meaning, and which is available to us only when we set our interests aside, sounds very like an experience of value. Indeed, Kant argued, that is what it is. The ideas of obligation and right, which grow from the exercise of freedom, inhabit our aesthetic perceptions. An 'interest of reason' cannot be content with less. Moreover, if my attitude towards the aesthetic object is truly disinterested, I refer its value to no desire or need, to no empirical predicament of my own: it has its value intrinsically. I discern aesthetic value by exercising the universal faculty of reason. You too could discern it, to the extent that you laid your interests aside. Hence you *ought* to discern it. If you do not, it is because your judgement is clouded by self-interest. You do not attend to the object as it really is – or rather, as it really appears. Aesthetic interest leads to the 'judgement of taste' which, like the categorical imperative, issues in a statement of law. All rational beings ought to feel as I do: if they do not, it is because they are not attending to what is there. (This looks most plausible, of course, if you consider extreme examples – whether of bad taste (*American Psycho*) or of good (*The Iliad*).)

If we accept that, however, we are landed with a paradox. Kant called this paradox the 'antinomy of taste', and derived it as follows. On the one hand aesthetic experience is 'immediate': its meaning is contained indissolubly within it, and cannot be detached from the experience. Hence, we cannot say, in universal terms, why the experience matters. There is no universal law of the kind: 'all paintings which show idolatry as a human weakness are good'. The features that make one painting great will make another one absurd: everything depends upon the context. That is why, Kant observed, you can never make your aesthetic judgements at second hand. You may be able to *guess* that Tintoretto's *Crucifixion* is a great painting, from the critic's description. But you cannot *know* until you have seen it. The description will always fall short of the meaning. This suggests that there can be no reasoned argument for an aesthetic conclusion: the judgement will always be in some sense 'groundless'. Hence there can be no right and wrong in matters of taste: *all* that matters is the immediate pleasure of the observer. So we derive the opposite conclusion from the one above. Both conclusions are compelled by our theory of aesthetic experience. Aesthetic experience *cannot* issue in an objective judgement; and it *must* issue in an objective judgement.

That is not the way in which Kant derived the contradiction; but it will do for our purposes. As with all paradoxes, we have several ways out: we can try to show that the contradiction is only apparent; we can dispute the premises; and we can dispute the reasoning that leads from them. And since, in this case, all three components are far from compelling, there are many opportunities to escape. But Kant is clearly on to something. For the paradox emerges whenever we try to discuss aesthetic choices. People who loathe modern architecture do not rest their case with the words 'it is all a matter of taste; you to yours and me to mine'. Their characteristic attitude is one of outrage: 'How *dare* you impose on me, against my will, the sight of something so loathsome!' And they will strive hard to justify their judgement, adding that the offending object is inhuman in its scale and materials; that it is alienating; that it destroys some fragile corner of the human world, or presents us with an image of public life that is meaningless, mechanical, overbearing. At the same time, they feel a pressure in the opposite direction, knowing that their reasons fall far short of proof, and that they depend upon the forlorn attempt to persuade the opponent to *see* the offending object in a way that he is not disposed to see it. 'Look,' comes the reply, 'you say that concrete is a brutal material; but it is concrete that forms the dome of the Pantheon, your favourite classical building. You say that the scale is inhuman; but it is nothing beside the scale of Chartres Cathedral, which you admit to be the greatest building in Europe. You say that it obliterates the street, and refuses to align itself with its neighbours; but that is precisely what is done by your favourite mosque in Isfahan.' And so on. Only the *particular* building, with its particular materials, scale and orientation, can persuade your opponent; and it is precisely in his experience of this particular thing that he is disposed to disagree with you.

And yet the matter cannot be allowed to rest. We even have laws which make it a crime to put up buildings of a certain kind, and these laws have a purely aesthetic justification. The city of Venice has been governed by such laws since the middle ages, with wonderful results. Their absence caused the total destruction of American cities by money-making vandals. Here, then, is something that we care about; a matter in which we know that there is a right and a wrong, and in which we can never win the argument. That is why aesthetic experience is so distressing, and why many modern people do their best to avoid it.

But can they avoid it? Kant said no. Interests of reason are inescapable. You can pretend to evade the moral law, but you will never live happily with the attempt. You can pretend to ignore the degradation of music, the spoliation of the landscape, the destruction of the town; but in your heart you will gradually be subdued and even destroyed by these things. Aesthetic judgement is a part of practical reason, and our truest guide to the environment. It is by aesthetic judgement that we adapt the world to ourselves and ourselves to the world. Take it away, and we shall be homeless: in the place of comfortable solitude will come irritated loneliness. Schiller went further, arguing that the 'aesthetic education' of man is his one true preparation for rational life, and the foundation of any ordered politics.

Those are bold ideas; but modern life has done something to confirm them.

4. Ends and Values

But perhaps we ought not to give so prominent a place to the aesthetic in our view of human fulfilment. For the plausibility of the Kantian position stems partly from the fact that aesthetic experience is the point of intersection of two other rational interests: the ludic and the religious. It is not only the object of aesthetic interest that is valued for its own sake. The same is true of people, of activities, of institutions and ceremonies. When I work, my activity is a means to an end:

making money, or 'producing value', as the Marxians prefer. When I play, however, my activity is an end in itself. Play is not a means to enjoyment: it is the very thing enjoyed. And it provides the archetype of similar activities: sport, for instance, conversation, socialising, maybe art itself. Schiller noticed this, and went so far as to exalt play into the paradigm of intrinsic value. With the useful and the good, he remarked, man is merely in earnest; but with the beautiful he *plays*. (*Letters on the Aesthetic Education of Man*.)

There is an element of paradoxism in Schiller's dictum. But you can extract from it a thought that is far from paradoxical, namely this: if every activity is a means to an end, then no activity has intrinsic value. The world is then deprived of its *sense*. If, however, there are activities that are engaged in for their own sake, the world is restored to us, and we to it. For of these activities, we do not ask what they are *for*; they are sufficient in themselves. Play is one of them; and its association with childhood reminds us of the essential innocence and exhilaration that attends such 'disinterested' activities. If work becomes play – so that the worker is fulfilled in his work, regardless of what results from it – then work ceases to be drudgery, and becomes instead the 'restoration of man to himself'. Those last words are Marx's, and contain the core of his theory of 'unalienated labour' – a theory which derives from Kant, via Schiller and Hegel.

The parallel between art and play has been emphasised by more recent thinkers – notably Kendall Walton, in his *Mimesis as Make-believe*, in which he argues that an artistic representation is like a prop in a game of make-believe. The idea is a fertile one, and reinforces the link that we have made, between aesthetic interest and imagination. But it also suggest how deep is the connection between art and leisure, and how art is only one of many activities in which we are at home in the world and at ease with ourselves.

Consider conversation: each utterance calls forth a rejoinder; but in the normal case there is no direction towards which the conversation tends. Each participant responds to what he hears with a matching remark, and the conversation unfolds unpredictably and purpose-lessly, until business interrupts it. Although we gain much information from conversation, this is not its primary purpose. In the normal case, as when people 'pass the time of day', conversation is engaged in for its own sake, like play. The same is true of dancing.

In order to understand such activities, we must distinguish purpose from function. A socio-biologist will insist quite rightly that play has a function: it is the safest way to explore the world, and to prepare the child for action. But its function is not the purpose. The child plays because he wants to play: play is its *own* purpose. Indeed, if you make the function into a purpose – playing for the sake of learning, say – then you cease to play. You are now, as Schiller puts it, 'merely in earnest'. Likewise, the urgent man who converses in order to gain or impart some piece of information, to elicit sympathy or to tell his story, has ceased to converse. Like the Ancient Mariner, he is the death of dialogue.

Conversation is a fleeting thing; but what I have said applies even more evidently to those long-term relations of love and friendship which are among the highest of human goods. Aristotle distinguished three kinds of friendship. There is the friendship founded in pleasure, as when children play together, men drink together, or women gossip. Such friendships show the primitive form of mutuality: they are valued for their own sake, but they are fleeting, and the companion can easily be replaced by another who will 'do just as well'. They contrast with friendships of utility: as in a business partnership, or a joint operation for a common goal. Here friendship is subordinate to a purpose, and may dissolve when the purpose is frustrated or fulfilled. Aristotle's third form of friendship is friendship as we should normally conceive it: in which the other is valued for his own sake, regardless of the pleasure or utility that we derive from him. Aristotle believed that such a friendship is possible only between virtuous people. That may not be true; nevertheless, we can recognise in Aristotle's discussion an important insight into the varieties of human commerce: there are relationships which are valued for their own sake, and those which are valued as a means. And among those valued for their own sake there is a lower, less formed, more infantile variety, and a higher, more conscious and more lasting kind, in which the other person is not just enjoyed but also valued.

Friendship brings us into the kingdom of ends, though it is a warmer and livelier realm than the one described by Kant. In friendship everything is an end, nothing a means only. I strive to please you; I do things for your sake, and not for any interest of mine (except my 'interest of reason' which is to do things for another's sake). The friendship itself is my purpose, and I treat you not as a means to that purpose but as the end of it. In friendship the quest of reason for the why of things comes to a final end: and the end is you.

Friendship has a function: it binds people together, making communities strong and durable; it brings advantages to those who are joined by it, and fortifies them in all their endeavours. But make those advantages into your purpose, and the friendship is gone. Friendship is a means to advantage, but only when not treated as a means. The same is true of almost everything worthwhile: education, sport, hiking, hunting, and art itself.

Those are difficult thoughts to unravel: but it is not absurd to believe that, if we could unravel them, we should have a clue to the meaning of life. Meaning lies in intrinsic value; we understand it by discovering the thing that interests us for its own sake; and such an interest must be disinterested, in the manner of aesthetic experience, friendship, and every other activity where we are not 'merely in earnest'.

Nor is it only in the realm of subjective spirit that these activities are important. Oakeshott has argued that the greatest mistake in political thinking is to envisage the *polis* on the model of an 'enterprise association' – a partnership for some common purpose. Yet that is how we view the state, whenever we subordinate the natural growth of peaceful congregation to some overmastering plan. 'Civil association' is to be compared less with an enterprise than with a conversation. (Aristotle, for similar reasons, described it as a species of friendship, in the wide sense encapsulated in the Greek *philia*.) Many have found inspiration in Oakeshott's idea: others have criticised it for ignoring Hegel's distinction between civil society and state. However, whether political organisation is to be seen as end or as means, it will never answer to our deepest needs, if it makes no room for associations that *are* ends in themselves.

5. The Religious Experience

Some light is cast on the Kantian position by setting aesthetic experience in another context, and studying another general phenomenon of which it is an instance. There are other states of mind besides the aesthetic, in which meaning and experience are intimately conjoined, and where questions of right and wrong seem forced on us. In particular, there are acts of worship, religious rites and ceremonies, and all the various 'graces'and 'epiphanies' that fill the literature of religious ecstasy. These experiences may involve an aesthetic component. But they are not purely aesthetic, since they are not disinterested. We do not participate in religious rites merely so as to contemplate their meaning in the detached way that we would contemplate a play or a painting. We are genuine *participants*, who are engaged for the sake of our salvation and with a view to the truth.

Nevertheless, there are interesting similarities with the aesthetic experience. Although the purpose of an act of worship lies beyond the moment – in the form of a promised salvation, a revelation, or a restoration of the soul's natural harmony – it is not entirely separable from the experience. God is *defined* in the act of worship far more precisely than he is defined by any theology, and this is why the forms of the ceremony are so important. Changes in the liturgy take on a momentous significance for the believer, for they are changes in his experience of God. The question whether to make the sign of the cross with two fingers or with three can split a church, as it split the Orthodox Church of Russia. So can the question whether or not to use the Book of Common Prayer or the Tridentine Mass.

The religious rite resembles the aesthetic experience in other ways as well. It is inexhaustible and endlessly renewable. The person who goes once to Mass, and comes away saying 'now I know what it means and there is no need to go again', has not seen the point of it. Even if he is able thereafter to remember every gesture and every word; even if he gives the most subtle and persuasive commentary on the associated theology and doctrine, he still has not seen the point of the ritual. The meaning of the Mass is inseparable from the experience and must be constantly renewed. You must enter the frame of mind in which you 'cannot have enough of it'; not because you look forward to it – on the contrary, you might, like Amfortas, dread it to the point of preferring death – but because it has become a spiritual necessity.

There is another, and more elusive, comparison between the aesthetic and the religious. The subjective nature of the aesthetic experience goes hand in hand with an implied idea of community: in thinking everything away except this unique and present object, and in addressing myself to it with all my interests discounted, I am also opening my mind to the thought of its meaning – not for me, only, but for the kind of which I am a member. The aesthetic experience is a lived encounter between object and subject, in which the subject takes on a universal significance. The meaning that I find in the object is a meaning that it has for all who live like me, for all whose joys and sufferings are mirrored in me. This is reflected in the antinomy of taste. As Kant put it, aesthetic judgement makes appeal to a 'common sense': it frees me from the slavish attachment to my own desires. I come to see myself as one member of an implied community, whose life is transcribed and vindicated in the experience of contemplation.

In the religious experience too there is an implied but absent community: not the community of the living only, but a community that extends to the dead and the unborn. The religious stories concern people long since gone beneath the earth; the ceremony of participation can be repeated unchanged from age to age; it unites me with my own dead, and also with those who are yet to be. Changes in the ritual are disturbing, since they suggest that the community may be cut short by time, that the words and gestures that I employ are no better than provisional, and that we shall all be forgotten.

This absent community is the real meaning of solitude, the justification of the hermit and the anchorite, and also the 'real presence' of the Sacred Service or the Mass. It is there in our thoughts, just as soon as we detach them from the fashions and appetites that set us against one another. And that is why religious and aesthetic experience are therapies for social life.

In *The Birth of Tragedy*, Nietzsche speculates on the religious origin of tragedy, and comes up with the following suggestion. The worshippers of Dionysus cast off their worldly concerns and join in a dance. This dance is an invocation of the god and he is present in it. All music derives from this desire to dance together, in a community that embraces each of us, and cancels our separation. The chorus that we form tells us the story of the god; and also the story of those who separate themselves, as we all must separate ourselves, from the pure communion, so as to embark on some fatal project of our own. Out of the dance there steps the tragic hero, whose fate appals and fascinates his fellow dancers. He acts apart, affirms himself, and is destroyed, sinking back into the unity from which he briefly emerged. There lies the consolation of the tragic dance, that the individual transgression is enacted at a distance, accepted and at last overcome.

The same experience can be repeated in the theatre. The audience dances by proxy, through the chorus of the play. The tragic hero is the centre of a represented action. The god himself has been quietly hidden away. But, Nietzsche suggested, it is essentially the same experience. And maybe it is the same experience when a priest recounts the tale of Christ's passion, reminds his congregation of their sins and the separation from God that sin engenders, and then invites them to a common 'sacrifice'. In each case the same story is told: the ideal community, the act that separates us (whether error or sin), and the ultimate restoration as the community is restored – not now a community of the living, but one of the living, the dead and the unborn. The tragic hero who passes over to the dead is like the worshipper who joins them in his worship.

6. Imagination and Design

Those ideas are fanciful, and this is not the place to develop them. But they help us to explain another of Kant's thoughts about the aesthetic. Disinterested contemplation sets all purpose aside. Not only its *own* purpose; but the purpose of the object too. The beautiful work of art, or the sublime landscape, are not understood by finding a purpose for them, and judging whether they are suited to it. (That would be to confuse, in Collingwood's terms, art with craft.) But just as the disinterested attitude is at the same time the expression of an interest (namely an interest of reason), so is the purposelessness of the aesthetic object a kind of purposefulness. We see the object as designed in every particular, brought together by a controlling intention, yet lacking any purpose beyond itself. This 'purposefulness without purpose' is characteristic of the aesthetic; it shows the world as designed, and ourselves as part of its design. We sense a kind of harmony between our own faculties and the phenomenal world to which they are directed. The world was designed to be contemplated, and it reflects the purposeful nature that we also find in ourselves.

Hence the aesthetic attitude leads naturally into the religious attitude, which sees all of nature as an expression of will. We cannot possibly know *what* God's intention was in creating the world. But we can at least see *that* he created it, by recognising the sublime purposefulness that flows through all things, and which ensures that even creatures as strange and paradoxical as we – free beings who are compelled at every moment to see themselves as apart from nature – can find their home in the natural order.

As a version of the argument from design this is most defective. Even if it is true that the aesthetic experience *sees* its object as purposeful, it does not follow that the object really *is* purposeful, any more than that there really *are* people dancing at the place where Poussin's *Golden Calf* now hangs. But Kant's argument makes an important contribution to religious anthropology, and it is worth pondering it. Kant is suggesting that we can extend to the whole of nature those attitudes which get their sense from interpersonal relations. When we see someone smile we see human flesh moving in obedience to impulses in the nerves. However, we understand the smile not as flesh but as spirit. It is an expression of freedom and will, and not merely a natural process. A smile is always more than flesh for us, even if it is only flesh.

And this is how we should understand such concepts as the

miraculous and the holy. A miraculous event is one which wears for us a personal expression. We may not notice this expression, just as someone may stare at a portrait, see all the lines and colours which compose it, and fail to see the face. Similarly, a sacred place is one in which personality and freedom shine forth from what is contingent, dependent and commonplace – from a piece of stone, a tree, or a patch of water. There is an attitude that we direct to the human person, and which leads us to see, in the human form, a perspective on the world that reaches from a point outside it. We may direct this very attitude, on occasion, to the whole of nature, and in particular to those places, things and events where freedom has been real. The experience of the sacred is the sudden encounter with freedom; it is the recognition of personality and purposefulness in that which contains no human will. In this way, our experience may be understood as a revelation of the divine. It is our hunger for that revelation which causes us, in an age without faith, to invest so many hopes in aesthetic values.

7. Epistemology Naturalised

One of the problems for religious faith is the epistemological difficulty, of discovering some ground in thought or experience that could possibly justify such a vast hypothesis. Epistemology is not what it was in Descartes's day. Under the influence of Wittgenstein and Quine, philosophers tend towards a third-person approach to epistemological questions, describing our epistemological capacities as features of the natural world. (See Chapter 5.) Those things really exist, they argue, which are referred to in the true *explanation* of our beliefs. For example, our beliefs about the physical world are best explained by the assumption that the physical world exists, in the form that science describes, and that we have evolved in such a way as to gain accurate information about its superficial contours. Beliefs about physical objects are therefore best explained on the assumption of their truth. We have seen the impact of that approach in the theory of knowledge and perception. Why not extend it to more difficult regions – morality, aesthetics and theology?

In the case of morality the result is obscure and difficult to interpret. In the case of theology, however, it is relatively clear. If the claims of faith are true, then God is transcendental. He is not part of nature, and not the possible object of a scientific inquiry. No scientific explanation of religious belief, therefore, could conceivably refer to him. It follows that, if there *is* an explanation, it will be 'naturalistic': it will explain religious belief in terms of forces and functions which make no reference to God. Precisely such an explanation has been assumed, both in this chapter, and in Chapter 11, when I discussed the nature of God. The belief in God is a response to our sense of awe and holiness, which in turn is a response to our need for a community, in which the dead and the unborn surround us with their faint but relentless affection. Does such a 'naturalised epistemology' destroy the claims of faith? Does it follow from the fact that the best explanation of our belief in the transcendental makes no reference to the transcendental, that the belief is unfounded? (Try the same for our belief in the existence of numbers.) The question is hard to answer. For the existence of such an explanation is precisely what God's existence would entail. Moreover, the explanation that we have given makes the belief (or something like it) a natural by-product of the social process whereby we come to believe or question anything. Hence St Augustine's remark, that 'our hearts are restless until they rest in Thee'.

Other ways of explaining religious belief make a more direct assault on the thing explained. Consider the explanation offered by Feuerbach, under the influence of Hegel, which provided Marx with one of his major ideas. (*The Essence of Christianity*.) Religion in general, Feuerbach argued, and Christianity in particular, can be seen as an elaborate device whereby man frees himself from the arduous task of self-improvement, by personifying his virtues and his communal life, and setting them up outside himself in a transcendental realm. Since access to that realm is forbidden to a merely empirical being, man thereby sunders himself from the possibility of his own improvement, and alienates himself from his 'species being'. Religion does not fulfil man's social nature, but thwarts it; it does not offer a path to virtue, but closes that path eternally; it does not provide our home in the world but alienates us from the world and ourselves.

This idea belongs to a spiritual and intellectual project which we must now consider.

30

The Devil

Could we envisage a negative ontological argument, proving the non-existence of the devil? On the model of Plantinga's argument for the existence of God, should we not define the devil as a being of 'minimal greatness', i.e. possessing all negative attributes, of which non-existence must certainly be one? That sounds absurd, partly because the temptation to regard *non*-existence as a predicate is virtually zero. How can non-existence be a *property* of anything?

But it is absurd, too, for another reason, namely, that evil is *increased* and not decreased by its reality: an evil deed is a greater evil than an evil thought. Hence non-existence would not be an addition to the devil's evil attributes, but a detraction from them. Only if we were to invent some new predicate, the analogue of existence, but attaching solely to destructive entities, could we propose an ontological proof of the devil – not as non-existing, but as existing negatively, so to speak. As a matter of fact there is one philosopher who has proposed such a predicate, namely Heidegger, who famously argued that there is something that is true of Nothing, namely that it 'noths' (Das Nichts *nichtet*). As always, he had a phenomenological idea in mind: there is a certain experience, which we might describe as the encroachment of nothingness: it is as though Nothing were doing something to us, as it steals across the landscape of the soul.

This experience is recorded in different words by Sartre, in his remarkable work *Being and Nothingness.* Nothingness, he tells us, lies 'coiled in the heart of Being, like a worm'. I can encounter Nothing at any juncture: for example, when I look for someone in a café where I expect to meet him, and he is not there. The world is suddenly coloured by his absence; and this negative fact has a peculiar reality all of its own. But, however strange this experience, it is surely not an archetype of evil. Entering Les Deux Magots to find that Sartre is not there is one of life's blessings.

1. Negation and the Moral Law

There is a kind of negation, however, which is captured neither by the negation sign in logic, nor by the phenomenology of absence. I mean the repudiation with which we rid ourselves of that which irks us. The human animal is distinguished by 'neoteny'. He is born long before the stage of development in which he is capable of surviving unaided. The growth of the human being does not parallel the growth of other animals, who can quickly strike out on paths of their own, after the first few months of dependence. Not only must the human being *learn* what he needs for survival over many years; his body itself reaches maturity only when a third of life is over, while his spirit passes through endless turmoil, and is never mature, even on the threshold of the grave.

Inevitably, therefore, human beings grow into a state of dependence on their parents which has no parallel in the animal kingdom. Unless there were some countervailing force, strong enough to rupture the primal bond, the young would never take that first step into the fearful world of competition and estrangement, so as once more to begin the cycle of reproduction. The passage from family to civil society, Hegel suggested, involves a dialectical negation. There is a sudden, even violent, repudiation of the 'comfortable' values of the home – a rejection precisely of that which renders home attractive. Solitude, risk and adventure magnetise the adolescent soul, and the journey outward into civil society, which is the realm of strangers, begins.

The journey into estrangement generates an intense longing for home; and in every human heart there remains an image of safety, of the final redemption which is also a return to the place from which one started, only to 'know it for the first time'. This – the third moment of Hegel's dialectic – is the goal of religion, and there is no image of consolation that is not conceived in such terms. Whether the plot takes the form of high romantic drama as in the *Flying Dutchman*; or a mystical summoning of the silent dead and their dead beliefs, as in *Four Quartets*; or a journey back into the landscape of childhood, as in Hölderlin's *Heimkehr*, the effect is the same: the crime of repudiation is expiated, and the soul is welcomed to its eternal home. The individual is shriven of original sin, which is, in Schopenhauer's words, 'the crime of existence itself' – that is, the *hubris* of individuality, and the alienation that springs from it.

The Dutchman notwithstanding, there need be nothing demonic in the state of repudiation. Some leave home with a mild rebuke to those who would keep them there; others look on their parents with humour and affection, and gradually discard their old dependence, usually by acquiring a new one. But there are situations – as described by Edmund Gosse in *Father and Son* – in which there is no way out of the family drama without the tragic rejection of the father's law. Such a rejection may lead at last to reconciliation, as the son learns to sympathise with the father, and to see the old law as a necessary precursor to his own self-affirmation. But it may lead to a lifelong posture of negation, a refusal to accept any external authority and a rejection of every value, every custom, every norm which impedes the 'liberation' of the self.

This arrest of the soul in the posture of negation is worthy of study, since it is at the root of much that passes for philosophy in a modern university: from Marxism to deconstruction, the modernist philosopher has occupied himself with the proof that there is no authority, no source of law, no value and no meaning in the culture and institutions that we have inherited, and that the sole purpose of thought is to clear the way for 'liberation'. Since modernism, so defined, is incompatible with the argument of this book, it is worth trying to understand its message.

The moral law, Kant argues, is addressed to all rational beings, in the form of a categorical imperative: it has not changed through the centuries, nor could it change. The same thought is expressed through the ten commandments, in the new law of Christ – thou shalt love thy neighbour as thy self – and in the newer law of Kant himself: thou shalt not treat humanity as a means only, but always as an end.

The spirit of rebellion resists the idea that there is a universal law of reason. It is the envious father who has written 'Thou shalt not' over the entrance to the temple of delights. So Blake tells us, and such is the message of Marx and his modernist disciples. A law should be understood, they argue, in terms of the interest that is advanced by it. This is how laws are to be explained. Society is a system of powers, which are held in place by the myth of their legitimacy. Law and morality are part of that myth, and their ability to survive is derived solely from the credulity of those who lose by obeying them.

Marxism tries to ground this repudiation of law in a science of history, according to which the institutions of a society are dependent on the relations of production which form its economic base. Each successive system of production relations – slavery, feudalism, capitalism, socialism and finally communism – is brought about by the development of the 'productive forces' of a society, and each (apart from the last, which is self-sustaining) is held in place by a superstructure of institutions and laws. The legal right of private property is explained by the fact that property rights reinforce bourgeois 'relations of production'. In doing so, however, they also contribute to the maintenance of a system that is incompatible with man's final liberation. This will come only with the advent of communism, when private property will be inconceivable, there being no relations of production that it could serve to maintain.

The theory is more complex than that. But it is often taken to show that there is no such thing as a valid law, or a valid system of morality, outside the economic order which requires it. In defending a law as absolute, you not only deny the 'historicity' of human institutions; you affirm the timeless validity of a particular order – usually the bourgeois order – and set your face against social change. It is almost as though morality itself requires the repudiation of morality. In attaching yourself to the moral law, you deny the hope for liberation.

The paradox expressed in those last sentences has been willingly embraced by revolutionaries, from Robespierre and St Just to Pol Pot and the *Sendera Luminosa*. It has been most clearly expressed in our time by the Hungarian communist György Lukács, a philosopher and literary critic who was one of the few members of Imre Nagy's government to escape execution after 1956. 'The question of legality or illegality reduces itself . . . for the Communist Party to a *mere question of tactics*,' he wrote in *History and Class-Consciousness*, one of the favourite books of the sixties radicals. 'In this wholly unprincipled solution lies the only possible practical and principled rejection of the bourgeois legal system.' (You can see here the exultant paradoxism which grips the revolutionary, who, like the mystic, is intoxicated by the thought of his distinguished destiny.) What is true of the legal system is true of every other feature of the bourgeois world: economic practices, social relations, emotions, ambitions, even morality itself. Thus Lukács was able to assert that 'Communist ethics makes it the highest duty to accept the necessity to act wickedly,' adding that 'this is the greatest sacrifice revolution asks from us'.

We are familiar with that sacrifice; and it was not paid by the revolutionary intellectuals themselves, but by the countless millions of their victims. As Lukács also wrote, 'it is not possible to be human in bourgeois society', so that 'the bourgeoisie possesses only a semblance of a human existence'. It is not so hard to exterminate people, when you describe them in such terms. The bourgeois is the Father, whose law is to be swept away; and if he himself must be swept along with it, then has he not asked for just such a destiny?

The scientific pretensions of Marxism have been extremely important as an exonerating device. If revolution is inevitable (being the result of an unavoidable clash between productive forces and the production relations that impede their development), so too are its crimes – which are not crimes at all, but merely birth-pangs of the new social order. Those who seek to facilitate the birth must act quickly and violently. But the small spillage of blood that results is the price of life and liberation.

2. The Genealogy of Morals

Marxist materialism is one example of the 'genealogy of morals', as Nietzsche called it, in a famous book of that title. It tries to explain the institution of morality, by showing how it arises in society, what function it fulfils, and why people are induced to subscribe to it. The explanation owes its appeal largely to the fact that it seems to undermine the thing explained. There is very little truth in the Marxian theory of ideology; but there is a great deal of Mephistophelian charm. If the world is like this, then surely morality has no claim to our endorsement? Indeed, we may have a kind of shadow obligation to jettison morality. The philosophy which undermines morality also creates a shadow morality of its own.

Nietzsche's genealogy has the same effect. He tries to show that moral values, in the Christian tradition, have been endorsed because they give strength to those who are inherently weak, so permitting the natural slaves to overcome their natural masters. Morality, by exalting the sentiments of pity and meekness, unites the herd against the hero, and excludes him from the fruits of his energy and power.

Nietzsche's genealogy has little in common with Marx's from the point of view of content. But it has a similar charm. If you happen to believe that you are, deep down, an *Übermensch*, whose powers, energies and talents are frustrated by the little people who tie you with their Lilliputian web of obligations, you will gladly endorse Nietzsche's vision, and believe that, by so explaining morality, you also explain it *away*.

The massive excuse prepared by Marx for the crimes of the

international socialists, was prepared by Nietzsche for their nationalist opponents. The same breathless cry of liberation can be heard in the propaganda of the Nazis, this time sounding in Nietzschean accents. Morality was scorned, as a mere artefact of the old society; not as an instrument of oppression, but as a disease by which the poorer specimens neutralise those who would rightfully oppress them.

Both philosophies raise an important question: When does the explanation of a belief also undermine it? Would the truth of the Marxian or Nietzschean genealogies in either case justify the rejection of moral claims? This parallels the question raised about religion. But it is, as I said, far harder to answer. Perhaps the best that can be said is that we ought not to believe that morality has been undermined, until the proof is before our eyes. The mere fact that someone has thought of a way to explain morality without reference to its truth should not license our liberation.

3. Alienation

Before proceeding, mention must again be made of a concept which, since Hegel, has played a critical role in defining the condition of modernity: the concept of alienation. Hegel's dialectic implies that all knowledge, all activity and all emotion exist in a state of tension, and are driven by this tension to enact a primeval drama. Each concept, desire and feeling exists first in a primitive, immediate and unified form – without self-knowledge, and inherently unstable, but nevertheless at home with itself. Its final 'realisation' is achieved only in a condition of 'unity restored', a homecoming to the primordial point of rest, but in a condition of achieved self-knowledge and fulfilled intention. In order to reach this final point, each aspect of spirit must pass through a long trajectory of separation, sundered from its home, and struggling to affirm itself in a world that it does not control. This state of alienation – the vale of tears – is the realm of becoming, in which consciousness is separated from its object and also from itself. There are as many varieties of alienation as there are forms of spiritual life; but in each form the fundamental drama is the same: spirit can know itself only if it 'posits' an object of knowledge – only if it invests its world with the idea of the other. In doing this it becomes other to itself, and lives through conflict and disharmony, until finally uniting with the other – as we unite with the object of science when fully understanding it; with the self when overcoming guilt and religious estrangement; with other people when joined in a lawful body politic.

Alienation was not, for Hegel, an evil, but a necessary part of every spiritual good, whether theoretical or practical. But the concept soon took a turn for the worse. Already in Feuerbach we find the idea that religion is *essentially* alienating, since it divides man from his essence in social life; religion is therefore something to be *overcome*. In Marx, the same is said of private property and the institutions of a 'capitalist' society. These sunder us from our social nature and must therefore be overcome if we are to 'realise' our freedom as social and rational beings. Feuerbach was a 'left Hegelian', as was the young Marx. They understood the dialectic as a diagnosis of human misery, and also as a promise of its cure. Both of them believed that history is inexorably proceeding towards a solution of the problems that history creates.

Other philosophers, equally impressed by the idea of alienation, have taken a bleaker view of it. For the existentialists, alienation is part of our condition as free beings. We see the world as other than ourselves, sundered from us, since it is a world of pure things, in which we are strangers. All we have is our freedom of choice, and if there is a cure for alienation it is for each of us severally to provide it, by exercising his freedom, and becoming what he truly is.

Even for those who are unimpressed by the existentialist 'solution', the problem remains. We are not at home in the world, and this homelessness is a deep truth about our condition. Is there anything we can do to alter it? Two things are usually identified as contributing to alienation: 'capitalism' and scientific thought. (Husserl emphasised the second of those, as did Ruskin and his followers in England.) Capitalism is a rude name for the market – i.e. for the system of economic relations established by consent. And science is a name for the sum of propositions thought to be true. Science and the market are the two fundamental forms of man's relation to an objective order: the two ways of recognising the world's objectivity, whether as thing (and therefore object of knowledge and use) or as person (and therefore subject of consent). But that is precisely what Hegel's original argument suggests: any attempt to recognise an objective reality is also an estrangement from reality. Here, indeed, is the root of original sin: through consciousness we 'fall' into a world where we are strangers.

Some of the emotional pull of these ideas can be explained by neoteny and the trajectory of adolescence. But their full intensity is best understood, once again, in terms of Durkheim's theory of religion. We are by nature social beings, who are born into a condition of dependence. In our 'angel infancy' we are at one with the world, protected and embraced by it. The separation between self and other is hardly apparent in the primordial experience of the tribe. We seek to recapture this experience through religion, but in the higher, conscious form, that can be purchased only by a knowledge of our separation. To affirm ourselves is to transgress; to worship is to be restored. In short, the sense of alienation stems from the loss of 'membership', and from the desire to be accepted once again by our tribal deities.

Now the relations established by a market, like those created by science, have a *universal* character. A contract requires no bond, no anterior attachment between the parties, and its meaning is exhausted by its terms. Moreover, terms are dictated 'impersonally', by the rational self-interest of all who have access to the market. The alienating quality of the market reflects the fact that the 'alien' has utter equality with the friend. Similarly with scientific thought. The categories of science arise directly from our rational interest in truth. Science is therefore *common* to all rational beings, and the peculiar possession of none of them. When I engage in scientific inquiry I free my perception of the world from intentional concepts, and from categories whose sense derives from relations of membership and obligation. I no longer see the world under the aspect of 'belonging'; I am not 'at home' in the world of science, for precisely the reason that I am just as much at home there, and just as little, as everyone else.

Thus arises the condition that Kierkegaard described as 'despair': the condition in which I cannot find the meaning of the world, since wherever I turn I see only the absence of God (and of the community which sheltered him). The work of the devil is to persuade us to remake the world in the image of this despair, while at the same time exciting us to the heights of religious passion, so that we believe that, in doing so, we are acting for the greatest good. We should bear this in mind when considering the great events of the modern world, and in particular those extraordinary revolutions which, beginning from the despair of intellectuals, ended in the despair of everyone.

4. Them

Conscious despair is the enemy of unconscious contentment. We should not be surprised to find, therefore, that intellectual alienation has gone hand in hand with a suspicion of ordinary humanity. This suspicion attains a kind of metaphysical dignity in the writings of Sartre and Heidegger, for whom Others are the source of inauthenticity, and therefore of the only crime that the modernist can commit: the crime against the self. My freedom is my essence and my salvation: I cannot lose it without ceasing to be. But it is everywhere threatened: I live as a subject among objects, and the danger is that I might 'fall' into that world of objects, and become one with them. In reaction, I may hide from myself, bury myself in some predetermined role, contort myself to fit a costume that is already made for me, so crossing the chasm that divides me from objects only to become an object myself. This happens, according to Sartre, when I adopt a morality, a religion, a social role that has been devised by others and which has significance for me only in so far as I am objectified in it. The result is 'bad faith' – the 'inauthenticity' of Heidegger, which is the normal condition of our bourgeois order. And bad faith is at the same time a state of alienation. Thus, according to Sartre, 'my Being-for-others is a fall . . . towards objectivity', and 'this fall is an alienation' (*Being and Nothingness*, pp. 274–5).

The false simulation of the object by the subject (of the in-itself by the for-itself, to use Sartre's adaptation of the Hegelian language) is to be contrasted to the authentic individual gesture: the free act whereby the individual creates both himself and his world together, by casting the one into the other. Don't ask *how* this is done, since the process cannot be described. (To describe it is to use the concepts of everyday morality, and so to be imprisoned once again by Them.) The end-point of the authentic gesture is what matters, and this Sartre describes as commitment. The freedom of the liberated self is expressed in commitment: but commitment to what?

There is, of course, no answer to that question which is not an exercise in bad faith. Any adoption of a system of values which is represented as objectively justified involves an attempt to transfer my freedom into the world of objects and so to lose it. The desire for an objective moral order is inauthentic, a loss of that freedom without which moral order of any kind is inconceivable. Sartre's justification of a self-made morality is therefore inherently contradictory, a fact which he energetically embraces:

> I emerge alone and in dread in the face of the unique and first project which constitutes my being: all the barriers, all the railings, collapse, annihilated by the consciousness of my liberty; I have not, nor can I have, recourse to any value against the fact that it is I, who maintain values in being . . .

Despite this nihilistic posture, however, Sartre insists that commitment must, in our time, be to revolutionary politics, of a broadly Marxist – or at any rate anti-bourgeois – kind. The reason for this seems to be that no *other* relation to people has intrinsic sense. Using his own version of the master and slave argument, Sartre attempts to show that human relations are rotten with contradiction. He introduces Heidegger's 'being for others', as a description of the state in which I, as a self-conscious being, inevitably find myself. I am at once a free subject in my own eyes, and a determined object in the eyes of others. When another self-conscious being looks at me, I know that he searches in me not just for the object but also for the subject. The gaze of a self-conscious creature has a peculiar capacity to penetrate, to create a demand. This is the demand that I, as free subjectivity, reveal myself to him. At the same time, my existence as a bodily object creates an opacity, an impenetrable barrier between my free subjectivity and the other who seeks to unite with it. This opacity in another's body is the origin of obscenity, and my recognition that my body stands to another as his does to me is the source of shame.

If I desire a woman, this is not simply a matter of lusting to gratify myself on her body. If it were no more than that, then any suitable object, even a doll, would do just as well. My desire would then unite me with the world of objects, as I am united with and dragged under by slime (*le visqueux* which, for Sartre, is repulsive because it is the image of a *meta*physical, rather than a physical, dissolution). I would experience the extinction of the 'for-itself' in the nightmare of obscenity. In true desire what I want is the *other, himself*. But the other is real only in his freedom, and is falsified by every attempt to represent him as an object. Hence desire seeks the freedom of the other, in order to appropriate it as its own. The lover, who wishes to possess the body of the other only as, and only in so far as, the other possesses it himself, is therefore tied by a contradiction. His desire fulfils itself only by compelling the other to identify with his body – to lose the for-itself in the in-itself of flesh. But then what is possessed is precisely not the freedom of the other, but only the husk of freedom – a freedom abjured. In a remarkable passage, Sartre describes sadism and masochism as 'reefs upon which desire may founder'. In sadomasochism one party attempts to force the other to identify with his suffering flesh, so as to possess him in his body in the very act of tormenting him. Again, however, the project comes to nothing: the freedom that is offered is abjured in the very offer. The sadist is reduced by his own action to a distant spectator of another's tragedy, separated from the freedom with which he seeks to unite himself by the obscene veil of tortured flesh.

Sartre's phenomenology of desire is a sincere description of existential horror. And it brings home an important observation: that his attempt to retain a pure and absolute freedom which surrenders to no moral law sets him at an impassable distance from the human world. There is no passage out of that freedom into the realm of human comfort. Every entry into that realm is also a surrender. If this is the price we pay for freedom, then perhaps freedom is too costly. And why describe this constant repudiation of the world as freedom, when it is so hedged round with the fear of others as to be a prison in itself?

If there is transcendental freedom, then it is surely right to see it as Kant sees it: as the foundation of a morality of law. We do not retain our freedom through a posture of constant disobedience; nor do we retain it by a gesture of 'commitment' to a cause that we refuse on principle to justify. We retain it by obedience. Obedience to the moral law means recognising the rights of others, and treating them as ends rather than means. It launches us down that path towards the 'bourgeois' order on which finicky intellectuals are so reluctant to tread.

It is useful here to recall Hegel's argument about the master and the slave. Personal existence, for Hegel, is achieved only in the condition of mutual recognition; and that in turn requires submission to the moral law, in which the other is no longer regarded as an alien competitor for the possession of my world, but as a sovereign will existing freely within it, the possessor of rights and duties which are the mirror of my own. For Sartre, it seems, we cannot reach that stage of mutuality. The very demand for a radical freedom *excludes* the other from my world, and if he is nevertheless to be found there, it is in the first instance as an enemy. Sartre illustrates this with one of his vivid examples. I am in a park, whose objects organise themselves around me, as I project my purposes towards them. This bench is to be sat upon; that tree is hidden but demands my gaze. Then suddenly I see another man; at once the park loses its unique distribution according to the principles of my desire, and begins to group itself around his purposes too. The bench becomes a bench that he avoids; the tree a tree that he approaches. 'The other is . . . a permanent flight of things towards a goal which . . . escapes me inasmuch as it unfolds about itself its own distance'; the other 'has stolen the world from me'. (*Being and Nothingness*, p. 255.) In short, there is no safety in

Them, and our mutual condition is one of war. The dialectic returns me always to its first stage: the life-and-death struggle with the other, from which the postulate of freedom can never release me. This is the real meaning of Sartre's celebrated remark, in the play *Huis Clos*, that 'Hell is other people': namely, that other people are Hell.

5. The Banalisation of Evil

Kant was confident that the personal morality that he had justified would provide a sufficient foundation for politics and jurisprudence. Yet it was the Kantian philosophy of the person that was turned against itself by Marx, in order to cast doubt on the bourgeois order. Wage labour, Marx argued in the early 1844 manuscripts, turns a person into a thing – a mere instrument or means. It denies the worker's reality as a person or end in himself. When a man's labour is extracted from him by a wage-contract, and when another owns the product of his labour, then the worker is alienated from his labour and so from himself. Marx's attempt to prove this is riddled with metaphor, and the argument does not survive its translation into consecutive steps. Nevertheless, it was this argument that inspired the most fervent of modern intellectual revolutionaries, and which was always offered during the sixties as the final indictment of the capitalist order.

What do we mean, by the alienation of the subject, or the translation of a person into a thing? It could be argued that Marx himself was tempted to see human beings as 'thing-like', in his desire to expose the 'mystifications' through which we observe the social world. There is, in his theory of history, a kind of scientistic attitude to our ordinary ways of thinking, an eagerness to dismiss as 'ideology' the concepts through which we construct our social world and endow it with a meaning. Although Marx believed that these concepts are riddled with falsehood, and prevent us from seeing man in his true nature and dignity, the effect of his assault on them was to leave nothing in their place besides a disenchanted spectacle of pure power. The veil of illusion was torn from the world by the science of history, to reveal the 'laws of motion' of human society. Those features of the world which constitute its personal face – rights and duties, laws and values, institutions of membership and religion – were effectively wiped away from the surface, by the impetuous desire to peer into the economic depths and discover the 'truth' of human history.

This assault on the human world in the name of science is not a Marxian monopoly. An equally important instance is provided by modern sexology. This too is more pseudo-science than science, and rejoices in its bald, unmoralised image of 'what we really are'. What we really are from the scientific point of view is precisely what we really aren't, and if anything shows this it is the study of sexual behaviour. Sexual desire is a response to the other as perceived, not as anatomically described. It is the response to a *person*, perceived as the bearer of rights, responsibilities and awareness. The other is not a means to my pleasure, still less something against which to scratch the itch of my lust. The point was made by Socrates, as recorded by Xenophon: 'Socrates . . . said he thought Critias was no better off than a pig if he wanted to scratch himself against Euthydemus as piglets against a stone.' We recognise in that picture of Critias' lust not true desire, but one of its infantile perversions. Yet this is the way in which desire is invariably represented by sexologists – in 'functional' terms which totally misrepresent its intentionality. The goal of sexual desire is not orgasm; nor is it 'sex', however described. It is possession of the other – the very thing that Sartre, in his bleak refusal to relinquish his freedom, deemed to be impossible. (Yet possession is easy, provided you do not recoil from being possessed.)

The moral of sexology is the same as that of Marxism. We belong to the human world, and to surrender that world to science is to lose sight of personality, and all the meaning that derives from it. We can improvise our way through the desert left by scientistic ways of thinking, but with no prospect of happiness. Moreover, once the veil of meaning has been torn away, everything is permitted. When all that I see of the other is 'the skull beneath the skin', there is no absolute interdiction – no 'thou shalt not' – which speaks to me from his features. The interesting results can be witnessed in the totalitarian governments that have arisen in our century – usually inspired by some spurious 'science' of man, whether biological (as in Nazi genetics) or social (as in the Marxist theory of history). Systematic murder becomes a bureaucratic task, for which no one is liable, and for which no one in particular is to blame. Hannah Arendt wrote in this connection of a 'banalisation of evil'; it would be equally appropriate to speak of a 'depersonalisation', a severance of evil from the network of human responsibility, which is strung through the human world, but which has no counterpart in the world of science. The totalitarian system, and the camp which is its most sublime expression, are without the marks of individual care. In such a system, human life is driven underground, and the ideas of freedom, responsibility and right receive no public recognition, since they have no place in the administrative process. The machine which is established for the efficient production of Utopia has total licence to kill. Nothing is sacred, and its killings are not murders (for which persons alone could be liable) but 'liquidations'.

With the depersonalisation of the world there comes a disintegration of language. The phenomenon of 'Newspeak', described with such extraordinary prescience by George Orwell in *1984*, should always be borne in mind by those who ask how the crimes of totalitarian politics have been possible. By emptying language of every vestige of a moral idea, we change the way in which the world is perceived. Those crucial words around which our aspirations congregate - words like 'liberty', 'truth', 'right', 'democracy', 'peace' – are either banished from the language, or used to mean both of two opposing things, as in Robespierre's invocation of the 'despotism of liberty', the communist slogan 'fight for peace', or Lenin's description of totalitarian government as 'democratic centralism'. When the murder of twelve million people (the kulaks) is described as 'the liquidation of a class', or of six million (the Jews) as the 'final solution to the Jewish question', all reference to the human reality is expunged from discourse. The terms are abstract, bureaucratic, almost without reference. Vocabulary, syntax, logic and style take on a new purpose which is neither to describe the world nor to interpret it, but to deprive it of its meaning, so that the dictates of power can be inscribed without resistance on the resulting blank.

6. The Personality of Institutions

We should remember in this connection that human persons are not the only persons that there are. Institutions too may have personality. This is so, not only in the legal sense, according to which a 'corporate person' is an entity with rights and liabilities at law, but also in the moral sense. Institutions can have moral rights and duties; they can invite our respect and consideration, not as means only, but as ends in themselves; and they are rational agents, with their own changing goals and their hopes and fears for the future. This is a strange fact, but it is a fact none the less. Indeed, it is one reason why institutions are so important to us: for they are the objective counterparts of our experience of membership, and can be loved as persons through every change in the persons who compose them. A corporation can be the object of praise, loyalty, pride, as well as of anger and resentment; it can possess habits of mind, virtues and vices. A corporation has even been the central character in a drama (Wagner's *Die Meistersinger*), in which the phenomenon of membership is powerfully vindicated.

There are difficult metaphysical questions here, and I propose to ignore them. For what is important to our present theme is the fate of institutions under the rule of 'liberation'. For Sartre and Foucault, institutions are inherently suspect: they are the sources of social power, the means whereby the free individual is conscripted into purposes that are not his own. Liberation can be achieved only through the total destruction of institutions. And this assault on institutions has been part of the revolutionary programme since 1789. Wherever they came to power, the Communist and Nazi Parties expropriated, infiltrated or abolished institutions, so as to destroy every trace of corporate personality, just as they had declared war on the personality of individuals. When the Polish trade union Solidarity, drawing on Hegel, called for a restitution of the distinction between civil society and state, this is what it meant: a return to a society of free institutions, each with its own personality, and each able to confer the benefit of membership. For the advocate of liberation, however, the experience of membership is inherently degrading: it involves a surrender to rules, hierarchies, uniforms; it sets a value on obedience and conformity; it brings people into line with one another, and confers legitimacy on Them.

Hence one of the prime motives of modernist thinking has been to undermine institutions, by showing that they are essentially arbitrary, replaceable, and in every instance ripe for a radical reform in some 'democratic' direction. The philosophy of Foucault owes its appeal largely to its ability to redescribe what seem like harmless gatherings of like-minded citizens as sinister structures of power. By seeing all institutions in this way, you also justify the political project which subverts them – which regards the control of civil society as legitimate. For if the one reality is power, then the only questions, as Lenin said, are 'Who?' and 'Whom?' Who exercises that power, over whom? We are entitled to seize control of every institution, to extinguish its personality, and reassign it to the class which it had been used to oppress. Indeed, it becomes the duty of the ruling party in totalitarian systems to exert this control. No part of civil society should escape its vigilance, since no power except its own can be permitted.

We should again make the distinction, however, between two kinds of free association:
(a) Those which come about as the unintended result of voluntary actions. The paradigm case is the free market; but this is only one instance of an institution arising by an 'invisible hand' and affecting the conduct of those who take part in it. (Other instances are costumes, manners, festivals, and – to some extent – religions.)
(b) Associations that are deliberately created, and which depend upon some criterion of membership: such as clubs, schools, universities, parliaments and churches.

In order to control civil society it is not enough to destroy the personality of corporations. You must also control the 'invisible hand'. The planned economy is therefore a natural result of the totalitarian project. So too is the destruction of private property, since private property permits the two fundamental relations from which all free associations ultimately derive: contract and gift. Rights too must be abolished, whether the rights of individuals or those of corporations, since rights draw limits to what the ruling power might do.

Control of civil society requires the creation of Potemkin institutions, and a Potemkin economy. The highest posts in every civil institution must be assigned to party members or to those under their control, while the people lower down must be harassed until their conformity can be guaranteed. In order to achieve this kind of control it is necessary not merely to destroy all property rights (which are the main vehicle of corporate freedom), but also to undo the laws of association. In normal societies an association is permitted until it is forbidden, and when permitted its actions are subject to the law, under which it has rights and liabilities. In totalitarian societies associations are forbidden unless expressly permitted – usually by some written dispensation from the party machine. Once permitted, however, their actions lie beyond the law: for the real decisions taken in their name are not taken by the corporations themselves, but by the party, which is immune from prosecution; while the responsibilities which fall on the corporations can never be honoured, unless the party itself requires it.

In place of the old forms of civil society, therefore, there is created a new kind of social unity: a conscript unity, foreshadowed in the *levée en masse* of the French Revolution. The people are pulled from the path of compromise and set marching side by side toward the future. Their regimented steps admit of no deviation, and their lingering backward glances are subject to the sternest reproof. Henceforth it is the *future* that counts, and those who recoil from it betray not the revolutionary vanguard, but the whole of society – indeed, humanity itself.

But the future has, from the point of view of our present obedience, a singular defect: it is unknowable. Those frail institutions which enable us to lay down tentative tracks from what has been to what might be – institutions like law and the market, which embody the mutual dealings of many generations – are destroyed by the totalitarian project. Society enters on its great march in total ignorance, cut off (as Hayek has powerfully argued) from the only sources of social knowledge. Burke rightly described society as a partnership between those who are living, those who are dead, and those who are yet to be born: implying thereby that those who repudiate the dead and discount their claims on us repudiate also, in that very act, their unborn successors. Their lives are lived for the present only, and just as they lay waste the accumulated wealth and wisdom of former generations, so they mortgage the entire future of their country, in the interests of policies that have no foundation other than ideology. In short, the totalitarian project leads not only to the disestablishment of the dead, but also to the disenfranchisement of the unborn. The very philosophy which tells us to march into the future, consumes the future in the fires of its present emergency.

7. The Holy Trinity

God, says the Christian, has three natures, and we come to him by three separate paths: when we worship him as transcendental lawgiver; when we encounter him incarnate; and when the Holy Spirit moves through us in its work of concord. It follows that there are three modes of rebellion against God: the repudiation of law; the assault on the sanctity of the human person; and the desecration of the work of the Spirit.

Our brief survey of the philosophy of liberation enables us to understand the rebellion against law. You yourself, the liberator tells us, are the author of these laws that bind you: change your way of thinking, and the 'mind-forged manacles' will drop away. Look into the basis of law and what do you find? Nothing but a transient human interest, and an interest, moreover, which is not an interest of yours. Such is the message of all the great debunkers, from Rousseau through Marx to Marcuse and Fromm.

The categorical imperative which tells us to respect the incarnation of reason is another impediment. Nevertheless, something in us resists the call to trample on it. For we live in the human world. We see the gaze of another person as a sign of his freedom. In respect and affection we encounter the incarnation in nature of a force that is not really at home in this world. We see the human form and the human face as sacred; another's life and happiness are not to be weighed, therefore, in the balance of our individual profit. Our calculations stop at the threshold of the place which the other occupies, and refuse to transgress.

But the call to liberation overcomes this squeamishness. The human world, it says, is your creation: ideas of personality and right, of the sacred and the untouchable, are human artefacts. They play no part in the science of man, which reveals us to be dispensible parts of the history machine; our rights and freedoms can be overridden whenever the better functioning of the whole requires it. Individual freedom and the flesh which harbours it are the raw materials of revolution. They can be placed in the balance of calculation and discarded for the sake of the 'liberation' of mankind. Revolution leads to murder, for the reason that it destroys the ideas and conceptions through which we endow the human world with personality.

We encounter the divine, not only in the incarnate person, but also in counsel, association and institution building. In countless ways people combine in a spirit of conciliation, willing to renounce even their dearest ambitions for the sake of agreement with their fellows. In the true council people are prepared to accept a corporate decision which corresponds to nothing that they previously desired, for the reason that the council itself is vested with authority. The spirit of co-operation issues in decisions which may coincide with the will of no participant; and this corporate will in turn implies a corporate liability, corporate rights and corporate duties – in short the kind of personality which is the expression of membership in the human world. From our experience of the 'little platoon' great emotions arise in us: our sense of duty is spread more widely than the circle which inspired it, to embrace other places and other times. We come at last to respect the dead and the unborn, and this is the experience on which free and stable government is founded.

But, says the liberator, your institutions are human artefacts, creators and distributors of power. Through them you bind yourself, by making permanent what is transitory, and absolute what is merely relative to a human interest – and again an interest that is not yours. With the advent of revolution, therefore, the work of institution-building comes to an end; so too does charity; so too does every other form of combination which lies outside the 'People's' control. There results a peculiar society, devoid of counsel, in which decisions have the impersonality of a machine. In this respect too, revolution turns against the world: it leads to the destruction of corporate persons, just as it leads to the murder of individuals, since it has abolished the experience of sanctity which conditions our respect for them.

The three modalities of the divine are the three pillars of community. By representing each as a form of oppression, the liberator conscripts us to the work of destruction. Yet for what end? The remarkable fact about all the philosophies of liberation, is that they have nothing coherent to say about their goal. Indeed, a kind of laughing paradoxism inspires their statements of conviction. Whether uttered in the abstract language of the French Revolution, or in the contradictory vision of a commitment to something whose value cannot be described, the result is the same: a kind of nothingness – a struggle to destroy, but for the sake of Nothing. Not a single advocate of liberation has troubled to tell us in concrete terms what his ideal consists in: we are to liberate ourselves *from* law, institutions and councils; but not *to* anything. The General Will of Rousseau, the People of Robespierre, the commune of Marx, the *fascio* of the Italian anarchists, the *groupe en fusion* of Sartre: all express the same contradictory idea, of a free society without laws or institutions, in which people spontaneously group together in life-affirming globules, despite the centuries-old proof that they are capable of no such thing. The aim is for a 'society without obedience', a 'unity in disobedience' where conflict, competition and subservience are all unknown. In pursuit of this contradictory phantom, the 'liberators' have sought to pull down actual institutions, to uproot actual relations between people, to destroy all that is merely negotiated, compromised and half-convinced. The real reference to the transcendental, which is there in the humble forms of human life, is cancelled, on behalf of a transcendental freedom that cannot be obtained. This idolised freedom destroys human relations, by measuring men by a standard which they cannot attain. If there is anything in the world which noths, it is the idea of liberation, and those phoney conceptions of authenticity and good faith with which the existentialists embellish it.

8. Deconstruction

By demonstrating that all law and interdiction, all meaning and value, all that troubles, contains or limits us, is our own invention, the devil fosters the belief that everything is permitted. In particular, revenge is permitted against the society from which you feel excluded and against the Father who created it. If you have reached the stage of repudiation, and are unable to advance beyond it to the reconciliation and forgiveness which are the signs of moral maturity, then the temptation of the liberator is irresistible. 'Ruin the sacred truths', as Marvel said; pull down the order that surrounds you; not only do you affirm yourself against it, you also liberate your fellows and will be rewarded by their admiring love.

The problem, however, is that too much of what has been achieved in the name of community is preserved in intangible form. The sacred text, the sublime harmony, the forms of art, poetry and liturgy: these come back to us, whispering over again the same troubling message. The old order is sacred, they tell us, and its meanings secure. When all institutions have been exposed as fraudulent; when all laws have been re-cast as human interests; when the human person himself has been depersonalised, and set down unprotected in the sterile void of science, those ancestral voices still murmur in the void, and prevent us from being content with our liberation.

There is a solution to this problem, which is mockery. Faust briefly wavered from the path of liberation; the response of Mephistopheles was Walpurgisnacht – the representation of the world as meaningless. No beauty or truth exists in the human world that does not have its mocking double in Walpurgisnacht; and any attempt to cling to the fragments of meaning that a culture or civilisation have bestowed on us, is an attempt to clothe ourselves in dear illusions: the *cari inganni* of Leopardi, which cannot hide *l'infinita vanità di tutto*.

The ruin of meaning would never be sanctioned by a philosopher who is merely modern; but it lies on the agenda of those modernists and post-modernists, from Sartre to Rorty, whose world is bereft of all authority. Art, however, poses a threat to the modernist; for in art we can display the victory of a moral idea, and by this method come to understand, in however imperfect a manner, the restoration of the good. That is why the devil is so interested in art, and seeks to use its divine attractions to 'do dirt on life', in D.H. Lawrence's famous phrase. The goal can be achieved by showing that there is no more meaning in *King Lear* or the *Odyssey* than in *American Psycho*.

Different modernist philosophers adopt different approaches to the problem. This, briefly, is the strategy of Jacques Derrida: first to identify the target, which is Western culture as such – that is, the sum total of the artefacts through which Christianity formed and administered our human world. Secondly, to identify a fault in this culture: this fault is 'logocentrism', meaning two things – the privileging of speech over the written word, and, more importantly, the belief that the world really is as our concepts describe it. Then the mining of that fault, until all that is built on it collapses.

Logocentrism, in the second reading, means realism. And the best way to attack realism is to marshall some of those nominalist arguments which, from William of Ockham to Nelson Goodman, have cast doubt on the view that concepts are anything more than human artefacts. Nominalist arguments loom large in Derrida, as they do in his masters, Barthes and Foucault. For they lend themselves to the proof that the true purpose of our concepts is not to describe the world but to fortify our power over it. If you don't like the prevailing power, then ruin its concepts!

How can we do this? Derrida borrows from Saussure the notion of '*différence*'. This was used by Saussure to emphasise the way in which signs are arbitrary. It does not matter *which* sounds we choose to mean hot and cold; what matters is that they are used to mark a certain difference, and this structural property is the true carrier of meaning. The French *différer* also means to 'defer', in the sense of postpone: and by a play of words Derrida decides that Saussure has proved that meaning is always deferred by the text: no one word bears a meaning until related to the next, which must be related to the next and so on. (That is why Derrida spells 'difference' with an 'a': *différance* is supposed to signify both difference and deferral.) The process of meaning something never gets started: or rather, it starts just when we want it to. Texts do not have a single authoritative meaning; there is a 'free play of meaning' and anything goes. In short, we are *liberated* from meaning.

Similarly the text is emancipated from authorship. Once written, the work is a public object, behind which the author disappears into the privacy and particularity from which he briefly emerged. It is for us to decide what the text means. And we are free to decide as we wish, since 'all interpretation is misinterpretation'. Alternatively: no reading is privileged. This is affirmed with great vehemence, even though – indeed because – it has the consequence that no text really says anything, including the text which says so. Deconstruction deconstructs itself, and disappears up its own behind, leaving only a disembodied smile and a faint smell of sulphur.

None of that stands up to a moment's scrutiny; but this *makes no difference whatsoever.* The advocate of deconstruction cheerfully accepts the disproof of his own philosophy; for what matters to him is not the truth of an utterance, but the interest that is being advanced through it. Your criticism shows you to be a denizen of the 'traditional criticism', a parasitical resident in the house of Western culture; and by criticising you merely display the invincible ignorance that leads you helplessly to reaffirm a vanished authority. To think that you can get rid of deconstruction by disproving it, is to fail to recognise that proof itself has been deconstructed. Proof is what Others do; and it's Them that we're against. (You find this kind of response in Marxism too, and in Foucault: not, what are you saying? but *d'où parles-tu*?)

Perhaps the most curious claim made by the disciples of Derrida is that the 'subject' has been finally abolished by the deconstruction of meaning. The self has been shown to be a fiction, by the argument that reference is impossible. In fact, however, nothing looms larger in the practice of deconstruction than the self – the destruction of meaning is in reality the destruction of the *other*, the final revenge against Them. All that remains thereafter is the subject, who can choose what to think, what to feel and what to do, released from external constraints, and answerable to nothing and to no one. There are indeed arguments, as I shall show, for the conclusion that the 'self' – at least as construed by Descartes, and perhaps in its subsequent manifestations too – is some kind of grammatical illusion. But these arguments nowhere occur in the literature of deconstruction, and depend on principles (that there really is meaning in language, and that words really do refer) which deconstruction denies. The suggestion that the subject is a fiction is itself a fiction, part of the attempt to claim over the objective world the kind of absolute sovereignty that attaches to a purely subjective view of things. One of the most important themes of recent philosophy has been the relation between object and subject, and between the objective and the subjective perspective. In exploring that theme, in the final chapter, I hope to provide an answer to deconstruction.

9. Home Again

You don't have to believe in God to see the sense of what Durkheim says about religion. Religion is the affirmation of a first-person plural. It tells us that, actually or possibly, we are members of something greater than ourselves, which is the source of consolation and continuity. Nor do you have to believe in the devil to accept the corollary: that communities may dissolve under the stress of disaffection, and that the force of dissolution can become active and wilful in the manner of a god. The devil has one message, which is that there *is* no first-person plural. We are alone in the world, and the self is all that we can guarantee against it. All institutions and communities, all culture and law, are objects of a sublime mockery: absurd in themselves, and the source of absurdity in their adherents. By promising to 'liberate' the self, the devil establishes a world where nothing *but* the self exists.

This is the final triumph of the demon. We are back with Descartes, surrounded by an ocean of doubt, but all the more solitary for our fruitless venture outwards. In response, we should reaffirm the factor which is ignored by the deconstructionists: our existence as social beings, who do not merely create a human world, but fill it with their culture. Does it matter that *we* are the source of that world and the originators of its meaning? After all, God himself could have endowed it with meaning in no other way.

31

Self and Other

The devil has appeared twice in our discussions: first as the source of Descartes's doubt, secondly as the 'deconstructor' of the social world. The effect of his two appearances is roughly the same: to cause us to retreat into the first person, to suspend the idea of objective order and objective knowledge, and to propose the self alone as the test of reality. For Descartes, the retreat into the self was also, however, the prelude to metaphysics; armed within his fortress he hoped to sally out against the demon, and so to repossess the world. For the philosophers discussed in the last chapter the self is sufficient: it is all we have and all we need, and its realisation *against* the social order is a good in itself. The devil returns in triumphal aspect; his victims become his disciples, and their sarcastic laughter resounds in a self-created void.

To vanquish the devil is hard; for it involves renouncing our dearest illusion – the illusion of the self and its sovereignty. But modern philosophy has had much to say about this topic, and it is important to understand both the arguments, and the world-view (for it is no less than that) that is unfolded through them.

1. A Fragment of History

The theme of the self might be taken to be the central theme of modern philosophy, introduced at the outset by Descartes, with his argument from self-consciousness to the immaterial nature of the mind. But the theme acquired a new character with Kant, and an almost religious urgency in Kant's immediate successors. Kant argued that the self, as subject of consciousness, can never be the object of its own awareness. It is 'transcendental', a pure subject – no part of the empirical world, but the point of view from which that world is understood. I can become an object for myself only by ceasing to be a subject: only by losing my essential character as self-conscious observer. Hence I cannot bring the self under concepts (which apply only to the *objects* of experience), nor determine its position in the world of nature. Arguments like Descartes's, which begin from the premise of the self-conscious subject and lead to conclusions about its place in the natural order, are inherently without cogency, since no such conclusion can ever be drawn. The immediate knowledge that we have of our mental states implies nothing about their nature, and the point of view expressed in that knowledge is outside nature, and inaccessible to science.

Such, at any rate, is the view expounded with great brilliance and subtlety in the *Critique of Pure Reason*. In the *Critique of Practical Reason*, however, and in his other works on morality and aesthetics, Kant tended towards another conclusion, and one which is hard to reconcile with the argument of the first *Critique*. The transcendental self, Kant now argued, is indeed unknowable to the understanding – which knows what it knows always as *object*, and therefore as part of the natural order. But it is not unknowable to practical reason. When thinking morally, I address my thought precisely to a transcendental self – to a subject 'outside nature', that is posited as *free*. (See the sections on Kantian morality in Chapters 17 and 20.) In the moral life, therefore, I obtain an intimation of this transcendental self, although it is an intimation that resists translation into the language of the understanding (the language of concepts).

This view may be difficult to grasp; but it is also immensely attractive. For we are assailed by two irresistible but incompatible thoughts. We think of ourselves as self-conscious observers of our world, occupying a unique perspective upon it – the perspective summarised in those mysterious words 'I, here, now'. But we also suppose ourselves to be part of the world, changing and changed by it, observable to others, and bound not only by a common moral law, but by the natural order of the universe. And maybe Kant is right, that the practical question – 'What shall I do?' – belongs to the first of those conceptions, and resists translation into the second. In which case we seem to be forced to the conclusion that we must think of ourselves in two ways: as free subjects, and as natural objects, and that the second way of thinking can make no room for the knowledge contained in the first.

Subsequent philosophers made much of this idea of the transcendental self. Indeed, Fichte used it to set a new agenda for philosophy, and so furnished both the devil and his enemies with their major arguments. The task of philosophy, Fichte argued, is to discover the 'absolutely unconditioned first principle of human knowledge' – i.e., the principle upon which all knowledge can rest, but which itself rests on nothing. Logicians offer us an instance of necessary and indisputable truth, in the law of identity: A = A. But even in that law something is presupposed that we have yet to justify, namely the existence of A. This thought enabled Fichte to introduce a concept which changed the course of philosophy, and which I have already employed in describing Kant's theory of the self, the concept expressed by the verb 'to posit' (*setzen*). I can advance to the truth that A = A, Fichte argued, once A has been 'posited' as an object of thought. But what justifies me in positing A? There is no answer. Only if we can find something that is posited in the act of thinking itself, will we arrive at a self-justifying basis for our claims to knowledge. This thing that is posited 'absolutely' is the I; for when the self is the object of thought, that which is 'posited' is identical with that which 'posits': in the statement that I = I we have therefore reached bedrock. Here is a necessary truth that presupposes nothing. The self-positing of the self is the true ground of knowledge in all its forms.

Would Fichte have drawn such a conclusion had he expressed the law of identity as we should express it, and as it should be expressed, as $(x)(x = x)$ – in other words, without the use of a name ('A')? I do not know. What is certain, however, is that he did not regard either the law itself, or the philosophy of the self from which he derived it, as a trivial matter. On the contrary, it seemed to him that we could make no sense of our world, if we did not recognise the (transcendental) self as both subject *and* object in all our knowledge. By a series of hair-raising arguments he developed a philosophy which, in outline, foreshadowed not only the idealism of Schelling and Hegel, but also the eschatological materialism of Marx, the transcendental phenomenology of Husserl, and the existentialism of Heidegger and Sartre. According to this philosophy the free and self-producing subject is the source not only of knowledge but also of the thing known: in a profound sense, all knowledge is self-knowledge, even when the self is known as 'not-self' (i.e. as part of the natural world). The self knows itself by 'determining itself', that is, by setting limits to itself, so as to comprehend itself as object. The object is posited by the subject, but stands opposed to it as its negation. The relation between subject and object is dialectical – thesis meets antithesis, whence a synthesis (knowledge) emerges. Every venture outwards into the object (the not-self) is also a self-alienation; the self achieves freedom and self-knowledge only after a long toil of self-sundering. All art, religion, science and institutions are gathered into this process, expressing some part of the great spiritual journey, whereby the empty I = I takes on flesh, so as to know itself at last as an ordered and objective reality, and also as free.

If you look back at Hegel's Master and Slave argument (Chapter 20) you will see that dramatic picture in another version, just as you will see it in Marx's theory of alienation and Sartre's theory of freedom. I mention it as an example of the amazing conclusions that can be drawn, once the concept of a transcendental self is taken seriously. But *should* we take it seriously? Is it really necessary to suppose the *existence* of such a thing, just because our language casts, so to speak, this shadow into the unknowable?

2. The Grammar of Self-reference

We refer to the self-same car, to the street itself, to the horse's attempts to get itself over a fence. But surely we do not suppose that words like 'self-same' and 'itself' here refer to an entity – the self of the car, street or horse? Why is it, then, that, in our own case, we take the reflexive pronoun so seriously? Why not assume that words like 'self' have no other function than to express the concept of identity? Indeed, in many languages, words for self and identity are either closely related, as in Greek, or indistinguishable, as in Arabic (*nafs* – which also means 'soul').

The answer is that we are self-conscious beings, with a conception of our own place in the scheme of things. This conception is one of the things that we try to articulate in our 'I'-thoughts. But are we right to assume that these 'I'-thoughts refer to a particular entity – the self – and that reference to this entity is the function of 'I'? Perhaps not. For our 'I'-thoughts possess two peculiar features which set them apart from ordinary reference.

Wittgenstein distinguished (*The Blue and Brown Books*, pp. 66–7) two different uses of the word 'I': the 'subject' use and the 'object' use. If I am wrestling with someone, and, catching sight of a leg which is bleeding, announce that 'I am bleeding', then I have employed the word 'I' in its object use, to identify an object as myself. And I could be wrong: it could be that the bleeding leg belonged to my wrestling partner, and that I am not bleeding at all. If, however, I say that 'I am wrestling', or 'I am in pain', the word 'I' occurs in its subject-use, in order to express my first-person perspective on the world. In such a case it is impossible that I should have made a mistake as to *who* is wrestling or in pain. I identify myself correctly, and could not be wrong. In its subject use, Sydney Shoemaker says, the term 'I' makes statements that are 'immune to error through misidentification'. In this, the normal use of the first-person pronoun, I always identify precisely the person I intend to identify, namely myself.

Secondly, two people who understand the term 'I' in the same way, and understand each other's use of it, will invariably use it to refer to different objects. I use the word to refer to me; you use it to refer to you. In saying that I am cold I am saying what you would say by means of the sentence 'Roger Scruton is cold'. Yet surely 'I am cold' does not *mean* the same as 'Roger Scruton is cold'.

Certain conclusions are suggested by those observations. First, it would seem that 'I' does not function as a proper name – nor as any other kind of 'rigid designator'. (See the argument of Kripke, discussed in Chapter 13.) If it denotes at all, it is with maximum flaccidity, being applied without ambiguity to different objects in different possible worlds, and to different objects in the actual world. Secondly, we must recognise in words like 'I' a difficult problem for Frege's theory of sense. The sense of 'I' does not determine its reference; just *who* is referred to can be determined only when the context of use is known. This context-dependence of 'I'-thoughts underlies their immunity to error. To understand the sense of 'I' is to know that it refers to the person who is speaking. Hence no speaker who understands the word can use it to misidentify its referent. It is this transparently grammatical feature of 'I' that provided Descartes with his original answer to the demon.

But now, if this 'immunity to error', upon which philosophers have built so much, is a merely grammatical phenomenon, ought we not to become suspicious of those grandiose Fichtean theories, which construe the self as the ultimate reality, and locate it outside nature, in a transcendental realm? After all, the grammatical peculiarities of 'I' are displayed by other words in our language – words like 'here', and 'now', which seem to have no special connection with self-consciousness. The place identified by 'here', and the time identified by 'now' are determined by context, in precisely the manner in which the reference of 'I' is determined: they denote the place and the time at which the speaker is situated. 'Here'-thoughts and 'now'-thoughts are likewise (in the normal case) immune to error through misidentification. You cannot understand the word 'here', and use it to misidentify a place. (For some ingenious exceptions which prove the rule, see Colin McGinn's *The Subjective View*.) Here is where I am, and about that fact I cannot be in error. Of course, to say that the place from which I am speaking is here is not to say very much. Indeed, it is to convey a thought that is all but empty, since it does not indicate *where* I am in relation to anything else. The thought is, in the language of Fichte and Hegel, *indeterminate*, and incorrigible *for that reason*. But that is the point of the comparison. Maybe the thought that I am *I* (myself) is as empty and 'formal' as the thought that I am here: maybe it contains no information as to what, or what kind of thing, I am.

It has therefore seemed to many philosophers that our understanding of the self is subordinate to our understanding of 'indexicals' (terms like 'I', 'here' and 'now', which are 'indexed' to the speaker), and 'demonstratives' like 'this' and 'that' which function in a similar way. (These are classed together by Reichenbach as 'token-reflexive' terms.) Indeed, if you cast your mind back to McTaggart's argument for the unreality of time, which makes such fatal play with the logic of 'now' (or the present), you will see how indexicals generate metaphysical paradoxes. In a sense all of metaphysics is contained in the problem of token-reflexive terms; for they capture in language the situation from which metaphysics arises: namely *our* situation, as parts of a world upon which we also have a conscious perspective. In picking out a particular item in the world as myself, here, now, I express thoughts which are – according to Thomas Nagel (*The View from Nowhere*) – ineliminably subjective, and not translatable into propositions that could figure in the 'book of the world' as physics conceives it. Nevertheless, such thoughts express genuine facts, and metaphysics involves the attempt to make sense of them.

3. Secondary Qualities

Nagel may be right in suggesting that indexical facts (as we might call them) have no place in the physicist's book of the world. But what follows from that? Surely, nothing follows concerning the nature of the perceiving subject; all that we can deduce is that the subject is not outside the world but in it, and able so to identify himself. Once we understand the grammar of 'I', 'here' and 'now', we begin to see precisely that the subject is not, and cannot be 'transcendental' in any true sense of the word. There is somewhere where he is in the natural order, and also somewhen.

From the third-person point of view, indeed, there is nothing puzzling. If we look at the self from outside, we are drawn to the following picture: the subject of experience is a person, that is to say, a being with a certain kind of mental life. He exists in the world, as one object among others. And he exists in relation to his kind, forming his conception of the natural order through a language that is essentially shared with his fellows. Because he gains knowledge through experience, his knowledge is inevitably marked by his point of view: the world has a certain *appearance* for him, and it is on the basis of this appearance that he explores its reality. That a certain place is presented to his thinking as *here* and a certain time as *now* says nothing about the nature of that place and time: it merely relates them to the point of view from which he observes them. From the third-person perspective we locate the subject in space and time through his relation to other places and times: in terms of the temporal 'B'-series, in McTaggart's sense, and its spatial equivalent. Here-ness and now-ness, like this-ness and that-ness, simply disappear from the picture, as belonging to the perspective of the subject, rather than to the objective order itself.

There is a parallel here with the old idea of a secondary quality, as Gassendi and Locke defined it. Locke described the quality *red* as a 'power' in objects to produce a certain 'sensation' (the experience of seeing red). His thought was roughly this: if you describe the qualities of an object which concern its location and activity in space, you have described it entirely. For among its activities will be the power to produce certain perceptual experiences in us. And there is nothing more to being red than that. Of course, you have not described the *way* things look when they look red. But that is precisely not the kind of fact that can be listed in the book of the world. It is a purely subjective, or 'phenomenal', fact, and to know how things look when they look red is simply to know what it is like to see them. Indeed, on one view of the matter (see Chapter 22 section 6), there is no fact at all here: 'knowing what it is like' is not a kind of propositional knowledge.

This is not to say that objects do not have secondary qualities, but rather – as Locke himself argued – that secondary qualities do not resemble in any way the 'ideas' which they produce in us. What we see as red is a complex surface-structure of an object, which causes it to reflect light of a certain wave-length and absorb light of other wave-lengths. These qualities of the object are as far removed as could be from that 'immediate' experience which leads us to describe things as red.

Secondary qualities remind us, however, that appearances are *systematic.* That the world is decked out in colours is a fact of which we are immediately aware; and its being so enables us to make rapid and reliable discriminations, which are incorporated into our picture of reality. We know little about the underlying causes of experience; but so long as experience varies in a systematic way with its cause, we can use it to classify objects, as we use the experience of colours. The world from our point of view really *is* as we see it: even though the book of physics treats that point of view as one among the many things to be explained. Thus we reconcile two seemingly incompatible thoughts: first, that secondary qualities have an irreducibly subjective component, a component of 'pure seeming'; and secondly that there really are facts about secondary qualities. So long as we keep the first-person and the third-person viewpoints separate in our thinking, we find nothing puzzling in this. Likewise, we should not be puzzled by the fact that particular places and times can come before us as *here* and *now*, or a particular person as *I*.

4. The Grammar of Consciousness

Someone might not be satisfied with such an approach, however. For it seems precisely to avoid the peculiar thing that troubles us: the fact of consciousness, and the first-person perspective that embodies it. The parallel between 'I' and 'now' looks plausible only so long as we focus exclusively on the role of 'I' in identifying the speaker. But there is another and more interesting use of the term, in the self-attribution of mental states. When I say 'I am in pain', or 'I am thinking', it is true that I can make no error as to the identity of the sufferer or the thinker. But it is also true that I can – in the normal case – make no error as to the suffering and the thinking. I know immediately and incorrigibly that *this*, that I now have, is a pain, a thought, a desire – and so on. It was precisely to this fact of privileged knowledge that Descartes made appeal, in constructing his theory of consciousness. For my immediate awareness of *my* pain contrasts with the mediate, fallible and hypothetical belief that I have concerning *yours*. There is an epistemological asymmetry between first-person and third-person awareness. Hence – the Cartesian supposes – while I can know my sensations, I cannot really know yours. Your sensations are accessible to you, but not to me. Thus a sensation is an essentially 'private' item, something with an 'inner' or 'phenomenal' essence, revealed to no one besides the subject. It is this which entitles us to speak of subjective facts, which have no place in the book of the world; and also to identify ourselves as things *outside* the physical world, the centres of awareness where seeming congregates.

I have already rehearsed in Chapter 5 the famous private-language argument of Wittgenstein, which is, on one reading, a response to that Cartesian way of thinking. But there is something left out of Wittgenstein's argument, which is the premise from which the Cartesian begins. How can we explain the privileged knowledge that I have of my own mental states? Certainly – and if the private language argument shows nothing else, it at least shows this – this privileged knowledge is not explained by the hypothesis that my mental states are private and knowable only to me. But if we can find no explanation, the puzzle of consciousness will remain with us, and the temptation will always exist, to situate the mind outside nature, and to grant to it the kind of metaphysical isolation in which the devil most rejoices.

We must explain, what no merely phenomenological (i.e. first-person) study *can* explain, namely the fact that whoever says sincerely 'I am in pain' speaks the truth – provided that he understands what he says. We are dealing here with necessities: the particular kind of absurdity involved in the suggestion that I should be mistaken, in my present belief that I am in pain, could be accounted for in no other way. It is the same kind of absurdity as attaches to the suggestion that there might be a non-spatial physical object, or a non-temporal experience. And here is the crucial point: no purely first-person account of consciousness could explain such a necessary truth, since it is a truth that is expressed in a public language, and guaranteed – if at all – either by the rules of that language or by the 'real essence' of the objects to which that language refers, objects which (by the private language argument) must be publicly identifiable if they are to be referred to at all.

Whoever utters the sentence 'I am in pain' sincerely, understanding the words, is in pain. From this it follows that whoever utters 'I am in pain' sincerely, when not in pain, does not understand what he says. In short, first-person privilege reflects a rule of language – a condition upon understanding the sentence 'I am in pain'. A person understands such a sentence only if his uses of it are (with permitted exceptions) true.

The utterance 'I am in pain' contains four words; which, if any, has been misunderstood by the person who uses it to make a false assertion? Suppose that he ascribes pain accurately to others: that is surely sufficient evidence that he understands the word 'pain', given the place of this word in a public language. So is the offending word 'I'? But how could that be so? How can it be made a condition of understanding this word that we should make no error of fact when it occurs as the subject term in a sentence attributing a present mental state?

It is fair to say that this is one of the most important questions to which modern philosophy is addressed. I shall briefly suggest an answer to it, which will also be an answer to the devil.

Wittgenstein asserts that, if a lion could speak, then we should not understand him (*Philosophical Investigations*, Part I, section 244, Part II, section xii). Consider what would have to be the case if we were to accept the speech issuing from the lion's mouth as an expression of the lion's mentality. We recognise two possibilities from the outset. Either the speaker *is* the lion, or it is not. If it is not, it might yet appear to us to possess a mental identity of its own. In such a case, it would appear to speak *out* of the skin of the lion, inhabiting the lion as a dryad inhabits a tree. We readily imagine such spirits in the objects that surround us, and it is natural to primitive people actually to believe in them, to fear and to worship them.

Now the lion, unlike the tree, has an independent mentality of his own. Whether or not the voice is his, the lion still has his own desires, sensations and satisfactions. We can therefore ask ourselves what must be true if the speech that issues from him is to be an expression of the lion's mentality, rather than the voice of some alien spirit that possesses him. Consider the lion of Androcles. He roars, and a voice issues from his mouth, saying 'I roar. Moreover there is a thorn in my paw. I seem not to be able to stand on the paw in question. Indeed, my behaviour exhibits the kind of disorganisation which is characteristic of pain. Therefore I must be in pain.' Suppose too that all the lion's 'self-ascriptions' are of that nature, and suppose also that many of them are simply wrong, even when emphatically asserted. In such a case, the voice is clearly describing the mental state of the lion just as it would describe the mental state of any other thing, using the common public basis, and neither claiming nor achieving any special immunity from error or doubt. The lion's voice is that of an 'observer', and speaks from *outside* – the sole difference being that, where an observer would normally use 'he', the voice uses 'I'. But 'I' in such a case really *means* 'he': the voice is not speaking in the first person, but merely borrowing the first-person pronoun in order to express third-person thoughts. The lion is possessed, but not inspired.

In order to combine the lion with the voice, so that it becomes *his* voice, we must grant to the voice just those powers of privileged self-attribution that I have been discussing. The voice must have a special kind of authority: it cannot, except occasionally and for special reasons, make mistakes about the lion's mentality, and its knowledge must be 'immediate', based on no observation. In other words, the voice must obey the rule of authority which I sketched above. This is not just a rule that the voice follows; it is a rule that it also *obeys*. That is how the *voice* must understand its own 'self-attributions'. It must treat the suggestion that it might be in error as absurd. Once it does so, body and soul are united in speech. The voice now expresses a self – itself – and does not merely refer to an animal organism upon which it maintains an external perspective. It now understands the word 'I', not simply as a reflexive pronoun, but as an instrument in the communication of mental life.

This is borne out by another important case of first-person privilege: the expression of intention. If someone says that he will do something, understanding what he says and speaking sincerely, then – provided he intends what he says as an expression of intention – he will indeed do the thing in question, when the occasion arises. Or, if he does not do so, it is because he has changed his mind. The rule of authority here is of course vastly more complicated than the rule that I gave for sensations. It can be seen as a kind of elaborate restriction upon the explanations that may be offered for the falsehood of first-person statements about the future. They might be insincere; they might be misunderstood; they might express predictions rather than decisions; they might have been superseded by a change of mind. Perhaps 'weakness of will' is a fifth escape route; and perhaps self-deception is a sixth. (Hence the interest of modern philosophers in these two subtle phenomena.) What is not permitted, however, is a sincere expression of intention that is neither cancelled nor fulfilled.

Let us look at the third 'escape route' – the much discussed distinction between predicting and deciding – which provides a vivid illustration of why a concept answering to the given rule of authority should be so useful to us. In forming an intention, I must see myself as playing an active and determining role in the future. I must, to put it simply, 'identify' with my future self, and take responsibility *now* for what that future self will do. This attitude contrasts with another, that of 'alienation' from my future self, in which I see myself as the passive victim of external forces and of my past, driven by causes that issue from outside my will. In the second case I may predict my future conduct; but I withhold my decision, and do not endorse what I assume I shall do. In the first case, by contrast, I must try to do as I say; in so far as I support my assertion, it is not with predictions (reasons for believing that such-and-such will happen), but with practical reasons – reasons for ensuring that such-and-such is done. Hence, in the first case, I have a peculiar certainty that I shall indeed try to do as I say: not to be certain is to be insincere. It is by virtue of this certainty that expressions of intention can be understood as obeying their own rule of first-person authority.

It follows that, if I have intentions, I must also have practical reason. Suppose that someone expresses the intention to do *x*, and realises that the only way to do *x* is by doing *y*, and yet denies that he intends to do *y*. He must regard himself as committed to the truth of the following propositions: 'I do *x*'; 'I do *x* only if I do *y*'; 'I do not do *y*'. In other words, he is committed to contradictory beliefs. So that, if he has theoretical reason – which leads him to reject such contradictions – he has practical reason too. He is able to reason about the means to his ends. Since the ability to understand ordinary inferences is essential to understanding language, we can see that a connection has been forged between the possession of speech and the possession of rational agency. This is one link in the chain of connection that joins intention, rational agency, language, self-consciousness and the first-person perspective into a single idea, and which forms the full elaboration of the concept of the person.

The concept of intention, as I have characterised it, can gain application only because of certain matters of fact about human beings. It is a matter of fact that those utterances about the future which are singled out as expressions of intention are generally followed by the agent's attempt to realise what they describe. But on this matter of fact rests an important practice. Given the general truth that a person will at some time attempt to realise his sincere expressions of intention, those expressions will provide us with a peculiar means of access to his future conduct. We can now, in effect, argue against what he plans to do. We can change his behaviour by persuading him to change his declarations of intention, and he will change these just so long as he is rational and our reasons are good. Hence we have direct access, through discourse, to the core of activity from which his behaviour springs. When this means of access fails –

either because the agent is unable to accept reasons (the case of irrationality), or because the matter-of-fact connection between declaration and performance breaks down (the case of insanity) – then we have no way of dealing with the person except through the science (such as it is) of human behaviour. The agent has become a patient. (Cf. the argument of Chapter 17.)

In supposing someone to express his intentions, we are permitting ourselves to trust his word, both now and in the future. Once again, we are holding him answerable – this time for his actions. It follows that reasons given to change what he says, will now also change what he does. Language becomes the means of access, both to his present mental states and to his future activity. Our attitude to him may now single him out as the focal point in a network of intentions, as an agent, capable of committing himself to his future, and taking responsibility for his past, as a creature with a perduring 'self-identity'. Towards such a being I may reasonably feel gratitude and resentment, admiration and anger. He is the possible object of a whole variety of 'interpersonal' responses, through which our lives as moral beings are conducted.

When I am interested in someone as a person, then his own conceptions, his reasons for action and his declarations of resolve are of paramount importance to me. In seeking to change his conduct, I seek first of all to change *these*, and I accept that he may have reason on his side. If I am not interested in him as a person, however – if, for me, he is a mere human object who lies in my path – then I shall give no special consideration to his reasons and resolves. If I wish to change his behaviour, then I shall (if I am rational) take the most efficient course. For example, if a drug is more effective than the tiresome process of persuasion, I shall use a drug. To put it in the language of Kant: I now treat him as a means and not an end. For his ends, his reasons, are no longer sovereign for me. I am alienated from him as a rational agent, and do not mind if he is alienated from me.

5. The Social Construction of the Self

Suppose we could explain first-person privilege in that way, as the result of those rules which make language into an *expression* of human thought and will, and which must be obeyed if the person or rational agent is to emerge as a part of the natural order. We can then draw an interesting conclusion, namely that from the third-person perspective, to be a self *is* simply to be a person. It is to be a language-using member of a community, in which interpersonal relations are the norm. Those features of our condition upon which the devil relies for his arguments – privileged access and 'transcendental' freedom – are not features of some private and self-vindicating entity outside the natural order. On the contrary, they are the gifts of community, which could not exist in the moral void to which the devil tempts us. There is privileged access to my own mental states only because I speak a public language which confers that privilege upon me, by enabling me to formulate and reflect upon my thoughts, feelings and desires. And I can see myself as free only because I have intentions, which I acquire through the practice of public criticism and argument, whereby I am situated in a moral community and made answerable for what I do.

It would seem, then, that many of the assumptions behind the radical critique of the moral order must be rejected. Communities are not formed through the fusion or agreement of rational individuals: it is rational individuals who are formed through communities. Membership comes first, and is the precondition of the outlook that would reject it. Of course, it does not follow that any *particular* form of community is ideally suited to the development and the flourishing of the rational individual. Maybe existing societies could be usefully amended so as to accommodate some higher ideal of the individual life. But we should beware of a philosophy which tempts us, as the existentialist tempts us, toward a posture of absolute rejection of the Other, in the interests of a transcendental freedom that can never be obtained.

But this sets a new task for philosophical psychology – a task that involves the study of politics, ethics and aesthetics, in addition to those questions in the philosophy of mind that I have touched on in the present chapter. However complete we may feel in our solitude, the community is latent in our concepts and in the experiences that are shaped by them. The subjective viewpoint, as we encounter it in aesthetic or religious experience, is not a retreat from others, but a search for the community in which we are truly at one with them. All that is most 'inward', 'private' and 'holy' in our experience is in reality most outwardly directed, most urgent and plaintive in its search for the order to which we belong, and in which we may lose ourselves.

Philosophy must study this latent social intentionality; for it is the source of the greatest disparities between the human world and the world of science. The mind's 'disposition to spread itself upon objects', as Hume expressed it, is at the same time a search for home.

Home is where my companions are. We see landscapes as peaceful or hostile, situations as comic or tragic, and hills as places of escape or refuge. Those 'tertiary' qualities of objects are grounded in our own responses. But the responses are like intimations of social life: experiences which, even though received in solitude, point to a world of fellow-feeling and consolation. This is what we learn from aesthetic experience.

Consider laughter, that peculiar reaction to which rational beings are susceptible, and which Hobbes described as a 'sudden glory'. It is not only I who change when I burst into laughter: the world changes with me. It is as though I become reconciled with the object that amuses me – it is no longer a threat to me, no longer something that cuts me off or limits my desire. And behind this attitude lies another – the sense that I am 'laughing with' my true companions. Even if they lie out of reach, these companions are present in imagination, supporting me. All fellowship is fortified by laughter, which is why social awkwardness is overcome when people begin to laugh. A truly solitary laugh – one which intimates no community – is not a laugh at all, but a snarl of isolation.

There is no way of understanding such a state of mind as laughter, except in the social context which it implies. And the same goes for the other distinctly human states of mind. The human world is a social world, and socially constructed. This is not to say that it must be constructed in only one way. But nor can it be constructed as we please. There are constants of human nature – moral, aesthetic and political – which we defy at our peril, and which we must strive to obey. Perhaps the principal task for philosophy in modern conditions is to vindicate the human world, by showing that the social intimations that underlie our understanding are necessary to us, and part of our happiness. Through understanding our concepts, as they inform and are informed by our social experience, we may find a path back to the natural community. And if it is a path that is closed to us, it is imperative to know why that is so.

Study Guide

Study Guide

The purpose of this Guide is to enable the reader to grasp the nature of recent discussions, and, where possible, to reach the frontier of the subject, at least as English-speaking philosophers conceive it. In the main text I have presented the subject in a way that would not meet with the approval of all anglophone philosophers, and which is in certain respects sceptical towards the analytical tradition. In this guide, however, I follow the recent literature.

I have provided reading, and occasional questions, for most of the topics touched on in the text. Where the main text is either conventional in its approach or straightforward, I have added little in the Guide, other than brief comments on the reading material. Occasionally, however, I have made extended comments, in order either to balance the emphasis of the text, or to cover the necessary ground that still remains if the reader is to come up to the frontier of the subject. This is especially so in the Guide to Chapter 16 ('The Soul'), and that to Chapter 28 ('Objective Spirit'), both of which concern areas of constant and rapid development which could not be properly treated in the main body of this work.

Some of the chapters are less surveys of recent philosophy than contributions to it, attempts to introduce considerations that are normally ignored or misrepresented in the anglophone tradition. This is especially so of Chapters 29 and 30; not surprisingly, therefore, the Study Guides for these chapters are rather thinner than those offered for the more pedestrian topics.

The reader should not take this guide as definitive, or as a substitute for his own researches. It records one person's attempt to select from and to summarise the vast and often none-too-readable literature in a subject that continues to grow in every direction. Moreover, modern philosophy bristles with technicalities and jargon, much of which has to be mastered in order to make headway. It is best to read sparingly but closely, and to write essays, especially about the matters that are hard to understand. There is no harm in consulting dictionaries and encyclopaedias. Indeed, it is a good idea to make use of Antony Flew's *A Dictionary of Philosophy*, London 1979, which, while no better than second-rate, will prevent the reader from using terms in complete ignorance of their meaning. The reader should also consult the impressive *Encyclopedia of Philosophy*, New York and London 1967, edited by Paul Edwards, while tackling some short introduction to the subject. I recommend, with reservations, *What Philosophy is*, Harmondsworth 1985, by Anthony O'Hear, and also Bertrand Russell's classic *Problems of Philosophy*, London 1912. A.J. Ayer's *The Central Problems of Philosophy*, London 1969, is also of considerable value, though extremely narrow. Roger Scruton's *Short History of Modern Philosophy*, London 1981, is at least short.

Blackwells are in the course of publishing a series of Companions to areas of philosophy. So far three have appeared: *A Companion to Epistemology*, ed. J. Dancy and E. Sosa, *A Companion to Ethics*, ed. P. Singer, and *A Companion to Aesthetics*, ed. D. Cooper. These also could be consulted, and give valuable insight into the richness and fertility of philosophical thinking.

Chapter 1 The Nature of Philosophy

For those who wish to explore the themes of this chapter at a higher level, here is how it might be done, section by section:

Modernity, Modernism and Post-modernism

These are fashionable labels, and will acquire a settled meaning only long after the smoke of contemporary controversy has cleared. First, the concept 'modern', as a component in historical explanation. Everything depends upon when you think the 'modern' world began, which in turn depends upon which currents of human life seem decisive to you. Some (e.g. Hedley Bull: *The Anarchical Society: a Study of Order in World Politics*, 1977) trace it back to the seventeenth century, and in particular to the Treaty of Westphalia, 1648, which brought the religious conflicts of Europe into equilibrium. Others (e.g. Martin Wight, in *Systems of States*, 1977) go back even further (in this case to the papal bull, *Inter Caestera Divinae*, issued by Pope Alexander VI in 1493). Others bring the date forward into the nineteenth century, as does Paul Johnson, in his *The Birth of the Modern World Society, 1815–1830*, London 1991. In the *intellectual* sphere, however, it is hard to get away from the belief that modernity is connected with enlightenment, with individualism, and with the emancipation of thought from religious and theological dogma. Whenever this emancipation began, it was certainly well under way by the mid-seventeenth century, and the scientist-philosophers of that period (Galileo, Gassendi, Boyle, Descartes, Locke, Leibniz and Spinoza) stand out as the manifest heralds of a new approach to abstract thinking.

At a certain point in this process was unleashed the celebrated 'battle between the ancients and the moderns'. This was probably the first occasion that the term 'modern' was used in its modern sense, to draw a contrast between us and our predecessors. This battle, conducted by the writers of the French *Encyclopédie*, which began appearing in the first half of the eighteenth century, involved all the major thinkers of Europe, from Hume and Lord Kames in Scotland, through Dr Johnson and his circle in England and Diderot and d'Alembert in France, to Vico in Italy and Wolff and Baumgarten in Germany. It was a battle simultaneously about the nature of philosophy, about the place of science in determining the nature of the world, about the authority of religion, and about the curriculum, and what should be studied by those who wished really to *know*. Few of its participants denied the value of classical learning; but all were suspicious of the medieval accretions that encumbered it, while the 'moderns' advocated discovery and experiment as the media of intellectual advance, and the demotic languages as the primary vehicles of literary and scientific discourse. Needless to say, to the inestimable misery of mankind, the moderns won.

The reader will see from that hopelessly potted history that modernity and its concept are both defined in Euro-centric terms. This is no accident; for it is only European civilisation (the civilisation that grew from the amalgamation of Roman law, Greek philosophy and Christian religion) that has possessed the necessary dynamism to endure through such a cataclysmic transformation.

For the historian of ideas, the period known as the Enlightenment (*siècle des lumières, Aufklärung*), is one of the richest objects of study. To date it precisely is difficult, but because it seems naturally to culminate in the philosophical and political ideas that animated the leaders of the French Revolution, there is a broad consensus that the second half of the eighteenth century is the core episode within it. The 'Scottish Enlightenment', which produced the sceptical conservatism of Hume and Adam Smith is, however, a far cry from the French Enlightenment as animated by Rousseau (himself a citizen of Calvinist Geneva), whose cult of natural man, and posture of chafing rebellion against every traditional constraint, had little appeal in Northern climates. In Germany the greatest thinker of the Enlightenment was undoubtedly Kant, who formulated – and, moreover, *justified* – the most lasting intellectual achievement of the age: the system of universal morality which lays down laws for all rational beings, regardless of their circumstances. The inspiration of countless political programmes, the Kantian morality remains the outstanding symbol of 'enlightenment universalism': it points to a universal order of society in which man's essential freedom will be realised, not in anarchy, but in consent under the rule of law. The Founding Fathers of the American Revolution were not influenced by Kant: their inspiration came from British empiricism, and especially from Locke, while their essentially conservative and law-abiding attitude to political reality echoed that of Smith and Hume. But they were men of the Enlightenment nevertheless, and accomplished the second most important achievement of the age.

'Modernism' is perhaps a little easier to define, since the phenomenon denoted by this word derives its existence precisely from the attempts to define it. The word 'modern' was introduced to describe a slow but accelerating social and cultural transformation, by those who were in the midst of it. The word 'modernist', by contrast, was introduced in order to *create* such a transformation, with a kind of authoritarian absolutism that would make itself heard. It began life as a manifesto word. The first manifestos appeared at the beginning of our century, and the whole structure was fully in place by the mid-twenties. Whether in music (Schoenberg), in painting (Marinetti, Braque), in poetry (André Breton, Apollinaire), in politics (Lenin, Mussolini), or in architecture (Loos, Le Corbusier), the ruling idea was simple: the past has *exhausted* itself; and as a result we are *free* of it. The future is our faith, and the sphere of our hopes and achievements.

As the phenomenon developed, however, there emerged a new kind of modernism – and especially in the arts. Poets, painters and musicians, while enjoying their regimen of experiment, began to seek a discipline that would give form to it. And they sought for this discipline in the art of the past. Gradually, the leading modernists began to envisage themselves as more committed to the past than to the future. Their new forms and procedures, they insisted, were the legitimate continuation and revivification of the tradition from which they seemed (but seemed only to the naïve observer) to depart. Schoenberg ceased to herald the music of the future, but wrote instead of the continued life of German musical culture, resting his faith in 'Brahms the progressive'; T.S. Eliot became the century's most articulate defender of literary tradition, and – almost to a man – the greatest of the literary modernists in the English-speaking world rejected revolutionary politics and embraced the tried old ways of constitutional government. The same was true of Matisse and Henry Moore. Only in architecture and politics did the ebullient invocation of a new order of things continue to dominate the modernist psyche: with castastrophic results (see Chapter 30).

This is what makes the *history* of modernism so complicated. Modernism began with a repudiation of the past; but some modernists repudiated this very project, or tried to win through it to a reaffirmation of that which they had once rejected. Other modernists regarded this new repudiation as a 'betrayal': and that is especially true of the Marxists among them, in particular of the school of philosopher-critics founded by Max Horkheimer in Frankfurt, and which included the musician-philosopher Theodor Adorno and the social theorist Herbert Marcuse. The last gasp of the Frankfurt School can be observed in the writings of Jürgen Habermas, whose *Philosophical Discourse of Modernity*, tr. F.G. Lawrence, Oxford 1987, shows a final turning away from the ideology of progress and emancipation, towards a bemused acceptance of Western culture, as the only culture we are likely to have.

'Post-modernism' entered the vocabulary in yet another way, rather as the terms 'Baroque' and 'Renaissance' entered the vocabulary of the art historian. It was fairly obvious by the early seventies that 'modernism' as an artistic and cultural creed was stone dead. It was also obvious that the world of culture was being transformed beyond recognition by two conflicting influences: a widespread pessimism about the modern world, and a flood of momentary artefacts (television, pop-culture, travel) which remove attention from every period of human life besides the present. The word 'post-modern' was used to describe the new cultural conditions; while the word 'post-modernism' was used to refer either to those who affirmed them, or to those who sought for a philosophical or aesthetic doctrine that would replace the sterile future-worship of the early modernists.

'Post-modernism' can therefore describe two distinct intellectual phenomena: the documentation of post-modern society, as conducted, for example, by the French sociologist Jean Baudrillard; or the attempt to advocate a new and post-modern stance towards that society, which will be one-up in sophistication and iconoclasm on the stodgy futurism of the modernists. The principal representative of this latter way of thinking is Jean-François Lyotard, whose *The Postmodern Condition*, Minnesota 1984, attempts a comprehensive summary of the post-modern attitude to knowledge. Lyotard argues (though 'argue' is not quite the word for his style of thinking) that the post-modern condition has been brought about by two vast revolutions: first, the collapse of the 'narratives of legitimation', including those of modernism, with which Western societies have justified themselves since the Enlightenment; and secondly, the ascendancy of information technology, which has replaced traditional culture, as a title to knowledge and distinction, with a new kind of expertise. Knowledge is now the property of machines, and whatever cannot be summarised as 'data' is destined for extinction. The society of the future will store the information that Shakespeare wrote so many plays, of so many thousand words, with so many characters. But it will not retain a trace of the meaning of those plays, which slips through the grid of information and disperses into the void. The stance implicitly recommended by Lyotard can be conveyed in a nutshell as 'the philosophy of inverted commas'. Since, in the new conditions, only unsophisticated people can have beliefs, values and meanings, all such must be placed in quotation marks by the philosopher. This way we use the post-modern condition to achieve a kind of emancipation from the narratives of power.

What is Philosophy?

It is difficult to explain what is meant by abstraction. But there are some formidable examples to choose from. I recommend J.G. Fichte's *Science of Knowledge* (ed. and tr. Peter Heath and John Lacks, Cambridge 1982), which is abstract to the point of unintelligibility, but which certainly gives a sense of just how far philosophy can go in removing itself from the real world of human events.

Aristotle's theory of the categories occurs in his *De Interpretatione* and *Prior Analytics*, and the appropriate passages are excerpted in J. Ackrill's *New Aristotle Reader*, Oxford 1987.

The argument from Kant occurs in the section of the *Critique of Pure Reason* entitled 'The Antinomies'. My version of it is contentious, and the interested reader should consult:
C.D. Broad, 'Kant's mathematical antinomies', *Proceedings of the Aristotelian Society*, 1955.
R.G. Swinburne, 'The Beginning of the universe', *Proceedings of the Aristotelian Society*, 1966.
Pamela Huby, 'Kant or Cantor? That the universe, if real, must be finite in both space and time', *Philosophy*, 1971.
Bertrand Russell, *Our Knowledge of the External World*, London 1956, pp. 170f.

Nietzsche's 'perspectivism' is not the straightforwardly paradoxical thing that I have described: although it has elements of wilful paradox. For a more sympathetic reading, consult Arthur Danto, *Nietzsche as Philosopher*, London 1965, and the unbelievably long-winded and boring *Nietzsche* by Richard Schacht (London 1983), in the Arguments of the Philosophers series, published by Routledge. The relevant passages from Nietzsche occur in the book translated as *The Gay Science*, by Walter Kaufmann (New York 1974), and which ought perhaps to be called, as it once was, *Joyful Wisdom*. (See, for example, section 265, and compare Nietzsche's *Beyond Good and Evil*, section 2.)

What is the Subject-Matter of Philosophy?
As an example of the first approach, the reader might study Plato's *Parmenides*; as an example of the second, Wittgenstein's *Philosophical Investigations*, Oxford 1952, and as an example of the third Schelling's *System of Transcendental Idealism* (tr. Peter Heath, intr. Michael Vater, Charlottesville 1978), which remains the most accessible work of German idealism.

Does Philosophy have a Distinctive Method?
(a) *Thomism*. Jacques Maritain's *Introduction to Philosophy*, London 1932, gives the flavour of modern Thomism. But of course there is very much more to Thomism than can be contained in so short a book.
(b) *Linguistic or 'conceptual' analysis*. Gilbert Ryle's celebrated work, *The Concept of Mind*, London 1949, exemplifies the method. But linguistic philosophy is essentially dynamic, and is never at one time what it was a moment before. An attempt to capture this kind of

philosophy as it was in 1979 was made by Ted Honderich and Myles Burnyeat, in *Philosophy as it is*, Penguin, Harmondsworth 1979, a collection of celebrated articles and extracts, most of which repay careful study.

(c) *Critical philosophy*. Short of tackling the *Critique of Pure Reason*, the reader might try Roger Scruton, *Kant*, London 1981, which is very far short of the *Critique* in every sense.

(d) *Phenomenology*. There is a good introduction by M. Hammond, called *Understanding Phenomenology*, Oxford 1991. Edmund Husserl wrote a survey, entitled 'Phenomenology', which was published in the eleventh edition (1911) of the *Encyclopaedia Britannica*, the greatest edition of that work, and the last (apart from additional volumes) before the right to publish the *Encyclopaedia* was acquired by American vandals.

The a priori *and the Empirical*

I return to this subject in Chapter 13. The idea of philosophy as an *a priori* discipline has been increasingly questioned, by W.V. Quine and his followers (who on the whole do not believe that the concept of the *a priori* is coherent), and also by philosophers impatient with the assumption that you can assign a question to the *a priori* category before the hard work of answering it. (See, for example, David Wiggins, 'Moral Cognitivism', in *Proceedings of the Aristotelian Society*, 1991.)

That is only one instance of a general truth: everything that I have said in this first chapter, just like most other things in philosophy, is disputed by someone, and often on very good grounds. And if it seems at times as though I am ignoring real complexities, do not be deceived: I am.

Branches of Philosophy

Here are some introductions:

(a) Logic: Mark Sainsbury, *Logical Forms*, Oxford 1991.

(b) Epistemology: Jonathan Dancy, *Introduction to Contemporary Epistemology*, Oxford 1985.

(c) Metaphysics: José Bernadete, *Metaphysics*, Oxford 1989 (sophisticated and over-confident.)

D. W. Hamlyn, *Metaphysics*, London 1984 (pedestrian, but instructive.)

(d) Ethics and aesthetics. See Chapters 20 and 29.

History
The important point to clarify is the distinction between history of philosophy and history of ideas. See the first chapter of Roger Scruton, *A Short History of Modern Philosophy*, especially pp. 11–13.

For a profile of the figures mentioned in this section, consult Edwards's *Encyclopedia*, or else Sir Anthony Kenny, ed., *An Illustrated History of Western Philosophy*, Oxford 1994.

Chapter 2 Scepticism

Preliminary reading:
Descartes, *Meditations of First Philosophy*, I.
Other works referred to:
Norman Malcolm, *Dreaming*, London 1954.
Gilbert Harman, 'The inference to the best explanation', in *Philosophical Review*, vol. 74, 1965, pp. 88–95.
Hilary Putnam, *Reason, Truth and History*, Oxford 1984.
George Berkeley, *Three Dialogues between Hylas and Philonous.*

The understanding of scepticism has been heavily influenced by Barry Stroud's *The Significance of Philosophical Scepticism*, Oxford 1984. The first chapter contains a patient exposition of Descartes's argument about dreaming, while the second chapter discusses what was for a while the favoured response to that argument: namely, that Descartes's sceptical conclusion (that we do not know that the external world exists) arises simply because he is re-defining the verb 'to know', placing upon it conditions so strict that they could not be satisfied. The *locus classicus* of that approach is J.L. Austin's celebrated essay 'Other Minds', contained in his *Philosophical Papers*, Oxford 1961. It is worth studying this paper, and Stroud's reply to it, since it gives a clear indication, first of the method of 'linguistic analysis', and secondly of the reasons why recent philosophers have been dissatisfied with that method. It should be said, however, that Stroud's book is immensely long-winded, and packed with redundant sentences. Philosophers have become so nervous of their nit-picking colleagues, that they dot every i and cross every t, lest they be accused of slap-dash thinking. Stroud's book is by no means the worst example; but reading it is as likely to stir an appetite for philosophy as a visit to the slaughter-house is likely to stir an appetite for meat.

The following sources give a picture of modern philosophy's response to the demon:

(1) G.E. Moore, 'Proof of an External World', in *Philosophical Papers*, Allen and Unwin, London 1959. This famous paper (which is the focus of Stroud's argument in his third chapter) will be discussed in Chapter 10, as will be:
(2) Ludwig Wittgenstein, *On Certainty*, Blackwell, Oxford 1969.
(3) P.F. Strawson, *Skepticism and Naturalism*, Methuen, London 1985. This develops one standard response, to which I shall occasionally refer in subsequent chapters.
(4) Robert Nozick, *Philosophical Explanations*, Oxford 1981. A good example of the exuberant logorrhea of American philosophy today, with a novel response to scepticism.
(5) Peter Unger, *Ignorance: a Case for Scepticism*, Oxford 1975. A radical reassertion of the sceptic's argument, using every trick in the game. Unger's argument is usefully discussed by Stroud, in a review article in *The Journal of Philosophy*, 1977.
(6) Hilary Putnam, *Reason, Truth and History*, Oxford 1984: for the re-statement of the demon argument, in terms of 'brains in a vat'. This unpleasant fantasy has gripped the imaginations of recent philosophers; it is doubtful that it constitutes any advance over Descartes, however.
(7) Stanley Cavell, *The Claim of Reason: Wittgenstein, Skepticism, Morality and Tragedy*, Oxford 1979. Cavell offers a response to the sceptic which resembles that which I advance in Chapter 5, although it is presented in a highly sophisticated way. Cavell's approach is endorsed by Stroud, in the final chapter of his book.
(8) Finally three articles from the British technocrat school:
(i) John McDowell, 'Criteria, Defeasability and Knowledge', *Proceedings of the British Academy*, 1982.
(ii) Crispin Wright, 'Facts and Certainty', *Proceedings of the British Academy*, 1985.
(iii) Crispin Wright, 'Imploding the Demon', *Mind*, 1991.

Some general questions:
1. What is the best response to the dreaming argument of Descartes? (See Stroud's and Wright's discussion of this.)
2. Is it a sufficient response to scepticism that it is a fruitless position? (This is, in essence, Strawson's response in (3).)
3. 'Perhaps the best scepticism-rebutting argument in favour of the existence of body is the quasi-scientific argument . . . that the existence of a world of physical objects having more or less the properties which current science attributes to them provides the best

explanation of the phenomena of experience.' (Strawson, in (3).) Do you agree?
4. Do you have an answer to the demon? If not, what do you propose to do about it?

Chapter 3 Some More -isms

To attempt to give the history of philosophy in one chapter is madness. That is why I start with Descartes and end with Plato. Here are a few supplementary thoughts, together with some reading and some questions:

1. Idealism

What I say about Berkeley is controversial, as you will see from consulting some of the papers in *Essays on Berkeley*, edited by J. Foster and H. Robinson, Oxford 1985. See in particular 'Berkeley's Central Argument' by A.D. Smith.

On idealism generally see *Idealism Past and Present*, published by Cambridge University Press in 1982, being lectures to the Royal Institute. This contains useful contributions – especially Myles Burnyeat on Greek philosophy.

The comparison and distinction between Kant and Berkeley is made by R.C. Walker, in his contribution to the volume edited by Foster and Robinson. 'Transcendental idealism' is also explained in my short book entitled *Kant*. There is no easy way to come to grips with the 'objective idealism' of Hegel, though I return to it in Chapter 11.

Some questions to consider:
(1) How did Descartes make Berkeley's idealism possible?
(2) How do Kant's 'appearances' differ from Berkeley's 'ideas'?

2. Verificationism

The chapter devoted to this in J.A. Passmore's *A Hundred Years of Philosophy*, London 1957, is thorough and sufficient. Far better than anything by Ayer are the essays by Feigl and Hempel, in H. Feigl and W. Sellars, *Essays in Philosophical Analysis*, New York 1949. These give the real flavour of the movement, as well as the sense of what it achieved.

It is always useful to write an essay summarising the verification principle, arguments for and against.

3. Reductionism

An excellent history by J.O. Urmson, called *Philosophical Analysis*, Oxford 1956, gives the background to this, and should enable the reader to answer the most important question:

What is a logical contruction? Are there any surprising examples?

4. Empiricism and 5. Rationalism

It is best not to tackle these subjects at this stage, but to consider the individual philosophers, one by one.

6. Realism

The question that will concern us in time, will be the relation between 'semantic' anti-realism, and anti-realism of other kinds. In the introduction to his fairly impenetrable *Realism, Meaning and Truth*, Oxford 1986, Crispin Wright has explored this issue, though in a manner that is at best provisional. Once again, it is best to postpone the subject until later.

7. Relativism

This subject is fairly clearly expounded in chapter 4 of John Passmore's *Philosophical Reasoning*, London 1961. The discussion in Plato's *Theaetetus*, Hackett 1990, has been provided with a superb commentary by Myles Burnyeat. See also his 'Protagoras and self-refutation in Plato's *Theaetetus*', *Philosophical Review*, vol. 85, 1976, and pp. 168–71 of John McDowell's edition of Plato's text (Oxford 1987). Burnyeat's commentary is published by Hackett, 1990, and the crucial pages are pp. 28–30.

For more recent debates, the reader should consult Bernard Williams, *Ethics and the Limits of Philosophy*, London and Cambridge Mass., 1985; Richard Rorty, *Philosophy and the Mirror of Nature*, Princeton 1979, and Hilary Putnam, *Renewing Philosophy*, Cambridge Mass. 1992, ch. 4.

Some questions:

(1) 'The content of such assertions [i.e. relativist theories] rejects what is part of the sense or content of every assertion and what accordingly cannot be significantly separated from any assertion' (Husserl, *Logical Investigations*, New York 1970). Do you agree?

(2) Is there any defensible form of moral relativism?

(See Gilbert Harman, 'Moral Relativism Defended', *Philosophical Review* vol. 84, 1975; and David Wiggins, 'Moral Realism, Motivating Beliefs, etc.', *Proceedings of the Aristotelian Society*, 1991.)

Chapter 4 Self, Mind and Body

Preliminary reading:
Descartes, *Meditations*, II.

Understanding the logic of Descartes's '*cogito*' is by no means an easy task. There are three recent commentaries on Descartes that are helpful: one by Anthony Kenny (Random House, 1968), which is dull, dutiful and depressed; one by Margaret Wilson (London 1978), which is worthy, worried and well argued, and one by Bernard Williams (Harmondsworth 1978), which is lively, luminous and likely to last. Important articles by Jaakko Hintikka (*Philosophical Review*, 1962, 1963) and others, are sufficiently discussed in Williams's book, which will repay continual study.

I shall return to the Cartesian theory of the mind in Chapter 15. Meanwhile it is worth reflecting on the following questions:
(1) What view of the mind is conveyed by Meditation II?
(2) What exactly is the argument of the '*cogito*'? Is it valid?
(3) Descartes argues that 'I exist' is true every time that I think it. He also believes that 'I have a perceptual experience' is true every time I think it. But are these sentences true for the same reason?

It is useful to look at the way philosophers introduce their discussion of the mind and mental states, and to ask yourself whether they are not covertly assuming, in doing so, that the mind is a separate sphere from the physical world, and inherently private to the one who possesses it. Look, for example, at Locke's *Essay of the Human Understanding*, the first major text of modern empiricism, Book II, ch. 1, where he introduces 'ideas'; or at Leibniz's reference to the soul in his *Monadology*, a short but powerful work which contains the seeds of Leibniz's philosophy. (It is available in N. Rescher, ed., *Leibniz: Selections*, second edition, London 1991. The first edition of this work contains misleading translations and should be avoided.) Phenomenology is harder to pin down. But the introduction to Maurice Merleau-Ponty's *Phenomenology of Perception*, second edn, London 1989, is quite revealing, as are the *Cartesian Meditations* of Husserl (Nijhoff, Amsterdam 1960). The introduction by Hammond (see Study Guide to Chapter 1) is a good place to begin; better still is the first chapter of David Cooper's *Existentialism*, Oxford 1990.

I return to the topic of intentionality in Chapter 18.

Chapter 5 The Private Language Argument

Preliminary reading:
L. Wittgenstein, *Philosophical Investigations*, Oxford 1952, sections 243–351.
Also referred to:
W.V. Quine, 'Epistemology naturalised', in *Ontological Relativity and Other Essays*, Cambridge Mass. 1961.

This argument has attracted so much commentary, that it is impossible to give an impartial survey of the literature. O.R. Jones edited a volume (now out of print) entitled *The Private Language Argument*, London 1971, which brought together the most important material published in the fifties and sixties. But the debate has moved on since then, and there is no great need to refer to it. There are three recent accounts which support the interpretation that I offer:
(1) James Hopkins, 'Wittgenstein and Physicalism', *Proceedings of the Aristotelian Society*, 1975.
(2) Roger Scruton, *Sexual Desire*, London and New York 1986, Appendix I.
(3) M. J. Budd, *Wittgenstein's Philosophy of Psychology*, London 1989.

These three texts share the view that the logic of first-person 'avowals' is a by-product of the public language, and cannot be used, either as proof of an 'inner' private realm, or as a blueprint for some supposedly 'private' language. (See further the discussion in Chapter 31.) In addition, Budd conclusively refutes the view that first-person awareness involves the observation of inner states.

All such arguments raise the question how first-person authority is to be explained. This question has been explored by Donald Davidson, in 'First Person Authority', *Dialectica*, 1984, and by Crispin Wright, in 'On making up one's mind: Wittgenstein on intention' (*Proceedings of the 11th International Wittgenstein Symposium*, Kirchberg 1987.) It is fair to say that the first person remains one of the great outstanding mysteries of modern philosophy. (See again Chapter 31.)

More recent discussion has been profoundly affected by S. Kripke, *Wittgenstein on Rules and Private Language*, Oxford 1984. Kripke's interpretation is challenged by Roger Scruton, in a review article in *Mind*, 1985, and by Crispin Wright in 'On Making up One's Mind', *op. cit.* There is also a school of Wittgenstein interpretation founded in Oxford by G. Baker and P.S. Hacker, which, while broadly in agreement with the conclusions of this chapter, is radically opposed to the way of reaching them.

Some questions:
(1) What is the 'beetle in the box' argument? Is it valid?
(2) If a solitary Robinson Crusoe invented a language, would his language be 'private' in Wittgenstein's sense?
(See A.J. Ayer, 'The Private Language Argument', in *The Concept of a Person* and other essays, London 1963.)
(3) Wittgenstein asserts that it doesn't make sense to say of myself, that I *know* that I am in pain. Do you agree?
(4) If Descartes were deceived by the demon, could he know what he means by '*cogito*'?

Chapter 6 Sense and Reference

Preliminary reading:
G. Frege, 'On Sense and Reference,' in P.T. Geach and M. Black, eds., *Philosophical Writings of Gottlob Frege*, Oxford 1960.
'The Thought', in *Mind*, vol. 65, 1956, pp. 287–311.

It is important to master the idea of quantification, and to understand its role in modern logic. To this end the reader should consult the article on quantifiers in M. Dummett's *Truth and Other Enigmas*, London 1978, pp. 8–33. Dummett traces the history of scholastic logic, and the difficulties it faced in explaining inferences involving sentences of multiple generality (e.g. 'Everyone loves someone'). 'The difficulty,' Dummett writes (p. 10) 'arose out of trying to consider a sentence such as the above as being constructed simultaneously out of its three components, the relational expression here represented by the verb, and the two signs of generality'. Frege showed how the sentence could be constructed in steps which correspond to the different signs of generality within it. From 'Peter loves Jane' I can abstract the 'existential generalisation', 'Peter loves someone'; and from this I can move to the next level of generality, so arriving at 'Everyone loves someone'. This way we can distinguish that sentence from 'There is someone whom everyone loves', since the order of construction is in each case different.

Frege is a growth industry in modern philosophy, and it is easy to get lost in the literature. It is lamentable that a philosopher who wrote so lucidly, accurately and well should now be buried beneath a mound of ill-written scholarship. At some stage you should become acquainted with the Frege described in Dummett's work – and in particular, in *Frege: Philosophy of Language*, London 1973. For although this Frege is in part an invention of Dummett's, he has become an important presence in modern philosophy. Dummett has returned to the fray several times, notably in *The Interpretation of Frege's Philosophy*, London 1981. For rival views it is best to consult the collection of papers edited by Crispin Wright, entitled *Frege: Tradition and Influence*, Oxford 1984, especially the articles by D. Bell and David Wiggins.

Frege's theory of sense incorporates several strands: there is the purely 'semantic' idea of sense, as the conditions that the referent must satisfy if the term is to be correctly applied to it (as in Frege's *Basic Laws of Arithmetic*, tr. M. Furth, Los Angeles 1964, p. 85); and there is also the 'epistemic' idea, of sense as the 'route to' reference, which dominates the article 'On Sense and Reference', in *Philosophical Writings*, ed. P.T. Geach and M. Black, Oxford 1960. In rehearsing the theory both the commentators and Frege himself have frequent recourse to metaphors; sense is the 'route to reference', the 'mode of presentation' of reference, the 'way of thinking' of reference; and so on. Attempts to discard these metaphors seem to lead to the purely 'semantic' idea, mentioned above. There is a growing literature, also, on the question whether sense determines reference, or vice versa. In some cases – the predicate 'chair', say – it seems quite plausible to say that reference is fixed by sense. We understand the predicate, and that enables us to identify the items which it applies to. But in other cases, it seems as though the relation goes the other way. I understand the name 'John Major' by knowing that it is the name of *this* man. By learning the reference of the name, I come to understand the name. So do we say that, in this case, reference determines sense? All discussions of these questions are greatly indebted to Hilary Putnam, whose views are neatly summarised in *Reason, Truth and History*, Oxford 1984. I discuss the matter further in Chapters 13 and 19. Two trendy but difficult papers on the topic are:
(1) Gareth Evans, 'Understanding Demonstratives', in *Collected Papers*, Oxford 1985.
(2) John McDowell, '*De re* senses', in C. Wright, ed., *Frege: Tradition and Influence*, Oxford 1984.

Some questions:
(1) Why did Frege distinguish sense from reference?
(2) Why should we think of the reference of a sentence as its truth value?

(3) What is a quantifier? Give a logical analysis of the sentence 'Everyone loves someone', using quantifiers.

There is a vast number of books explaining the revolution in logic. No better history of the subject has been written than the massive study by W. and M. Kneale, called *The Development of Logic*, Oxford 1962. But it is not the history that is important, so much as the subject as it is now understood. The preface to W.V. Quine's *Methods of Logic*, London 1952, is helpful; but you will also need to work through a textbook of logic. The strange thing is that almost nobody who teaches the subject is prepared to recommend one, unless he is himself the author of it. My own recommendation is that by Mark Sainsbury, entitled *Logical Forms*, Oxford 1991.

Chapter 7 Descriptions and Logical Form

Preliminary reading:
Bertrand Russell, 'On Denoting', in *Logic and Knowledge*, London 1956.
P.F. Strawson, 'On Referring', in *Logico-Linguistic Papers*, London 1971.

G. McCulloch's *The Game of the Name*, Oxford 1990, provides a general introduction to the issues discussed by Russell, beginning with Frege and working through to the most recent 'developments', which means roughly what is being said about the subject in Oxford. McCulloch analyses Russell's Problem, and the way in which the theory of descriptions was supposed to solve it. He also connects the theory with the general question of naming, which lies beyond the scope of this book.

Russell's arguments for the theory of descriptions are to be found in various places, besides 'On Denoting': for example, *The Problems of Philosophy*, Oxford 1905, p. 29; the 'Lectures on Logical Atomism' in *Logic and Knowledge*, London 1956, and the introduction to *Principia Mathematica*, by Russell and A.N. Whitehead, Cambridge 1913. They are inconclusive, but well summarised by Mark Sainsbury, in his dry but thorough book on Russell (London 1979).

Strawson's response occurs also (though in an amended form) in his book *Introduction to Logical Theory*, London 1952. In neither version does he make clear whether the sentence 'The king of France is bald' should be given no truth-value, or a third truth-value, in those cases where it 'makes no statement'. But there are other problems too. How, for example, are we supposed to construe the utterance 'The murderer of Mozart was insane', when it is unknown whether Mozart was murdered? Does this sentence only say something when we actually know that Mozart was murdered? Here we see an important point. Russell's problem was to account for 'empty singular terms' (i.e. singular terms which lack a referent). Strawson's theory merely postpones this problem, and does not answer it.

The article by K.S. Donnellan ('Reference and Definite Descriptions', *Philosophical Review*, vol. 75, pp. 284–304) has been influential, for reasons discussed by McCulloch. (See also Donnellan's 'Proper Names and Identifying Descriptions' in D. Davidson and G. Harman, eds., *Semantics of Natural Language*, Dordrecht 1972.) Saul Kripke has also influenced the debate, with his 'Speaker's Reference and Semantic Reference', in P. French, ed., *Contemporary Perspectives on the Philosophy of Language*, Minnesota 1977, arguing for Russell against Donnellan. The debate goes on, not only because it is central to the philosophy of language, but also because, ridiculous though it may seem, the truth about metaphysics really does depend on the meaning of 'the'. (Though not quite in the same way as the truth about love depends upon the meaning of 'and' – see *Tristan und Isolde*, Act II.) An illuminating recent discussion is that by P. Millikan, 'Content, Thought and Definite Descriptions', *Aristotelian Society, Supplementary Volume*, 1990.

Some questions:
(1) What does Russell mean by a 'definite description'? How does he account for the meaning of sentences involving such descriptions?
(2) Does ordinary language function as Russell's theory of descriptions says it should?
(3) How should a philosopher settle the question, whether this or that exists?
(4) What exactly does Strawson mean by a 'statement'? Are there any statements?

The topic of logical form has again come to the fore, as a result of Donald Davidson's approach to the theory of meaning. I return to this topic in Chapter 19.

Chapter 8 Things and Properties

Preliminary reading:
Bertrand Russell, chapter on universals, in his *Problems of Philosophy*, London 1910.

This chapter is extremely controversial, and it is well to rehearse some of the points at which other philosophers would raise a protest.

2. Universals: Primary and Secondary Qualities

This is a subject of such debate, not only among scholars of Locke and Berkeley, but also among contemporary philosophers of science, that almost nothing that is said about it will remain unchallenged. The chapter in J.L. Mackie's *Problems from Locke*, Oxford 1976, is useful; so too is the article on the subject by A.D. Smith, in *Philosophical Review*, 1990. The subject will, I hope, be clearer after Chapter 15.

The concept of a sortal has become important in recent writings about substance and identity, particularly through the work of David Wiggins, who argues that 'A is identical with B if there is some substance concept F such that A coincides with B under F' ('On being in the same place at the same time', *Phil. Rev.* 1968, p. 90f). 'Coinciding under a substance concept' is a pleasant passtime: you and your past self coincide under the concept human being (or do you? See Chapter 21); Decartes's hard, cold, fragrant lump coincides with the soft, warm, odourless mush under the substance concept: wax. But this is *not* a sortal. That is, we do not *count* things as 'waxes'; we speak instead of a certain amount, quantity of wax. Identity then becomes a tenuous idea: 'what happened to that lump of wax?' 'It's smeared all over the table.' 'You mean the *same* lump of wax?' 'No: the same *wax*, but not the same *lump*, since it is no longer a lump.' We get into deep water here, because we have no clear way of counting waxes: we measure wax, but we do not count it. That is why we might deny that *wax* is a sortal concept.

3. The Problems of Universals: Plato

Did he believe in the theory of Forms? The 'third man argument' against the Forms (which I mention in the main text of this chapter) is in fact used against them by Plato himself, in the *Parmenides*. (The argument is well expounded by Gail Fine, 'Aristotle and the more accurate arguments', in Martha Nussbaum and E. Schoefield, eds., *Language and Logos*, Cambridge 1982.)

Here is the argument as Plato gives it:

> Again there is another question.
>
> What is that?
>
> How do you feel about this? Imagine your ground for believing in a single Form in each case is this: when it seems to you that a number of things are large, there seems, I suppose, to be a certain character (*idea*) which is the same when you look at them all; hence you think that Largeness is a single thing.
>
> True, he replied.
>
> But now take Largeness itself and the other things which are large. Suppose you look at all these in the same way in your mind's eye, will not yet another unity make its appearance – a Largeness by virtue of which they all appear large?
>
> So it would seem.
>
> If so, a second Form of Largeness will present itself, over and above Largeness itself and the things that share in it; and again, covering all these, yet another, which will make all of them large. So each of your Forms will no longer be one, but an indefinite number.
>
> (*Parmenides* 131E–132B.)

The name 'Third Man' is due to Aristotle (*Metaphysics* 990b, 15), although there were a variety of arguments alive in Plato's day to which Aristotle could have been referring. (See F.M. Cornford, *Plato and Parmenides*, London 1939, pp. 88f.) The general point is this. If individual men are men by virtue of their participation in the Form of Man, by virtue of what is the Form itself a Form of Man? Does it not have something in common with its instances (manhood), and does this not mean that both it and they participate in a further Form, which determines what they have in common? In which case, are we not at the beginning of an infinite regress? The question here is whether 'man' is predicated of the Form in the same way as it is predicated of the individual instances.

For a traditional approach to Plato's theory of universals, see W.D. Ross, *Plato's Theory of Ideas*, Oxford 1951. Ross assumes that the Idea or Form of Bed, discussed by Socrates at *Republic* 597, is one among the many Forms, of which there is one for every general term. This does not square with Plato's own doubts (expressed, however, in the *Parmenides*, which was probably written later) as to whether there are Ideas of man, fire, water, hair, dirt, mud or other 'undignified and trivial objects' (130 b–e). In a thorough examination of this matter, Charles Griswold ('The Ideas and the Criticism of Poetry in Plato's *Republic*, Book 10', *Journal of the History of Philosophy*, vol. XIX no. 2, 1981, pp. 135–50) concludes that the remarks about the Idea of bed in the *Republic* are a kind of joke on Socrates' part.

In 'Teleology and the good in Plato's *Phaedo*', *Oxford Studies in Ancient Philosophy*, vol. IV, Oxford 1986, David Wiggins argues that the Forms are determinations of the good, which strive jointly to realise the Good. The theory of Forms is not, in the normal sense, a theory of universals at all. For a surprising view of the theory by one of Plato's contemporaries, see *Xanthippe's Republic*, 45D, collected in R. Scruton, ed., *Xanthippic Dialogues*, London 1993.

On the question of abstract particulars – whether there are such things, and, if so, what precisely *are* they? – see Nelson Goodman's classic assaults on them: 'On Abstract Entities', *Journal of Symbolic Logic* 1947, and 'Nominalism', *Philosophical Review*, 1957; you might also consult M. Dummett, *Frege: The Philosophy of Language*, ch. 4, and Crispin Wright, *Frege's Conception of Numbers as Objects*, Aberdeen 1983, ch. 1. This issue will be clearer after Chapter 26, on mathematics.

4. Realism and Nominalism

My statement of this position is heavily influenced by Wittgenstein and Goodman. The reader will not find it stated like this in standard works on the subject, such as David Armstrong's *Universals and Scientific Realism*, Cambridge 1978, which is a robust defence of hard-headed realism against a kind of nominalism that is timid enough to give way. A nominalist who really sits it out can always win the argument: the only question is whether he is prepared to pay the price.

Other discussions of universals distinguish many more positions than I have mentioned. For example, there are the following distinct kinds of nominalism:

(i) Predicate nominalism (which makes predication into the fundamental generalising device). See: J.R. Searle, *Speech Acts*, Cambridge 1969, pp. 105, 120, and David Armstrong, *Nominalism and Realism*, Cambridge 1978, pp. 11–25;

(ii) Class nominalism (which takes class membership as the fundamental idea), explained by Armstrong, *ibid.*, pp. 28–43;

(iii) Conceptualism (which puts mental in the the place of linguistic generalisation), as in Hume, *Treatise*, Bk I, Part I, sec. 7.

There are also various resemblance theories, which are difficult to classify, since they so disastrously miss the point of the problem. (Resemblance is what has to be *explained* by a theory of universals, and surely cannot be used to construct that theory?) Armstrong discusses these in *ibid.*, pp. 44–57; as does Goodman, in a strong contribution to J. Foster, ed., *Experience and Theory*, Cambridge Mass. 1970 (see especially pp. 24–5). There is also the 'family resemblance' 'solution', which I discuss in Chapter 14.

The most radical kind of nominalism recognises *only* individuals, arguing that nothing exists except concretely, and that therefore we should re-express all general statements in terms of the individuals which make them true. On this view, even to refer to classes or sets can be ontologically criminal, since it involves a departure from the pure ontology of individual objects. It is possible that William of Ockham was an extreme nominalist in this sense. It is certain that Nelson Goodman is. See his: 'A World of Individuals', in *Problems and Projects*, Indianapolis 1972. Goodman, in earlier writings (notably *The Structure of Appearance*, Cambridge Mass. 1951), actually tried to produce a calculus of individuals, which would permit the replacement of all references to predicates, classes and sets, with sentences which quantify over individuals alone. See also the article by Carl Hempel in *Philosophical Review*, 1960.

Further reading:

F.P. Ramsey, 'Universals', in *Foundations of Mathematics*, London 1927.

P.F. Strawson, 'Particular and General', in *Logico-Linguistic Papers*, London 1971.

M. Loux, ed., *Universals and Particulars*, a useful collection of analytical readings.

David Lewis, 'New Work for a Theory of Universals', *Australasian Journal of Philosophy*, 1983.

David Pears, 'Universals', in A. Flew, ed., *Logic and Language*, vol. 1, Oxford 1951.

5. Substance

Again an immensely controversial area, and one to which I return in Chapters 12 and 15. A proper understanding of the idea of substance, David Wiggins suggests, leads us to see that the opposition between nominalism (or rather, its mental variant, conceptualism) and realism has been wrongly stated. We choose the way in which we conceptualise reality, but we do not choose what is there to be conceptualised. The conceptualist is right in opposing the view that reality comes to us already classified; but he is wrong in believing that our concepts are deeply arbitrary. There are right and wrong ways to classify; and one way of going wrong is by carving against the joints of nature.

All this is set out in *Sameness and Substance*, Oxford 1980, and also in the paper 'On Singling out an Object Determinately', in John McDowell, ed., *Subject, Thought and Context*, Oxford 1986. I shall return to it when discussing natural kinds.

Some questions:
(1) What is the problem of universals?

(See H.H. Price, *Thinking and Experience*, London 1953, ch. 1, for an old-fashioned view, and also David Pears, in Antony Flew, ed., *Logic and Language*, vol. 1, Oxford 1951; and D.W. Hamlyn, *Metaphysics*, Cambridge 1984, pp. 95–101.)

(2) What does Strawson mean by a 'basic particular'?

(See *Individuals*, ch. 1.)

(3) Is there any sense in which a cat is more substantial than a dead cat in aspic?

(If you can answer that, you can answer anything.)

Chapter 9 Truth

Preliminary reading:
P.F. Strawson, 'Truth' and 'A Problem about Truth', in *Logico-Linguistic Papers*, London 1971.

If you read the following, you will understand the traditional positions reasonably well:
(1) D.W. Hamlyn, *The Theory of Knowledge*, London 1970, ch. 5.
(2) Susan Haack, *Philosophy of Logics*, Cambridge 1978, ch. 7.
(3) George Pitcher, ed., *Truth*, Prentice Hall, 1964. This contains all the standard texts from the modern period, up to the date of its publication.
(4) J.L. Austin, *Philosophical Papers*, Oxford 1961, pp. 117–33.
(5) Ralph Walker, *The Coherence Theory of Truth*, London 1990.
(6) Simon Blackburn, *Spreading the Word*, Oxford 1984, chapters 5, 6 and 7. An excellent introduction to the modern philosophy of language, which also shows how the problem of truth has been profoundly affected by the theory of meaning.
(7) D.J. O'Connor, *The Correspondence Theory of Truth*, London 1975, which contains a useful discussion of Austin and Wittgenstein.

When considering pragmatism it is important to know that Peirce, who first thought of it, did not formulate it as a theory of truth, but as a theory of scientific method. See his paper 'How to make things clear' in the Dover edition of *The Philosophical Writings of C.S. Peirce*, New York 1955. In this article he writes pithily of truth as 'the opinion which is fated to be ultimately agreed to by all who investigate'. There is, condensed into this sentence, a whole theory of science, language and discovery which would need a chapter in itself. Sometimes Peirce's approach is criticised for trying to embody too many philosophical ambitions within a single theory – see, e.g. Richard Rorty, 'Pragmatism, Davidson and Truth', in *Objectivity, Relativism and Truth*, Cambridge 1991. This may not be just; but in any case, the classical text is not by Peirce, but by his disciple William James: *Pragmatism*, New York 1907. See also: 'The Pragmatist Theory of Truth,' by Susan Haack, *British Journal for the Philosophy of Science*, vol. 27, 1976, for a clear exposition and partial defence of the theory, in its modern form.

The most recent form of pragmatism, defended by Richard Rorty, has a socio-political character. The quest for truth (and, Rorty adds, for goodness too) is simply the quest for a community of free inquiry and open encounter (the kind of community that appeals to Rorty). Call this the pursuit of truth if you like: but there is no practical difference between pursuing truth and pursuing such a community. And where there is no practical difference, the pragmatist will say, there is no theoretical difference either. The inspiration for this approach lies in the writings of the American popular sage John Dewey. Rorty defends his views in the collection of essays entitled *Objectivity, Relativism and Truth*, Cambridge 1991, and in particular in the essay 'Science as Solidarity', in which he leans heavily on the proven ability of coherence theories to disarm their opponents, by showing that there is, in Wittgenstein's words, 'no way of using language to get between language and the world'. Rorty is a sophisticated thinker, well trained in the methods of his opponents, and a convert to pragmatism, which he defends with all the subtlety and all the *odium theologicum* of a medieval schoolman. The big question is this: What difference does it make to the rest of philosophy, that you agree with him or not? Rorty thinks that the difference is very great. The reply mounted by Fred Sommers is as yet largely unpublished. (But see Fred Sommers, 'The Enemy is us', in Howard Dickman, ed., *The Imperilled Academy*, New Brunswick, NJ 1993.)

More recent discussions of truth emphasise its role in the theory of meaning, and the problem of epistemological constraints: Could we acquire a concept of truth in which the world is represented as something that transcends our capacity to discover the truth about it? (I return to this in Chapter 19.) Here the classic text is the article 'Truth' by M. Dummett, in his *Truth and Other Enigmas*, London 1978.

Some questions:
(1) 'Truth is correspondence to the facts'. Could an advocate of the coherence theory agree with that statement?
(2) It is a fact that the earth is not made of cheese. Does this mean there are negative facts?

(This question has exercised many of the defenders of the correspondence theory, who have often envisaged facts as real, solid and 'out there', just like things and their properties. But if the non-existence of Mr Pickwick is a fact, in what way is it real and hard and solid and 'out there'? Once again this is ammunition to the defender of the coherence theory.)
(3) What precisely is at issue between Austin and Strawson?

(The debate here is summarised in Hamlyn, *The Theory of Knowledge*, pp. 132–6; the papers of Austin and Strawson, and a few other contributions to the controversy, are all reprinted in Pitcher's volume.)

The redundancy theory is expounded by F.P. Ramsey, 'Facts and propositions', in *The Foundations of Mathematics*, Cambridge 1931.

The 'semantic conception of truth' is defended by Tarski, in an article with that title (reprinted in H. Feigl and W. Sellars, eds., *Readings in Philosophical Analysis*, New York 1949). It is fair to say that Tarski's purpose in this article was for a long time misunderstood (in a manner exemplified by Hamlyn, pp. 126–9). His purpose was (a) to reaffirm the conception of truth as correspondence; (b) to give what he thought to be the only coherent analysis of that idea; (c) to show that it is this very idea that is required, if truth is to have the semantic role required by the Fregean approach to logic and language; (d) building on those thoughts, to show how 'theories of truth' can be designed, which show the precise and systematic relations between language and world. I return to these matters in Chapter 19, when the concept of truth will feature in the theory of meaning. Tarski composed a layman's guide to the semantic theory, under the title 'Truth and Proof', which appeared in *Scientific American* for 1969.

The 'minimalist' theory of truth, which retains no more than what is essential to safeguard the role of truth in passing from words quoted to words used, has been ably defended in Paul Horwich, *Truth*, Oxford 1990. Many of those who defend Tarski's approach to truth on formal grounds, deny that it commits us to the correspondence theory. Donald Davidson, for example, famous for his application of Tarski to virtually every philosophical problem, finds himself accepting the coherence theory, while Quine, also a disciple in this matter of Tarski, is a pragmatist.

(See Donald Davidson, 'A Coherence Theory of Truth and Knowledge', in E. LePore, ed., *Truth and Interpretation: Perspectives on the Philosophy of Donald Davidson*, Oxford 1986.)

Chapter 10 Appearance and Reality

Preliminary reading:
Locke, *Essay*, Book II, ch. 8.
F.H. Bradley, *Appearance and Reality*, Oxford 1893, chs. 1 and 2.

Bradley is worth studying, as an example of a philosopher who, by emphasising the limitations of our perspective, tries to show that our common-sense world is a world of 'mere appearance'. The same result is reached by 'phenomenalism', a theory that I discuss in Chapter 23. There is a good, though difficult, contemporary discussion of the problems encountered by such a view in Gareth Evans, 'Things Without the Mind', *Collected Papers*, Oxford 1986. All philosophers who distinguish primary from secondary qualities must answer the question 'What do secondary qualities inhere in?' Bradley's answer, in effect, is nothing. And this answer pushes him towards a similar conclusion for all other qualities as well. His absolute reality seems to be entirely *without* qualities: and also without individuality, since there is no coherent means of distinguishing it from anything else.

On the 'absolute conception of reality', which I discuss at the end of the chapter, see Bernard Williams, *Descartes*, already referred to, and Thomas Nagel, *The View from Nowhere*, Oxford 1986, which, for all its deliberate exaggeration, has acquired a place as one of the central documents of recent philosophy: engaging, serious and – as always with Nagel – well written. But throughout the book there is much play on the word 'objective', which tends to change meaning several times in the course of any one proof.

The common-sense rejoinder is given by G.E. Moore, in 'Proof of an External World', in *Philosophical Papers*, London 1959, and, in a highly sophisticated version (so sophisticated that the term 'common sense' can hardly be applied to it) by Wittgenstein in *On Certainty*, Oxford 1971. Moore's 'proof' has received a forbidding number of commentaries, and for a while occupied the centre of the stage, at least among anglophone philosophers. Not everything written about it is now worth returning to; however, a certain kind of 'Wittgensteinian' philosophy made fruitful use of Moore's method, as a point of departure for more adventurous speculations. The most important member of this school was Norman Malcolm, and the following articles are worth studying:

(1) N. Malcolm, 'Moore and Ordinary Language', in P.A. Schilpp, ed., *The Philosophy of G.E. Moore*, New York 1952.

(2) N. Malcolm, 'George Edward Moore', in his *Knowledge and Certainty*, Englewood Cliffs NJ, 1963.

(3) N. Malcolm, 'Defending Common Sense', in *Philosophical Review*, 1949. Stroud's ponderous discussion of Moore's proof in chapter 3 of *The Significance of Philosophical Scepticism*, Oxford 1984, summarises the arguments of Malcolm and others, and gives an up-to-date assessment of the issue. The volume edited by Schilpp contains Moore's own interesting reply to objections raised by Malcolm and by Alice Ambrose.

Some questions:

(1) What is the distinction between primary and secondary qualities, and how important is it?

(See Locke, Book II ch. 8, and A.D. Smith, referred to above, p. 517. Also Colin McGinn, *The Subjective View*, Oxford 1983.)

(2) What is an 'appearance'?

(3) Examine any one of Bradley's arguments for the conclusion that some feature of our ordinary world-view relates only to appearances.

(4) Does Moore *know* that he has two hands?

Chapter 11 God

Preliminary reading:

Descartes, *Meditations*, III.

Kant, 'The Ideal of Reason', in *The Critique of Pure Reason*.

E. Durkheim, *The Elementary Forms of Religious Life*, Paris 1912.

There is an excellent reader on this topic, edited by John Hick, entitled *The Existence of God*, New York and London 1964, which contains all the relevant extracts from Plato, Aquinas, St Anselm, Hume,

Norman Malcolm, and others. The arguments are summarised and dismissed by J.L. Mackie, in his *Miracle of Theism*, Oxford 1982, a kind of plain man's guide to unbelief, which repays careful study.

1. God and gods

In advancing a Durkheimian account of the genesis of religious belief I am consciously simplifying. Hume, in his *Dialogues of Natural Religion*, London 1947, advances another explanation of the 'miracle' of theism; Freudians, Jungians and Feuerbachians would each make suggestions of their own. Maybe no theory captures the whole truth about religious belief; and certainly nobody has really been able to account for the fact that it is a human universal, which can be abolished in one form only to re-emerge in another – for example, as the militant atheism of modern revolutions, the fervour and conscious absurdity of which cannot be understood without recourse to religious categories.

As to the question whether the explanation of religious belief also undermines it, I return to this in the last chapter. A useful reminder of the experiential nature of religion is William James's remarkable *tour de force, The Varieties of Religious Experience*, London 1960.

3. The God of the Philosophers

There is an excellent overview of the medieval account of God, and the difficulties to which it gives rise, by Sir Anthony Kenny, entitled *The God of the Philosophers*, Oxford 1979.

4. Arguments for God's Existence

Aquinas's five ways are reproduced in the volume edited by Hick, and extensively criticised by Sir Anthony Kenny in *The Five Ways*, London 1969. Malcolm's paper on the ontological argument is included in Hick. Plantinga's argument occurs in his *The Nature of Necessity*, Oxford 1974. I discuss it in the next chapter. Also relevant to a study of the ontological argument is the classic discussion of the question whether existence is a predicate, by D.F. Pears and J. Thompson, reprinted in P.F. Strawson, ed., *Philosophical Logic*, Oxford Readings in Philosophy, Oxford 1967.

On the cosmological argument, see the following:

Aquinas and Hume, in Hick.

Paul Edwards, 'The Cosmological Argument', in *The Rationalist Annual*, 1959.

On the argument from design, see:

(1) W. Paley, *Natural Theology*, reprinted in the Library of Liberal Arts series, 1964.
(2) Hume, *Dialogues*, X and XII.
(3) J.L. Mackie, 'Evil and Omnipotence', *Mind*, vol. 64, 1955.
(4) Richard Dawkins, *The Blind Watchmaker*, London 1986.
(5) Stephen Hawking, *A Brief History of Time*, London 1988.

6. The Problem of Evil
It is almost impossible to get clear about this. If God created the world, then the world of his creation must in some sense be 'lesser' than him, dependent upon him, and deprived of some part of his infinite perfection. If evil is the same as deprivation, then it seems to follow that a created world will be in some measure evil. But in what measure? Furthermore, evil has a remarkable knack of turning into good: tragedy shows the most horrible events taking on a strange and compelling beauty, an inevitability which vindicates suffering and shows it to be integral to what is noblest in the human soul. Could the whole universe be like that? On the other hand, the worst things that happen are not tragic at all, but merely senseless or disgusting. True, some of them are the result of human cruelty, and so may be understood as the consequence of the great good of human freedom. But not all of them are like that. We approach here the mystery of death: Is there any way of regarding death itself as a good, or at any rate a non-evil? I return to this in Chapter 21.

Some questions:
(1) What, in your view, is the most plausible version of the cosmological argument? Is it valid?
(2) Is the ontological argument refuted by showing that existence is not a predicate?
(3) Why should we think that the prime mover of the world is also a person?

Other works referred to:
Dan Jacobson, *The Story of the Stories: the Chosen People and its God*, London 1982.
Chateaubriand, *Le Génie du christianisme*, 1802.

Chapter 12 Being

Preliminary reading:
Aristotle, *Metaphysics*, Books IV and VI.

Hegel, *The Science of Logic*, Book I.
Heidegger, *Being and Time*, Introduction and chs. 1 and 2.
J. Bernadete, *Metaphysics, the Logical Approach*, Oxford 1989.

1. Aristotle
Aristotle's key thoughts about the way in which reality is divided are to be found in the *Categories* and the *Metaphysics*; the relevant passages have been extracted in the *New Aristotle Reader*, ed. J. Ackrill, Oxford 1987. Aristotle's underlying thought in the *Categories*, which is the earlier of the two works, is that the universe is made up in the last analysis of individual substances like Socrates; anything else is a dependent property, and exists only in so far as it is instantiated in some individual substance. In *Metaphysics*, Aristotle argues that substance is form (i.e. that which gives to 'matter' its particular nature and identity), and this appears to conflict with his earlier view. See the article by J.A. Driscoll in *Studies in Aristotle*, ed. O.J. O'Meara, Washington DC 1981. The concisest definition of substance that Aristotle offers is at *Categories* C5: 'A substance – that which is called a substance most strictly, primarily, and most of all – is that which is neither said of a subject nor is in a subject, e.g. the individual man or house.' This is the idea which Strawson resurrects, in the form of his 'basic particular' (see Chapter 8).

2. The Analytical View
Most modern philosophers accept the credentials of concrete particulars, like tables and chairs, as occupants of reality. There is no problem about their existence, and no hesitation to quantify over the variables that pick them out. On the other hand, both science and metaphysics are suspicious of them. There are no tables and chairs in the 'book of the really real'. If the scientist writes the book, you will find only sub-atomic particles, or points in space and time; if the metaphysician writes it you will find perhaps organisms and people, but very few 'medium-sized dry goods', to use Austin's phrase.

But what else occurs in our inventory? Much modern philosophy is devoted to such questions as whether facts, events or propositions have a 'place in our ontology'. The very phrasing shows the influence of Quine. The subterranean thought is this: the question what exists concerns which theory to adopt, and the scope of the quantifiers in that theory. In his *Essays on Actions and Events*, Oxford 1980, Donald Davidson has vigorously defended an 'ontology of events', challenging Strawson's view of ordinary physical objects as the basic particulars of our universe, and putting events in the place of them. (See Chapter 14.) Others worry about whether numbers really exist (see Chapter 26), or such entities as states, corporations and institutions (See Chapter 28).

The crucial reading here is:

(1) W.V. Quine, 'Ontological Relativity', in *Ontological Relativity and Other Essays*, New York and London 1962.
(2) David Wiggins, *Sameness and Substance*, Oxford 1980.
(3) P.F. Strawson, *Individuals*, London 1959, ch. 1.

Some questions:

(1) How would you settle the question, whether entities of a certain kind exist?
(2) What does it mean to say that ordinary physical objects are 'more basic' than, or 'ontologically prior' to events, or vice versa?

Strawson and Davidson are relevant here.

3. Identity, etc.

A topic which leads everywhere. The best text is still Wiggins, *Sameness and Substance, op. cit.* An interesting suggestion is made by Quine in 'Identity, hypostasis, etc.' in *From a Logical Point of View*, Cambridge Mass. 1953. Moving on from the dispute between Geach and Wiggins, is Harold Noonan, in *Identity*, Dartmouth Publishers 1993. The papers in Geach's *Logic Matters*, Oxford 1972, are all worth reading. His *Reference and Generality*, Cornell 1962, which is the principal object of Wiggins's attack, is rather more eccentric.

4. Necessary Being

The topic is difficult, and I therefore postpone discussion of it until Chapter 13.

5. Hegel

It is good to get to grips with Hegel at an early stage. It does not matter if you feel bewildered by it: you will not be alone. (There is a chapter on Hegel in Roger Scruton, *A Short History of Modern Philosophy*, London 1982.) The most important questions are these:

(1) What is Hegel's 'dialectic'? Are there any examples which do not depend for their plausibility on our accepting Hegel's metaphysics?
(2) What does Hegel mean by the 'going over' or 'passage' from being to not-being, and how is it accomplished?

Some of the essays in Robert Solomon's *From Hegel to Existentialism*,

Oxford 1987, are helpful. However, I would not recommend many of the commentators: see 'Understanding Hegel', in Roger Scruton, *The Philosopher on Dover Beach*, Manchester 1990.

6. Heidegger

The commentators on Heidegger are also dreadful. Very rarely does one of them ask whether Heidegger's utterances are true or false: and by avoiding that crucial question, you also avoid the question what Heidegger *means*. (Remember the connection between sense and truth-conditions.) (Among respectable philosophers, only Rorty, *Objectivity, Relativism and Truth*, Cambridge 1991, and elsewhere, is able to refer to Heidegger with a clean conscience, since only he – or almost only he – believes that the question of truth is irrelevant.) The shining exception is David Cooper, whose *Existentialism*, Oxford 1990, while not devoted to Heidegger by any means, gives a trenchant and persuasive account of the anxious theory of Being. The following question is worth thinking about:

What is the distinction between being and Dasein? Why is it important?

Chapter 13 Necessity and the *a priori*

The fundamental texts are these:

(1) Kant, *Critique of Pure Reason*, Introduction.

(2) W.V. Quine, 'Two Dogmas of Empiricism', in *From a Logical Point of View*, Cambridge Mass. 1953.

(3) H.P. Grice and P.F. Strawson, 'In Defence of a Dogma', *Philosophical Review* 65, 1956. (A reply to Quine.)

(4) Gilbert Harman, 'Quine on Meaning and Existence', *Review of Metaphysics*, 1967, a defence of Quine.

(5) Saul Kripke, *Naming and Necessity*, Oxford 1980.

(6) A. Quinton, 'The *a priori* and the analytic', in P.F. Strawson, ed., *Philosophical Logic*, Oxford 1967.

(7) A. Plantinga, *The Nature of Necessity*, Oxford 1974.

On possible worlds, you might read the defence of a realist position (i.e. that possible worlds really – but not actually – exist) given by David Lewis, *Counterfactuals*, Oxford 1973, pp. 84–91, and *The Plurality of Worlds*, Oxford 1986, ch. 1.

Russell's defence of the 'logically proper name' occurs in his 'Lectures on Logical Atomism', contained in *Logic and Knowledge*,

London 1956. The modal logic of C.I. Lewis is explained and discussed in Casimir Lewy, *Meaning and Modality*, Cambridge 1976, while Kripke's 'Semantical considerations on modal logic' (*Journal of Symbolic Logic*, 1963) is explained in G.E. Hughes and M.J. Creswell, *An Introduction to Modal Logic*, London 1968, pp. 74–80.

Some questions:

(1) What does Kant mean by 'analytic'?

Look at Kant's definition in the introduction (taken from Aquinas's definition of 'self-evident', *Summa Theologica*, article 3). Compare it with Frege, *Foundations of Arithmetic*, tr. J.L. Austin, Oxford 1950, p. 4.

(2) Are there any arguments for thinking that the distinctions: analytic/synthetic; *a priori/a posteriori*; necessary/contingent, are really one distinction?

See Quinton, in P.F. Strawson, *op. cit.*

(3) What is a necessary truth? What is it to know something *a priori*? Examine the claim that all and only necessary truths are knowable *a priori*.

See P. Edwards and A. Pap, *A Modern Introduction to Philosophy*, section VII, introduction.

A. Plantinga, *The Nature of Necessity*, ch. 1.

Quinton, *op. cit.*

S. Kripke, *Naming and Necessity*, esp. pp. 34–9; 53–7; 97–105.

(4) What is a rigid designator?

Kripke, pp. 48–9, 55–60.

(5) Is Quine right?

(6) What are Kripke's arguments for the view that there are (a) contingent *a priori* truths, (b) necessary *a posteriori* truths?

(7) Critically assess Plantinga's version of the ontological argument.

See A. Plantinga, *The Nature of Necessity*, Oxford 1974, ch. 10, and J.L. Mackie, *The Miracle of Theism*, Oxford 1982, ch. 3.

Chapter 14 Cause

Hume made two attempts at a theory of causation, one in *Treatise*, Book I, ch. 5, xiv (p. 170 of the Selby-Bigge edn, Oxford 1978); the other in the first of his two *Enquiries*. Hume was taken by Kant, and many others since, to be arguing for a sceptical position, according to which our causal beliefs are without justification, and possibly even fundamentally confused. In fact he was more sophisticated than that, and modern commentators – notably Barry Stroud (*Hume*, London 1977), and David Pears (*Hume's System*, Oxford 1990), emphasise his 'naturalism' – meaning his desire to *explain* our beliefs in terms of human nature. His main aim was not to open or to close the gap between evidence and conclusion in respect of causality, but to describe how we reach our beliefs about causes. Pears argues that Hume is successful in this, despite his restrictive psychology. Of course, it is another question whether there is any point in being successful, in a task which leaves all of epistemology, metaphysics and philosophy of science exactly where it was. Still, that is how modern philosophers see poor old Hume.

The older interpretation of Hume, as primarily concerned to emphasise the gap between evidence and conclusion abut causality and to propose a 'regularity theory' as his sceptical solution, is maintained by J.L. Mackie, in *The Cement of the Universe*, Oxford 1974. Mackie gives an updated version of what he takes to be Hume's analysis of causation, arguing that the counter-factual conditional must feature in the analysis of causal regularity, but can be accounted for in terms that harmonise with Hume's empiricist premises. He describes a cause as an 'insufficient but necessary part of a sufficient but unnecessary condition' – an INUS condition, for short. His book contains impressive and argumentative surveys of all the major questions about causality, and is an excellent guide for the more advanced student. His reluctance to countenance real necessities is, however, hard to reconcile with his counterfactual analysis. See the criticisms in José A. Bernadete, *Metaphysics: the Logical Approach*, Oxford 1989, ch. 22.

G.E.M. Anscombe's paper is in her *Metaphysics and the Philosophy of Mind*, Oxford 1981. It is also reprinted (together with Davidson's article, which is Anscombe's principal target) in Ernest Sosa, ed., *Causation and Conditionals*, Oxford Readings in Philosophy, Oxford 1975, which contains several other useful papers. David Lewis, in his 'Causation' (*Philosophical Papers*, vol. 2, Oxford 1986), thinks that the counterfactual analysis of causation can overcome the difficulties facing a Humean theory in terms of regular connection. This is emphatically denied by Dorothy Edgington, in a series of papers, some in an obscure Mexican journal called *Critica*, others unpublished.

(The great advantage of unpublished papers is that you don't have to read them, because you can't. So what's the point of referring to them?)

Hume was no fool. There *is* a gap between evidence and conclusion, and all the usual attempts to close it are doomed to failure. Perhaps we should accept, as he does, that our causal beliefs are 'audacious', involving a leap beyond the evidence that is both unwarranted and indispensable.

Some questions:
(1) Can the problems that confront the regularity theory be overcome? (See Dorothy Edgington, 'Explanation, Causation and Laws,' *Critica*, vol. XXII, no. 66, 1990; also Hume, Davidson, Anscombe, Lewis, all in E. Sosa, *Causation and Conditionals*, Oxford Readings, Oxford 1975.)
(2) Do singular causal statements imply causal laws?
(Anscombe and Davidson.)
(3) 'Causal statements differ from statements of coincidence, because they support counterfactuals'. Is that true?
(Lewis, Mackie, and Jaegwon Kim in Sosa, *op. cit.*)
(4) Causality is a relation between events. Is that true?
Davidson, Mellor (in *Matters of Metaphysics*, Cambridge 1991.)

Dummett's article on 'Bringing about the past' occurs in his *Truth and Other Enigmas*, London 1978.

Chapter 15 Science

A useful and up-to-date introduction to the philosophy of science is that by Anthony O'Hear, *An Introduction to the Philosophy of Science*, Oxford 1989. Older but highly influential works that should be consulted are:
K. Popper, *The Logic of Scientific Discovery*, London 1959.
T. Kuhn, *The Structure of Scientific Revolutions*, London 1962, second edn 1980.
C. Hempel, *Aspects of Scientific Explanation*, 1965, New York 1970.
Hume sets out the problem of induction (which he discovered in its corrosive modern form) in *Treatise*, I, 5, vi and *Enquiries*, s. 4, part ii, 30. He argues that any attempt to justify induction will be circular, since it can *be* a justification only if deductive – in which case it will need a major premise which must be justified by induction – or inductive – in which case it will assume the principle of induction in the course of the proof. There is just no way round this. That is why modern philosophers try to embrace the circularity and show that it is not vicious. (It is very difficult to embrace what you believe to be

vicious: but see *Phryne's Symposium*, 22d, in R. Scruton, ed., *Xanthippic Dialogues*, London 1993, for a counter-example.) J.S. Mill in *System of Logic*, London 1943, Bk III, ch. 21, argued for this position, and Braithwaite, in *Scientific Explanation*, Cambridge 1953, gave a determined and probably incomprehensible defence of it. His pupil D.H. Mellor revives it in a novel form in 'The Warrant of Induction', in *Matters of Metaphysics, op. cit.* Mellor's view is that induction works because it is a reliable form of inference; it is reliable because induction shows it to be so.

Some questions:
(1) Can there be a warrant for induction?
(2) 'A wise man proportions his beliefs to the evidence'. Does this solve the problem of induction?
(This involves discussing probability.)
(3) Does Strawson have a solution to the problem of induction, or does he merely refuse to see it as a problem?
(Look at *Introduction to Logical Theory*, London 1952, ch. 9.)

The 'solution' proposed by Quine and his followers has been carefully examined by Barry Stroud, in the last chapter of his *Significance of Philosophical Scepticism*, who concludes that sceptical problems cannot be overcome by a 'naturalised epistemology', which must always, in the last analysis, presuppose precisely what the sceptic is calling into question.

Hempel's paradox was presented first in 'Studies in the logic of Confirmation', *Mind*, 1945. But see also Hempel's *Aspects of Scientific Explanation*, New York 1965, and J.L. Mackie, 'The Paradox of Confirmation', *British Journal for the Philosophy of Science*, 1963, reprinted in P.H. Nidditch, ed., *The Philosophy of Science*, Oxford 1968. Mark Sainsbury discusses the paradox in ch. 4 of his *Paradoxes*, Cambridge 1988.

A question:
Is there a satisfactory solution to Hempel's paradox?

Goodman's paradox, presented in *Fact, Fiction and Forecast*, Cambridge Mass. 1979, is discussed by B. Skyrms in his *Choice and Chance: an Introduction to Inductive Logic*, California 1986, and also in Sainsbury, *op. cit.*, ch. 4, and by S. Blackburn in his *Reason and Prediction*, Cambridge 1973, ch. 1, and S. Kripke in *Wittgenstein on Rules and Private Language*, Oxford 1984. The question to ask is the same as the one asked abut Hempel's paradox, namely: Does it have a solution? However, the difficult thing is to learn how to state the paradox without begging any questions, either for or against it.

The dispute over the nature of theories and theoretical terms is well explored in E. Nagel, *The Structure of Science*, London 1961, Mary Hesse, *The Structure of Scientific Inference*, London 1974, and I. Hacking, *Representing and Intervening*, Cambridge 1983. There are two kinds of instrumentalist: some argue that theories should not be interpreted (or at least, not literally), but merely used; others argue that theories should be interpreted, but that we should not believe what they say. See: B. Van Fraassen, *The Scientific Image*, Oxford 1980, and the papers in P. Churchland and C. Hooker, eds., *Images of Science*, Chicago 1985. More radical still is the kind of relativism inspired by Kuhn's *Structure of Scientific Revolutions*, Chicago 1962. Kuhn (and also Paul Feyerabend in his *Against Method*, London 1975) uses the term 'incommensurability' to express his view that there is no common standard against which competing scientific theories can be measured. For illuminating contributions to the debate inspired by Kuhn and Feyerabend, see I. Hacking, ed., *Scientific Revolutions*, Oxford 1981. Craig's theorem was originally presented in William Craig, 'On Axiomizability within a System', *Journal of Symbolic Logic*, 18, 1953, pp. 30–32; its philosophical application was developed in William Craig, 'Replacement of Auxiliary Expressions', in *Philosophical Review* 65, 1956, pp. 38–55. A useful introduction to the theorem, which also provides a fairly plausible sceptical response to it, is the article on 'Craig's Theorem' by Max Black, in Paul Edwards, ed., *The Encyclopedia of Philosophy*.

Perhaps the most important question in the philosophy of science today is that of the place in scientific thinking of the concept of probability. Several philosophers (notably Mellor, in *op. cit.*, and Patrick Suppes, in *A Probabilistic Theory of Causality*, Amsterdam 1970) have argued that concepts like *cause*, *law*, and *induction* cannot really be understood without recourse to ideas of evidence, where *p* is evidence for *q* only if *p* makes *q* more probable.

The most appropriate recent texts are:

C. Howson and P. Urbach, *Scientific Reasoning: the Bayesian Approach*, La Salle Ill. 1989.

B. Skyrms, *Choice and Chance, op. cit.*

P. Horwich, *Probability and Evidence*, Cambridge 1982.

D.K. Lewis, 'A Subjectivist Guide to Objective Chance', in *Philosophical Papers*, Oxford 1983, vol. 2.

Older approaches, such as those of Keynes, Jevons and Reichenbach, are discussed in W. Kneale, *Probability and Induction*, Oxford 1949. J.R. Lucas's defence of objective probabilities occurs in his *The Concept of Probability*, Oxford 1991.

The literature on criteria and symptoms is extensive, and not always very illuminating. The leading texts are Wittgenstein's *Philosophical Investigations*, Oxford 1952, and *Blue Book*, Oxford 1958. See also R. Scruton and Crispin Wright, 'Truth Conditions and Criteria', in *Proceedings of the Aristotelian Society, Supplementary Volume*, 1976.

On the concept of a natural kind, the reader should consult:

(1) J.S. Mill, *A System of Logic*, London 1943, Book I, ch. 7, section 4.

(2) C.S. Peirce: 'Natural Classes' in *Collected Papers*, vol. 1, ed. Charles Hartshorne and Paul Weiss, Cambridge Mass. 1960, pp. 83–99.

(3) H. Putnam, 'Is semantics possible?' in his *Collected Papers*, vol. 2: *Meaning and Metaphysics*, Cambridge 1975.

(4) S. Kripke, *Naming and Necessity*, Oxford 1980.

The paradoxes of quantum mechanics are discussed by many authors. Perhaps the most accessible text is John Polkinghorne, *The Quantum World*, Harmondsworth 1986. Among more difficult texts, the following are particularly important:

(1) Nancy Cartwright, *How the Laws of Physics Lie*, Oxford 1983.

(2) Bernard d'Espagnat, *Conceptual Foundations of Quantum Mechanics*, 2nd edn, 1972.

(3) A.J. Leggett, *The Problems of Physics*, Oxford 1987, which contains, on pp. 168–170, a convenient summary of Schrödinger's argument about the cat.

(4) N. Bohr, *Atomic Physics and Human Knowledge*, New York 1949 (see especially his 'Discussion with Einstein').

(5) Richard Healey, *The Philosophy of Quantum Mechanics*, Cambridge 1989: more technical than philosophical.

(6) Einstein, Podolsky and Rosen: this occurs in *Physical Review*, no. 47, 1935, p. 777.

(7) J.S. Bell, in *Physics*, I, 1964, p. 195. Bell's theorem and its implications have been conveniently summarised in N. David Mermin: 'Is the moon there when nobody looks at it?', *Physics Today*, April 1985, pp. 38–47, Bernard d'Espagnat, 'The Quantum Theory and Reality', in *Scientific American*, vol. CCXL no. 5, Nov. 1979, pp. 158–181, and Abner Shimony, 'An Exposition of Bell's theorem', in *Search for a Naturalistic World View*, vol. 2, Cambridge 1993. See also

J.S. Bell, 'Bertlemann's socks and the nature of reality', in *J. Phys.* (Paris), 42, 1981.
(8) David Bohm, *Wholeness and the Implicate Order*, London 1980.
(9) W. Heisenberg, *Physics and Philosophy*, New York 1962.
(10) Michael Redhead, *Incompleteness, Nonlocality and Realism*, Cambridge 1987: very difficult.
(12) Bas C. Van Fraassen, *Quantum Mechanics: an empiricist view*, Oxford 1992. Perhaps the best work on the subject.

Chapter 16 The Soul

This topic has proved to be of great interest to academic philosophers in recent years, although the word 'soul' has been discarded, on the assumption that 'mind' or 'consciousness' are clearer and less contentious. Recent literature is copious, difficult and often written so badly that the student will be tempted either to abandon the topic, or to conclude that these unintelligible writers must have something really deep to say. Neither attitude should be adopted. But, because help is necessary, I shall give a rather more extended guide to the area.

A way in to the modern philosophy of mind is provided by two classic texts: Gilbert Ryle's *The Concept of Mind*, London 1949, and John Wisdom's *Other Minds*, Oxford 1952, which contains an important criticism of Ryle. The agenda has changed radically since Ryle's day; and on the Continent it has always been different, on account of the domination of phenomenology, as exemplified in Maurice Merleau-Ponty's classic study, *The Phenomenology of Perception*, tr. Colin Smith, London 1962. Rather than attempt to cover all the many options in the philosophy of mind, I shall once again confine myself to guiding the reader through recent discussions in the anglophone tradition. One useful approach to those discussions is to read first the introduction by P. Smith and O.R. Jones, entitled *The Philosophy of Mind*, Cambridge 1986, and then to tackle some of the writings of Colin McGinn, one of the more lucid and imaginative contributors to recent debates: try, for example, *The Character of Mind*, Oxford 1982, and the papers in *The Problem of Consciousness*, Oxford 1991.

I begin from Aristotle, because his concept of *psuche* – the animating principle in all living things – presents a radical challenge to our modern assumptions, and forces us to confront the question what we really mean by 'mind' and 'mental'. Whatever the mind is, it must be part of what animates the thing that has it: in which case, what is the distinction between mental and other forms of animation? The comparison between the concept of *psuche* and the modern philosopher's concept of mind is illuminatingly explored by Kathleen Wilkes, '*Psuche* versus the Mind', in Martha C. Nussbaum and Amelie O. Rorty, eds., *Essays on Aristotle's De Anima*, Oxford 1992, an important book which also contains an exchange between Myles Burnyeat ('Is an Aristotelian Philosophy of Mind still Credible?') and Martha Nussbaum and Hilary Putnam ('Changing Aristotle's Mind'), to which I return below. Aristotle's view is sometimes called 'hylomorphism' – from *hule*, matter, and *morphe*, form. For Aristotle argued that the soul and body are related as form and matter. The soul is the principle which organises the matter of the body into a single and self-activating life. Hylomorphism (expounded in the *De Anima*, a work unsurpassed in its succinctness and penetration) is not a form of dualism; nor is it clearly physicalist in the modern sense – though Nussbaum and Putnam (see below) consider it to be a precursor, and maybe even a version, of functionalism. Burnyeat's view is that Aristotle's distinction between matter and form, and the concept of matter that derives from it, belong to a superseded stage of scientific thinking, and that no theory phrased in these terms is credible today.

For other attempts to set Aristotle's theory in the context of modern debate, see Charles Kahn, 'Aristotle on Thinking', and M. Frede, 'On Aristotle's Conception of the Soul', both in Nussbaum and Rorty, *op. cit.*; and also Hugh Lawson-Tancred's introduction to his translation of the *De Anima*, Penguin, Harmondsworth 1987.

1. The Cartesian Theory of Mind

It is necessary, as we have already seen (Chapter 4), to distinguish Descartes's particular theory of the mind, as a thinking substance, from the theories that we now call Cartesian, for the reason that they perpetuate Descartes's view that the mind is, wholly or in part, distinct from the physical world. Descartes himself is best studied through his two expositions of the argument for the 'real distinction' between mind and body – in *Principles of First Philosophy*, 60, and *Meditations*, VI – and through the commentaries on this argument. Perhaps the most illuminating commentary, which brings out the hidden assumptions and the modal structure of Descartes's reasoning, is that given by Bernard Williams, in *Descartes: the Project of Pure Enquiry*, Penguin, Harmondsworth 1978, ch. 4.

Descartes thought of the mind as a substance, in the specific sense common to the rationalist philosophers: it is a bearer of properties, which depends for its existence upon no other thing besides God. (God, for Descartes, is the only truly self-dependent entity, and is therefore a substance in another sense from created substances.) The mind is also not extended in space, and all of its properties are modes of thinking. It is often held to be a consequence of the rationalist conception of substance, that separate substances cannot interact (since causal interaction is a form of mutual dependence). Since there *must* be causal relation between mind and body (else how are perception, knowledge and action possible, and how are we *in* the world at all?) this has been construed as a *reductio ad absurdum* of Descartes's position. Indeed, there are those (Davidson, see below, is one of them) who regard the fact of causal relations between mind and body as a conclusive refutation of dualism in any form. As I note in the main text of this chapter, however, Descartes believed that mind and body *do* interact; the difficulty lies in explaining how this could be so, in a way that is consistent with the suppositions of a rationalist science.

Philosophers still exist who are prepared to defend dualism in something like the form that Descartes gave to it, although purged of any dependence on the rationalist conception of substance. One such is Richard Swinburne, who persists in regarding the mind as an entity (if not quite a substance in the rationalist sense), which is non-physical and the bearer of non-physical properties. Of particular interest is his *Evolution of the Soul*, Oxford 1986, and his defence of dualism against Sydney Shoemaker's attacks on it in their joint publication *Personal Identity*, 1986. Shoemaker's attack on Swinburne's position seems fairly decisive; Swinburne, however, maintains that he has yet to be refuted. (It is a useful exercise to write an essay, taking sides in this debate.)

The dualism of Descartes is a 'substance dualism': it argues for two kinds of *thing*, rather than two kinds of property. Even Strawson's position in *Individuals*, London 1956, ch. 3, is a kind of substance dualism, since Strawson wishes to argue that persons are the sole bearers of 'P-predicates' as he calls them, and have separate conditions of identity from their bodies. He is even prepared to allow, in an attenuated sense, the survival of the person, as a pure 'subject of consciousness', after the death of the body. What more is required to be a full Cartesian dualist? John Foster, in his *The Immaterial Self*, London 1991, argues that anti-dualist theories of the mind are incoherent: in this extremely well-argued book a new kind of case is put, for a new kind of dualism.

Cartesian theories of the mind (as characterised in Chapter 4), do not always countenance a duality of substances. Maybe a duality of properties is enough: maybe we are Cartesians, to the extent that we think of mental properties and states as being non-physical states, even if they are also states of a physical thing. (As we might regard the aspect in a picture as a property of the physical canvas, even though it is not a physical property.) In this sense, those philosophers who believe in a realm of 'subjective' properties, 'raw feels' or '*qualia*' often seem to commit themselves to the belief that there are properties of the person (or maybe even of the brain) that are non-physical properties, and which would not be listed in the 'book of the world' as physics conceives it. This minimal Cartesianism seems to survive in the writings of Thomas Nagel (specifically 'What it's like to be a bat', and 'Subjective and objective' in *Mortal Questions*, Oxford 1982, and the expansion of the argument of those articles in *The View from Nowhere*, Oxford 1986).

The Cartesianism of the phenomenologists – and in particular of Husserl – is more difficult to define: this is a topic to which I return in Chapter 31. In all its forms, Cartesianism seems to remain committed to the view that there are facts which are essentially private, observable to one person alone, and describable, if at all, only in a language which is private to the speaker. The argument against such a language therefore tells against the Cartesian theory. Some philosophers, however, wish to retain the idea that there are 'purely subjective' facts – facts accessible only from the first person point of view – while rejecting the Cartesian picture of first-person *knowledge*. The error of the Cartesian, they say, is in supposing that we know the 'inner realm' by a kind of observation. And it is this fiction of inner observation (the internal theatre, in which the I is audience) that is the malign legacy of Descartes's real distinction. Abolish that picture, and the private language argument fails to get off the ground. Hence we can retain the idea of purely subjective facts, while denying the existence of any private language. This kind of position is taken by Nagel, and also by Kathleen Wilkes, in her discussion of Aristotle.

By contrast, there are philosophers who have wished to revitalise the Cartesian argument for a real distinction – at least between mental and physical states. The most challenging is Kripke, who, in lecture 3 of *Naming and Necessity*, Oxford 1980, presents a modal argument for the conclusion that pain cannot be identical with any physical process, since pain has an essential feature (namely, that it is felt painfully) which no physical process could have. The argument has a question-begging appearance, and has been much discussed, for example in:

W.G. Lycan, 'Kripke and Materialism', *Journal of Philosophy* 1971, 18: 677–89.
W.G. Lycan, *Consciousness*, Cambridge Mass. 1987.
G. Sher, 'Kripke, Cartesian Intuitions and Materialism', *Canadian Journal of Philosophy*, vol. 7, 1977.
J.R. Searle, *The Rediscovery of the Mind*, Cambridge Mass. 1992, ch. 2.

2. The First-person Perspective

The topic of this section can be approached in many ways. There is a distinction between those who believe that the topic concerns a special phenomenon – namely the phenomenon of *consciousness* (see, for example, the above-mentioned book by Lycan) – and those (such as Colin McGinn, in *The Subjective View* and elsewhere), who believe that the topic concerns rather a certain point of view on the world: the point of view of the 'subject of experience'. Philosophers of the first kind ask whether consciousness might be identical with, reducible to, or at any rate explicable in terms of physical features of an organism, and if not, whether this gives grounds for some kind of dualism. Philosophers of the second kind tend to consider whether the first-person viewpoint can be accomodated within a third-person account of the world, and if not, whether this signifies anything in particular about the metaphysical condition of the 'subject'. But despite the difference of emphasis and approach, the questions concerned are ultimately the same. These questions include the following:

(1) Are there facts about experience, sensations, and other mental states, which are accessible only to the subject – for example, facts about 'what it's like', 'how it feels'? This is sometimes expressed as the question of *qualia*, a word used to denote the inner characteristics (supposing there are such) of our mental life. A philosophy that is committed to denying, or explaining away, the existence of *qualia* is often dismissed (for example by Searle, in *The Rediscovery of the Mind op. cit.*) as wrong from the start. (See also G. McCullough, 'On the Very Idea of the Phenomenological', *Proceedings of the Aristotelian Society*, 1992.)

(2) What is consciousness, and is it related to, identical with, or entirely distinct from self-consciousness? Some forms of Cartesianism seem to regard consciousness as a kind of inner observation – a view damagingly criticised by Ryle among many others, as involving what Searle calls the 'homunculus fallacy': the fallacy of believing in the existence of an 'inner observer' who, since he too must have a mind, must contain another such observer so as to be conscious of his

own mental states. (And so on *ad infinitum.*) Others argue that consciousness is a natural phenomenon, a property of many higher organisms including dogs, cats and elephants, which is sharply to be distinguished from the self-consciousness that comes with the use of language, and the first-person case. But then, which is it that creates the philosophical problem – consciousness, self-consciousness, or both? And how is the problem to be resolved?

(3) Can consciousness be understood in terms of the 'subjective view': the view that is necessary if there is to be experience at all? Or is the subjective view to be understood as the product of consciousness? Is there any advantage in either of these positions that the other lacks? (In many recent philosophers, notably Nagel and Searle, there is a tendency to treat the subjective view and consciousness as equivalent, and to confound them both with self-consciousness.)

(4) Specific puzzles, generated by our intuitions about consciousness. One such is that of the 'inverted spectrum'. Could we not imagine people who are exactly like us in their language and behaviour, who classify objects just as we do, and who make all the discriminations to which we are accustomed, but whose 'inner experience' of red objects is like our inner experience of green objects, and vice versa? If so, must we not admit that there is a crucial *fact of the matter* that is left out of any account of the mind in terms of what is publicly observable and accessible to physical science? For an entertaining discussion of this problem, see D.C. Dennett, in *Consciousness Explained*, London 1992. Dennett argues that the suggestion of an inverted spectrum is incoherent: that there just *could not be* such people. If not, why not? The argument is a kind of reply to Nagel *et al.*, and is meant to suggest that there are no subjective facts that are left out of the account by physicalist theories. Something similar seems to be implied by Wittgenstein, in his many incursions into this territory, notably in *Zettel*, Oxford 1967. The Wittgensteinian position seems to be that all such speculations are empty, and that they can never amount to more than a 'temptation to say' something, which in fact has no real role in the 'language game' of the mental. (See further the discussion of functionalism, below.)

3. Intentionality

For a survey of this topic, as recently discussed, see the second appendix to Roger Scruton, *Sexual Desire*, London 1986. Brentano's path-breaking account occurs in *Psychology from an Empirical Standpoint*, vol. 1, ed. Oscar Kraus, tr. Linda McAlister *et al.*, London

1976. Brentano identified intentionality as a feature of mental *phenomena*, by which he meant a sub-class of those things that are 'before the mind' when we think, desire or feel. This adds a complication which subsequent philosophers have, on the whole, chosen to ignore, considering intentionality as a property of the mental as such, rather than of the mental as a 'phenomenon'. Brentano's approach introduces a vertiginous doubling of the intentional realm, which was welcome to Husserl, in his extended account of the matter, but which has seemed to others already to beg the principal question against physicalist theories.

According to Husserl, intentionality introduces the need to 'bracket' the object of every mental state, in order to obtain a philosophical understanding of the state itself, conceived as a mode of consciousness. The procedure of 'phenomenological reduction' that results from this is glancingly discussed in chapters 4 and 31; there is no evidence to my mind that discussion needs to be more than glancing. (See the dry but dutiful study by David Bell, in his *Husserl*, London 1990.) It would take us far afield, however, to examine the full basis for the phenomenological method. Husserl's own recognition in his later writings that he had reached an impasse, together with Wittgenstein's strictures against the very possibility of a pure phenomenology (in his *Remarks on Colour*, Oxford 1976, and, by implication, in the private language argument), do something to confirm a robust scepticism towards this much-vaunted method, so rich in technicalities, and so poor in results.

Among those who have recently regarded intentionality as an obstacle to physicalism, perhaps the most persistent and pugnacious has been Roderick Chisholm, starting with his 'Sentences about Believing', *Proceedings of the Aristotelian Society*, 1956, and *Perceiving: a Philosophical Study*, Ithaca, New York 1957. There are three stock responses to this kind of argument. First, one might accept that intentionality is an obstacle to a physicalist account of mental concepts, and use this as an argument for dismissing such concepts as flawed, or destined to be replaced by a non-intentional science. This thought underlies the attack on 'folk psychology' which has been mounted most systematically by S.P. Stich, in *From Folk Psychology to Cognitive Science: The Case against Belief*, Cambridge Mass. 1983. It was also accepted by D.C. Dennett in some of his earlier writings. The second response is more characteristic of the later writings of Dennett, who, in such papers as 'Intentional Systems' in *Brainstorms: Philosophical Essays on Mind and Psychology*, Cambridge Mass. 1978, and in *The Intentional Stance*, Cambridge Mass. 1987, has argued that there is nothing in principle absurd in the attribution of intentional states to physical systems, even while denying that those systems exhibit consciousness or any other of the marks of the 'mental'. The third, and most radical response, is that explored by Donald Davidson, in his 'Mental Events', in his *Essays on Actions and Events*, Oxford 1980, according to which the existence of intentionality provides an argument against dualism, and for a kind of 'anomalous monism'. Davidson argues that there can be no *laws* connecting events under mental (i.e. intentional) descriptions with events described as physical. However, since there are causal connections between events in the two categories, it follows that such laws must exist, and therefore that there is another, and intentionality-free way of describing the mental events. Events are mental (and intentional) only *as described*. A view close to that of Davidson is developed by Spinoza, in his *Ethics*, where it is argued that *everything* belongs to two complete but incommensurable systems of description, mental and physical, and that the very interconnectedness of events argues for the existence of a single and universal substance, which can be described now as mental (as Thought) and now as physical (as Extension).

The topic of intentionality has been explored by J.R. Searle, in his *Intentionality: an Essay in the Philosophy of Mind*, Cambridge 1983, and in *The Rediscovery of the Mind*, ch. 8. Searle's position, very roughly, is that we cannot make sense of intentionality if we divorce it from the other mark of the mental – consciousness. (The 'connection' thesis.) All attempts to describe intentionality as a property of non-conscious systems (e.g. that of Dennett), fail to capture its leading characteristic, which is its *representational* or *aspectual* nature. An intentional state presents its object *in a certain way*, and that means *to a certain centre of consciousness*. For a contrasting approach, in which 'aboutness' is presented as a biological characteristic, see Ruth Millikan's *Language, Thought and Other Biological Categories*, Cambridge Mass. 1984.

4. The Unconscious

This topic is really two: first, is consciousness, or the 'subjective view' an essential part of the mental? Secondly, is there an *unconscious* in the Freudian sense – a realm of the mental which is inaccessible to the subject, and which can be brought into consciousness by the therapeutic process? In a trivial sense there are of course unconscious mental states: consider beliefs, for example. Mary believes that platinum is chemically inert. But is that belief 'before her mind' as she plays chess with John? Since we all of us have a vast store of beliefs, the suggestion that they *must* be before the mind is surely absurd. On the other hand, this belief of Mary's is clearly not unconscious in the sense required by Freudian theory (according to which Mary may very well entertain an unconscious belief that her father was murdered by her mother, and did not die, as she regularly tells the story, of a surfeit of lampreys). In another sense, however, Mary's belief about platinum is conscious, in that she knows immediately and without having to find out, that *this* is something she believes, just as soon as the proposition comes before her mind.

In a classic attack on the Freudian theory, Sartre has argued (*Being and Nothingness*, tr. Hazel Barnes, London 1943) that mental states are essentially the states of a centre of consciousness, and that Freud is forced by his theory of the unconscious to duplicate that 'centre of consciousness', and to postulate another and hidden self, which by definition is not *me*. In other words, Freud introduces unconscious mental states, only to ascribe them to something other than the subject. In which case they are not unconscious states of that subject, but – if they exist at all – conscious states of another subject. Sartre's position is, however, paradoxical, since it relies upon an entirely 'transcendental' view of the subject, as something which is totally unconnected with the surrounding world, a pure 'I', which enjoys absolute sovereignty over the unreal sphere of its own existence, but relates to nothing else. (See Chapter 30.) (Though Sartre himself, pp. 50–54, tries to argue directly against such a view.)

The position adopted by Wittgenstein in his *Lectures on Aesthetics, Freud and Religious Belief*, Oxford 1966, is rather more subtle. According to Wittgenstein mental states are such that, in the normal case, one can ascribe them to oneself on no basis, and *that* is what consciousness amounts to. Sometimes, however, mental states may be attributed to me by another, even though I refuse to 'confess' to them. Are they nevertheless mine? The suggestion that they are, Wittgenstein argues, makes sense only against the background of the psychoanalytic process. And what the process consists in is precisely this, that it *brings* me to accept these mental states as mine, to confess to them and to ascribe them to myself on no basis. The 'bringing into consciousness' provides the criterion for the truth of the psychoanalytic hypothesis, which relies, therefore, on precisely the criterion of consciousness that our ordinary mental attributions employ. This position is also essentially that adopted by Searle, in ch. 7 of *The Rediscovery of the Mind*.

Others take a more radical view, arguing for the existence of mental states wherever there is intentionality, while denying that intentionality is either actually or potentially conscious. Searle believes this to be confused, on account of the 'connection' thesis (see above). To clarify this difficult topic is no easy matter. First, we must distinguish the presence of an intentional state (an unconscious belief or desire, say) from the mere fact that a teleological explanation can be provided of someone's behaviour. For instance, much of my behaviour could be explained by the fact that I have certain goals – wealth, food and sexual satisfaction, say. It may be that the pursuit of sexual satisfaction can be invoked to explain some particular piece of behaviour: my purchase of a sports car, for example. In a sense I bought the car 'in order to' impress the ladies. But I was not aware of any such motive: on the contrary, I liked the look of the car, was impressed by its performance, and bought it for all the reasons that were spelled out to me by the salesman. If it is said that I bought the car 'in order to' impress the ladies, this would be true only in the sense that a flower turns to the sun 'in order to' receive the light that it needs. A general disposition of the organism is invoked through the postulation of a 'goal', and the resulting 'teleological' explanation has a structure not unlike that of an explanation in terms of desire. But of course, there is no suggestion that a flower has a *desire* for sunlight, conscious or otherwise. Nor is it the case that I must be motivated by a desire for sexual satisfaction, just because my possession of this general goal can be invoked to explain my purchase of a particular car. I had no such *desire* on this occasion. My desire was for just such a car as the one I purchased.

The topic of teleological explanation, its legitimacy or otherwise, is a complex one. But the example suffices to show that we need something more if we are to introduce the concept of an unconscious motive. It is precisely the question 'what more?' that is puzzling, and which prompts philosophers like Wittgenstein and Searle to believe that the theory of the unconscious mind is at some level relying upon and reaffirming the thesis that mental states are essentially conscious.

On the other hand, there are philosophers who believe that consciousness is, in general, an *addition* to the mental, and that our modern tendency to build it into the very concept of the mental is responsible for recurring philosophical errors: not the least being the error of Cartesianism itself, which, on account of its emphasis on consciousness, is tempted towards the 'inner theatre' view of the mind. This, in brief, is the argument of Kathleen Wilkes in *Real*

People, Oxford 1988. According to Wilkes, our over-emphasis on consciousness is partly responsible for the temptation to dualism, a temptation to which the Greeks (and especially Aristotle) were less prone, since they lacked the view of the mind as an 'inner space', seeing it instead as an active principle, manifest in action, perception and response. This view is also maintained by Richard Rorty in *Philosophy and the Mirror of Nature*, Oxford 1980. (But a caution is in order: the Greeks were not fairly represented by Aristotle, and had many other views of the mind than those encountered by modern philosophers. See Ruth Padel, *In and Out of the Mind, Greek Images of the Tragic Self*, Princeton 1992.)

5. Physicalisms

A general sympathetic survey is contained in Kathleen Wilkes, *Physicalism*, Atlantic Highlands, NJ, 1978. A passionate rejection can be found in Searle's *The Rediscovery of the Mind*. It is possible to disagree with Searle, and lean towards one or another form of physicalism, while sharing Searle's sense that the philosophy of mind has been narrowed, distorted and, ultimately misrepresented, by concentrating on the single question, whether the mind is physical, and by the assumption that the terms of that question make obvious sense. Nevertheless, for the time being, this question occupies the central place in the modern philosophy of mind.

(i) Behaviourism.

The classical statement of the first kind of behaviourism is J.B. Watson, *Behaviourism*, New York 1925. Philosophical behaviourism is more difficult to pin down, since it has never been more than a distant hope of certain verificationists and empirically-minded philosophers (notably A.J. Ayer, in *Language, Truth and Logic*, London 1936). However, the nearest approach to a reasoned defence of a quasi-behaviourist position is *The Concept of Mind*, by Gilbert Ryle. Some people have detected traces of behaviourism in the later Wittgenstein, notably in the private language argument, and the remarks that surround it. In functionalism, too, there survives the behaviourist's intuition, that something that behaves *just as we do*, and whose behaviour is connected with its environment as our behaviour is connected, is surely endowed with a mind. (And if science fiction is only the flimsiest authority for such an intuition, the reader should not ignore those genuine works of art which also confirm it: Mary Shelley's *Frankenstein*, and Karel Čapek's play *R.U.R.*, in which the word 'robot' appeared for the first time.)

(ii) The Identity Theory.

The revival of this theory by Australian philosophers in the fifties and sixties is documented in the articles contained in C.V. Borst, *The Mind/Brain Identity Theory*, London 1970 – the articles by Smart are particularly important. The position is spelled out comprehensively in David Armstrong's *A Materialist Theory of Mind*, London 1968. A sophisticated contribution to the debate can be found in David Lewis, 'An Argument for the Identity Theory', in his *Collected Papers*, vol. 1, Oxford 1983.

(iii) Functionalism.

This theory has been the focus of most recent discussions – at least, if we suppose the various attempts at a 'cognitive science' to be natural outgrowths of it. A good introduction is that by Ned Block, 'What is Functionalism?', in Ned Block, ed., *Readings in the Philosophy of Psychology*, vol. 1, London 1988. The leading statement of the position in recent literature is that by Hilary Putnam, 'The Nature of Mental States', in the same volume, and also in Putnam's *Mind, Language and Reality*, Cambridge 1975, which contains several other essays which elaborate the author's position. In the article co-authored with Martha Nussbaum, referred to above, Putnam links functionalism to Aristotle's hylomorphism. Putnam and Nussbaum discuss the original motivation behind Aristotle's distinction between matter and form, and suggest why this motivation is absent from the mind-body debate as we know it. Whereas the mind-body debate begins from the question of the nature of mental awareness, Aristotle is concerned with the more general question of the relation between organisation and composition. Putnam and Nussbaum connect Aristotle's approach not only with functionalism (which tries to detach mental organisation from the composition of the entity that exemplifies it), but also with the approach of contemporary philosophers like Chisholm and Wiggins, who make use of Aristotelian ideas of essence. By taking an Aristotelian approach to the mind-body problem, Putnam thinks, we can see the problem evaporate before us: 'The soul is not an "it" housed in the body, but a functional structure in and out of matter. Likewise matter is not a "thing" to which the structures of life can be reduced.'

Functionalism is sometimes described as the 'black box' view of the mind. It does not explore what is 'inside', but contents itself with the observable connections between input and output, and locates the mental in the system of connections, while remaining neutral as to the mechanism which, in this or that organism (or artificial intelligence), serves to make them. It rejects behaviourism, on the ground that the latter theory neglects the causal nature of the mental, and fails to see that the whole point of the concept of a mental state lies in its role in the *explanation* of behaviour. But it wishes to describe this explanatory role without invoking any particular 'inner' process. This has naturally led to the accusation that the mental has precisely been *left out of account* by the functionalist. What, for example, does the functionalist say about the 'inverted spectrum'? If it is a possibility, then surely this signifies a fact about the mental that cannot be reduced to the input-output relations that are of interest to the functionalist? This question is discussed in detail by Ned Block and Jerry Fodor, in 'What Psychological States are Not', *Philosophical Review*, vol. 81, 1972, pp. 159–181. Similar attention has been paid by functionalists to the problem of *qualia*: see especially Sydney Shoemaker's 'Functionalism and Qualia' in his *Identity, Cause and Mind*, Cambridge 1984.

The original inspiration for functionalism was the 'Turing Machine', described by Alan Turing, 'Computing Machinery and Intelligence', *Mind*, 1950, reprinted in Margaret Boden, ed., *The Philosophy of Artificial Intelligence*, Oxford Readings in Philosophy, Oxford 1990. Turing's conjecture is that mental operations are sufficiently like those performed by computational systems to warrant explanation in the same way. In particular, mental questions seem to be iterative, leading to results by the repeated application of algorithmic devices. Maybe the brain *is* a Turing machine. Turing proposes a test for artificial intelligence (the Turing test), which is that the machine should be able to match any given human performance: if it can do that, what grounds have we for withholding the description 'intelligent' from the machine?

Out of these speculations arose not only functionalism, but the whole discipline (or, some would say, pseudo-discipline) of cognitive science, which aims to explore the properties of systems that satisfy the Turing test. What has to be true of a system in order to say that it matches our capacity to perceive, to think, to reason, to believe, to feel anger, and so on? Indeed, there are philosophers who regard the philosophy of mind as little more than the groundwork to a cognitive science, and who believe that the questions that we have been considering become fully clear only when posed in the course of speculating about artificial rather than natural minds. See, for example, P.M. Churchland, *Matter and Consciousness: A Contemporary Introduction to the Philosophy of Mind*, Cambridge Mass. 1984, and the articles in W.G. Lycan, ed., *Mind and Cognition*, Oxford 1990. The assumption behind this approach has been characterised as Strong Artificial Intelligence (Strong AI) by Searle: it is the assumption that 'the mind is to the brain as the programme is to the hardware'. In other words, we should see the mind as a system of software, and the brain as a computer that is programmed by it. Searle (*The Rediscovery of the Mind*) objects to this approach on several counts – notably on the ground that nothing is intrinsically computational. A computational role can be assigned to a system, but only if it is used as such. And that merely poses the question: *Who* is using this brain in such a way? Less interesting but more notorious is Searle's 'Chinese Room' argument, referred to in many places by its author, but considered question-begging by his critics. The argument is this: I am alone in a room with a set of instructions which tell me which cards to pass out of the room to those outside in response to cards that are passed in to me by them. The cards carry Chinese characters, and the instructions in fact cause me to pass out excellent answers in Chinese to well-formed questions in that language. But I do not understand Chinese. Likewise, a computer could be programmed to give all the correct responses to a given input, but this would not amount to *understanding* that input in the precise sense of that term which implies a mental capacity on the part of the subject. What is wrong with that argument? One answer is that it assumes that the distinction between understanding and not understanding cannot be analysed in terms of more, and more flexible, connections between input and output, of the kind exhibited in a conversation.

Searle's is one of increasingly many expressions of discontent with functionalism, on account of its inability to give a clear account of intentionality. The mind does not simply respond to input from the outer world; it represents that input, and responds to the representation. It is not clear that computers do this, or that it even makes sense to say that they do it. See J. Fodor, 'Banish Discontent', in W.G. Lycan, *Mind and Cognition, op. cit.*

For David Lewis's subtle version of functionalism, see his two papers in the volume edited by Block.

Advocacy of a kind of functionalism has led Jerry Fodor to his distinctive idea of a 'language of thought'. He believes that we should understand the structure of the mind as being like that of a language, and mental interaction as a kind of ongoing semantic interpretation. See his *Psychosemantics: The Problem of Meaning in the Philosophy of Mind*, Cambridge Mass. 1987. Fodor adopts this view partly because he does not believe that intentionality can be accepted as a primitive property of mental states; as he puts it: 'if aboutness is real, it must really be something else' (p. 97). Only through semantic theory can we make sense of such an idea.

6. Emergent Properties and Supervenience
This topic is technical and controversial. The concept of a 'supervenient' feature has been explored by many recent philosophers – normally in the context of ethics. The thesis of the supervenience of the mental is introduced by Davidson's argument in 'Mental Events', *op. cit.* But the leading discussions of the idea are those of Jaegwon Kim, 'Causality, Identity and Supervenience in the Mind-Body Problem', *Midwest Studies in Philosophy*, no. 4, 1979, pp. 31–49; and 'Psychophysical supervenience', *Philosophical Studies*, vol. 41, 1982, pp. 51–70.

See also J. Haugeland, 'Weak Supervenience', *American Philosophical Quarterly*, vol. 19, 1982, pp. 93–104.

7. The Self
I take up this topic in Chapter 31. For the purpose of the present discussion, however, it is important to note the problem created by the self and self-consciousness for the standard physicalist views. First, there are non-self-conscious creatures, which nevertheless have minds: horses, for instance. So the self is not simply the mind in general, but at best a certain *kind* of mind. What kind? Secondly, when I attribute mental states to myself, I do so with a particular authority. What explains this fact, and to what precisely *am* I attributing those mental states? Is the 'self' part of the physical world, and if so which part?

One response to these questions is to take a 'reductionist' view of the self, as does Derek Parfit, in *Reasons and Persons*, Oxford 1984. According to Parfit the existence of a person just consists in the existence of a brain, a body and a set of inter-related mental and physical events. (This is supposed to be a reductive view of the person, and *therefore* of the self, which is simply the person in its self-describing aspect.) Parfit's position is criticised by Shoemaker in his review of *Reasons and Persons*, in *Mind*, 1985, pp. 443–53, and by Quassim Cassam, 'Reductionism and First-person Thinking', in David Charles and Kathleen Lennon, eds., *Reduction, Explanation and Realism*, Oxford 1992. Cassam argues that we cannot capture what it is to be a subject of thinking simply by examining (as Parfit, Evans and others would wish) the logical properties of I-thoughts. Cassam offers what he calls a weak reductionist reply to Parfit. The weak reductionist accepts that the existence of a person consists of a brain and a body and the relevant mental and physical events, but denies that these events can be adequately described in impersonal terms. 'It is distinctive of the life of a person or subject to be self-conscious, and hence to include the thinking of first-person or I-thoughts and it is constitutive of I-thoughts that they are ascribed to a particular person or subject'. Cassam, like Parfit, seems not to distinguish clearly the person from the self. Maybe the two concepts are coextensive, but surely they are distinct?

The questions here are very difficult: yet surely it is important for any account of the mind to make sense of self-consciousness, and in particular of the unified self that seems to be, in our case at least, the fundamental starting point of all mental life and knowledge? For this reason, many philosophers in the phenomenological tradition remain dissatisfied with the whole tendency of recent analytical philosophy, which proposes theories of the mental which make no mention of the self, or alternatively simply remove the idea to the sidelines, and deal with mental states in isolation, as separate atoms, detached from the entity to which they are attributed whenever we are aware of them.

Chapter 17 Freedom

1. Determinism

The following are useful:

(1) The article by Richard Taylor in Paul Edwards's *Encyclopedia of Philosophy*.

(2) Richard Taylor, 'Freedom and Determination', in *Metaphysics*, Englewood Cliffs NJ 1974.

(3) C.D. Broad, 'Determinism, Indeterminism and Libertarianism', in *Ethics and the History of Philosophy*, London 1952.

(4) A.C. MacIntyre, 'Determinism', *Mind*, 66, 1957, 28–41.

(5) G.J. Warnock, 'Every Event has a Cause', in A. Flew, ed., *Logic and Language*, vol. 1, Oxford 1955.

There is also a dialogue on *Free Will and Determinism*, by Clifford Williams, published by Hackett, Indianapolis, in the hope that American undergraduates will not only read it but also go on to answer some of the many questions at the end. This dialogue is worth studying: however, it does not make any attempt to connect its with the literature listed in the bibliography.

Some questions on determinism:
(1) What is determinism? Is it true?
(2) Has quantum physics shown that some kinds of occurrence are uncaused?
(3) Are reasons for action causes?

(This is touched on in Williams's dialogue, and dealt with in more detail by Donald Davidson, in 'Reasons as Causes', in *Essays on Actions and Events*, Oxford 1980, replying to A.I. Melden, *Free Action*, London 1961.)

Responsibility

This is a vast field, but the crucial issues can be grasped by studying:
(1) J.L. Austin, 'A Plea for Excuses', in *Philosophical Papers*, Oxford 1961.
(2) C.L. Stevenson, 'Ethics and Avoidability', in P. Schilpp, ed., *G.E. Moore*, library of living philosophers, Open Court 1968.
(3) H.L.A. Hart, *Law, Liberty and Morality*, Oxford 1963, and *Essays in Jurisprudence and Philosophy*, Oxford 1983.
(4) H.L.A. Hart, 'Defeasability', in A. Flew, *Logic and Language*, vol. 2, Oxford 1951.

Some questions to consider:
(1) Am I to blame for doing that which I was coerced into doing?
(2) What is an excuse? Do animals ever have excuses?
(3) Can you escape blame by proving that your action was caused by factors beyond your control?
(4) 'You are responsible for that child': what does this mean?

The Kantian Position

There are no good commentaries, so far as I know, on Kant's philosophy of freedom, though the vast and dull book by Henry E. Allison (*Kant's Theory of Freedom*, Cambridge 1990) is not seriously misleading. The best way to study Kant's text is to read it in conjunction with Strawson's 'Freedom and Resentment', and to reflect on the concept of the person, as this emerges from Kant's Categorical Imperative. (See Chapter 20.)

An interesting reconstruction of a quasi-Kantian position is associated with the name of Harry Frankfurt, who locates the freedom of the self-conscious subject precisely in his self-consciousness: specifically in 'second-order desires'. I do not merely desire things, as a dog does; I also desire to desire them, and desire not to desire them. A fully responsible choice springs not merely from desire, but from desire reinforced by the second-order desire to have that desire. This approach is discussed by Frankfurt, Gary Watson and Charles Taylor, in a book of essays edited by Gary Watson, entitled *Free Will*, 1982.

The Kantian theory raises the whole question of action: What is the relation between an action and a movement? What are reasons for action? To what extent am I responsible for my actions? The literature here is extensive, and I have skated over the subject in the chapter. Here are some of the basic texts:

(1) A.R. White, ed. *The Philosophy of Action*, in the Oxford Readings series (Oxford 1968): contains important early articles from recent discussions: see especially A.C. Danto.
(2) Brian O'Shaughnessy, *The Will*, 2 vols., Oxford 1981.
(3) T. Honderich, ed., *Essays on Freedom and Action*, London 1973.

Some questions:
(1) What is a reason for action?
(2) Do reasons purport to explain an action, to justify it, or both?
(3) What is a 'basic action'? Are there any?
(4) What is left over when I subtract the fact of my arm's rising from the fact of my raising my arm?

(This question is raised by Wittgenstein, in his *Remarks on Philosophy and Psychology*, Oxford 1980, section 452, and discussed in Malcolm Budd, *Wittgenstein's Philosophy of Psychology*, London 1989.)

Chapter 18 The Human World

Following Wilfrid Sellars, modern philosophers often make a contrast between the 'scientific image' of the world, and the 'manifest image' – the world as it appears to us. And this provides an alternative idiom for discussing the issues raised in this chapter. However, it is fair to say that recent discussions have for the most part been very narrow, focussing on the distinction between primary and secondary qualities, or that between real and nominal essence, or between natural and functional kinds, or between the objective and the subjective view. (See Chapter 31.) The suggestion that the human world may be through and through the product of unscientific ways of thinking, and yet at the same time a true representation of an objective reality, has rarely been made by English-speaking philosophers in modern times – though it has often been endorsed in other terms by cultural and literary critics. (For example, by Matthew

Arnold, in his *Culture and Anarchy*, by John Ruskin, in *Modern Painters* and elsewhere, and by F.R. Leavis, in his brilliant if intemperate attack on C.P. Snow, entitled 'The Two Cultures', and reprinted with embellishments in *Nor Shall My Sword*, London 1972.)

Indeed, the standard response of the analytical philosopher to the suggestion that we conceptualise the world in ways that defy the procedures of scientific explanation is to rule our ordinary concepts defective, and to draw the conclusion that there is no such thing as the reality that they purport to describe. This approach (applied to the concept of mind in general and belief and other intentional states in particular) is illustrated by Steven Stich, *From Folk Psychology to Cognitive Science: the case against belief*, Cambridge Mass. 1983. Among those modern philosophers who are prepared to uphold the claims of the human world against the imperialist stance of science, the following deserve special mention:

Bernard Williams, in *Ethics and the Limits of Philosophy*, London and Cambridge Mass. 1985. (Williams does not resist the conclusion, however, that the human world falls short of the 'absolute' conception that is the goal of science, and is in a deep sense relative to the cultural and communal identity of those who perceive it.)

Hilary Putnam, *Renewing Philosophy*, Cambridge Mass. 1992: an attempt to place some parts of the human world, at least, on a level footing with science.

Anthony O'Hear, *The Element of Fire: Science, Art and the Human World*, London 1988: an account which reflects the traditional (Arnoldian) conception of a 'culture', as a form of knowledge of human life in its concrete conditions.

David Cooper, *Existentialism*, Oxford 1990: especially chs. 3 and 4.

The interested reader might also consult Roger Scruton, *Sexual Desire*, London and New York 1986, in which the intentional concept of sexual desire is contrasted with the scientific theory of sexuality.

The concept of the *Lebenswelt* was introduced by Husserl in *Phenomenology and the Crisis of the Human Sciences*, Northwestern University 1954. This fairly impenetrable work does not really fulfil its promise, and leaves the reader as uncertain at the end as he is at the beginning, as to whether philosophy really can vindicate the human world in the face of scientific scepticism. Other writers have given elaborate and often highly plausible descriptions of our intentional concepts, the most notable being those phenomenological sociologists who were directly influenced by Husserl: Alfred Schutz (see his two volumes of *Collected Papers*, tr. M. Natanson, The Hague 1967,

and *Life Forms and Meaning Structure*, tr. H.R. Wagner, London 1982) and Max Scheler (especially *The Nature of Sympathy*, tr. P. Heath, London 1954 and 'On Knowing, Feeling and Valuing', in *Selected Writings*, Chicago 1992). Unfortunately, such writers do nothing to vindicate the concepts that they study. The same is true of the extremely suggestive works of Helmuth Plessner, Martin Buber and Rudolf Otto. See especially:
Martin Buber, *I and Thou*, Edinburgh 1984.
Rudolf Otto, *The Idea of the Holy*, tr. J.W. Harvey, Oxford 1923.
Dilthey's concept of *Verstehen*, from which this school of thought originally derives, can be studied through W. Dilthey, *Selected Writings*, tr. and ed. H.P. Rickman, Cambridge 1976.

Some questions:
(1) What is the relation between intentionality and intensionality? (See the second appendix to Roger Scruton, *Sexual Desire, op. cit.*, and the references cited therein.)
(2) Why should we try to save the appearances? (Compare the very different arguments mounted by Anthony O'Hear, in *The Element of Fire, op. cit.*, and Colin McGinn, in *The Subjective View*, Oxford 1983.)
(3) Are there kinds other than natural kinds? (See Roger Scruton, *op. cit.*, ch. 1.)
(4) What is the distinction between intention and desire? (See the argument of G.E.M. Anscombe, *Intention*, Oxford 1957.)
(5) Are there efforts of will? (See Brian O'Shaughnessy, *The Will*, especially vol. II, chs. 11 and 12.)

Chapter 19 Meaning

Stevenson's causal theory of meaning is set out in his *Ethics and Language*, Yale 1944. Paul Grice's account appears in 'Meaning', *Phil. Rev.* vol. 66, 1957, reprinted in *Philosophical Logic*, ed. P.F. Strawson, Oxford 1967. Grice deals with counter-examples to his theory in 'Utterer's Meaning and Intentions', *Phil. Rev.* vol. 78, 1968, and in 'Meaning Revisited', in Neilson Smith, ed., *Mutual Knowledge*, London 1982.

Strawson draws the distinction between communication-intention approaches to meaning and truth-theoretic semantics in his 'Meaning and Truth', reprinted in *Logico-Linguistic Papers*, London 1971. Jonathan Bennett defends Grice in his *Linguistic Behaviour*, Cambridge 1976. See also David Lewis, *Convention*, Cambridge Mass.

1969, ch. 4. Donald Davidson questions the cogency of Lewis's approach to convention in 'Communication and Convention', *Synthese* vol. 59, 1984. The extension of Grice's theory to provide a general account of speech acts is attempted in J.R. Searle, *Speech Acts*, Cambridge 1969.

Gilbert Harman gives an overview of the relations between language-meaning, thought and pragmatics in 'Three levels of Meaning', in D.D. Steinberg and L.A. Jacobovits, *Semantics: an Interdisciplinary Reader in Philosophy, Linguistics and Psychology*, Cambridge 1971. (Pragmatics is, roughly, the study of those aspects of meaning which attach to the context of language-use, rather than to semantic rules.)

Tarski's paper on truth, 'The Concept of Truth in Formalized Languages', in *Logic, Semantics, Metamathematics*, Oxford 1956, is very difficult, but its main ideas are expounded in 'The Semantic Theory of Truth' referred to in the notes to Chapter 9, and also by Quine, in his *Philosophy of Logic*, Cambridge Mass. 1986, ch. 3. Davidson's truth-theoretic account of meaning is set out in 'Truth and Meaning', reprinted in *Essays in Truth and Interpretation*, Oxford 1984. The Davidson industry is now in full swing. Bjorn T. Ramberg, in *Donald Davidson's Philosophy of Language*, Oxford 1989 provides a sympathetic, though one-sided account, while Simon Evnine, in *Donald Davidson*, Oxford 1991 gives a useful commentary on all aspects of Davidson's philosophy, and an excellent appendix on 'the Frege argument'. Some useful articles are to be found in G. Evans and J. McDowell, *Truth and Meaning*, Oxford 1976.

Some questions:
(1) Is there a connection between meaning and truth?
(2) What role is played by Tarski's convention T in Davidson's theory of meaning?
(3) What is the connection between meaning as studied by Grice, and meaning as studied by Davidson?
(4) What are the most important objections to Grice's theory? Can they be overcome?

The literature concerning the debate between realism and anti-realism is vast and forbidding. A comprehensive overview is to be found in the introduction to Crispin Wright's *Realism, Meaning and Truth*, Oxford 1986, a book with a most attractive cover. Wright expounds the 'manifestation' and 'acquisition' arguments of Dummett, whose views are to be found in 'Truth', 1959, reprinted in *Truth and Other*

Enigmas, London 1978, and in 'What is a Theory of Meaning?', in S. Guttenplan, ed., *Mind and Language*, Oxford 1974, and 'What is a Theory of Meaning II', in Evans and McDowell, *op. cit.* Dummett's views are also repeated in *The Logical Basis of Metaphysics*, London 1991, and thoroughly discussed by Crispin Wright in *Truth and Objectivity*, Oxford 1993.

Crispin Wright's views, like those of Dummett, are constantly circling around the carcass of anti-realism, indifferent to the stench that rises from it. From the anthropological view, the debate is most instructive.

Some questions:
(1) How cogent are Dummett's arguments against the realist?
(2) Can the meaning of a sentence be given in terms of its assertibility conditions?
(3) What constraints are placed on a theory of meaning, by the idea that meaning is what is understood by a speaker of the language?

The externalist approach to meaning is defended by Hilary Putnam in *Reason, Truth and History*, Oxford 1984, while a causal theory of reference is developed by Jerry Fodor in *A Theory of Content*, Cambridge Mass. 1990. Putnam is not persuaded by this theory, or by the attempt to reduce representation to some more primitive relation. See his extensive criticisms of Fodor's position in *Renewing Philosophy*, Cambridge Mass. 1992, ch. 3. Tyler Burge's criticisms of psychological individualism can be found in his 'Individualism and Psychology', in *Philosophical Review*, 1986.

On the rule-following argument, the reader should consult S. Kripke, *Wittgenstein on Rules and Private Language*, Oxford 1982, which is, however, very suspect on the private language argument. Crispin Wright connects the rule-following argument with anti-realism, in his 'Rule-following, Meaning and Constructivism', in C. Travis, ed., *Meaning and Interpretation*, Oxford 1986, and 'Rule-following, Objectivity and the Theory of Meaning', in S.H. Holzman and C.M. Leich, eds., *Wittgenstein: to Follow a Rule*, London 1981. Wright believes that Kripke's interpretation of Wittgenstein, inspired by the beginning of section 201 in the *Philosophical Investigations*, would not have survived had he read closely to the end of the section.

Chapter 20 Morality

Introductions to ethics abound, very few of them impartial or comprehensive. Bernard Williams's *Ethics and the Limits of Philosophy*, Fontana, London 1985, which is more a statement of position than a survey of the subject, is worth reading nevertheless, since it plunges the reader into the middle of modern moral philosophy. The Penguin introduction by J.L. Mackie: *Ethics: Inventing Right and Wrong*, Harmondsworth 1977, is clear enough to lay good intellectual foundations. Its central argument is disposed of by H.P. Grice, in his *Carus Lectures on the Conception of Value*, Oxford 1991. The best introduction to the subject is probably still Kant's *Groundwork of the Metaphysic of Morals*, together with the same author's *Critique of Practical Reason*.

The Naturalistic Fallacy

First stated in G.E. Moore's *Principia Ethica*, Cambridge 1903, it is better expressed in his unfinished draft of a preface to the second edition of that book. This draft was never printed, but is summarised and reconstructed in C. Lewy: 'G.E. Moore on the Naturalistic Fallacy', *PBA* 1964, reprinted in P.F. Strawson, ed., *Studies in the Philosophy of Thought and Action*, Oxford 1968. There are useful discussions of the fallacy by W. Frankena and others, in P. Foot, *Theories of Ethics*, Oxford 1967, a selection of papers in the Oxford Readings series.

Some questions:

(1) What is the naturalistic fallacy? Is it a fallacy?

(2) Is there a useful comparison to be made, between Moore's argument for a naturalistic fallacy, and Hume's remarks concerning the gap between 'is' and 'ought'?

(3) What does Moore mean by a 'non-natural' property?

Emotivism

First seriously defended by Spinoza, and then by Hume, its modern statement has been hampered by naïve theories of meaning: e.g. the statement by I.A. Richards and C.K. Ogden in *The Meaning of Meaning*, London 1927, and by A.J. Ayer, in *Language, Truth and Logic*, London 1936, and also by C.L. Stevenson, *Ethics and Language*, Yale 1944. A more up-to-date defence is given in R. Scruton, 'Attitudes, Beliefs and Reasons', in J. Casey, ed., *Morality and Moral Reasoning*, London 1971.

Some questions:
(1) 'Moral judgements express attitudes'. What might be meant by the term 'express' in that statement?
(2) Can emotions be justified?

(See R. Scruton, 'Emotion and Common Culture', in *The Aesthetic Understanding*, Manchester 1983; and E. de Sousa, 'The Rationality of Emotion', in A. Rorty, ed., *Explaining Emotions*, Berkeley California 1980.)

Prescriptivism
Hare's views are to be found in *The Language of Morals*, Oxford 1952, *Freedom and Reason*, Oxford 1963, *Practical Inferences*, London 1971, *Applications of Moral Philosophy*, London 1972 and *Essays on the Moral Concepts*, London 1972, which contains the essay on 'universalisability' of 1955. His later works, especially *Moral Thinking*, Oxford 1981, show a shift of emphasis towards questions of practical reasoning, and away from 'the analysis of moral judgement'. Useful discussions, by Geach et al., are reprinted in Foot's *Therories of Ethics, op. cit.*

Some questions:
(1) Is there a proof, from the premise of prescriptivism, that there is no deriving an 'ought' from an 'is'?
(2) Can a prescriptivist account for weakness of will?

(There is a question whether *anybody* can account for it. See D. Davidson, 'Weakness of Will', in *Actions and Events*, Oxford 1980. But read Hare, *Freedom and Reason*, ch. 5.)

(3) Can a prescriptivist believe that moral judgements are objective?
(See Hare in *Moral Thinking*.)
(4) What, according to Hare, is wrong with 'descriptivism'? Is he right?

(See the essay on 'Descriptivism' in *Essays on the Moral Concepts*.)

Moral Realism
There is a valuable collection, ed. T. Honderich: *Morality and Objectivity*, London 1985: see especially the articles by S.L. Hurley, John McDowell, Bernard Williams, and Simon Blackburn. Blackburn's anti-realism is also expounded in his *Spreading the Word*, Oxford 1984; Hurley's qualified realism is developed at length in her *Natural Reasons*, Oxford 1989. The best statement of the position is perhaps that given by David Wiggins, in a series of tightly argued and

often excruciatingly dense articles, some of them reprinted in his *Needs, Values, Truth*, London 1987, to which should be added 'Moral Cognitivism, Moral Relativism, and Motivating Moral Beliefs', *Proceedings of the Aristotelian Society*, 1990. Wiggins makes an attempt to explain what he means by a 'motivating belief', but this is – ultimately – the crux for the moral realist. What is to be said to someone who agrees with the moral argument, but feels no inclination to act on it? That he does not really agree with it, because he hasn't understood it? See also R. Scruton, 'Reason and Happiness', *Royal Institute of Philosophy Lectures*, 1980.

Some questions:
(1) What is a 'motivating belief'?
(2) Could you disagree with the moral realist, and still believe in an objective morality?
(3) What are the principal arguments against moral realism?

Utilitarianism

The classic texts are Jeremy Bentham, *Introduction to the Principles of Morals and Legislation*, London 1789, and J.S. Mill, *Utilitarianism*, London 1861. Modern discussions are well summarised in J.J.C. Smart and Bernard Williams, *Utilitarianism, for and against*, Cambridge 1973, and A.K. Sen and B. Williams, eds., *Utilitarianism and Beyond*, Cambridge 1982. Good criticisms of the theory are contained in Williams's *Ethics and the Limits of Philosophy*, and in Phillipa Foot's 'Utilitarianism and the Virtues', *Mind*, 1990, which offers arguments against the view that consequentialism could define the moral viewpoint. See also Samuel Scheffler, ed., *Consequentialism and its Critics*, Oxford Readings, Oxford 1988: a fairly parochial collection of articles, mostly written in horrible jargon by liberal Americans. (But with good contributions from T. Scanlon and A. Sen.)

Some questions:
(1) Is absolutism untenable?
(2) Can utilitarianism escape the charge that it justifies injustice?
(3) What is happiness? Is there any sense in the suggestion that it can be measured?

Kant

The best translations are by L.W. Beck, in his *Selections* from Kant, London 1988. The *Critique of Practical Reason* is in many ways clearer than the *Foundations*. Commentaries are useless, although there is an inaugural lecture by David Wiggins, entitled 'Categorical Requirements . . .', and published in *The Monist* for 1991. This brings Kant and Hume together, to make Wiggins.

With or without Wiggins, it is always best to discuss Kant and Hume together: the relevant texts for the latter being the third book of the *Treatise* and the second *Enquiry*.

Some questions:
(1) 'Empirical principles are not at all suited to serve as the basis for moral laws' (Kant). What does he mean, and is he right?
(2) How many categorical imperatives are there?
(Commentators give answers ranging from 1 (H.J. Paton, *The Categorical Imperative*) to 11 (Bruce Aune, *Kant's Moral Theory*).)
(3) Could there be an objective morality without a synthetic *a priori* basis?
(4) Why does Hume think that reason is and ought only to be the slave of the passions. Is he right?

Hegel and Side-constraints
On Hegel's Master and Slave argument it is useful to read Robert Solomon, *In the Spirit of Hegel*, Oxford 1983, especially pp. 448ff. Nozick's philosophy of side-constraints is contained in his *Anarchy, State and Utopia*, Oxford 1974. The relevant passage of Hegel's *Phenomenology of Spirit* is 'Lordship and Bondage', pp. 111–118 of the Miller translation, Oxford 1977. The best-known exposition of this complex work is the rhapsodic text by Alexandre Kojève, entitled *Introduction to the Reading of Hegel*, tr. J.H. Nichols Jr., Ithaca and London 1980. This makes no effort whatsoever to transcribe Hegel's argument into terms that could be used in serious discussion with an opponent, and is intellectually entirely worthless. This may be why it is widely read.

Among recent attempts to derive a morality of side-constraints, in which respect for persons and the sanctity of rights are given justifications in terms of the requirements of collective rationality among self-interested but cooperating agents, two in particular stand out: David Gauthier, *Morals by Agreement*, Oxford 1986, and Loren Lomasky, *Persons, Rights and the Moral Community*, New York 1987. These are very far from the 'spirit of Hegel', but tend, by surprising routes, towards the conclusions of the argument about the master and the slave.

Some questions:
(1) What compels the master to recognise the slave as a person?
(2) Could there be a rational being who treated all other rational beings as means only?
(3) What is the difference between a person and a thing?

Aristotle and Nietzsche
There are many collections of essays on Aristotle's ethics, including one by A. Rorty, *Essays on Aristotle's Ethics*, California 1980, containing important modern discussions (see especially articles by Williams, Pears and Wilkes). The *Nicomachean Ethics* (tr. W.D. Ross, Oxford 1954) is complemented by the *Eudemian Ethics*, with which it shares a chapter or two, and various subsidiary texts, not all of them certainly by Aristotle. The two key works by Nietzsche are *The Genealogy of Morals*, New York 1969, to which I return in Chapter 30, and *Joyful Wisdom* (*Fröhlische Wissenchaft*), translated by Walter Kaufmann as *The Gay Science*, New York 1974. Commentaries on Nietzsche are mostly terrible – either sterilising, like Arthur Danto's, or appropriating, like Foucault's *Folie et déraison*. Among the exceptions is Erich Heller, *The Disinherited Mind*, Harmondsworth 1961, which has a powerful chapter on Nietzsche.

Recent work on the virtues helps to clarify some of the issues – notably Martha Nussbaum's long-winded *Fragility of Goodness*, Cambridge 1985, and John Casey's short-winded *Pagan Virtue*, Oxford 1990. There are those who find important ideas in Alasdair MacIntyre's *After Virtue*, Notre Dame 1981; not being one of them, I recommend instead those of the *Enneads* of Plotinus which deal with virtue (e.g. 1.2. (19), 1.4. (46), and 1.5. (36)), together with other ancient texts like *Phryne's Symposium* (in R. Scruton, ed., *Xanthippic Dialogues*, London 1993).

Some questions:
(1) What is a virtue?
(2) Is happiness the final end?
(3) Why should I be courageous?
(4) Why should I be just?
(5) Can I aim at excellence while despising the *Übermensch*?
(6) What is Aristotle's strategy in the *Nicomachean Ethics*?

(See e.g. R. Scruton, *Sexual Desire*, ch. 11; David Wiggins, *Needs, Values, Truth*, Oxford 1986, and Philippa Foot, 'Utilitarianism and the Virtues', *Mind* 1989.)

(7) Why, and with what consequences, does Nietzsche contrast the 'will to truth' with the 'will to power'?

(See *Beyond Good and Evil*, opening sections.)

Chapter 21 Life, Death and Identity

There are three topics covered in this chapter, and it is normal to deal with them separately: life, personal identity and death. It is the second that has received the most sustained attention from modern philosophers.

Life

The defence of vitalism given by Henri Bergson in *Creative Evolution*, London 1914, is not without interest. Broadly there are two approaches to life among philosophers. There is the attempt to understand life as a special process in the natural world, and to describe its place in the great chain of being. This is the approach adopted by Aristotle in his biological writings, and also by Spinoza. And there is the attempt to derive from our concept of life some paradigm of existence, to which all else aspires. This is the approach of Leibniz, with his theory of the *vis viva*, and of those thinkers like A.N. Whitehead who argue that life reveals the nature of the world as process. Whitehead's *Process and Reality*, New York 1978, and *Adventures of Ideas*, Cambridge 1933, are still worth reading; as are the works of the 'process theologians' whom he inspired. (See Chapter 25.)

Personal Identity

Much has been written on this subject, and many gripping thought-experiments devised which push our concept of the person to its limits. However, the questions considered are not always the same. Hume, for example, in *Treatise*, Book I, Part IV, 5–6, is concerned with the question of what, if anything, unites the ideas and impressions of a particular individual. He is not concerned with the question of identity – either in the sense of what distinguishes one person from another, or in the sense of sameness over time. His attempt at the problem he poses and his heroic failure are well summed up by David Pears in his *Hume's System*, Oxford 1990.

The problem of personal identity as it is discussed today grew from an early paper of Bernard Williams: 'Personal Identity and Individuation', *Proceedings of the Aristotelian Society*, 1956–7. This was followed by other papers, of which 'Imagination and the Self' (*Proceedings of the British Academy*, 1966) and 'The Self and the Future', *Phil. Rev.* 1970 are the most important. These are reproduced in Williams's collection *Problems of the Self*, Cambridge 1973. Other important players in the game are Sydney Shoemaker (*Self-knowledge and Self-identity*, New York 1963), and Derek Parfit, beginning with 'Personal Identity' in the *Phil. Rev.* for 1971, and ending with the book *Reasons and Persons*, Oxford 1986. Parfit's radical approach, designed to disestablish the concept of personal identity altogether, is rejected by David Wiggins: see especially his 'Locke, Butler and the Stream of Consciousness' in A. Rorty, ed., *The Identities of Persons*, Berkeley 1989. Mention should also be made of John Perry, whose papers on this subject occur in John Perry, *Personal Identity*, California 1975, and whose *Dialogue on Personal Identity and Immortality*, Indianapolis 1978, is an exemplary introduction to some of the issues discussed in this chapter. Interesting discussions are also contained in A. Peacocke and G. Gillett, eds., *Persons and Personality*, Oxford 1987, especially the article by David Wiggins, entitled 'The Person as object of science, as subject of experience, and as locus of value'.

The historical debate between Locke, Butler and Reid is contained in:

Locke: *An Essay Concerning Human Understanding*, Book II, ch. xxvii.
Butler: 'Of Personal Identity', obtainable in A. Flew, ed., *Body, Mind and Death*, London 1964.
Thomas Reid: *Essays on the Intellectual Powers of Man*, Essay 3, chs. 4 and 6. (London 1941.)

The debate between these three established the question of personal identity across time, which goes to the heart of what we mean by 'person'.

Some questions:
(1) Is the charge of circularity against Locke effective? (See Perry, 'Personal Identity, Memory and the Problem of Circularity', in his *Personal Identity*, *op. cit.*, and Wiggins, 'Locke, Butler and the Stream of Consciousness', in Rorty, *op. cit.*)
(2) What is wrong with Locke's definition of a person? (See *Essay*, II, xxvii, ii, and Wiggins, *op. cit.*)
(3) Some philosophers – e.g. Strawson (*Individuals*, ch. 3) and Wiggins – argue that the concept, *person*, is a primitive concept. What does this mean, and is it true?
(4) Is anything proved by thought-experiments concerning brain-

transplants and the rest? (See Wiggins, *op. cit.*, who is most sceptical about this.)
(5) Could there be a Parfitian person?
(6) Are persons members of a natural kind? If so, is the concept *person* the concept of that kind?
(7) Is the solution to the problem of personal identity to be discovered or stipulated?

Existence and Essence
The reader interested in existentialism will find an excellent introduction in David Cooper, *Existentialism*, Oxford 1990. The relevant texts are all cited by Cooper; the quotations I give are from Sartre, *Existentialism and Humanism*, London 1966; Heidegger, *Being and Time*, and José Ortega y Gasset; 'Man the technician', in *History as a System*, Norton 1962.

Death
The principal readings on this subject are:
Lucretius, *De Rerum Natura*, Book 9.
Epicurus, Letter to Menoeceus, in *Epicurus, the Extant Remains*, ed. C. Bailey, Oxford 1926.
Thomas Nagel, 'Death' in *Mortal Questions*, Cambridge 1979.
D.Z. Phillips, *Death and Immortality*, Cambridge 1970.
Fred Feldman, *Confrontations with the Reaper*, Oxford 1992.
C.J. Ducasse, *Nature, Mind and Death*, London 1951.
John Donnelly, ed., *Language, Metaphysics and Death*, New York 1978.
Paul Edwards, 'My Death', in *Encyclopedia of Philosophy* vol. V, pp. 416–9.
Stephen Rosenbaum, 'The Symmetry Argument: Lucretius Against the Fear of Death', *Phil. and Phenomenological Research*, 1989.
(A defence of the Lucretian view that fear of not existing in the future is no more rational than fear of not having existed in the past.)
H. Silverstein, 'The Evil of Death', *Journal of Philosophy*, 1980.
Antony Flew, *The Logic of Mortality*, Oxford 1987.
(A vigorous critique of the belief in life after death.)
S.T. Davis (ed.), *Death and the Afterlife*, London 1989.

Important too are Plato's original arguments for the immortality of the soul, in the *Phaedo*, 67A, and 102A–107A.

Bernard Williams's reflections on the *Makropoulos Case* occur in his *Problems of the Self*, *op. cit.*; Schopenhauer's defence of suicide occurs in Book IV of *The World as Will and Representation*, and also in an essay on Suicide, collected in the Penguin edition of Schopenhauer's essays and elsewhere; Nietzsche's defence of timely death (voluntary death) occurs in *Thus Spake Zarathustra*, tr. R.J. Hollingdale, Harmondsworth 1961, pp. 97–9; Heidegger's account of 'Being-towards-death' is in Division II, ch. 1 of *Being and Time*; pp. 290–311 of the English language edition, tr. J. Macquarrie and E. Robinson, Oxford 1962.

Chapter 22 Knowledge

Basic reading:
J. Dancy, *Introduction to Contemporary Epistemology*, Oxford 1985, a lucid summary of modern positions, which is a useful textbook.
T. Nagel, *The View from Nowhere*, Oxford 1986, ch. 5.
A. Phillips Griffiths, ed., *Knowledge and Belief*, Oxford Readings, Oxford 1967: contains the article by Gettier (which originally appeared in *Analysis*, vol. 23, 1963), as well as other useful material.
R. Nozick, *Philosophical Explanations*, Oxford 1981. A difficult text, whose argument is more clearly set out in ch. 3 of Dancy.

It is also useful to study the following:
Plato: *Theaetetus*, ed. Burnyeat, Cambridge 1991, together with Burnyeat's commentary.
J.L. Austin, 'Other Minds', in *Philosophical Papers*, Oxford 1961: the first major statement of the reliability theory.
A. Goldman, 'A Causal Theory of Knowledge', *Journal of Philosophy*, vol. 64, 1967.
A. Goldman, 'Discrimination and Perceptual Knowledge', *Journal of Philosophy*, vol. 73, 1976.
W.V. Quine, 'Epistemology naturalised', in *Ontological Relativity*, New York 1969.
Colin Radford, 'Knowledge by Examples', *Analysis*, vol. 29, 1969.
Colin McGinn, 'The Concept of Knowledge', unpublished.

McGinn argues that local analyses of knowledge fail since they do not accommodate the fact that, in order to know that *p*, a subject must be reliable with respect to a whole *range* of propositions, not just the proposition that *p*. Suppose there is a bent stick, placed in water, in which condition straight sticks normally look bent. I believe that it *is* bent, on the basis of looking at it; and on Nozick's analysis this would be a cause of knowing that it *is* bent. But I do not *know*, for in these circumstances I cannot be deemed to be reliable with regard to propositions of that kind.

Some questions:
(1) What is the lesson of the Gettier examples?
(2) Why do we say that 'knowing how' is a form of knowledge?
(3) Can a theory of knowledge be used to combat the arguments of the sceptic? (See Stroud, *The Significance of Philosophical Scepticism*, Oxford 1984.)
(4) What is a belief? (See Phillips Griffiths, in Phillips Griffiths, *op. cit.*)
(5) What are the main difficulties for Nozick's 'tracking theory'? (See Dancy, *op. cit.*)

Chapter 23 Perception

The most important articles are included in two of the Oxford Readings, that entitled *The Philosophy of Perception*, edited by G.J. Warnock, Oxford 1967, and that entitled *Perceptual Knowledge*, edited by Jonathan Dancy, Oxford 1988. The second is more up-to-date, and reflects developments that have occurred since Warnock's collection was published. Also important are the essays contained in Tim Crane, ed., *The Contents of Experience*, Cambridge 1992.

Three questions have troubled recent philosophers of perception. First, the character of perceptual experience: can any kind of experience be or become a perception? What is the difference between having a perceptual experience and *inferring* something about the world, from an experience that is not itself perceptual? There seems to be a real difference here, but how is it to be characterised? Is it part of the *phenomenology* of the perceptual experience? Or does it follow from the circumstances in which the experience occurs? Maurice Merleau-Ponty's book, *The Phenomenology of Perception*, tr. Colin Smith, London 1962, tries to give a phenomenological account of the perceptual experience, which will show its peculiarities, and also refute the empiricist theory that perception involves on the one hand a datum, and on the other an interpretation. Interpretation and experience interpenetrate, and experience itself marks the world with its own intentional character. Moreover this character reflects our nature as active beings: we see things under concepts whose sense derives from activity, planning, intention and desire, and not merely from intellectual speculation. The human world is essentially 'ready for action', and is already seen in just that way.

The second problem concerns the nature of the information obtained through perception. What I see depends not only on my sensory 'input', but also on my prior knowledge. For this reason,

Fred Dretske, in *Seeing and Knowing*, London 1969, describes perceptual information as essentially incremental: to perceive is to ascend to a new information state from an old one. Dretske also gives a thorough analysis of the process of perception which, while questionable in many ways, is so incredibly dry and boring that most philosophers are prepared to take it on trust.

The third field of speculation concerns the connection between perception and what some philosophers (e.g. John McDowell, 'Criteria, Defeasability and Knowledge', in Dancy, *op. cit.*, and '*De Re* senses', in Crispin Wright, ed., *Frege, Tradition and Influence*, Oxford 1984) call *de re* belief. If I see Mary, and so come to believe that there is a woman standing by the front door in a state of distress, then it would be normal to say that I believe *of* Mary, that she is standing by the door in a state of distress. But I may not know that this woman is Mary. Here the object of my belief is identified through my perception, which is in turn identified through the causal connection between my mental state and an item in the world. By identifying the object of belief in this way, I have identified it as a 'material', rather than an 'intentional' object. I can substitute equivalent descriptions of Mary *salva veritate*; and indeed, in order to give a precise account of my belief, it is necessary to identify it in precisely this way. There is no way of determining what my belief is about, simply by looking 'inwards' to the mental representation. In identifying the belief I make *essential* reference to the thing in the world. (This is parallel to the *de re* modalities discussed in Chapter 13.) *De re* belief is interesting partly because some philosophers – Putnam, for example, in 'Is Semantics Possible?', in *Mind, Language and Reality, op. cit.*, – believe it to be the normal case, and believe also that this fact forces us entirely to revise our picture of the mind, as somehow constituted by the *way things seem* to the subject. (See further Chapter 19, section 6.)

Some questions:

(1) If John sees Mary, is there anything that he must believe about what he sees?

(2) What is a sense-datum? Are there any?

(3) What is proved by the argument from illusion?

(4) What are the arguments for the causal theory of perception? How plausible are they?

(5) Is perception the foundation of knowledge?

(6) What is a perceptual experience, and in what way is it like, and unlike, a sensation?

(7) Is there any convincing argument either for or against phenomenalism?

(See Russell, *Lectures on Logical Atomism*, in *Logic and Knowledge*, London 1988; R. Scruton, 'Objectivity and the Will', in *Mind*, 1971; and Gareth Evans, 'Things Without the Mind', in *Collected Papers*, Oxford 1985.)

Chapter 24 Imagination

This lecture explores a topic that is normally considered to be a philosophical by-way: though wrongly, in my view. Much confusion is caused by the fact that philosophers have used the word 'imagination' and its cognates in radically different ways – often confounding the two distinct ideas to which I refer in the chapter, that of mental imagery, and that of creative thought (whether or not accompanied by or expressed in imagery). In her historical survey, *Imagination*, London 1976, Mary Warnock makes little or no effort to hold these two notions apart, to the detriment of the argument. (Her principal interest is to show the place of imagination in the epistemologies of major historical figures, such as Hume and Kant, both of whom argued that perception and common-sense belief involve exercises of something that they called imagination.)

The view that imagination is involved in everyday perception is defended also by modern philosophers, for example by P.F. Strawson, in 'Imagination and Perception', in L. Foster and J.W. Swanson, eds., *Experience and Theory*, Cambridge Mass. 1970. This view is countered by Roger Scruton, in *Art and Imagination*, London 1974, especially Part II.

Other sources include the following:

J.P. Sartre, *L'Imagination*, Paris 1936.

J.P. Sartre, *L'Imaginaire*, Paris 1940, tr. as *The Psychology of the Imagination*, New York 1948.

Hidé Ishiguro, 'Imagination', in B. Williams and A. Montefiore, eds., *British Analytical Philosophy*, 1966.

Hidé Ishiguro, 'Imagination', *A.S.S.V.*, 1967.

L. Wittgenstein, *Philosophical Investigations*, Part II, section xi, for the classic discussion of 'seeing as'.

Two ancillary topics arise out of that of the imagination: the nature of representation (and especially of representation in art); and the nature of fictions and of our responses to fictions.

Representation

(1) Nelson Goodman, *Languages of Art*, London 1969, chs. 1 and 2: for a clear statement of the semantic theory.

(2) Richard Wollheim, *Painting as an Art*, London 1987, which contains incidental criticisms of Goodman, as well as the development of the distinction adumbrated in the same author's 'Seeing as and Seeing in', contained in the second edition of his *Art and its Objects*, Cambridge 1980.

(3) Roger Scruton, *Art and Imagination, op. cit.*, ch. 13.

(4) Roger Scruton, *The Aesthetic Understanding*, Manchester 1983, for a discussion of representation in relation to music and photography.

Fictions

Kendall Walton, *Mimesis as Make-Believe*, Cambridge Mass. 1990.

Colin Radford and Michael Weston, 'How can we be moved by the fate of Anna Karenina?', *Proceedings of the Aristotelian Society, Supplementary Volume*, no. 69, 1975, pp. 67–93.

Frank Palmer, *Literature and Moral Understanding*, Oxford 1992.

Some questions:

(1) What is the connection between imagining and having an image?

(2) What happens when I switch from seeing one aspect to seeing the other in an ambiguous picture?

(3) What is the connection between imagining that *p* and believing that *p*?

(4) What is depiction?

(5) When I respond to the grief of a character in a play, do I feel real sympathy?

Chapter 25 Space and Time

Reading:

D.W. Hamlyn, *Metaphysics*, Cambridge 1984, ch. 7.

H. Reichenbach, *The Philosophy of Space and Time*, New York 1957.

Kant, 'Transcendental Aesthetic' from the *Critique of Pure Reason*.

R.M. Gale, ed., *The Philosophy of Time*, London 1968. A useful collection which contains extracts from the major writers, including all that the reader needs of McTaggart.

The study of space and time is complicated by the intrusion of physics; philosophically minded physicists (notably Einstein, in his *Relativity: the Special and the General Theory*, London 1960, and

Stephen Hawking in his *Brief History of Time*, London 1988) have certainly changed the emphasis of recent discussions. To come to terms with modern physics is hard, and it is best to be guided by a physically-minded philosopher, such as Bas Van Fraassen, whose *An Introduction to the Philosophy of Time and Space*, New York 1970, remains one of the best introductory texts on the subject.

1. Euclid and Visual Geometry

The debate continues as to whether there is such a thing as visual, or phenomenal geometry (as opposed to the physical geometry of things in space). And the suggestion is still sometimes made (for example, by P.F. Strawson, in *The Bounds of Sense*, London 1966, endorsing Kant, in 'The Transcendental Aesthetic' from *The Critique of Pure Reason*), that visual geometry *is* essentially Euclidean. The argument is countered in an illuminating article by G.J. Hopkins, 'Visual Geometry', *Phil. Rev.*, 1976, reprinted in Ralph Walker, ed., *Kant on Pure Reason*, Oxford 1982.

2. Hilbert and Axiomatic Systems

The best way to come to grips with the subject of this section is through a study of axiomatic systems generally. See, for example, W.V. Quine, *Methods of Logic*, Cambridge Mass. 1982, and, for the application to the study of space, Graham Nerlich, *The Shape of Space*, Cambridge 1976.

3. Non-Euclidean Space

The most comprehensive (though far from inviting) discussion is that of A. Grunbaum, *Philosophical Problems of Space and Time*, Dordrecht 1973, although Reichenbach, *op. cit.*, remains useful and accessible. See also:

H. Minkowski, 'Space and Time', in Albert Einstein *et al.*, *The Principle of Relativity*, New York 1923.

M. Jammer, *Concepts of Space*, New York 1960 (a good historical survey).

G. Nerlich, *The Shape of Space, op. cit.*

4. Relative and Absolute Space

The correspondence between Leibniz and Clarke (ed. H.G. Alexander, Manchester 1956) remains an excellent introduction to the issues concerned. See also:

Isaac Newton, *Mathematical Principles of Natural Philosophy and his*

System of the World, Berkeley California 1934, Scholium to the definitions.
A. Einstein, *Relativity, the Special and the General Theory*, London 1960.
Kant's argument about the left and right hand has attracted much commentary. See, for example, Chris Mortensen and Graham Nerlich, 'Spacetime and handedness', *Ratio*, 25, 1983, pp. 1–13.

5. How Many Spaces are There?
A. Quinton, 'Spaces and Times', *Philosophy* 37, 1962.
The argument has been discussed in:
Jonathan Bennett, *Kant's Analytic*, Cambridge 1966.
T.E. Wilkerson, *Kant's Critique of Pure Reason*, Oxford 1976.

6. The Mystery of Time
See the intriguing discussion in H. Reichenbach, *The Direction of Time*, Berkeley California 1956.
St Augustine, *Confessions*, Harmondsworth (Penguin Edition) 1970.
W.C. Salmon, ed., *Zeno's Paradoxes*, Indianapolis 1970.
Bertrand Russell, 'On the Experience of Time', *Monist*, vol. 25, 1915, pp. 212–33.
J. Butterfield, 'Seeing the Present', in *Mind*, vol. 93, 1984, pp. 161–76.
P. Coveney and R. Highfield, *The Arrow of Time*, London 1985: gives a good account of the irreversible nature of time and physical process.

7. The Unreality of Time
Aristotle, *Physics*, IV. 10–14. 217b–244a.
Plotinus, *Enneads*, III. 7. (45), 'On Eternity and Time'.
J. McT. E. McTaggart, *The Nature of Existence*, ed. C.D. Broad, Cambridge 1921 and 1927, vol. 2, ch. 33.
P.T. Geach, *Love, Truth and Immortality*, London 1979. (Contains a sympathetic account of McTaggart's argument.)
W. Sellars, 'Time and the World Order', in H. Feigl and G. Maxwell, eds., *Minnesota Studies in the Philosophy of Science*, Minneapolis 1962, pp. 527–616.

8. Responses to the Argument
Michael Dummett: 'The Unreality of Time', in *Truth and Other Enigmas*, London 1976.
D.H. Mellor, *Real Time*, Cambridge 1980. This book began life as a

series of radio talks, and is therefore uncommonly lucid, without debasing the argument in any way.
R.M. Gale, *The Language of Time*, London 1968: an earlier attempt in the same direction.

9. Time and the First Person
Henri Bergson, *Time and Freewill*, tr. F.L. Pogson, London 1911.
Maurice Merleau-Ponty, *The Phenomenology of Perception*, tr. Colin Smith, London 1962.
See also the discussion in Chapter 31.

10. Process and Becoming
A useful introduction to process philosophy is the collection edited by D. Browning, entitled *Philosophers of Process*, Cambridge Mass. 1972. This contains extracts from all the major writers, including Whitehead, Dewey, James and Hartshorne. For more detailed discussion, see:
A.N. Whitehead, *Process and Reality*, New York 1978.
Charles Hartshorne, *Divine Reality: a Social Conception of God*, New Haven 1982.

11. Eternity, and 12. The Music of the Spheres
These sections deal with matters that have received little attention from modern philosophers. There is an excellent summary of Schopenhauer's theory of music, however, in Malcolm Budd, *Music and the Emotions: the Philosophical Theories*, London 1985. For an account of musical movement, see Roger Scruton, 'Understanding Music', in *The Aesthetic Understanding*, London and Manchester, 1983.
See also Plotinus, *op. cit.*, above, under 7.

Chapter 26 Mathematics

Reading:
Frege, *Foundations of Arithmetic*, tr. J.L. Austin, Oxford 1951.
Charles Parsons, 'Mathematics', in the *Encyclopedia of Philosophy*, ed. by Paul Edwards.
The following are also useful:
Bertrand Russell, *Principles of Mathematics*, New Edition, London 1992.

Bertrand Russell, *Introduction to Mathematical Philosophy*, London 1919.
P. Benacerraf and H. Putnam, eds., *Philosophy of Mathematics*, Cambridge 1984.
(This remains the most accessible and philosophically central collection of materials dealing with this topic.)
David Bostock, *Logic and Arithmetic*, vol. 1, Natural Numbers, Oxford 1974; vol. 2, Rational and Irrational Numbers, Oxford 1979.
(A painstaking and meticulous account.)
There is an accessible introduction to Gödel's theorem by E. Nagel and R. Newman: *Gödel's Proof*, London 1990.

Beyond those essential references, I do not propose to guide the reader further through this topic, which rapidly becomes too technical to be dealt with in an introductory text.

Chapter 27 Paradox

There is no easy access to further study in this area. The only way to proceed is to take the paradoxes one by one, and attempt to solve them, referring to the best of the recent literature. This is what R.M. Sainsbury does in his exemplary book on the subject – *Paradoxes*, Cambridge 1988.

There is no better guide to the *arguments* of the pre-Socratics than Jonathan Barnes, *The Pre-Socratic Philosophers*, London 1979, 1982. On the second Achilles, see Lewis Carroll in *Mind*, vol. 4, 1895, pp. 278–80. The many paradoxes of infinity are illuminatingly discussed by A.W. Moore, *The Infinite*, London 1990.

Chapter 28 Objective Spirit

This is another growth area, like the philosophy of mind, in which modern philosophy has begun to change the way in which the topic is conceived. The reader will therefore benefit from a more detailed guide to the territory, and to the constantly growing literature in the subject.

Introductory textbooks are few and disappointing. One major problem is that political philosophy has not until recently been regarded by analytical philosophers as a central area of concern, and the methods of modern philosophy have therefore been applied only in a spasmodic way to the analysis of political institutions and arguments. The first important work of analytical philosophy devoted to the subject – T.D. Weldon's *The Vocabulary of Politics*, London 1953 – seems largely to be motivated by the desire to show that there *is* no such subject, or at least, that clarity in our use of language would dispose of its major problems. Weldon remains important, although nobody is now persuaded either by his method or his conclusions.

I recommend the introduction by A. Quinton to *Political Philosophy* Oxford Readings, Oxford 1967, a book which also contains important papers and extracts, all of which are worth reading. Roger Scruton's *Dictionary of Political Thought*, London 1982, contains thumb-nail sketches of basic political concepts and theories, philosophies, philosophers and ideologies. This may be of assistance in one of the major tasks in political philosophy, which is to use terms consistently, knowing what they mean.

Modern literature is copious, and, as already noted, far from satisfactory. There is a short introduction by D.D. Raphael called *Problems of Political Philosophy*, London 1980, which is rather dry and sketchy, although useful in parts. More substantial, though somewhat dated, is S.I. Benn and R.S. Peters, *Social Principles and the Democratic State*, London 1959. The best collections of articles are those edited by Quinton, *op. cit.*, and by Jeremy Waldron (*Theories of Rights*, Oxford 1985) in the Oxford Readings in Philosophy series, and the five series of papers entitled *Philosophy, Politics and Society*, Oxford 1956–79, edited by P. Laslett, W.G. Runciman *et al.*

I begin from the prisoner's dilemma, though this is, properly treated, no subject for the beginner. Standard game-theoretical approaches to the topic can be found in:

A. Rapoport and A. Chammah, *Prisoner's Dilemma: a Study in Conflict and Cooperation*, Michigan 1965; and

R. Duncan Luce and Howard Raiffa, *Games and Decisions: Introduction and Critical Survey*, New York 1957.

The prisoners' dilemma poses a problem; the question is whether there is some strategy available to the parties that would be a solution to that problem, in the sense of being both rationally preferred, and also maximally beneficial. The theory of games distinguishes zero-sum games (in which gains to one party are losses to the other, as in a duel), from games in which both parties may gain or lose simultaneously (as in this case). The search for strategies that conform to various *a priori* requirements of rationality, consistency etc., provides a fascinating branch of mathematics, whose applications to other areas are many and surprising.

1. The Social Contract

The key historical texts are these:

Hobbes, *Leviathan*, ed. M. Oakeshott, Oxford 1957.

Locke, *Two Treatises of Civil Government.*

Rousseau, *The Social Contract.*

Kant also entertained a version of the social contract, though in hypothetical terms. A law is legitimate, he argued, if rational beings could envisage the *possibility* of their all agreeing to be bound by it. This conception of a hypothetical contract has become extremely important in recent philosophy, on account of the work of Rawls (see below).

The problem of the 'free rider' owes its name to Mancur Olson.

2. Traditional Objections

These occur, respectively, in:

(1) Hume, 'Of the Original Contract', in *Essays Moral, Political and Literary.*

(2) Edmund Burke, *Reflections on the Revolution in France*, Oxford 1993.

(3) Hegel, *Philosophy of Right*, tr. T.M. Knox, Oxford 1952.

3. Collective Choice and the Invisible Hand

The theory of collective choice is fascinating and increasingly technical. A lively and comprehensive introduction is that by Duncan Black, *The Theory of Committees and Elections*, Kluwer US 1987. The second part of this book contains a useful historical survey. In the same tradition is Brian Barry's *Political Argument*, Brighton 1990.

On the invisible hand, the classic text by Adam Smith, *The Wealth of Nations*, London 1982, should be supplemented by the subtle restatement of its central argument by F.A. von Hayek, especially in his 'Cosmos and Taxis', in *Law, Legislation and Liberty*, London 1982.

4. Paradoxes of Social Choice

The paradox of democracy has been stated many times, but given precise form in the context of modern political philosophy by Richard Wollheim, in 'The Paradox of Democracy', in P. Laslett and W.G. Runciman, *op. cit.*, vol. 1.

Arrow's theorem is discussed in Kenneth Arrow, *Social Choice and*

Individual Values, Yale 1990, and in A.F. Mackay, *Arrow's Theorem: the Paradox of Social Choice*, Yale 1980.

Condorcet's voting paradox is discussed in Duncan Black, *op. cit.*

5. General Will, Constitution and the State

This vast topic is in one sense the whole of political philosophy. The theory of the constitution is uppermost in Aristotle's *Politics*, and in J.D. Mabbott's *The State and the Citizen*, New Edition, London 1967, and these texts remain extremely valuable. On representative government, the study of that title by J.S. Mill is also still without serious competitor. The thesis of the separation of powers, introduced by Locke in his *Second Treatise of Civil Government*, was given its modern form by C. de Montesquieu, in *The Spirit of the Laws*, Cambridge 1989. These three topics, much discussed by political theorists, have received comparatively little attention from recent political philosophers.

A good survey of the philosophy of the state is that by R.A.D. Grant, 'Defenders of the State', in G.H.R. Parkinson, ed., *An Encyclopedia of Philosophy*, London 1988.

6. Justice

This is the area in which recent discussion has been most heavily concentrated, due largely to the seminal work, *A Theory of Justice*, by John Rawls, Oxford 1971, and the strongly framed reply by Robert Nozick, *Anarchy, State and Utopia*, Oxford 1978. However, these two thinkers do not have a monopoly of the subject by any means, as the reader may ascertain from the useful, if partisan, survey by J.R. Lucas, *On Justice*, Oxford 1989, and by referring to the following classics:

Aristotle, *Nicomachean Ethics*, 1131f;
Politics, 1280f.
Hume, *Treatise*, Book III, section 2.
H.L.A. Hart, *The Concept of Law*, Oxford 1976, ch. 8.
F.A. von Hayek, *The Constitution of Liberty*, London 1969.

Nevertheless, it is necessary to understand the dispute between Rawls and Nozick, if you are to acquire a grasp of recent political philosophy. Rawls is concerned, very roughly, to work out the philosophical foundations for a 'welfare state', while Nozick updates and sharpens the libertarian attacks on state intervention in the distribution of rewards.

It is impossible to grasp Rawls's theory without discomposing it. It is useful to begin by reading Rawls's article 'Justice as Fairness' in Laslett and Runciman, *op. cit.*, second series, and *A Theory of Justice* sections 1 to 13, and section 39, and then turning to the criticisms contained in the following:

Norman Daniels, ed., *Reading Rawls*, Oxford 1975. (Contains useful articles, especially those by Nagel, Dworkin and Hart.)
Brian Barry, *The Liberal Theory of Justice*, Oxford 1973.
Michael Sandel, *Liberalism and the Limits of Justice*, Cambridge 1982.

The above writers see the distinctive features of Rawls's theory in its liberalism. Nozick, however, is more concerned by the *socialist* aspect of the theory, and the ultimate restrictions placed on liberty in the name of a 'just distribution'. Rawls himself returns to the theme of his book in his Harvard lectures, printed in *J. Phil.*, 1981, and in an important article in *Philosophy and Public Affairs*, 1985. In the first of these he comes to grips, in a spectral way, with the metaphysical presuppositions of his approach, and acknowledges that his theory is one of 'Kantian constructivism' – i.e. an attempt to derive principles of morality from reflecting on the idea of a 'pure practical reason' – practical reason applied in isolation from the 'empirical conditions' of everyday choice. In the later article Rawls repudiates the idea of justice as founded in a 'hypothetical contract' and opts for an actual contract instead.

The theory can be broken down as follows:

(i) The original position. Socially conferred distinctions and advantages are eliminated from the premise of social choice by:

(ii) The veil of ignorance. This is drawn over all socially conferred goods, but not only over those. Problems are posed for Rawls by his attempt to make his original choosers *ignorant* of all those features of their situation which he believes are (from the point of view of justice) *irrelevant* – e.g. their sex, and even their particular 'conceptions of the good'.

(iii) Primary goods. These are the goods that all human beings need for a minimally tolerable life in society, and which cannot therefore be hidden behind the veil of ignorance. It is to be supposed that these, at least, will already be objects of desire, even without reference to the particular 'conceptions of the good' that we acquire through our social membership.

(iv) Social contract. The preferred principles of distribution are arrived at by a procedure of collective choice which has the form of a social contract (i.e. a choice over which each member of society exerts a veto). Only this will guarantee what Rawls sometimes thinks to be the fundamental requirement of a theory of justice, namely, that its recommendations must be acceptable to every member of society, whatever his position.

(v) Rational choice. In fact, however, because all social differences have been hidden behind the veil, the reasoning underlying the contract need not refer to others, but only to the basic principles of rational choice in conditions of risk and uncertainty. Hence Rawls frames his resulting theory in terms taken partly from decision theory, in which the individual and his interests are alone of concern.

(vi) The Liberty Principle. This is the first principle of the two chosen, and it holds roughly that each person is to enjoy the maximum liberty compatible with an equal liberty for everyone else. It is subsequently revised, refined and qualified, but it is meant to capture two ideas: (a) the priority of liberty in the political order; (b) the individual right to an equal respect.

(vii) The Difference Principle. This says roughly that differences in social advantage are justified (a) to the extent that they benefit everyone, and (b) to the extent that they are attached to offices and positions open to all. (Again this is substantially refined in subsequent discussions.)

(viii) Maximin. Condition (a) in (vii) is interpreted in this way: that people in the original position will seek to maximise the minimum pay-off (terms taken from decision theory and game theory): in other words, the prime concern will be with the position of the worst-off member of society.

(ix) Lexical ordering. The liberty principle is lexically prior to the difference principle: i.e. it must be satisfied before the difference principle applies, so that 'liberty can be sacrificed only for the sake of liberty'. Further applications of the idea of a lexical order among principles lead to later refinements of the difference principle itself.

(x) Intuitionism. Rawls does not go so far as Kant, and claim *a priori* validity for his principles. He says instead that they are to be measured against our intuitions concerning justice, with which they enter, at last, into 'reflective equilibrium'.

Rawls's theory is meant to correspond to the intuitions behind the social contract theories of political obligation, the 'Original Position' being a rational reconstruction of the 'State of Nature', and the principles of justice the outcome of a Social Contract, leading to a kind of general Will. He is concerned therefore to build in the condition of liberty (corresponding roughly to the Lockean idea of Natural Rights) as a fundamental component of a theory of justice, and not as a political value which is in some sense additional to, and in potential conflict with, the value of justice. The 'lexical' priority of liberty can be thought of as generating a theory of rights, in the sense made familiar by Dworkin ('Rights as Trumps', in *Taking Rights Seriously*, London 1978). (For a more perspicuous account of what this sense actually amounts to, see J. Raz, 'The Nature of Rights', in *Mind*, vol. 93, 1984.)

Some questions:
(i) Why should we believe that reflection about choice in the original position could cast light on our ideas of justice in an actual social arrangement?
(Rawls 3, 4, 22, 29; Nagel and Dworkin in Daniels.)
(ii) How thick is the veil of ignorance, and how thick should it be if Rawls's principles are to emerge?
(Rawls 24, 35 and Hare in Daniels.)
(iii) How can a hypothetical contract be binding?
(Dworkin in Daniels; on the Kantian position see Roger Scruton, 'Contract, Exploitation and Consent', in Howard Williams, ed., *Kant's Political Philosophy*, Aberystwyth 1992.)
(iv) What are the grounds for choosing Rawls's two principles?
(Rawls, 26; Nagel and Scanlon in Daniels.)
(v) Why is Rawls not a utilitarian? Is there any reason why averaging utilitarianism (e.g.) should not be just as rational a choice of principle for those choosing behind a veil of ignorance?
(Rawls, 27; Hare and Lyons in Daniels; Barry, *The Liberal Theory of Justice*.)

Nozick:
The principal contention behind Nozick's argument is that individuals have rights, these rights are absolute moral obstacles to those who would disregard them, and that, among the rights that people have, are rights of property, free action and non-interference in the peaceful pursuit of their projects. (Rights are, in a sense, freedoms.) The underlying vision is reminiscent of Locke, in his Second Treatise and letter on toleration. However, the only argument that Nozick considers for the indefeasibility of rights consists in a reiteration of Kant's second formulation of the categorical imperative (which says that people must be treated as ends and not as means only). The idea of a justice-preserving transaction looms large, and the main complaint against redistributive theories (such as Rawls's is supposed to be) is that, because they can be applied only by coercion, they cannot

achieve their aim without violating the procedural requirements of justice.

Reading:

Jeffrey Paul, *Reading Nozick*, Oxford 1982.

T. Scanlon, 'Nozick on Rights, Liberty and Property', in *Philosophy and Public Affairs*, 1976.

Some questions:

(i) Do 'pattern theories' of justice always licence unjust treatment of individuals? Is Rawls's theory such a theory? (Nozick, *ASU*, ch. 7.)

(ii) Is Wilt Chamberlain really entitled to his profit?
(Cohen and Nagel, in Paul.)

(iii) Does Nozick distinguish liberties from rights?

(iv) Is there any version of the 'end state' conception of justice that would escape Nozick's criticisms?

7. *Law*

Analytical philosophy of law has its roots in two traditions: the utilitarian theories of Jeremy Bentham, and the defence of legal positivism by John Austin (*The Province of Jurisprudence Determined*, 1832). The major works are these:

H.L.A. Hart, *The Concept of Law*, Oxford 1961. (A sophisticated defence of a new kind of legal positivism.)

Ronald Dworkin, *Taking Rights Seriously*, London 1978: a kind of reply to Hart, which adumbrates a new theory of the common law.

Ronald Dworkin, *Law's Empire*, Cambridge Mass. 1986, the same author's unsuccessful attempt to complete that theory.

Joseph Raz, *The Authority of Law*, Oxford 1983, perhaps the most sophisticated recent attempt to say what law is, and what distinguishes it from other kinds of reasoning.

8. Freedom

Reading:

J.S. Mill, *On Liberty*, in *Three Essays: On Liberty, Representative Government, and the Subjection of Women*, Oxford 1975.

Sir James FitzJ. Stephen, *Liberty, Equality, Fraternity*, Chicago 1991.

The dispute between Mill and Stephen has been re-enacted in recent times in:

H.L.A. Hart, *Law, Liberty and Morality*, Oxford 1968; and

P. Devlin, *The Enforcement of Morals*, Oxford 1968.

In addition the following texts are important:

I. Berlin, *Four Essays on Liberty*, Oxford 1958.
R.S. Peters, ed., *Of Liberty*, Royal Institute of Philosophy Lectures, 1981–2.
David Miller, ed., *Liberty*, Oxford Readings in Philosophy, Oxford 1991.

To a great extent modern discussions depend upon Mill's exposition for an account of the problem, if not for a solution to it. For Mill the question is the extent to which the individual may be granted freedom against, or conversely may be constrained by, 'society'. The loose use of the term 'society' by Mill (to mean sometimes civil society, sometimes the state, and sometimes mere 'social pressure'), and the failure to distinguish those social entities (such as the state) which are also corporate agents from those (such as a crowd) which are not, makes his exposition difficult to assess, despite its superficial clarity. As we might now see the problem, it has the following form:

We can distinguish, following Sir Isaiah Berlin ('Two Concepts of Liberty', in *op. cit.*), two ideas of political freedom, a negative and a positive. According to the negative idea I am free to the extent that my projects and intentions are unimpeded by others. According to the positive idea, I am free to the extent that I realise myself as a power in the world. In the first instance the problem of political freedom is concerned with the negative kind of freedom; in other words, with securing, for each individual, a sphere of 'non-interference' from others. (A sphere of negative rights.) The question is how far this sphere should extend, and who has the right and the authority to secure it. There is also the question of what is to be done with that sphere of negative freedom; in what way can I benefit from it, and use it to realise my nature as a free (rational) being? That is the question of positive freedom, which has a corrosive effect on liberal doctrines generally, since liberalism wishes to construct both politics and morality from the negative concept alone: the idea that I can become free in a positive sense, by expanding my power over the world, implies that negative freedoms (permissions) might have to be sacrificed for a higher goal. (This is the root of the Marxist view that negative freedoms are really 'bourgeois' freedoms, which secure the privileges of those able to take advantage of them, but contribute thereby to the continuing servitude of the remainder, as well as being 'not true freedoms'. True freedom comes with the full emancipation of human nature from its historical bondage: only then do we realise ourselves as free beings. It is therefore not an assault on human freedom, in its true sense, to cancel property rights, or any other right to non-interference, in the interests of that higher state.)

The standard liberal approaches can be divided into three broad categories:

(i) Those that make liberty (in one or another sense) into the supreme political value, and which therefore see the maximisation of liberty as the underlying purpose of the law. For such theories, the law may legitimately place constraints on our conduct, but only for the purpose of maximising liberty overall. (Sometimes the emphasis on liberty is qualified by a concern for justice, as a concomitant value, so that the maximisation of liberty is displaced in favour of some optimal combination of liberty and justice. In Rawls liberty is *absorbed* into justice; in Nozick, justice is absorbed into liberty.)

(ii) Those that recognise a plurality of political values, and which see the law as an instrument for the realisation of those values, but which regard liberty as a *sine qua non*. On this view, a certain basic minimum of liberty must be guaranteed to the individual, before the pursuit of other political objectives can be embarked upon. Such a view explains the emphasis on 'rights' in the liberal tradition: a right being the individual's veto against intrusion. One objection to the positive idea of freedom – especially as employed by the Marxist – is that it generates no conception of individual rights. It is therefore no accident, according to Sir Karl Popper (*The Open Society and its Enemies*, London 1952) that such a conception of freedom leads to tyranny.

(iii) Those which attempt to combine the liberal emphasis on negative freedom with a utilitarian theory of justification: the approach adopted by Mill (in some frames of mind) and by Henry Sidgwick (*Principles of Politics*, London 1891). The argument is roughly this: when people are negatively free they are, *ceteris paribus*, unhindered from doing what they decide to do; and people are by and large the best judges of what is most likely to satisfy them; hence negative freedom tends to maximise utility overall. The naïveté of the second premise needs no comment.

In all three forms, liberal theories encounter two major problems:

(a) What exactly is meant by freedom? Freedom to fulfil desires, whatever their objects? Or freedom to exercise autonomous rational choice? If the latter (freedom in the Kantian sense), then we must admit the legitimacy of all those constraints which further the development of the autonomous individual. And this might mean recognising the legitimacy of a highly illiberal political order. If the

former, then we need to say why freedom is a value, and to show how far other values should be sacrificed in order to obtain it. (See Minogue and Scruton in Peters, *op. cit.*)
(b) Should the law permit what morality forbids? If so, what are the limits to legal permission? The first kind of liberal thinker will argue that everything should be permitted until it can be shown to limit the liberty of someone. The second kind is more cautious, and usually recognises some *a priori* limit on what it is plausible to permit. Mill writes of 'harm' in this context: everything should be permitted unless this should produce harm to others. But what does harm consist in? In particular, am I harmed by those things which disturb and upset me, and which perhaps tempt me away from the path of righteousness?

Those questions are raised and dropped with alarming rapidity, both in the dispute between Mill and Stephen, and in that (which shows no real intellectual advance over its predecessor) between Hart and Devlin.

Some questions:
1. 'The sole end for which mankind are warranted, individually or collectively, in interfering with the liberty of action of any of their number, is self-protection.' (*On Liberty*.) What does Mill mean by 'self-protection', and how does he justify this claim?
2. What is the distinction between positive and negative freedom, and of what importance is it to political philosophy?
3. Is Mill right to criticise the 'despotism of custom' as he does? (Mill, Burke in *Reflections on the Revolution in France*, Scruton in Peters.)
4. What is the value of negative freedom? With what values, if any, does freedom compete?
5. How far can political reasons for a legislative decision be distinguished from moral reasons?

9. Property

This too is a growth area: the aftermath of the sixties saw renewed attacks on the institution of private property, and a renewed sympathy for Marxist theories – whether in their scientific or humanistic form. The collapse of communist government has again changed the perspective, and the patient arguments in defence of capitalism and private property presented by F.A. von Hayek are now also gaining a hearing (see especially the essays in *Law, Legislation and Liberty*, London 1982).

The classical dispute, between the defenders of communal property, and the defenders of private property, is elegantly rehearsed by Plato (*Republic*, Book IV, opening section, and Book V, which also defends the abolition of the household), and by Aristotle (*Politics* 1262b–1266a, where Aristotle expressly takes issue with Plato, and defends private property as an integral part of marriage, the family and the household).

The modern dispute was given canonical form by Locke, in ch. 5 of the *Second Treatise of Civil Government*, in which, however, Locke draws heavily on arguments developed by Aquinas, and by such theorists of natural law as Hooker and Grotius. Locke's argument is weak and intuitive. Nevertheless its basic conceptions have remained influential, inspiring on the one hand the socialist critique of private property (as creating the conditions of 'exploitation': see Karl Marx, *Capital*, vol. 1 part 3), and on the other hand, the modern defence of private property as the necessary outcome of procedural justice (as in Robert Nozick's *Anarchy, State and Utopia*, ch. 7). It is still best to begin the study of the arguments about property by considering Locke. Recent discussions include the following:

L. Becker, *Property Rights*, Cambridge 1982, ch. 4.
K. Olivencrona, in *The Journal of the History of Ideas*, 1974.
Jeremy Waldron, in *Phil. Quarterly*, 1976.
J.P. Day, in *Phil. Quarterly*, 1976.
Nozick, *ASU*, pp. 174ff.
Jeremy Waldron, *The Right to Private Property*, Oxford 1990.

The commentators mostly agree that Locke has three separate arguments at the back of his mind in discussing the original acquisition of private property:

(i) By mixing my labour with something, I come to own it in the same way as I own my labour.

(ii) By mixing my labour with something I add value to it; and value that is created solely by me is mine by right.

(iii) In a condition of need, provided others are in no way disadvantaged by my helping myself to the fruits of the earth, I have a right to do so.

The proviso in (iii) is more complicated than that implies, dividing into two parts: 'non-spoliage' and 'enough and as good left over'. Moreover, Locke means it to apply generally, qualifying the other arguments as well. The following questions will lead the student into some of the intricacies of Locke's argument:

(1) Is Locke's proviso ever satisfied? (See Nozick, pp. 178ff.)

(2) When I mix my labour with something that belongs to no one, do I gain the thing (Locke) or just lose my labour (Nozick)?
(3) What does Locke mean by 'in common', when he says that God gave the earth to mankind in common?
(4) Is it a fair criticism of Locke that his idea of a 'natural right' is simply an attempt to move from an 'is' to an 'ought' without even raising the question?
(5) Which rights of property does Locke set out to justify? Does he succeed in justifying any?

A useful counterbalance to Locke is Hegel who, while also defending private property, is much more conscious of the nature of property as an *institution*, with a history and personality over and above the natural rights of the individual. The following should be read:

Philosophy of Right, sections 41–71 (property); 169–172 (household and family); 189ff. (the 'system of needs').

Marx, 'Alienated Labour' and 'The Relationship of Private Property', both in the 1844 manuscripts, contained in David McLellan ed., *Selections from Marx*, London 1988. (The classical left-Hegelian critique of Hegel.)

Dudley Knowles, 'Hegel on Property and Personality', *Phil. Quarterly*, vol. 33 no. 130, 1979.

Jeremy Waldron, *The Right to Private Property, op. cit.*, chapter on Hegel. (Waldron is tempted to accept the broad tenets of Hegel's position, while bending the conclusion in a more socialist direction.)

Hegel's argument is phrased at first in the language of Locke (labour, natural right, etc.), partly because he wishes to justify a 'right of property' as pre-political (even though radically opposed to the possibility of a 'state of nature' as Locke envisages it). But it becomes clear in due course that he wishes to build far more into the concept of labour than Locke acknowledged. For Locke I own my labour as I own my body (the 'sweat of my brow'). For Hegel labour and bodily activity are distinct. Only a rational being (a self-conscious subject) is capable of *labour*, since labour involves the intention to produce value. This intention is not only limited to rational beings; it is essential to them, since without it no rational being can realise his potential, and become what he truly is. Moreover, the intention can persist only in those circumstances which permit private property in the object produced. Property must be private, since it is an expression of the individual will, and must stand in an exclusive relation to the will that covets it.

The language of the argument is largely metaphorical, as is that of Marx's imaginative reply. The reader should try to translate both into other terms, and present them as though they might be believed by us, here, now.

Hegel's argument is probably no more valid than Locke's. However, it has one or two interesting features:
(i) It justifies the institution of property, but no particular property right. (On the contrary, particular rights are determined by history.)
(ii) It easily takes on board the multifarious and composite character of private property, and is quite compatible with arguments favouring, for example, welfare legislation (which, indeed, Hegel was one of the first to advance).
(iii) It shows that the right of property, if it exists, is connected with rational activity *as such*, and therefore is intrinsic to humanity, even though it can exist only in the full social context which permits the realisation of the individual person.

Some questions:
(1) Why does Hegel distinguish (section 53) three 'modifications' of property?
(2) 'Since my will, as the will of a person, and so as a single will, becomes objective to me in property, property acquires the character of private property' (section 46). Explain and assess the underlying argument for that remark.
(3) Could there be a family without the need for private property?
(4) Is there any plausibility in Hegel's view that I define myself in property and realise my freedom through the use of it?
(5) What does Marx mean by 'alienation'? What, precisely, does he think the connection is, between alienation and private property?

10. Institutions
The major questions here are metaphysical: for example, to what extent does it make sense to think of individuals existing in a 'state of nature' without institutions? The Hegelian argument is that individual personality is an artefact, and is brought into being only with the institutions that produce it. The argument is summarised and partially defended in Roger Scruton, 'Corporate Persons', *Proceedings of the Aristotelian Society, Supplementary Volume*, 1989. I return to this question in Chapter 31.

Chapter 29 Subjective Spirit

The main text is Kant's *Critique of Judgement*, concerning which there is no really satisfactory commentary. For a summary of modern aesthetics, see R. Scruton, 'Aesthetics', in the current *Encyclopaedia Britannica*. Rather than attempt to guide the reader through this subject – which is highly controversial, and which seldom attracts the attention of the best minds among philosophers – I content myself with a few suggestions.

Aesthetics and the Philosophy of Art

Modern philosophers have had little to say about the nature of aesthetic interest; almost nothing to say about its relation to moral, religious and scientific interests. The concentration has been on the philosophy of art, and in particular on puzzles created by boring impostors like Duchamp: is this signed urinal a work of art? etc. This makes for an exceedingly dull literature, devoted to questions which can be answered in any way while leaving everything important exactly as it is.

Some philosophers, however, have tried to link the philosophy of art to central questions concerning meaning, understanding, and value. The most noteworthy are Richard Wollheim, in *Art and its Objects*, 2nd edn, Cambridge 1980, and *Painting as an Art*, London and Princeton 1987; Nelson Goodman, in *Languages of Art*, 2nd edn, Indianapolis 1976 and *Ways of Worldmaking*, Indianapolis 1978, and Stanley Cavell, in *Must We Mean What We Say?*, 2nd edn., Cambridge 1976, and *Disowning Knowledge: In Six Plays of Shakespeare*, Cambridge 1987. The principal topics that can be extracted from the works of those and similar philosophers (by which I mean, philosophers who share their rare combination of aesthetic sensibility and philosophical competence) are these:

The Ontology of Art

What kind of thing is a work of art? Where, or when, is art? Do works of music, works of literature, paintings, sculptures and buildings all occupy a like place in our ontology?

Leading works in this area are:

Goodman and Wollheim, *op.cit*.

Roman Ingarden, *The Literary Work of Art*, Evanston Ill. 1973.

Jerrold Levinson, *Music, Art and Metaphysics*, Ithaca NY 1990.

Nicholas Wolterstorff, *Works and Worlds of Art*, Oxford 1980.

Arthur Danto, *The Transfiguration of the Commonplace: a Philosophy of Art*, Cambridge Mass. 1981.

Representation
What does it mean to say that a painting represents a battle scene? Is depiction in painting the same property as description in imaginative literature? Does music represent things? I have touched on this topic in Chapter 24. The principal texts are mentioned in the study guide to that chapter; to them should be added Flint Schier's *Deeper into Pictures*, Cambridge 1986, and the essays on photography and film in Roger Scruton, *The Aesthetic Understanding*, London and Manchester 1983.

Expression
Representational works of art present us with imaginary worlds, in which we can identify objects, people, episodes and actions of the same kind that we encounter in the real world. But abstract (i.e. non-representational) works of art are also meaningful, while representational works of art have a meaning that is not exhausted by their representational content. These facts lead philosophers to postulate another dimension of aesthetic meaning, for which they use the term expression, derived from Croce, whose *Aesthetic*, 1902, first made the distinction between representation and expression, in order to dismiss the first as irrelevant to the aesthetic enterprise, and to elevate the second as the essence of art. On this fascinating topic, the authorities include Croce's brilliant disciple, R.G. Collingwood (*The Principles of Art*, Oxford 1938), as well as those already referred to.

Aesthetic Judgement, Aesthetic Value, and Criticism
When we interpret a work of art do we also evaluate it? And if so what kind of value does it have, and why should we be interested in that value? In 'Of the Standard of Taste', in *Essays, Moral Philosophical and Political*, 1745, Hume tries to establish a criterion of aesthetic judgement, in terms of the 'joint verdict of men of strong sense, united in delicate sentiment, improved by practice, perfected by comparison, and cleared of all prejudice'. He believes that you can establish such a criterion, even though aesthetic excellence (the object of aesthetic judgement) is not a 'real quality' of the things in which it is discerned. This is the forerunner of many modern views, which try to reconcile the subjective character of the aesthetic (the fact that beauty

is in the eye of the beholder), with a real distinction between good and bad taste. (See for example Roger Scruton, *The Aesthetics of Architecture*, London 1979.) Interesting light is cast on this by the argument of Anthony Savile, *The Test of Time*, Oxford 1982, and also by Mary Mothersill, in *Beauty Restored*, Oxford 1982.

Imagination
This topic has already been covered in Chapter 24. In the context of aesthetics, the theory of imagination figures prominently in:
R. Scruton, *Art and Imagination*, London 1974.
Arthur Danto, *The Politics of Imagination*, Lawrence 1988.
Kendall Walton, *Mimesis as Make-believe*, London 1990.

Recent articles of interest have been collected in the following:
J. Margolis, ed., *Philosophy Looks at the Arts*, Philadelphia 1978.
W. Elton, ed., *Aesthetics and Language*, Oxford 1954.
E. Schaper, ed., *Pleasure, Preference and Value*, Cambridge 1983.
R. Shusterman, ed., *Analytic Aesthetics*, Oxford 1990.

The argument of this chapter has been influenced by Hegel, whose *Aesthetics, Lectures on the Philosophy of Art*, tr. T.M. Knox, Oxford 1974, is one of the seminal texts of modern philosophy, and also by the *Letters on the Aesthetic Education of Man*, by F. Schiller, tr. E. Wilkinson and L.A. Willoughby, Oxford 1967. Those further interested in disentangling this topic should consult David Cooper, ed., *A Companion to Aesthetics*, Oxford 1993.

(Barthes's analysis of Sarrasine is entitled *S/Z*, and is published by Editions du Seuil, Paris 1970; English translation, Oxford 1990. L. Feuerbach's *Essence of Christianity* is translated by M. Evans (George Eliot), and published by Harper, New York 1957. Michael Oakeshott's distinction between civil and enterprise association is elaborated in his *On Human Conduct*, London 1974.)

Chapter 30 The Devil

The subject has been explored by R.G. Collingwood, in an essay on the existence of the devil: 'The Devil', in D.Z. Phillips ed., *Religion and Understanding* Oxford 1967. However, Collingwood's thought is a long way from the ancillary themes of this chapter, which are seldom discussed by modern philosophers. Again, there is little point in presenting a full study guide to the chapter. Those interested in pursuing some of the thoughts contained in it might consult some of the following:

Nietzsche, *The Genealogy of Morals*, New York 1988.
Cz. Milosz, *The Captive Mind*, Harmondsworth 1990.
R. Scruton, *The Philosopher on Dover Beach*, Manchester 1990, especially 'Man's Second Disobedience'.
F. Dostoevsky, *The Devils*, Oxford 1992.
Erich Heller, 'The Taking Back of the Ninth Symphony', in *In the Age of Prose*, Cambridge 1984.
Alain Besançon, *La Falsification du bien*, Paris 1986, on Orwell's insight into the devil's purpose.

On Sartre, see:
David Cooper, *Existentialism, op. cit.*
R. Scruton, *Thinkers of the New Left*, London 1986, ch. 11.
Cooper's reading of Sartre is sympathetic, and is a useful antidote to my remarks.

On Marx, by far the clearest exposition of the theory of history is G.A. Cohen, *Karl Marx's Theory of History: A Defence*, Oxford 1979.

On Foucault, see R. Scruton, *Thinkers of the New Left, op. cit.*, ch. 4.

On deconstruction, see R. Scruton, *Upon Nothing*, Swansea 1994.

Chapter 31 Self and Other

This chapter deals with difficult topics in a highly speculative way; not many anglophone philosophers would regard the modern approach to the self as providing an answer to a predicament that is primarily spiritual. But not many anglophone philosophers recognise such predicaments in the first place.

1. A Fragment of History
The historical material in this section is covered more extensively in my contribution to the *Illustrated History of Western Philosophy*, edited by Sir Anthony Kenny, Oxford 1993. Fichte's views are expounded in his *The Science of Knowledge, with the first and second introductions*, ed. and tr. Peter Heath and John Lacks, Cambridge 1982. Commentaries on this work are for the most part misleading, and usually far removed from the concerns of a modern philosopher. The same is not true of Kant's philosophy of the self, which has been interestingly discussed by several recent commentators, notably:

P.F. Strawson, *The Bounds of Sense*, London 1966, pp. 162–174.
Sydney Shoemaker, *Self-Knowledge and Self-Identity*, Cornell 1963, ch. 2.
Patricia Kitcher, *Kant's Transcendental Psychology*, Oxford 1991.

The crucial passages in Kant himself are contained in the chapter of the *Critique of Pure Reason*, entitled 'The Paralogisms of Pure Reason' (a chapter which was substantially rewritten for the second edition), and in the *Critique of Practical Reason*.

The transcendental self reappears in another role in the philosophy of Husserl, and here deserves independent study. For Husserl the pure subject can never be given as the object of its awareness: for then its essential character as subject would have been removed from it. If phenomenology involves the exploration of consciousness, then it must eventually abstract from all the objects of consciousness, so as to uncover the structure of consciousness as such, and this structure is transcendental. The important question is this: Did Husserl succeed in saying anything positive about this transcendental consciousness? And could he do so? See *Cartesian Meditations*, The Hague 1960, and the dry but methodical and illuminating discussion by David Bell, in his *Husserl*, London 1990.

2. The Grammar of Self-reference

This is one of the growth areas of modern philosophy, which brings together some of the hardest of philosophical problems: the nature of the subject; the problem of indexical reference and its place in a theory of meaning; the problem of time, and the asymmetry between the 'objective' and 'subjective' viewpoints. Any detailed study of the area must begin by breaking it down into its constituent parts. I tentatively suggest the following division:

(i) *The meaning of demonstratives*. Does a term like 'this' have sense as well as reference? If so, does the sense determine the reference, or must additional features, relating perhaps to the context of use, be supplied? The literature on this topic goes back to Frege himself – specifically to the essay entitled 'The Thought', reprinted in P.F. Strawson, ed., *Philosophical Logic*, Oxford Readings in Philosophy, Oxford 1967. Russell too was greatly interested in the analysis of demonstratives, since they seemed to provide a paradigm of meaning: the word 'this' is correctly used only in the presence of the thing referred to (the 'meaning', as Russell called it). See his *Lectures on Logical Atomism*, contained in his collection *Logic and Knowledge*, ed. R.C. Marsh, *op. cit.* More recent studies include:

David Kaplan, 'Dthat', reprinted in P. Yourgrau (ed.) *Demonstratives*, Oxford Readings in Philosophy, Oxford 1990.
John Perry, 'The Essential Indexical', also in Yourgrau.
Gareth Evans, 'Understanding Demonstratives', also in Yourgrau.

One question to ask is that of the relation between demonstratives like 'this' and 'that', and indexicals like 'I', 'here' and 'now'. In both cases the possibilities of error are restricted. But words like 'this' and 'that' can be used to identify objects that are taken to be real, but which are not real at all (hallucinations, for example). Could we use the words 'I', 'here' and 'now' in that way? In dreams perhaps?

(ii) *Indexicals*. Of particular importance is the comparison between 'I', 'here' and 'now'. If the peculiarities of 'I' are exactly duplicated by 'here' and 'now', then this might suggest that 'I-thoughts' are not, as Fichte thought, the clue to reality, but only the reflection of the speaker's viewpoint. (Alternatively, it might suggest that space and time are in some deep sense *subjective*, or even, as for McTaggart, unreal.) The comparison is interestingly made by D.H. Mellor, in *Real Time*, Cambridge 1981, and in 'I and Now', *Proceedings of the Aristotelian Society*, 1988–9. It is made, but perhaps not as seriously as it should be, by Thomas Nagel, in *The View from Nowhere*, Oxford 1986.

(iii) *'I'*. Of course, the principal object of study must be this word, and the context required for its successful deployment. In fact, it is not the *word* 'I' that interests us, but the first-person case, as represented by the *use* of that word – or at least, by *one* familiar use of it (the use which Wittgentein calls the 'as subject' use: *Blue and Brown Books*, Oxford 1958, pp. 66–7). There are languages which lack the first-person pronoun, or which use it sparingly. But all languages contain, and must contain, a 'first-person case'. (Why must they?)

Among recent discussions, the following are particularly illuminating, although there is very little agreement between them:
Elizabeth Anscombe, 'The First Person' in S. Guttenplan, ed., *Mind and Language*, Oxford 1975.
Zeno Vendler, 'A Note on the Paralogisms', in G. Ryle, ed., *Aspects of Philosophy*, London 1976.
Gareth Evans, *The Varieties of Reference*, Oxford 1982, ch. 7.
Colin McGinn, *The Subjective View*, Oxford 1983, ch. 2.
Sydney Shoemaker, 'Self-reference and Self-awareness', in *Identity, Cause and Mind*, Cambridge 1979.

One question to consider is this: 'Does the term "I" refer?' Anscombe seems to think that it does not; others, on the whole, that it does.

(iv) *The 'subjective' view*: what exactly is it? Is it a point of view on facts that are in themselves objective (and therefore accessible to anyone)? Or does it contain a revelation of specifically subjective facts: facts of which only one person can have knowledge? This is a point at issue between Nagel and McGinn, though it is difficult to obtain a clear answer from Nagel's discussion, either in *The View from Nowhere*, or in his classic papers 'Objective and Subjective', and 'What it's like to be a bat', both in *Mortal Questions*, Cambridge 1979.

3. Secondary Qualities

The connection between secondary quality concepts and indexicals has been illuminatingly made by Colin McGinn, in his *The Subjective View, op. cit.*, where he argues for the following conclusions: that indexical thoughts and ascriptions of secondary qualities may be objective; that there are, nevertheless, no *facts* which require indexicals or secondary quality terms for their description; that indexicals and secondary quality terms contain an essential relation to the point of view of a subject of experience; that we cannot dispense with indexical or secondary quality ways of thinking, since they form an essential link between the world in which we are situated and our actions towards that world. All these interesting theses are disputable, and all merit discussion.

Of great importance in the literature of secondary qualities are the following:

Pierre Gassendi, *The Selected Works of Pierre Gassendi*, tr. C. Bugh, London 1972.

Descartes, *Principles* I, lxviii; II, iv; in *Selected Philosophical Writings*, tr. J. Cottingham, Cambridge 1988, pp. 160–213.

N. Malebranche, *The Search after Truth*, tr. T. Lennon and P. Olscamp, Columbus Ohio 1980, pp. 55, 75 and 441.

Leibniz, *New Essays on the Human Understanding*, tr. P. Remnant and J. Bennett, Cambridge 1982, pp. 130–37.

John Locke, *Essay*, Book II, chs. 3, 8 and 18.

J.W. von Goethe, *Theory of Colours*, tr. Charles Eastlake, London 1840, reissued with intro. by D.B. Judd, Cambridge Mass. 1970.

Thomas Reid, *Works*, 7th edn, Edinburgh, 1872, ed. Sir William Hamilton, who quotes from Reid and others on the distinction in vol. II, p. 835; *Essays on the Intellectual Powers of Man*, Cambridge Mass. 1969, p. 254.

L. Wittgenstein, *Remarks on Colour*, ed. G.E.M. Anscombe, tr. L.L. McAlister and M. Schattle, Oxford 1977.

Among more recent discussions, the student might consult:
John Campbell, 'A Simple View of Colour', in J. Haldane and C. Wright, eds., *Reality, Representation and Projection*, Oxford, 1994.

4. The Grammar of Consciousness

This section takes up the questions of sections 2 and 3, in the context of the puzzling fact of first-person immunity. How is this immunity to be explained? And should we assume that it is always to be explained in the same way? Is there anything to *be* explained? Among recent contributions to this debate, the following are notable:
Donald Davidson, 'First-person authority', in *Dialectica*, 1984.
Sydney Shoemaker, 'First-person Access', in J. Tomberlin, *Philosophical Perspectives*, vol. 4.
Malcolm Budd, *Wittgenstein's Philosophy of Psychology*, London 1989.

The Wittgensteinian position which I defend briefly in the text of this lecture is spelt out at greater length in Chapter 3 of *Sexual Desire*, London 1986.

On the contrast between predicting and deciding, and its connection with the first-person view, see:
D.F. Pears, in P.F. Strawson, ed., *Studies in the Philosophy of Thought and Action*, Oxford 1968.
H.P. Grice, 'Intention and Uncertainty', *Proceedings of the British Academy*, 1976.

5. The Social Construction of the Self

In a sense the theme of this section is that of Hegel's *Phenomenology of Spirit* and *Philosophy of Right*. It is discussed in another way by the phenomenological sociologist Alfred Schutz, in his somewhat dry and jargon-infested writings: notably the two collections of his essays, and *The Structure of the Lifeworld*, The Hague 1974.

On laughter see:
R. Scruton, 'Laughter', in *The Aesthetic Understanding*, London and Manchester 1983, in which the traditional sources are surveyed and discussed.

Index of Subjects

Index of Names

Also available in paperback

JOHN KENNETH GALBRAITH

THE WORLD ECONOMY SINCE THE WARS

A PERSONAL VIEW

From World War I and the Russian Revolution to Reagan and the monetarist revolution of the eighties, John Kenneth Galbraith – one of the century's most revered and reviled economists – explores the history and consequences of eighty years of changing economic fashion, policy and reality around the world.

With the sharpness and clarity that have marked his writing for thirty books, this is the crowning achievement of Galbraith's remarkable career, an extensive and accessible view of twentieth-century economic and political history that will be read and referred to for years to come.

'At least as ambitious in scope, as appealing in simplicity and as alarming in its implications as Marx ever was'
The Times

'Galbraith has much to be immodest about . . . highly engaging . . . can be thoroughly recommended to those interested in the economic debate, but not quite sure where to start. They are promised immaculately sardonic prose, with not an equation or chart in sight'
New Statesman and Society

'He writes exceedingly well . . . richly informative and relevant'
New York Times

'Timely and much needed'
Irish Independent